Applied Structural Mechanics

Springer
Berlin
Heidelberg
New York
Barcelona
Budapest
Hong Kong
London
Milan
Paris
Santa Clara
Singapur
Tokyo

H. Eschenauer, N. Olhoff, W. Schnell

Applied Structural Mechanics

Fundamentals of Elasticity, Load-Bearing Structures, Structural Optimization

Including Exercises

With 179 Figures

Springer

Prof. Dr.-Ing. H. Eschenauer

University of Siegen
Research Center for Multidisciplinary Analyses
and Applied Structural Optimization FOMAAS
Institute of Mechanics and Control Engineering
D - 57068 Siegen / Germany

Prof. Dr. techn. N. Olhoff

Aalborg University
Institute of Mechanical Engineering
DK - 9220 Aalborg East / Denmark

Prof. Dr. Dr.-Ing. E. h. W. Schnell

Technical University of Darmstadt
Institute of Mechanics
D - 64289 Darmstadt / Germany

ISBN 3-540-61232-7 Springer-Verlag Berlin Heidelberg New York

Die Deutsche Bibliothek - CIP-Einheitsaufnahme
Eschenauer, Hans A.: Applied structural mechanics: fundamentals of elasticity, load bearing structures, structural optimization; including exercises / H. Eschenauer; N. Olhoff; W. Schnell. - Berlin; Heidelberg; New York; Barcelona; Budapest; Hong Kong; London; Milan; Paris; Santa Clara; Singapur; Tokyo: Springer, 1997
ISBN 3-540-61232-7
NE: Olhoff, Niels; Schnell, Walter

Printed in Germany

Typesetting: Camera-ready by editors
SPIN:10508157 61/3020-5 4 3 2 1 0 - Printed on acid -free paper

Preface

The present English-language work is a compilation of the two-volume 3rd edition (in German) of "Elastizitätstheorie" (1993, 1994) published by BI-Wissenschaftsverlag Mannheim, Leipzig, Wien, Zürich. Since the first edition of this book had appeared in 1983, the fundamental concept of this book has remained unaltered, in spite of an increasing amount of structural-analytical computation software (e.g. Finite Element Methods). The importance of computer-tools, may this be supercomputers, parallel computers, or workstations, is beyond discussion, however, the responsible engineer in research, development, computation, design, and planning should always be aware of the fact that a sensible use of computer-systems requires a realistic modeling and simulation and hence respective knowledge in solid mechanics, thermo- and fluiddynamics, materials science, and in further disciplines of engineering and natural sciences. Thus, this book provides the basic tools from the field of the *theory of elasticity* for students of natural sciences and engineering; besides that, it aims at assisting the engineer in an industrial environment in solving current problems and thus avoid a mere *black-box* thinking. In view of the growing importance of product liability as well as the fulfilment of extreme specification requirements for new products, this practice-relevant approach plays a decisive role. Apart from a firm handling of software systems, the engineer must be capable of both the generation of realistic computational models and of evaluating the computed results.

Following an outline of the fundamentals of the theory of elasticity and the most important load-bearing structures, the present work illustrates the transition and interrelation between *Structural Mechanics* and *Structural Optimization*. As mentioned before, a realistic modeling is the basis of every structural analysis and optimization computation, and therefore numerous exercises are attached to each main chapter.

By using tensor notation, it is attempted to offer a more general insight into the theory of elasticity in order to move away from a mere Cartesian view. An "arbitrarily shaped" solid described by generally valid equations shall be made the object of our investigations (Main Chapter A). Both the conditions of equilibrium and the strain-displacement relations are presented for large deformations (nonlinear theory); this knowledge is of vital importance for the treatment of stability problems of thin-walled load-bearing structures. When deriving the augmented equations as well as the corresponding solution procedures, we limit our considerations to the most essential aspects. All solution methods are based on the HOOKEAN concept of the linear-elastic solid. As examples of load-bearing structures, disks, plates and shells will be treated in more detail (Main Chapters B,C). Finally, an introduction into Structural Optimization is given in order to illustrate ways of determining optimal layouts of load-bearing structures (Main Chapter D).

In the scope of this book, the most important types of exercises arising from each Main Chapter are introduced, and their solutions are presented as comprehensively as necessary. However, it is highly recommended for the reader to test his own knowledge by solving the tasks independently. When treating structural optimization problems a large numerical effort generally occurs that cannot be handled without improved programming skills. Thus, at corresponding tasks, we restrict ourselves to giving hints and we have consciously avoided presenting details of the programming.

The authors would like to express their gratitude to all those who have assisted in preparing the camera-ready pages, in translating and proofreading as well as in drawing the figures. At this point, we would like to thank Mrs A. Wächter-Freudenberg, Mr K. Gesenhues, and Mr M. Wengenroth who fulfilled these tasks with perseverance and great patience. We further acknowledge the help of Mrs Dipl.-Ing. P. Neuser and Mr Dipl.-Ing. M. Seibel in proofreading.

Finally, we would also like to express our thanks to the publisher, and in particular to Mrs E. Raufelder, for excellent cooperation.

Hans Eschenauer
Siegen/GERMANY

Niels Olhoff
Aalborg/DENMARK

Walter Schnell
Darmstadt/GERMANY

April 1996

Contents

List of symbols

Note: The following list is restricted to the most important subscripts, notations and letters in the book.

Scalar quantities are printed in roman letters, vectors in boldface, tensors or matrices in capital letters and in boldface.

1. Indices and notations

The classification is limited to the most important indices and notations. Further terms are given in the text and in corresponding literature, respectively.

$i, j, k, \ldots$	latin indices valid for 1, 2, 3
$\alpha, \beta, \mu, \ldots$	greek indices valid for 1, 2
k	Index for a layer of a laminate
x_i	subscripts for covariant components
x^i	superscripts for contravariant components
(ii)	indices in brackets denote no summation
$'$	prime after index denotes rotated coordinate system e.g. $\sigma_{x'x'}$
$,$	comma denotes partial differentiation with respect to the quantity appearing after the comma, e.g. $u_{,x}$
$'$	superscript prime before symbol denotes deviator, e.g. $'\tau^i_j$
$\vert$	vertical line after a symbol denotes covariant derivative relating to curvilinear coordinates ξ^i, e.g. $v_i\vert_j$
$\bar{}$	bar over a symbol denotes virtual value, e.g. $\bar{F}_i$
$\hat{}$	roof over a symbol denotes the reference to a deformed body
$\sim$	tilde denotes approximation
*	asterisk right hand of a small letter denotes physical component of a tensor, e.g. a_i^*
*	asterisk right hand of a letter denotes extremal point
*	asterisk right hand of a capital letter denotes the *complementary* of work or energy, e.g. U^*
∇	nabla–operator
$\Diamond^4$	differential operator
$A \cap B$	intersection of A and B
$A \subset B$	A is a subset of B
$\forall$	for all

2. Latin letters

a	determinant of a surface tensor
a	radius of a spherical or a circular cylindrical shell
$\mathbf{a}_\alpha$, $\mathbf{a}^\alpha$	co– and contravariant base vector of a surface in arbitrary coordinates
$\mathbf{a}_3$	normal unit vector to a surface
$a_{\alpha\beta}$, $a^{\alpha\beta}$	co– and contravariant components of a surface tensor
a , b	semiaxis of an ellipse
b	determinant of the covariant curvature tensor
$b_{\alpha\beta}$, $b^{\alpha\beta}$, b^α_β	co–, contravariant and mixed curvature tensor
e	volume dilatation
$\mathbf{e}_i$	orthonormalized base (Cartesian coordinates)
e_{ijk} , e^{ijk}	permutation symbol
$\mathbf{f}$	volume force vector
f , $\mathbf{f}$	objective function, – vector
g	weight per area unit
g	determinant of the metric tensor
g_j , $\mathbf{g}$	inequality constraint function, – vector
$\mathbf{g}_i$, $\mathbf{g}^i$	co– and contravariant base (arbitrary coordinates)
g_{ij} , g^{ij}	co– and contravariant metric components, metric tensor
h	height of a boiler
h_i , $\mathbf{h}$	equality constraint function, – vector
h_C , h_{ku} , h_{kl}	core height, distance of the k. layer from the mid–plane
k	buckling value
$k = \dfrac{t^2}{12\,a^2}$	shell parameter
$\mathbf{n}$	normal unit vector
p	parabola parameter
$\mathbf{p}$	vector of external loads ; vector of control polygon points
$\mathbf{P}_j$	pseudo–load matrix
$p[\mathbf{f}(\mathbf{x})]$	preference function
p^α , p	circumferential and normal loads
p_x , p_ϑ , p	external loads of a cylindrical shell
r	distance perpendicular to axis of rotation
r_1	radius of curvature
r_2	distance to axis of rotation along the curvature radius
$\mathbf{r}$	load vector
$\mathbf{r}$	position vector to an arbitrary point of the mid–surface or a body

$\mathbf{r}_j, \mathbf{r}_k$	orthogonal vectors
s	coordinate in meridional direction
$\mathbf{s}_i$	vector of search direction
t, t_k	wall thickness, layer thickness ($k = 1, \ldots n$)
$\mathbf{t}$	stress vector
$t^i; t_x, t_y, t_z$	components of a stress vector
$\mathbf{u}$	state vector of a cylindrical shell, state variable vector
u, v	displacements in meridional and in circumferential direction
$\mathbf{v}$	displacement vector
v_α	displacements tangential to the mid-surface
w	displacement perpendicular to the mid-surface
w_i	weighting factors, penalty terms
w^*	approximation for deflection
$\mathbf{x}$	design variable vector
$x^i; x, y, z$	Cartesian coordinate system, EUCLIDIAN space
x_i	shape parameter
$\mathbf{y}, y_i$	transformed variables
z	complex variable
$\mathbf{z}_i$	state vector at point i of a shell of revolution
A	area, surface; concentrated force at a corner
$\mathbf{A}$	strain-stiffness matrix; matrix of A-conjugate directions
B_{ik}	B-spline base functions, BERNSTEIN-polynomials
$\mathbf{B}$	rotation matrix; coupled stiffness matrix
$\mathbf{C}_i, \mathbf{C}$	transfer matrix of a shell element, total transfer matrix
C^{ijkl}	elasticity tensor of fourth order
$\mathbf{C}$	elasticity matrix
$\mathbf{D}$	flexibility matrix
D	tension stiffness of an isotropic shell
$D_x, D_\vartheta, D_\nu, D_{x\vartheta}$	strain- or shear stiffnesses of an orthotropic shell
D_{ijkl}	flexibility tensor
$E, \mathbf{E}$	YOUNG's modulus, elasticity matrix
$E^{\alpha\beta\gamma\delta}$	plane elasticity tensor
$F, \mathbf{F}$	objective functionals
$F_i, \mathbf{F}$	concentrated forces; load vector
$F(x^i)$	implicit representation of a surface
$\mathbf{F}$	symmetrical flexibility matrix – mixed transformation tensor – system matrix

G	shear modulus
G_i	penalty function
G_j	operator of inequality constraints
$H^{\alpha\beta\gamma\delta}$	elasticity tensor of a shell
H	mean curvature
H_k	operator of equality constraints
$\mathbf{H}, \mathbf{H}_k$	HESSIAN matrices
I	integral function
I_1, I_2, I_3	invariants
$\mathbf{I}$	unity matrix, – tensor
$\mathbf{J}$	JACOBIAN matrix
K	compression modulus
K	bending stiffness of an isotropic shell
K	GAUSSIAN curvature
$K_x, K_\vartheta, K_\nu, K_{x\vartheta}$	bending and torsional stiffness of an orthotropic cylindrical shell
K_x, K_y, K_{xy}, H	stiffnesses of an orthotropic plate
$\mathbf{K}$	bending stiffness matrix
L	differential operators
L	LAGRANGE–function
M	boundary moment
$M^{\alpha\beta}$	moment tensor
$M_{xx}, M_{\vartheta\vartheta}, M_{x\vartheta}$	bending and torsional moments
$\widetilde{M}^{\alpha\beta}$	*pseudo*–resultant moment tensor
$N^{\alpha\beta}$	membrane force tensor
$N_{xx}, N_{\vartheta\vartheta}, N_{x\vartheta}$; N_{xx}, N_{yy}, N_{xy}	normal and shear components of membrane forces
$\bar{N}_{x\vartheta}$	effective inplane shear force
P_i	polynomials
Q^α	transverse shear forces
$\bar{Q}_x, \bar{Q}_\vartheta$	effective transverse shear force
R	boundary force
R_i	penalty parameter
R_j	polynomial approximations
R_1, R_2	radii of principal curvatures
$\mathbf{R}$	shape function of a shell surface
$\mathbb{R}^n$	n–dimensional set of real numbers
$\mathbf{S}$	stress tensor

$\mathbf{S}$	shear stiffness matrix
T	tensor of n–th order (n = 0 , 1 , 2 , 3 , 4 . . .)
T_i	CHEBYCHEV–polynomials or –functions
$\mathbf{T}$	transformation matrix
$\bar{U}$	specific deformation energy
$\bar{U}^*$	specific complementary energy
V	potential for field of conservative forces
V	volume
$\mathbf{V}$	tensor of deformation derivatives, deformation gradient
$\mathbf{V}_s$	strain tensor (symmetrical part of V)
$\mathbf{V}_a$	tensor of infinitesimal rotations (antisymmetrical part of V)
W	weight
W, W^*	external work, complementary work
X	feasible design space, subset

3. Greek letters

α	semi–angle of a cone
α_i	optimal step length
α_T	coefficient of thermal expansion
$\alpha_{\alpha\beta}$	strains of the mid–surface of a shell
α_i, β_j	LAGRANGE multipliers
$\beta_{\alpha\beta}$	distortions of the mid–surface of a shell
β^i_j	components of a rotation matrix
β^{ij}	thermal–elastic tensor
γ_{ij}	strain tensor
$\gamma_{xy}, \gamma_{xz}, \gamma_{yz}$	shear strains in Cartesian coordinates
${}^0\gamma_{\alpha\beta}, {}^1\gamma_{\alpha\beta}$	strains, distortions
γ_α	shear deformation
$\gamma_{s\vartheta}, \gamma_{\varphi\vartheta}$	shear strain
δ	variational symbol
δ^i_j, δ_{ij}	KRONECKER's tensor in curvilinear and Cartesian coordinates
δ_{ij}	MAXWELL's influence coeffients
$\varepsilon, \boldsymbol{\varepsilon}$	factor of the step length , strain vector
$\varepsilon_{ijk}, \varepsilon^{ijk}, \varepsilon_{\alpha\beta}$	permutation tensors
$\varepsilon_{xx}, \varepsilon_{yy}, \varepsilon_{zz}$ $\varepsilon_{ss}, \varepsilon_{\varphi\varphi}, \varepsilon_{\vartheta\vartheta}$	strains in Cartesian and spherical coordinates
$\boldsymbol{\varepsilon}_\Theta$	vector of free thermal strains
ζ	coordinate perpendicular to mid–surface

η_l	slack variable
ϑ	coordinate in latitude direction, latitude angle
κ_1, κ_2	decay factors
κ_1, κ_2	principal curvatures, variable exponents of an ellipse function
$\kappa_{\alpha\beta}$	tensor of curvatures
λ_i	eigenvalue
$\boldsymbol{\lambda}_i$	vector of auxiliary variables
λ, μ	LAMÉ constants
μ	decay factor for a conical shell
ν, ν_x, ν_y	POISSON's ratio
ξ^i, ξ^α	curvilinear coordinate system, GAUSSIAN surface parameters
ρ	mass density
$\rho_{\alpha\beta}$	tensor of curvatures (shallow shell)
$\sigma_{xx}, \sigma_{yy}, \sigma_{zz}$	normal stresses in Cartesian coordinates
$\sigma_I, \sigma_{II}, \sigma_{III}$	principal stresses
$\boldsymbol{\sigma}$	stress vector
σ_M	mean value of normal stresses
τ	time
τ^{ij}	stress tensor
$\tau_{xy}, \tau_{xz}, \tau_{yz}$	shear stresses in Cartesian coordinates
φ	coordinate in meridional direction, meridional angle
χ, ψ	physical components of the bending angles of a shell of revolution
χ	LOVE function
$\psi_\alpha, \psi_x, \psi_y$	slope of cross-sections, bending angle
ω, λ	eigenfrequency, eigenfrequency parameter
ω_1, ω_2	coordinates of a spherical shell (starting from the boundaries)
$\boldsymbol{\Gamma}$	GREEN-LAGRANGE's strain tensor
$\Gamma_{ijk}, \Gamma^k_{ij}$	CHRISTOFFEL symbols of first and second kind
Δ, Δ^*	LAPLACE-operator, modified LAPLACE operator
$\Theta(\xi^\alpha, \xi)$	temperature distribution
Θ_{kl}	thermal-elastic tensor
Π_e, Π_i	external, internal potential
Π^*_e, Π^*_i	external, internal complementary potential
Π, Π^*	total potential, total complementary potential
Φ	AIRY's stress function
$\boldsymbol{\Phi}_i$	modified function, penalty function

1 Introduction

The classical fundamentals of modern Structural Mechanics have been founded by two scientists. In his work "Discorsi", Galileo GALILEI (1564 – 1642) carried out the first systematic investigations into the fracture process of brittle solids. Besides that, he also described the influence of the shape of a solid (hollow solids, bones, blades of grass) on its stiffness, and thus successfully treated the problem of the *Theory of Solids with Uniform Strength*. One century later, Robert HOOKE (1635 - 1703) stated the fundamental law of the linear theory of elasticity by claiming that *strain (alteration of length) and stress (load) are proportional* ("ut tensio sic vis"). On the basis of this material law for the *Theory of Elasticity*, Edme MARIOTTE (1620 – 1684), Gottfried Wilhelm von LEIBNIZ (1646 – 1716), Jakob BERNOULLI (1654 – 1705), Leonard EULER (1707 – 1783), Charles Augustin COULOMB (1736 – 1806) and others treated special problems of bending of beams.

Until the beginning of the 19th century, the *Theory of Beams* had almost exclusively been the focus of the *Theory of Elasticity and Strength*. Claude-Louis-Marie-Henri NAVIER (1785 – 1836) developed the general equations of elasticity from the equilibrium of a solid element, and thus augmented the beam theory. Finally, he also set up a torsion theory of the beam. Hence, he may quite justly be seen as the actual founder of the Theory of Elasticity. NAVIER's disciple Barré de DE SAINT-VENANT (1797 – 1886) augmented the work of his teacher by contributing new theories on the impact of elastic solids. His contemporary, the outstanding scientist and engineer Gustav Robert KIRCHHOFF (1824 – 1887), derived with scientific strictness the plate theory named after him. The first mathematical treatments of shell structures were contributed by mathematicians and experts in the *theory of elasticity* as Carl Friedrich GAUSS (1777 – 1855), CASTIGLIANO (1847 – 1884), MOHR (1835 – 1918), Augustin Louis Baron CAUCHY (1789 – 1857), LAMÈ (1795 – 1870), BOUSSINESQ (1842 – 1929), and, as mentioned above, NAVIER, DE SAINT-VENANT and KIRCHHOFF. A complete bending theory of shells was derived systematically by Augustus Edward Hough LOVE (1863 – 1940) on the basis of a publication by ARON in 1847.

During the 19th century, numerous works have been published in the field of *Structural Mechanics* which cannot be described in detail here. However, based on the above-said one might assume that this discipline is an *old* one, the problems of which have largely been solved. As a matter of fact, this surmise may have been true until recently. However, the continuous development of the sciences and the technology, especially during recent years, calls for an increased exactness of computations, in particular in the construction of complex systems and plants and in lightweight constructions, respectively. Owing to the introduction of duraluminium and other advanced materials like composites, ceramics, etc. into the lightweight constructions, the number of publications in the field of shell and lightweight structures has witnessed a substantial increase. In [C.6] it is shown that the amount of publications has doubled per each decade since 1900. Proceeding from about 100 papers in the year 1950, one counted about 1000 publications in 1982, i.e. three per day. Thus, the references to this book can only comprise a very limited selection of textbooks and publications.

The still continuing importance of Structural Mechanics also stems from the fact that the relevance of structures that are *optimal* with respect to bearing capacity, reliability, accuracy, costs, etc., is becoming much more apparent than in former times. Especially in the field of *structural optimization*, considerable progress has been achieved during recent years and this has prompted increased research efforts in underlying branches of solid mechanics like fracture and damage mechanics, viscoelasticity theory, plastomechanics, mechanics of advanced materials, contact mechanics, and stability theory. Here, the application of computers and of increasingly refined algorithms allows treatment of more and more complex systems. In this

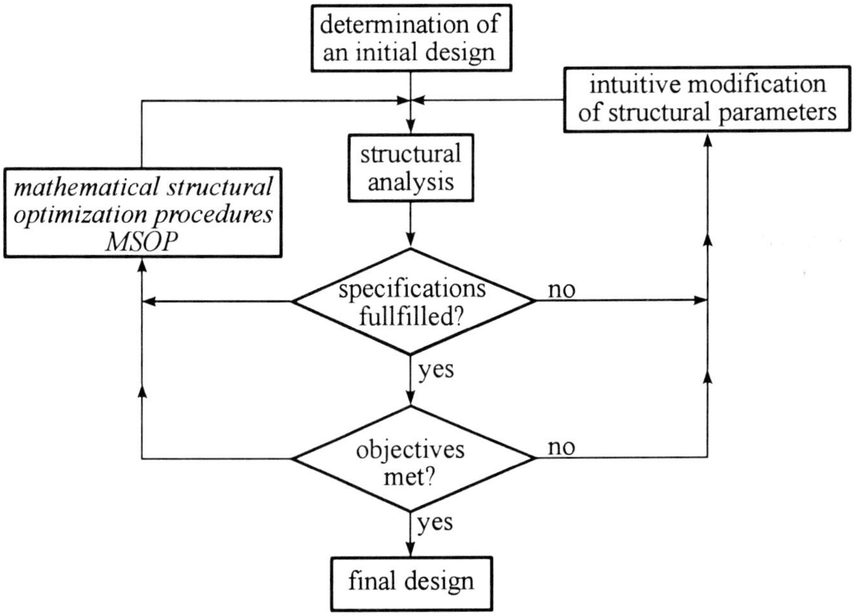

Fig. 1.1: Integration of *mathematical structural optimization procedures* (*MSOP*) into the design process

context, one should mention the large amount of novel finite computation procedures (e.g. Finite Element Methods [A.1, C.25]) as well as the Algorithms of Mathematical Programming applied in structural mechanics. One can thus justly claim that all of the above-named more *novel* fields and their solution approaches are all based on the fundamentals of elasticity without which the currently occurring problems cannot be solved and evaluated. The field of *Structural Optimization* increasingly moves away from the stage of a mere *trial-and-error* procedure to enter into the very design process using mathematical algorithms (Fig. 1.1). This development roots back to the 17th century, and is closely connected with the name Gottfried Wilhelm LEIBNIZ (1646 – 1716) as one of the last universal scholars of modern times. His works in the fields of mathematics and natural sciences may be seen as the foundation of analytical working, i.e. of a coherent thinking that is a decisive assumption of structural optimization. LEIBNIZ provided the basis of the differential calculus, and he also invented the first mechanical computer. Without his achievements, modern optimization calculations would yet not have been possible on a large scale.

Here, one must also name one of the greatest scientists Leonard EULER (1707 – 1783) who extended the determination of extremal values of given functions to practical examples. The search for the extremal value of a function soon led to the development of the variational calculus where entire functions can become extremal. Hence, Jakob BERNOULLI (1655 – 1705) determined the *curve of the shortest falling time (Brachistochrone)*, and Issac NEWTON (1643 – 1727) found the *solid body of revolution with the smallest resistance*. Jean Louis LAGRANGE (1736 – 1813) and Sir William Rowan HAMILTON (1805 – 1865) set up the principle of the smallest action and formulated an integral principle, and thus contributed to the perfection of the variational calculus that still is the basis of several types of optimization problems. Many publications on engineering applications over the previous decades utilize the variational principle. LAGRANGE, CLAUSEN and DE SAINT-VENANT had already treated the optimal shape of one-dimensional beam structures subjected to different load conditions. Typical examples here are the buckling of a column, as well as the cantilever beam for which optimal cross-sections could be found using the variational principle. This requires the derivation of optimality criteria as necessary conditions; these are EULER´s equations in the case of unconstrained problems. If constraints are considered, as, e.g., in solution of an isoperimetrical problem, LAGRANGE´s multiplier method is used.

A Fundamentals of elasticity

A.1 Definitions – Formulas – Concepts

2 Tensor algebra and analysis

2.1 Terminology – definitions

The use of the index notation is advantageous because it normally makes it possible to write in a very compact form mathematical formulas or systems of equations for physical or geometric quantities, which would otherwise contain a large number of terms.
Coordinate transformations constitute the basis for the general concept of tensors which applies to arbitrary coordinate systems. The reason for the use of tensors lies in the remarkable fact that the validity of a tensor equation is independent of the particular choice of coordinate system. In the following we confine our considerations to quantities of the three-dimensional EUCLIDEAN space. We introduce the following definitions:
A *scalar* characterized by *one* component (e.g. temperature, volume) is called a *tensor of zeroth order*.
A *vector* characterized by *three* components (e.g. force, velocity) is called a *tensor of first order*.
The dyadic product of two vectors, called a *dyad* (e.g. strain, stress), is a *tensor of second order* characterized by *nine* components.
Tensors of higher order appear as well.

Notation of *tensors of first order*:

a) Symbolic in matrix notation: $\mathbf{a} = \begin{bmatrix} a^1 \\ a^2 \\ a^3 \end{bmatrix}$.

b) Analytical: $\mathbf{a} = a_x \mathbf{e}_x + a_y \mathbf{e}_y + a_z \mathbf{e}_z$

or $\mathbf{a} = a^1 \mathbf{e}_1 + a^2 \mathbf{e}_2 + a^3 \mathbf{e}_3 = \sum_{i=1}^{3} a^i \mathbf{e}_i$

with $\mathbf{e}_x$, $\mathbf{e}_y$, $\mathbf{e}_z$ as base vectors in a Cartesian coordinate system. The superscripts are indices, and not exponents. In index notation the expression a^i (or a_i) ($i = 1, 2, 3$) denotes the total vector (see Fig. 2.1).

Notation of *tensors of second order*:

a) Symbolic in matrix notation: $\mathbf{T} = \begin{bmatrix} t_{11} & t_{12} & t_{13} \\ t_{21} & t_{22} & t_{23} \\ t_{31} & t_{32} & t_{33} \end{bmatrix}$.

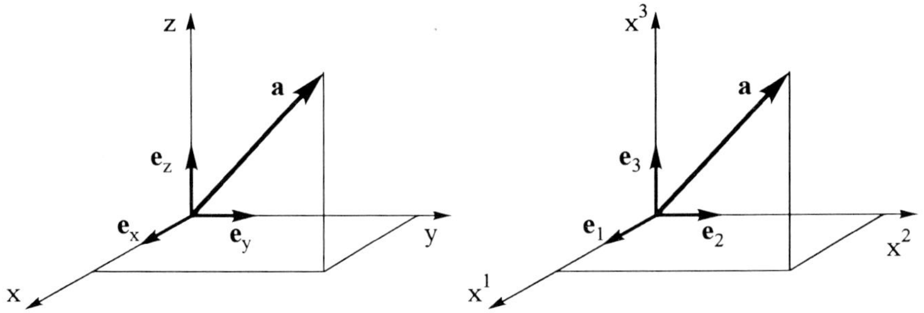

Fig. 2.1: Definition of the base vectors

b) Analytical:

$$\mathbf{T} = t_{11}\,\mathbf{e}^1\mathbf{e}^1 + t_{12}\,\mathbf{e}^1\mathbf{e}^2 + \ldots + t_{33}\,\mathbf{e}^3\mathbf{e}^3$$

or
$$\mathbf{T} = \sum_i \sum_j t_{ij}\,\mathbf{e}^i\mathbf{e}^j ,$$

where $\mathbf{e}^i\mathbf{e}^j$ is the dyadic product of the base vectors. In index notation the expression t_{ij} denotes the total tensor.

2.2 Index rules and summation convention

(i) Index rule

> If a letter index appears one and only one time in each term of an expression, the expression is valid for each of the actual values, the letter index can take. Such an index is called a *free index.*

(ii) EINSTEIN's summation convention

> Whenever a letter index appears twice within the same term, as subscript and/or as superscript, a summation is implied over the range of this index, i.e., from 1 to 3 in the three-dimensional space (Latin indices used), and from 1 to 2 in the two-dimensional space (Greek indices used). Such an index is called *dummy.*

(iii) Maximum rule

> Any letter index may never be applied more than *twice* in each term.

Examples of (i):

$$a^i + 2b^i = 0 \quad <=> \quad \begin{cases} a^1 + 2b^1 = 0 , \\ a^2 + 2b^2 = 0 , \\ a^3 + 2b^3 = 0 . \end{cases}$$

$$t_\beta = T_{,\beta} \quad <=> \quad \begin{cases} t_1 = T_{,1} = \dfrac{\partial T}{\partial x_1}, \\ t_2 = T_{,2} = \dfrac{\partial T}{\partial x_2}. \end{cases}$$

Note: Comma implies partial differentiation with respect to the coordinate(s) of succeeding indices. The rules *(i)* – *(iii)* apply for these indices as well.

Examples of (ii):

$$\mathbf{a} = a^i \mathbf{e}_i = a^1 \mathbf{e}_1 + a^2 \mathbf{e}_2 + a^3 \mathbf{e}_3, \qquad \text{three-dimensional space},$$

$$\mathbf{a} = a_\alpha \mathbf{e}^\alpha = a_1 \mathbf{e}^1 + a_2 \mathbf{e}^2, \qquad \text{two-dimensional space (surface)},$$

$$\mathbf{T} = t_{ij} \mathbf{e}^i \mathbf{e}^j = t_{11} \mathbf{e}^1 \mathbf{e}^1 + t_{12} \mathbf{e}^1 \mathbf{e}^2 + \ldots + t_{33} \mathbf{e}^3 \mathbf{e}^3,$$

$$t_i^i = t_j^j = t_1^1 + t_2^2 + t_3^3,$$

$$df = f_{,i}\, dx^i = \frac{\partial f}{\partial x^1} dx^1 + \frac{\partial f}{\partial x^2} dx^2 + \frac{\partial f}{\partial x^3} dx^3.$$

Attention: As it is of no importance which notation a doubly appearing index possesses, this so-called *dummy* index can be arbitrarily renamed:

$$\mathbf{a} = a^i \mathbf{e}_i = a^j \mathbf{e}_j = a^k \mathbf{e}_k = \ldots.$$

Exception: No summation over paranthesized indices, i.e.

$$a_i^* = a_i \sqrt{g^{(ii)}} \longrightarrow a_1^* = a_1 \sqrt{g^{11}} \quad \text{etc}.$$

Examples of (iii):

Following expressions are meaningless:

$$c_i t_i^i = 0 \quad , \quad b_\alpha^\alpha \cos \Phi_\alpha = 1.$$

The following expressions are also meaningless, as the free indices have to be the same in each term:

$$t_j^i + b_k^i = 0 \quad , \quad A^\alpha_{\beta,\alpha} = B^\alpha_\beta.$$

2.3 Tensor of first order (vector)

Base vectors (Fig. 2.2)

$\mathbf{e}_i$ = orthogonal base with the unit vectors $\mathbf{e}_1, \mathbf{e}_2, \mathbf{e}_3$,

$\mathbf{g}_i$ = base in arbitrary coordinates with the base vectors $\mathbf{g}_1, \mathbf{g}_2, \mathbf{g}_3$.

Measure or metric components

$$\mathbf{g}_i \cdot \mathbf{g}_j = g_{ij} = g_{ji} = \mathbf{g}_j \cdot \mathbf{g}_i. \tag{2.1a}$$

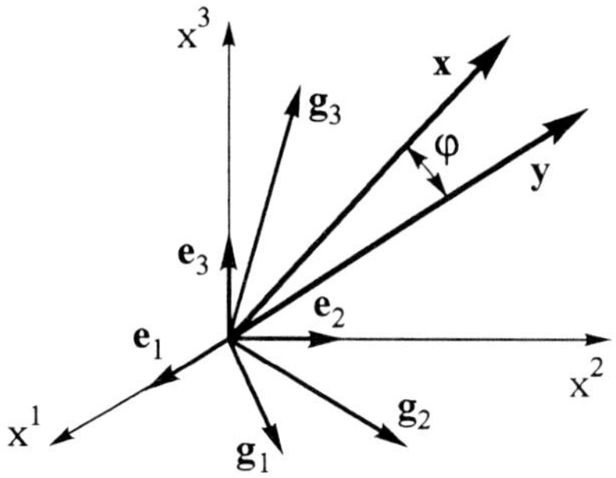

Fig. 2.2: Orthonormalized and arbitrary base

Metric tensor

$$(g_{ij}) = \begin{bmatrix} g_{11} & g_{12} & g_{13} \\ \cdot & g_{22} & g_{23} \\ \cdot & \cdot & g_{33} \end{bmatrix} \longrightarrow \quad \text{Due to } (2.1a) \text{ the metric tensor is } \textit{symmetrical}. \tag{2.1b}$$

Determinant of the metric tensor $\quad \det(\mathbf{g}) = |\mathbf{g}| = |g_{ij}|$. (2.1c)

Scalar product $\mathbf{x} \cdot \mathbf{y}$ of the vectors $\mathbf{x} = x^i \mathbf{g}_i$ and $\mathbf{y} = y^j \mathbf{g}_j$ (Fig. 2.2)

$$\mathbf{x} \cdot \mathbf{y} = g_{ij}\, x^i y^j \,. \tag{2.2a}$$

Length of a vector $\mathbf{x}$

$$d = |\mathbf{x}| = \sqrt{g_{ij}\, x^i x^j} \,. \tag{2.2b}$$

Angle φ between vectors $\mathbf{x}$ and $\mathbf{y}$

$$\cos\varphi = \frac{\mathbf{x} \cdot \mathbf{y}}{|\mathbf{x}| \cdot |\mathbf{y}|} = \frac{g_{ij}\, x^i y^j}{\sqrt{g_{mn}\, y^m y^n}\, \sqrt{g_{pq}\, x^p x^q}} \,. \tag{2.2c}$$

Covariant and contravariant base
An arbitrary base $g_i\,(i = 1,2,3)$ is given in the three-dimensional EUCLIDEAN space. We are searching a second base g^j so that the following relation exists between the base vectors :

$$\mathbf{g}_i \cdot \mathbf{g}^j = \delta_i^j \,, \tag{2.3a}$$

where KRONECKER's delta is defined by

$$\delta_i^j = \begin{cases} 1 & \text{for } i = j \,, \\ 0 & \text{for } i \neq j \,. \end{cases} \tag{2.3b}$$

If the base g_i is known, the base g^j can be determined by means of the nine equations *(2.3a)*. The base g_i is called the covariant base and g^j the contravariant base.

Covariant metric components

$$g_{ij} = \mathbf{g}_i \cdot \mathbf{g}_j = g_{ji} \, . \tag{2.4a}$$

Contravariant metric components

$$g^{ij} = \mathbf{g}^i \cdot \mathbf{g}^j = g^{ji} \, . \tag{2.4b}$$

Rule of exchanging indices

$$\mathbf{g}^i = g^{ij} \mathbf{g}_j \, , \tag{2.5a}$$

$$\mathbf{g}_i = g_{ij} \mathbf{g}^j \, , \tag{2.5b}$$

$$\delta^i_k = g^{ij} g_{jk} \, . \tag{2.5c}$$

Other determination of the contravariant base vectors

$$\mathbf{g}^1 = \frac{\mathbf{g}_2 \times \mathbf{g}_3}{[\mathbf{g}_1 , \mathbf{g}_2 , \mathbf{g}_3]} \, , \quad \mathbf{g}^2 = \frac{\mathbf{g}_3 \times \mathbf{g}_1}{[\mathbf{g}_1 , \mathbf{g}_2 , \mathbf{g}_3]} \, , \quad \mathbf{g}^3 = \frac{\mathbf{g}_1 \times \mathbf{g}_2}{[\mathbf{g}_1 , \mathbf{g}_2 , \mathbf{g}_3]} \, , \tag{2.6}$$

where $[\mathbf{g}_1 , \mathbf{g}_2 , \mathbf{g}_3]$ is the scalar triple product of the three covariant base vectors $\mathbf{g}_1$, $\mathbf{g}_2$, $\mathbf{g}_3$.

Transformation behaviour
A fundamental (defining) property of a tensor is its behaviour in connection with a coordinate transformation. In order to investigate this transformation behaviour, the following task shall be considered:
An initial base $\mathbf{g}_i$ or $\mathbf{g}^i$ ($i = 1,2,3$) is given together with a "new" base $\mathbf{g}_{i'}$ or $\mathbf{g}^{i'}$ ($i' = 1,2,3$) generated by an arbitrary linear transformation with prescribed transformation coefficients $\beta^j_{i'}$. Additionally, a vector be given in the initial base by its components v^i or v_i. Its components $v_{i'}$ or $v^{i'}$ shall now be determined in the "new" base.

Rules of transformation
Transformations of bases

$$\mathbf{g}_{i'} = \beta^j_{i'} \mathbf{g}_j \, , \quad \mathbf{g}^{i'} = \beta^{i'}_j \mathbf{g}^j \, , \tag{2.7a}$$

$$\mathbf{g}_k = \beta^{i'}_k \mathbf{g}_{i'} \, , \quad \mathbf{g}^k = \beta^k_{i'} \mathbf{g}^{i'} \, . \tag{2.7b}$$

The following relations are valid

$$\beta^j_{i'} \beta^{k'}_j = \delta^{k'}_{i'} \, , \quad \beta^{j'}_i \beta^k_{j'} = \delta^k_i \, . \tag{2.8}$$

Transformations of tensors of first order

$$v_{j'} = \beta^{i}_{j'} v_i \quad , \quad v^{j'} = \beta^{j'}_{i} v^{i} , \tag{2.9a}$$

$$v_j = \beta^{i'}_{j} v_{i'} \quad , \quad v^{j} = \beta^{j}_{i'} v^{i'} . \tag{2.9b}$$

Physical components of tensors of first order (vector)

$$a^{*i} = a^{i} \sqrt{g_{(ii)}} \quad \text{or} \quad a^{*}_{i} = a_i \sqrt{g^{(ii)}} . \tag{2.10}$$

2.4 Tensors of second and higher order

Definitions:

Two vectors $\mathbf{x}$ and $\mathbf{y}$ are given in the EUCLIDEAN space. With that we are forming the new product

$$\mathbf{T} = \mathbf{x}\,\mathbf{y} . \tag{2.11}$$

The notation without dot or cross shall indicate that it is neither a scalar product nor a vector product.

Depending on whether the covariant base vectors $\mathbf{g}_i$ or the contravariant base vectors $\mathbf{g}^i$ are applied here, one obtains four kinds of descriptions for a tensor of order two:

$$\mathbf{T} = t^{ij}\,\mathbf{g}_i\,\mathbf{g}_j = t^{i}{}_{j}\,\mathbf{g}_i\,\mathbf{g}^j = t_i{}^{j}\,\mathbf{g}^i\,\mathbf{g}_j = t_{ij}\,\mathbf{g}^i\,\mathbf{g}^j . \tag{2.12}$$

According to the position of the indices one denotes

t_{ij} as covariant components ,

t^{ij} as contravariant components ,

$t^{i}{}_{j}$ as mixed contravariant-covariant components ,

$t_i{}^{j}$ as mixed covariant-contravariant components of the tensor $\mathbf{T}$.

Formal generalization of tensors

$\mathbf{T}^{(0)} = t$	tensor of zeroth order (scalar)	$3^0 =$	1 base element,
$\mathbf{T}^{(1)} = t^{i}\,\mathbf{g}_i$	tensor of first order (vector)	$3^1 =$	3 base elements,
$\mathbf{T}^{(2)} = t^{ij}\,\mathbf{g}_i\,\mathbf{g}_j$	tensor of second order (dyad)	$3^2 =$	9 base elements,
$\mathbf{T}^{(3)} = t^{ijk}\,\mathbf{g}_i\,\mathbf{g}_j\,\mathbf{g}_k$	tensor of third order	$3^3 =$	27 base elements,
$\mathbf{T}^{(4)} = t^{ijkl}\,\mathbf{g}_i\,\mathbf{g}_j\,\mathbf{g}_k\,\mathbf{g}_l$	tensor of fourth order	$3^4 =$	81 base elements.

Transformation rules

For a transformation of a vector base $\mathbf{g}_i$ into a new vector base $\mathbf{g}_{i'}$ equations *(2.7a)* and *(2.7b)* are used:

$$\mathbf{g}_i = \beta_i^{j'} \mathbf{g}_{j'} \quad \text{and} \quad \mathbf{g}_{i'} = \beta_{i'}^{j} \mathbf{g}_j \ .$$

The tensor $\mathbf{T}$ can be given either in the old base $\mathbf{g}_i$ or in the new base $\mathbf{g}_{i'}$

$$\mathbf{T} = t^{k'l'} \mathbf{g}_{k'} \mathbf{g}_{l'} = t^{ij} \mathbf{g}_i \mathbf{g}_j \ . \qquad (2.13)$$

The transformation formulas read as follows

$$t^{ij} = \beta_{k'}^{i} \beta_{l'}^{j} t^{k'l'} \quad \text{or} \quad t^{i'j'} = \beta_k^{i'} \beta_l^{j'} t^{kl} \ . \qquad (2.14)$$

From $\mathbf{T} = t_{ij} \mathbf{g}^i \mathbf{g}^j = t_{k'l'} \mathbf{g}^{k'l'}$ *(2.15)*

follows $t_{ij} = \beta_i^{k'} \beta_j^{l'} t_{k'l'}$ or $t_{i'j'} = \beta_{i'}^{k} \beta_{j'}^{l} t_{kl}$. *(2.16)*

In a similar way one obtains the transformation formulas of the mixed components of the tensor $\mathbf{T}$.

Note: It is worth mentioning that tensors are actually defined by the rules by which their components transform due to coordinate transformations. Thus, any quantity $\mathbf{T}$ with $3^2 = 9$ components is then and only then a *second order tensor* if its components transform according to *(2.14)* or *(2.16)* in connection with an *arbitrary* coordinate transformation.

Physical components of a tensor of second order

The physical components for *orthogonal* coordinate systems can be determined as follows (for non-orthogonal coordinate systems see [A.8]):

$$\left.\begin{aligned} t^{*ij} &= t^{ij} \sqrt{g_{(ii)}} \sqrt{g_{(jj)}} \ , \\ t^{*i}{}_{j} &= t^{i}{}_{j} \sqrt{g_{(ii)}} \sqrt{g^{(jj)}} \ , \\ t^{*}{}_{i}{}^{j} &= t_{i}{}^{j} \sqrt{g^{(ii)}} \sqrt{g_{(jj)}} \ , \\ t^{*}{}_{ij} &= t_{ij} \sqrt{g^{(ii)}} \sqrt{g^{(jj)}} \ . \end{aligned}\right\} \qquad (2.17)$$

Symmetrical and antisymmetrical tensors of second order

Any tensor of second order can always be presented as a sum of a symmetrical and an antisymmetrical (or skew-symmetrical) tensor:

$$t^{ij} = \overset{(s)}{t}{}^{ij} + \overset{(a)}{t}{}^{ij} \qquad (2.18a)$$

with $\overset{(s)}{t}{}^{ij} = \frac{1}{2}(t^{ij} + t^{ji})$, *(2.18b)*

$$\overset{(a)}{t}{}^{ij} = \frac{1}{2}(t^{ij} - t^{ji}) \ . \qquad (2.18c)$$

Permutation tensor or $\boldsymbol{\varepsilon}$*-tensor*

As permutation tensor a tensor of third order is defined

$$\varepsilon_{ijk} = \sqrt{g}\, e_{ijk} \quad , \quad \varepsilon^{ijk} = \frac{1}{\sqrt{g}}\, e^{ijk} \tag{2.19}$$

with the permutation symbol

$$e_{ijk} = e^{ijk} = \begin{cases} +1 & \text{for } \{i,j,k\} \text{ cyclic} \quad , \\ -1 & \text{"} \quad \{i,j,k\} \text{ anticyclic} , \\ 0 & \text{"} \quad \{i,j,k\} \text{ acyclic} \quad . \end{cases} \tag{2.20a}$$

Permutation symbol in two dimensions

$$\begin{aligned} e_{11} &= 0 \quad , \quad e_{12} = +1 , \\ e_{21} &= -1 \quad , \quad e_{22} = 0 . \end{aligned} \tag{2.20b}$$

Vector product as application of the $\boldsymbol{\varepsilon}$*-tensor*

$$\mathbf{x} \times \mathbf{y} = \varepsilon_{klm} x^k y^l \mathbf{g}^m . \tag{2.21}$$

Eigenvalues and eigenvectors of a symmetrical tensor
– Principal axis transformation

Lemma: For any symmetrical, real valued, three–column matrix $\mathbf{T}$ there always exist three mutually orthogonal principal directions (eigenvectors) $\mathbf{a}$ and three corresponding real eigenvalues λ (which not necessarily have to be different from each other). These eigenvectors and eigenvalues are governed by the following algebraic eigenvalue problem, where $\mathbf{I}$ is the unit tensor:

$$(\mathbf{T} - \lambda \mathbf{I})\,\mathbf{a} = \mathbf{0} \quad \text{or} \quad (t_i^j - \lambda\, \delta_i^j)\, a_j = 0 . \tag{2.22a}$$

Determination of the eigenvalues:

$$\det(t_i^j - \lambda\, \delta_i^j) = \begin{vmatrix} t_1^1 - \lambda & t_2^1 & t_3^1 \\ t_1^2 & t_2^2 - \lambda & t_3^2 \\ t_1^3 & t_2^3 & t_3^3 - \lambda \end{vmatrix} = 0 . \tag{2.22b}$$

Characteristic equation of *(2.22b)*:

$$\lambda^3 - I_1 \lambda^2 + I_2 \lambda - I_3 = 0 . \tag{2.22c}$$

The roots $\lambda = \lambda_I$, λ_{II} and λ_{III} of this cubic equation are invariant with respect to transformations of coordinates. Substituting sequentially these eigenvalues into *(2.22a)* and solving for $\mathbf{a}$, we obtain $\mathbf{a}_I$, $\mathbf{a}_{II}$ and $\mathbf{a}_{III}$.

The quantities I_1, I_2, I_3 in *(2.22c)* are invariants defined by [A.8]:

$$I_1 = t^i_i \, , \qquad (2.23a)$$

$$I_2 = \frac{1}{2}(t^i_i t^j_j - t^i_j t^j_i) \, , \qquad (2.23b)$$

$$I_3 = \det(t^i_j) \, . \qquad (2.23c)$$

2.5 Curvilinear coordinates

Base vectors – metric tensor

In the three-dimensional space a vector $\mathbf{r}$ can be presented in Cartesian coordinates x^i and in curvilinear coordinates indicated by $\xi^i (i = 1, 2, 3)$ (Figs. 2.3 and 2.4).

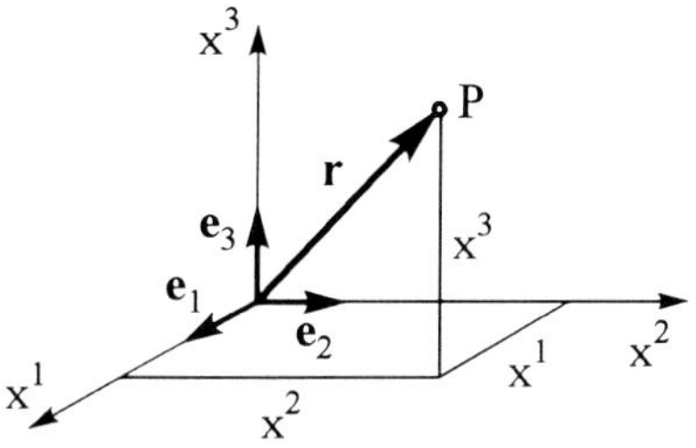

Fig. 2.3: Position vector in orthonormalized base

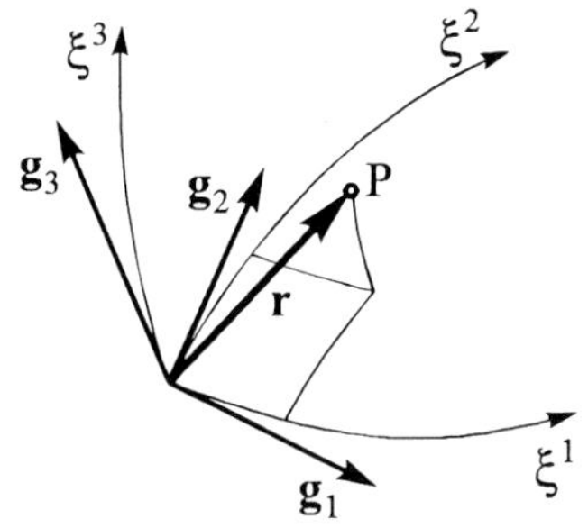

Fig. 2.4: Curvilinear coordinates and base vectors

Position vector $\mathbf{r}$ of a point P

$$\mathbf{r} = \mathbf{r}(x^i) \quad \Big| \quad \mathbf{r} = \mathbf{r}(\xi^i) \, . \qquad (2.24)$$

Base vectors

$$\mathbf{e}_i = \frac{\partial \mathbf{r}}{\partial x^i} = \mathbf{r}_{,i} \quad \Big| \quad \mathbf{g}_i = \frac{\partial \mathbf{r}}{\partial \xi^i} = \mathbf{r}_{,i} \, . \qquad (2.25)$$

Relation between base $g_i(\xi^i)$ and orthonormalized base $\mathbf{e}_i$

$$\mathbf{g}_k = \frac{\partial x^i}{\partial \xi^k} \mathbf{e}_i \, . \qquad (2.26)$$

Length of a line element

$$ds^2 = d\mathbf{r} \cdot d\mathbf{r} \longrightarrow \text{First fundamental form of a surface.} \qquad (2.27a)$$

Indicating the derivative with respect to the curve parameter t by a dot, the length of the curve between t_0 and t_1 is given by :

$$s = \int_{t_0}^{t_1} \sqrt{g_{ij} \dot{\xi}^i \dot{\xi}^j} \, dt \, . \qquad (2.27b)$$

Volume element

$$dV = \sqrt{g}\, d\xi^1\, d\xi^2\, d\xi^3 \,. \qquad (2.28)$$

Partial base derivatives – CHRISTOFFEL symbols

$$\mathbf{g}_{k,l} = \Gamma^{j}_{kl}\, \mathbf{g}_j \,, \qquad (2.29a)$$

$$\mathbf{g}^{i}_{,k} = -\Gamma^{i}_{kl}\, \mathbf{g}^{l} \,. \qquad (2.29b)$$

CHRISTOFFEL symbols of the first kind

$$\Gamma_{ijk} = \frac{1}{2}\left(g_{jk,i} + g_{ki,j} - g_{ij,k} \right) . \qquad (2.30)$$

CHRISTOFFEL symbols of the second kind

$$\Gamma^{m}_{ij} = g^{km}\, \Gamma_{ijk} \,. \qquad (2.31)$$

Rule: The CHRISTOFFEL symbols can be expressed alone by the metric tensor and its derivatives.

Note: The CHRISTOFFEL symbols do not have tensor character.

For the CHRISTOFFEL symbols of the first kind *(2.30)* the following relations hold:

1) $\Gamma_{ijk} = \Gamma_{jik}$ interchangeability of the *first two* indices , *(2.32a)*

2) $\Gamma_{mns} + \Gamma_{msn} = \dfrac{\partial g_{ns}}{\partial \xi^m}$ interchangeability of the *last two* indices . *(2.32b)*

For the CHRISTOFFEL symbols of the second kind, the following relations are derived from *(2.31)* using *(2.30)*:

1) $\Gamma^{k}_{ij} = \Gamma^{k}_{ji}$ interchangeability of subscripts (symmetry) , *(2.32c)*

2) $\Gamma^{i}_{ij} = \dfrac{1}{2} g^{ik} \dfrac{\partial g_{ik}}{\partial \xi^j} = \dfrac{\partial (\ln \sqrt{g}\,)}{\partial \xi^j}$. *(2.32d)*

Covariant derivatives

Tensor of first order

$$\mathbf{a}_{,j} = a^{i}|_{j}\, \mathbf{g}_i \qquad (2.33)$$

with $\quad a^{i}|_{j} = a^{i}_{,j} + \Gamma^{i}_{jk}\, a^{k} \,.$ *(2.34a)*

By anology $\quad a_{i}|_{j} = a_{i,j} - \Gamma^{k}_{ij}\, a_{k} \,.$ *(2.34b)*

Tensor of second order

$$a_{ij}|_k = a_{ij,k} - \Gamma^m_{ik}\, a_{mj} - \Gamma^m_{jk}\, a_{im} \,, \tag{2.35a}$$

$$a^{ij}|_k = a^{ij}_{\;,k} + \Gamma^i_{km}\, a^{mj} + \Gamma^j_{km}\, a^{im} \,. \tag{2.35b}$$

Gradient of a scalar funktion Φ

$$\mathbf{v} = \operatorname{grad} \Phi = \nabla \Phi = \Phi|_j\, \mathbf{g}^j \,. \tag{2.36a}$$

Gradient of a vector $\mathbf{v}$

$$\operatorname{Grad} \mathbf{v} = \nabla\, \mathbf{v} = v^j|_i\, \mathbf{g}^i \mathbf{g}_j \,. \tag{2.36b}$$

Divergence of a vector $\mathbf{v}$

$$\operatorname{div} \mathbf{v} = \nabla \cdot \mathbf{v} = v^j|_j = \frac{1}{\sqrt{g}} \frac{\partial}{\partial \xi^j} (\sqrt{g}\; v^j) \,. \tag{2.37a}$$

Divergence of a tensor $\mathbf{T}$ of second order

$$\operatorname{Div} \mathbf{T} = \nabla \cdot \mathbf{T} = t^{kl}|_k\, \mathbf{g}_l \,. \tag{2.37b}$$

Rotation of a vector $\mathbf{v}$

$$\operatorname{rot} \mathbf{v} = \nabla \times \mathbf{v} = v^j|_i\, (\mathbf{g}^i \times \mathbf{g}_j) \,. \tag{2.38}$$

LAPLACE operator

$$\Delta \Phi = \nabla^2 \Phi = \operatorname{div} \operatorname{grad} \Phi = \Phi|^i_i = \frac{1}{\sqrt{g}} (\sqrt{g}\, g^{jk}\, \Phi_{,k})_{,j} \,. \tag{2.39}$$

Bipotential operator

$$\begin{aligned} \Delta\Delta \Phi &= \nabla^4 \Phi = \nabla^2 (\nabla^2 \Phi) = \Phi|^{ij}_{ij} = \\ &= \frac{1}{\sqrt{g}} \left\{ \sqrt{g}\, g^{ij} \left[\frac{1}{\sqrt{g}} \left(\sqrt{g}\, g^{kl}\, \Phi_{,l} \right)_{,k} \right]_{,i} \right\}_{,j} \,. \end{aligned} \tag{2.40}$$

GAUSSIAN theorem

$$\iiint\limits_V v^j|_j \sqrt{g}\, d\xi^1\, d\xi^2\, d\xi^3 = \iint\limits_A v^j\, n_j\, dA \,. \tag{2.41}$$

Example: Application of the previous formulas to cylindrical coordinates
Single-valued relations between Cartesian coordinates x^i and cylindrical coordinates ξ^i read as follows (see Fig. 2.5):

$$x^1 = \xi^1 \cos \xi^2 \quad , \quad x^2 = \xi^1 \sin \xi^2 \quad , \quad x^3 = \xi^3 . \tag{2.42a}$$

Position vector

$$\mathbf{r}(\xi^i) = \xi^1 \cos \xi^2 \, \mathbf{e}_1 + \xi^1 \sin \xi^2 \, \mathbf{e}_2 + \xi^3 \, \mathbf{e}_3 . \tag{2.42b}$$

Covariant base vectors according to *(2.25)*

$$\mathbf{g}_i = \mathbf{r}_{,i} = \frac{\partial \mathbf{r}}{\partial \xi^i} \longrightarrow \begin{cases} \mathbf{g}_1 = \cos \xi^2 \, \mathbf{e}_1 + \sin \xi^2 \, \mathbf{e}_2 , \\ \mathbf{g}_2 = -\xi^1 \sin \xi^2 \, \mathbf{e}_1 + \xi^1 \cos \xi^2 \, \mathbf{e}_2 , \\ \mathbf{g}_3 = \mathbf{e}_3 . \end{cases} \tag{2.43}$$

Covariant metric components according to *(2.4a)*

$$g_{ij} = \mathbf{g}_i \cdot \mathbf{g}_j .$$

For example: $g_{22} = \mathbf{g}_2 \cdot \mathbf{g}_2 = (\xi^1)^2 \sin^2 \xi^2 + (\xi^1)^2 \cos^2 \xi^2 = (\xi^1)^2 .$

Covariant metric tensor

$$(g_{ij}) = \begin{bmatrix} 1 & 0 & 0 \\ 0 & (\xi^1)^2 & 0 \\ 0 & 0 & 1 \end{bmatrix} . \tag{2.44}$$

According to *(2.5c)*, because of $g_{ij} = \mathbf{g}_i \cdot \mathbf{g}_j = 0$ for $i \neq j$

$\longrightarrow$ Orthogonal base

Contravariant components from *(2.5c)*

$$g^{(ii)} g_{(ii)} = 1 \longrightarrow g^{(ii)} = \frac{1}{g_{(ii)}} .$$

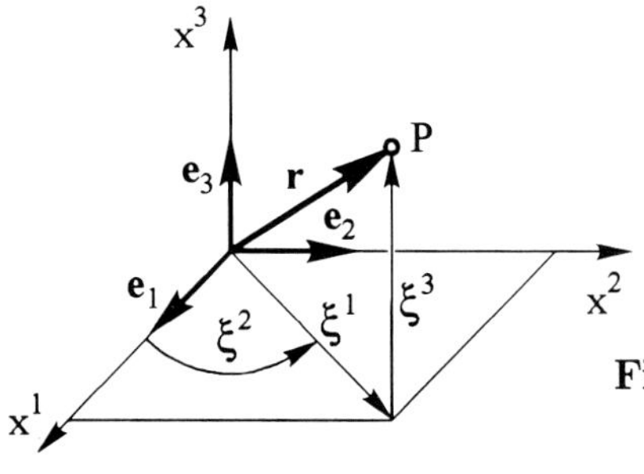

Fig. 2.5: Presentation of a position vector in cylindrical coordinates

Contravariant metric tensor

$$(g^{ij}) = \begin{bmatrix} 1 & 0 & 0 \\ 0 & (\xi^1)^{-2} & 0 \\ 0 & 0 & 1 \end{bmatrix} . \tag{2.45}$$

Determinant g of the covariant metric tensor

$$|\mathbf{g}| = |g_{ij}| = \begin{vmatrix} 1 & 0 & 0 \\ 0 & (\xi^1)^2 & 0 \\ 0 & 0 & 1 \end{vmatrix} = (\xi^1)^2 . \tag{2.46}$$

CHRISTOFFEL symbols of the first kind according to *(2.30)*

For example: $\Gamma_{221} = \frac{1}{2}(g_{21,2} + g_{12,2} - g_{22,1}) = \frac{1}{2}(0 + 0 - 2\xi^1) = -\xi^1$.

CHRISTOFFEL symbols in matrix notation

$$\left. \begin{aligned} (\Gamma_{ij1}) &= \begin{bmatrix} 0 & 0 & 0 \\ 0 & -\xi^1 & 0 \\ 0 & 0 & 0 \end{bmatrix} , \\ (\Gamma_{ij2}) &= \begin{bmatrix} 0 & \xi^1 & 0 \\ \xi^1 & 0 & 0 \\ 0 & 0 & 0 \end{bmatrix} , \\ (\Gamma_{ij3}) &= \mathbf{0} . \end{aligned} \right\} \tag{2.47}$$

CHRISTOFFEL symbols of the second kind according to *(2.31)*

For example: $\Gamma^1_{22} = g^{1k}\Gamma_{22k} = 1 \cdot (-\xi^1) + 0 \cdot 0 + 0 \cdot 0 = -\xi^1$.

CHRISTOFFEL symbols in matrix notation

$$\left. \begin{aligned} (\Gamma^1_{ij}) &= \begin{bmatrix} 0 & 0 & 0 \\ 0 & -\xi^1 & 0 \\ 0 & 0 & 0 \end{bmatrix} , \\ (\Gamma^2_{ij}) &= \begin{bmatrix} 0 & (\xi^1)^{-1} & 0 \\ (\xi^1)^{-1} & 0 & 0 \\ 0 & 0 & 0 \end{bmatrix} , \\ (\Gamma^3_{ij}) &= \mathbf{0} . \end{aligned} \right\} \tag{2.48}$$

LAPLACE operator according to *(2.39)*

$$\Delta\Phi = \frac{1}{\sqrt{g}}(\sqrt{g}\,g^{jk}\,\Phi_{,k})_{,j} = \frac{1}{\sqrt{g}}\left[\sqrt{g}\,(g^{j1}\Phi_{,1} + g^{j2}\,\Phi_{,2} + g^{j3}\,\Phi_{,3})\right]_{,j} =$$

$$= \frac{1}{\sqrt{g}}\left[(\sqrt{g}\,g^{11}\,\Phi_{,1})_{,1} + (\sqrt{g}\,g^{22}\,\Phi_{,2})_{,2} + (\sqrt{g}\,g^{33}\,\Phi_{,3})_{,3}\right] =$$

$$= \frac{1}{\xi^1}\left[(\xi^1\,\Phi_{,1})_{,1} + (\frac{1}{\xi^1}\Phi_{,2})_{,2} + (\xi^1\,\Phi_{,3})_{,3}\right] =$$

$$= \frac{1}{\xi^1}\left[\xi^1\,\Phi_{,11} + \Phi_{,1} + \frac{1}{\xi^1}\Phi_{,22} + \xi^1\,\Phi_{,33}\right]$$

$$\longrightarrow \quad \Delta\Phi = \Phi_{,11} + \frac{1}{\xi^1}\Phi_{,1} + \frac{1}{(\xi^1)^2}\Phi_{,22} + \Phi_{,33}\,. \qquad (2.49)$$

3 State of stress

3.1 Stress vector

The essential objective of structural analysis is the calculation of stresses and deformations of bodies. As shown in Fig. 3.1 we make a cut through the body, which is in equilibrium under external loads in the form of volume forces $\mathbf{f}_i$, surface tractions $\mathbf{p}_i$ and concentrated forces $\mathbf{F}_k$. A resulting force $\Delta\mathbf{F}$ is transmitted at every element Δ A of the cut.

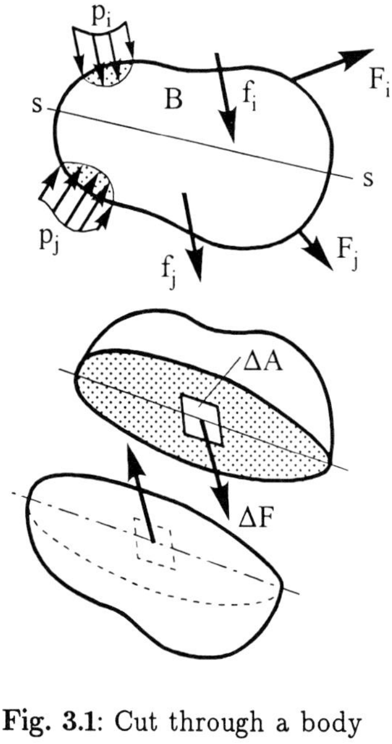

Fig. 3.1: Cut through a body

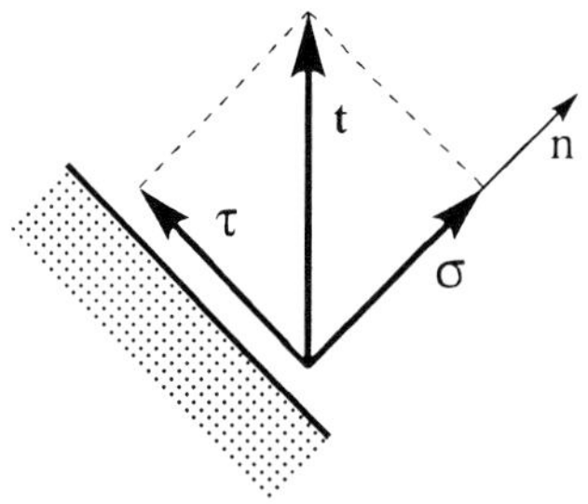

Fig. 3.2: Resolution of the stress vector

According to NEWTON's principle of "*actio* = *reactio*", reaction of the resulting force $\Delta\mathbf{F}$ is found on the same plane of the opposite part of the body, in the form of an opposite directed force of the same magnitude. We assume that the relation $\Delta\mathbf{F}/\Delta A$ in the limit of $\Delta A \longrightarrow 0$ tends to a finite value, and we call this limiting value

stress vector $$\mathbf{t} = \lim_{\Delta A \to 0} \frac{\Delta\mathbf{F}}{\Delta A} = \frac{d\mathbf{F}}{dA} . \qquad (3.1)$$

Here it is assumed that only *forces* (and no *moments*) are transmitted at any point of the cut.
The stress vector **t** can be resolved into a part perpendicular to the surface of the cut, the so-called *normal stress* of the value σ, and into a part tangential to the surface, the *shear stress* of the value τ (Fig. 3.2).

Sign convention: Stresses on cut planes with outward normals pointing in the positive (respective negative) coordinate directions, are taken positive in the positive (respective negative) coordinate directions (Fig. 3.3).

Stress vectors on the positive cut planes of the cubic element in Cartesian coordinates:

$$\mathbf{t}^x = \begin{bmatrix} \sigma_{xx} \\ \tau_{xy} \\ \tau_{xz} \end{bmatrix} , \quad \mathbf{t}^y = \begin{bmatrix} \tau_{yx} \\ \sigma_{yy} \\ \tau_{yz} \end{bmatrix} , \quad \mathbf{t}^z = \begin{bmatrix} \tau_{zx} \\ \tau_{zy} \\ \sigma_{zz} \end{bmatrix} . \qquad (3.2)$$

In this context, σ_{ii} $(i = x, y, z)$ are normal stresses and τ_{ij} $(i, j = x, y, z)$ are shear stresses.

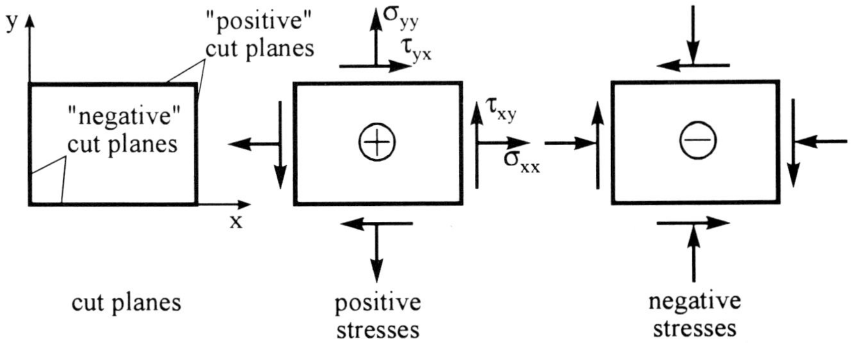

Fig. 3.3: Sign convention for the stresses

3.2 Stress tensor

The stress vectors can be assembled in matrix notation as the so-called stress tensor **S**. In Cartesian coordinates it reads

$$\mathbf{S}^T = [\,\mathbf{t}^x \,|\, \mathbf{t}^y \,|\, \mathbf{t}^z\,] = \begin{bmatrix} \sigma_{xx} & \tau_{yx} & \tau_{zx} \\ \tau_{xy} & \sigma_{yy} & \tau_{zy} \\ \tau_{xz} & \tau_{yz} & \sigma_{zz} \end{bmatrix}. \qquad (3.3)$$

The superscript T indicates transpose of a matrix.

The important CAUCHY's formula in arbitrary coordinates is written

$$\mathbf{t} = \mathbf{S}\,\mathbf{n} \qquad (3.4a)$$

or

$$t^i = \tau^{ji}\, n_j\,. \qquad (3.4b)$$

In words: This formula gives the stress vector **t** at a given surface or cut plane in terms of the stress tensor **S** and the unit outward normal vector **n** for the surface or cut.

The stress vector **t** acts on the infinitesimal area dA of the inclined cut plane characterized by the unit outward normal vector **n** (Fig. 3.4):

$$\mathbf{n} = \begin{bmatrix} n_x \\ n_y \\ n_z \end{bmatrix} = \begin{bmatrix} \cos\alpha \\ \cos\beta \\ \cos\gamma \end{bmatrix}, \quad \mathbf{t} = \begin{bmatrix} t_x \\ t_y \\ t_z \end{bmatrix}. \qquad (3.5a,b)$$

The remaining infinitesimal surfaces of the tetrahedron result from the projection of dA which can be written as follows in index notation with $x = x^1$, $y = x^2$, $z = x^3$:

$$\left.\begin{aligned} dA_x &= dA\cos\alpha = dA\, n_x \\ dA_y &= dA\cos\beta = dA\, n_y \\ dA_z &= dA\cos\gamma = dA\, n_z \end{aligned}\right\} \quad dA_i = dA\, n_i\,. \qquad (3.6)$$

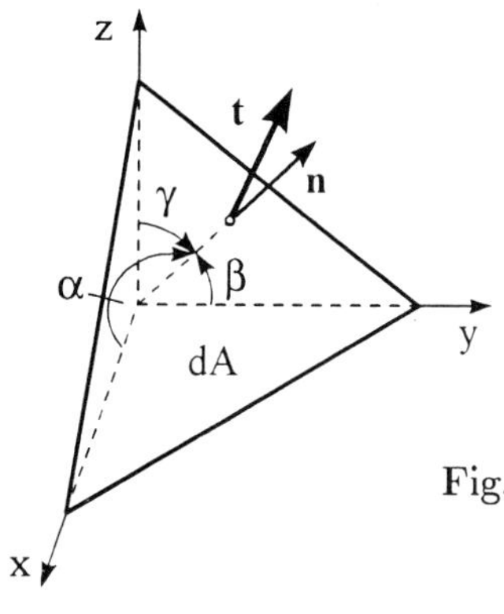

Fig. 3.4: Stress vector **t** at a tetrahedron in Cartesian coordinates

Relationships between Cartesian and other coordinates will be given later. In Cartesian coordinates there is no difference between a covariant and a contravariant base, and for this reason the indices can always be lowered.

Component notation in Cartesian coordinates

$$\left.\begin{aligned} t_x &= \sigma_{xx}\, n_x + \tau_{yx}\, n_y + \tau_{zx}\, n_z \,, \\ t_y &= \tau_{xy}\, n_x + \sigma_{yy}\, n_y + \tau_{zy}\, n_z \,, \\ t_z &= \tau_{xz}\, n_x + \tau_{yz}\, n_y + \sigma_{zz}\, n_z \,. \end{aligned}\right\} \qquad (3.7)$$

Note: Shear stresses are pairwise equal to one another, i.e., *the stress tensor* $\boldsymbol{S}$ *is symmetric* ⟶

$$\tau_{xy} = \tau_{yx} \quad , \quad \tau_{yz} = \tau_{zy} \quad , \quad \tau_{zx} = \tau_{xz} \,. \qquad (3.8)$$

The symmetry reflects satisfaction of moment equilibrium conditions.

3.3 Coordinate transformation – principal axes

We consider a Cartesian coordinate system x^i and a rotated system $x^{i'}$ (see Fig. 3.5).

Stresses in a rotated system according to *(2.14)*

$$\tau^{i'j'} = \beta^{i'}_k \, \beta^{j'}_l \, \tau^{kl} \,.$$

Symbolic notation

$$\mathbf{S}' = \mathbf{B} \cdot \mathbf{S} \cdot \mathbf{B}^T \,. \qquad (3.9)$$

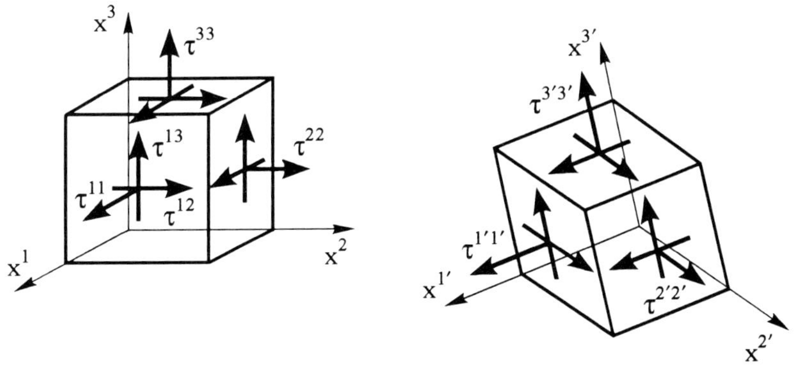

Fig. 3.5: Stresses in a rotated coordinate system

Arrangement of transformation coefficients in a rotation matrix **B**

$$\mathbf{B} = \begin{bmatrix} \beta_1^{1'} & \beta_2^{1'} & \beta_3^{1'} \\ \beta_1^{2'} & \beta_2^{2'} & \beta_3^{2'} \\ \beta_1^{3'} & \beta_2^{3'} & \beta_3^{3'} \end{bmatrix} . \qquad (3.10)$$

Principal stresses, principal axes

Principal stresses (see *(2.22)* and *(2.23)*)

$$(\tau_j^i - \sigma \delta_j^i)\, n_i^* = 0 . \qquad (3.11)$$

Characteristic equation

$$\sigma^3 - I_1 \sigma^2 + I_2 \sigma - I_3 = 0 \qquad (3.12)$$

with the invariants for any direction and for the principal stresses σ_i ($i = I, II, III$):

$$I_1 = \sigma_{xx} + \sigma_{yy} + \sigma_{zz} = \sigma_I + \sigma_{II} + \sigma_{III} \quad \text{sum of normal stresses ,} \qquad (3.13a)$$

$$I_2 = \begin{vmatrix} \sigma_{xx} & \tau_{xy} \\ \tau_{xy} & \sigma_{yy} \end{vmatrix} + \begin{vmatrix} \sigma_{yy} & \tau_{yz} \\ \tau_{yz} & \sigma_{zz} \end{vmatrix} + \begin{vmatrix} \sigma_{zz} & \tau_{zx} \\ \tau_{zx} & \sigma_{xx} \end{vmatrix} = \sigma_I \sigma_{II} + \sigma_{II} \sigma_{III} + \sigma_{III} \sigma_I , \qquad (3.13b)$$

$$I_3 = \begin{vmatrix} \sigma_{xx} & \tau_{xy} & \tau_{xz} \\ \tau_{xy} & \sigma_{yy} & \tau_{yz} \\ \tau_{xz} & \tau_{yz} & \sigma_{zz} \end{vmatrix} = \sigma_I \sigma_{II} \sigma_{III} . \qquad (3.13c)$$

Note: It can be shown that the three roots of *(3.12)* comprise the maximum and the minimum normal stress appearing on all possible cut planes through a given point. That is where the name *principal stresses* is coming from. For the symmetrical stress tensor the principal stresses are always real. The directions of principal stresses of different magnitudes are always unique and mutually orthogonal.

State of plane stress in Cartesian coordinates

Definition: $\sigma_{zz} = \tau_{xz} = \tau_{yz} = 0 .$ *(3.14)*

Stress tensor

$$\mathbf{S} = \begin{bmatrix} \sigma_{xx} & \tau_{xy} \\ \tau_{xy} & \sigma_{yy} \end{bmatrix} . \qquad (3.15)$$

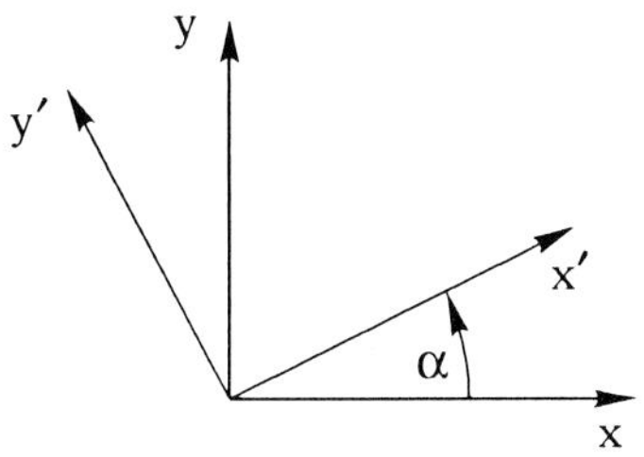

Fig. 3.6: Coordinate transformation

Transformation coefficients according to Fig. 3.6

$$\left.\begin{aligned} \beta_1^{1'} &= \cos\alpha \;, & \beta_2^{1'} &= \cos\left(\frac{\pi}{2} - \alpha\right) = \sin\alpha \;, \\ \beta_1^{2'} &= \cos\left(\frac{\pi}{2} + \alpha\right) = -\sin\alpha \;, & \beta_2^{2'} &= \cos\alpha \;. \end{aligned}\right\} \qquad (3.16)$$

Formulas of transformation for any rotation α of the coordinate system

$$\left.\begin{aligned} \sigma_{x'x'} &= \frac{1}{2}(\sigma_{xx} + \sigma_{yy}) + \frac{1}{2}(\sigma_{xx} - \sigma_{yy})\cos 2\alpha + \tau_{xy}\sin 2\alpha \;, \\ \sigma_{y'y'} &= \frac{1}{2}(\sigma_{xx} + \sigma_{yy}) - \frac{1}{2}(\sigma_{xx} - \sigma_{yy})\cos 2\alpha - \tau_{xy}\sin 2\alpha \;, \\ \tau_{x'y'} &= \frac{1}{2}(\sigma_{yy} - \sigma_{xx})\sin 2\alpha + \tau_{xy}\cos 2\alpha \;. \end{aligned}\right\} \qquad (3.17)$$

Principal stresses

$$\left.\begin{matrix}\sigma_I \\ \sigma_{II}\end{matrix}\right\} = \frac{1}{2}(\sigma_{xx} + \sigma_{yy}) \pm \sqrt{\left(\frac{\sigma_{xx} - \sigma_{yy}}{2}\right)^2 + \tau_{xy}^2} \;. \qquad (3.18)$$

The directions of the principal stresses follow from the extremal condition to be

$$\tan 2\alpha^* = \frac{2\tau_{xy}}{\sigma_{xx} - \sigma_{yy}} \qquad (3.19)$$

and from this the principal directions $2\alpha^*$ and $2\alpha^* + \pi$ or α^* and $\alpha^* + \frac{\pi}{2}$.

The principal directions are *orthogonal* to each other.

Maximum shear stress

$$\tau_{max} = \sqrt{\left(\frac{\sigma_{xx} - \sigma_{yy}}{2}\right)^2 + \tau_{xy}^2} = \frac{\sigma_I - \sigma_{II}}{2} \;. \qquad (3.20)$$

Direction $\quad \alpha^{**} = \alpha^* \pm \frac{\pi}{4}$.

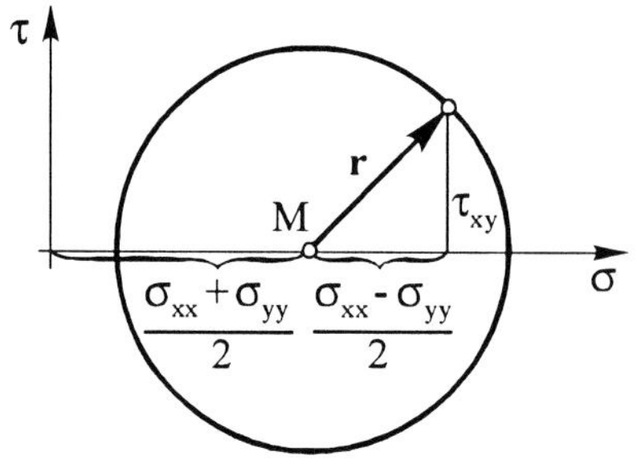

Fig. 3.7: MOHR's stress circle

MOHR's circle
The formulas for transformation of the plane state of stress lead to the MOHR's circle ($\sigma_{x'x'} \mathrel{\hat{=}} \sigma$, $\tau_{x'y'} \mathrel{\hat{=}} \tau$)

$$\left(\sigma - \frac{\sigma_{xx} + \sigma_{yy}}{2}\right)^2 + \tau^2 = (\sigma - \sigma_0)^2 + \tau^2 = r^2 \tag{3.21}$$

with the distance of the centre M on the σ-axis

$$\sigma_0 = \frac{1}{2}(\sigma_{xx} + \sigma_{yy}) \tag{3.22a}$$

and the radius of the circle

$$r = \sqrt{\left(\frac{\sigma_{xx} - \sigma_{yy}}{2}\right)^2 + \tau_{xy}^2} \mathrel{\hat{=}} \tau_{max} \,. \tag{3.22b}$$

3.4 Stress deviator

Definition: $${}'\tau^i_j = \tau^i_j - \sigma_M \delta^i_j \tag{3.23}$$

with the mean normal stress σ_M

$$\sigma_M = \frac{1}{3}(\sigma_{xx} + \sigma_{yy} + \sigma_{zz}) = \frac{1}{3}(\sigma_I + \sigma_{II} + \sigma_{III}) = \frac{1}{3}I_1 \,. \tag{3.24}$$

Physical interpretation:
The stress deviator ${}'\tau^i_j$ expresses the deviation of the state of stress from the mean normal stress.

Since ${}'I_1 = 0$, the principal values of the stress deviator follow in analogy to *(3.12)* from

$${}'\sigma^3 + {}'I_2\,{}'\sigma - {}'I_3 = 0 \tag{3.25}$$

with

$$\left.\begin{aligned} {}'I_2 &= I_2 - 3\,\sigma_M^2 \,, \\ {}'I_3 &= I_3 - I_2\,\sigma_M + 2\,\sigma_M^3 \,. \end{aligned}\right\} \tag{3.26}$$

3.5 Equilibrium conditions

The conditions of *force equilibrium* are stated with regard to the *undeformed* configuration of the body in this section [A.11, A.15, A.16, A.17].

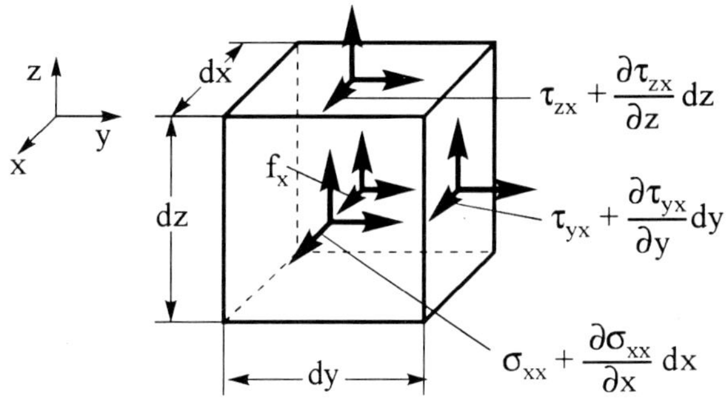

Fig. 3.8: Equilibrium for an infinitesimal volume element in Cartesian coordinates

1) Cartesian coordinates (Fig. 3.8)

$$\left.\begin{aligned} \frac{\partial \sigma_{xx}}{\partial x} + \frac{\partial \tau_{yx}}{\partial y} + \frac{\partial \tau_{zx}}{\partial z} + f_x &= 0\,, \\ \frac{\partial \tau_{xy}}{\partial x} + \frac{\partial \sigma_{yy}}{\partial y} + \frac{\partial \tau_{zy}}{\partial z} + f_y &= 0\,, \\ \frac{\partial \tau_{xz}}{\partial x} + \frac{\partial \tau_{yz}}{\partial y} + \frac{\partial \sigma_{zz}}{\partial z} + f_z &= 0\,. \end{aligned}\right\} \tag{3.27a}$$

f_i ($i = x, y, z$) are the components of the vector of volume forces.

Abbreviated notation

$$\tau^{ji}{}_{,j} + f^i = 0\,. \tag{3.27b}$$

2) Curvilinear coordinates

$$\tau^{ji}\big|_j + f^i = 0 \tag{3.28a}$$

or $$\mathrm{Div}\,\mathbf{S} + \mathbf{f} = 0\,. \tag{3.28b}$$

3) Cylindrical coordinates ($\xi^1 \mathrel{\widehat{=}} r$, $\xi^2 \mathrel{\widehat{=}} \varphi$, $\xi^3 \mathrel{\widehat{=}} z$) (Fig. 2.5)

$$\left.\begin{aligned} \frac{\partial \sigma_{rr}}{\partial r} + \frac{1}{r}\frac{\partial \tau_{\varphi r}}{\partial \varphi} + \frac{\partial \tau_{zr}}{\partial z} + \frac{1}{r}(\sigma_{rr} - \sigma_{\varphi\varphi}) + f_r &= 0\,, \\ \frac{\partial \tau_{\varphi r}}{\partial r} + \frac{1}{r}\frac{\partial \sigma_{\varphi\varphi}}{\partial \varphi} + \frac{\partial \tau_{z\varphi}}{\partial z} + \frac{2}{r}\tau_{\varphi r} + f_\varphi &= 0\,, \\ \frac{\partial \tau_{rz}}{\partial r} + \frac{1}{r}\frac{\partial \tau_{\varphi z}}{\partial \varphi} + \frac{\partial \sigma_{zz}}{\partial z} + \frac{1}{r}\tau_{rz} + f_z &= 0\,. \end{aligned}\right\} \tag{3.29a}$$

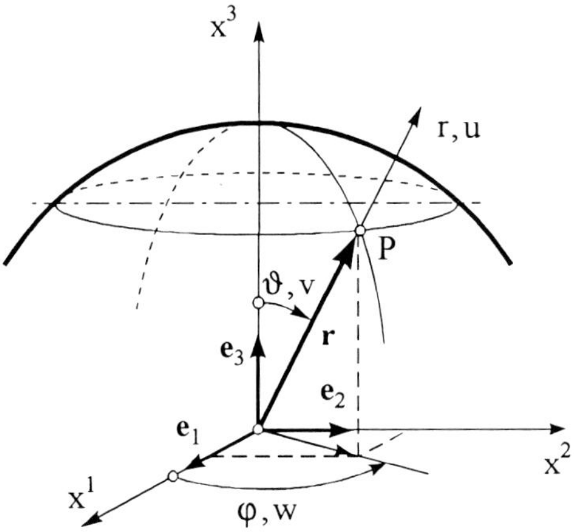

Fig. 3.9: Spherical coordinates
$(\xi^1 \mathrel{\widehat=} r,\ \xi^2 \mathrel{\widehat=} \vartheta,\ \xi^3 \mathrel{\widehat=} \varphi)$

Two-dimensional case: Polar coordinates r, φ

$$\left.\begin{aligned} &\frac{\partial \sigma_{rr}}{\partial r} + \frac{1}{r}\frac{\partial \tau_{\varphi r}}{\partial \varphi} + \frac{1}{r}(\sigma_{rr} - \sigma_{\varphi\varphi}) + f_r = 0 , \\ &\frac{\partial \tau_{\varphi r}}{\partial r} + \frac{1}{r}\frac{\partial \sigma_{\varphi\varphi}}{\partial \varphi} + \frac{2}{r}\tau_{\varphi r} + f_\varphi = 0 . \end{aligned}\right\} \quad (3.29b)$$

4) Spherical coordinates r, ϑ, φ (Fig. 3.9)

$$\left.\begin{aligned} &\frac{\partial \sigma_{rr}}{\partial r} + \frac{1}{r}\frac{\partial \tau_{r\vartheta}}{\partial \vartheta} + \frac{1}{r\sin\vartheta}\frac{\partial \tau_{r\varphi}}{\partial \varphi} + \frac{1}{r}\tau_{r\vartheta}\cot\vartheta + \\ &\qquad + \frac{1}{r}(2\sigma_{rr} - \sigma_{\vartheta\vartheta} - \sigma_{\varphi\varphi}) + f_r = 0 , \\ &\frac{\partial \tau_{r\vartheta}}{\partial r} + \frac{1}{r}\frac{\partial \sigma_{\vartheta\vartheta}}{\partial \vartheta} + \frac{1}{r\sin\vartheta}\frac{\partial \tau_{\varphi\vartheta}}{\partial \varphi} + \frac{3}{r}\tau_{\varphi\vartheta} + \\ &\qquad + \frac{1}{r}\cot\vartheta(\sigma_{\vartheta\vartheta} - \sigma_{\varphi\varphi}) + f_\vartheta = 0 , \\ &\frac{\partial \tau_{r\varphi}}{\partial r} + \frac{1}{r}\frac{\partial \tau_{\varphi\vartheta}}{\partial \vartheta} + \frac{1}{r\sin\vartheta}\frac{\partial \sigma_{\varphi\varphi}}{\partial \varphi} + \frac{3}{r}\tau_{r\varphi} + \frac{2}{r}\tau_{\varphi\vartheta}\cot\vartheta + f_\varphi = 0 . \end{aligned}\right\} \quad (3.30)$$

4 State of strain

4.1 Kinematics of a deformable body

Description of the deformation of a body with LAGRANGE's notation:

> The displacement of a material point of a body B is observed as a function of the initial state.

We distinguish between the initial state $t = t_0$ (without $\hat{}$) and the deformed state of the body $t = t$ (with $\hat{}$).

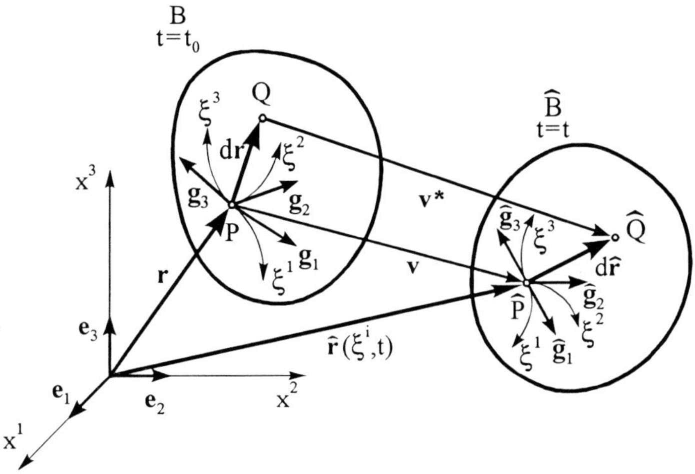

Fig. 4.1: Kinematics of a deformable body

Position vector $\hat{\mathbf{r}}$ of the material point $\hat{\mathrm{P}}$ of the deformed body $\hat{\mathrm{B}}$ (Fig. 4.1)

$$\hat{\mathbf{r}}(\xi^i) = \mathbf{r}(\xi^i) + \mathbf{v}(\xi^i) \tag{4.1}$$

with the position vector $\mathbf{r}(\xi^i)$ and the displacement vector $\mathbf{v}(\xi^i)$ of the same material point P of the undeformed body B.

Differential increase $d\mathbf{v}$ of the displacement vector $\mathbf{v}$

$$d\mathbf{v} = d\hat{\mathbf{r}} - d\mathbf{r} = \mathbf{V}\, d\mathbf{r}\ , \tag{4.2}$$

where $\mathbf{V}$ is the *tensor of the displacement derivatives.*

According to *(2.25)* the base vectors $\mathbf{g}_i$ and $\hat{\mathbf{g}}_i$ result from the total differential of the respective position vectors

$$d\mathbf{r} = \frac{\partial \mathbf{r}}{\partial \xi^i} d\xi^i = \mathbf{r}_{,i}\, d\xi^i = \mathbf{g}_i\, d\xi^i\ , \tag{4.3a}$$

$$d\hat{\mathbf{r}} = \frac{\partial \hat{\mathbf{r}}}{\partial \xi^i} d\xi^i = \hat{\mathbf{r}}_{,i}\, d\xi^i = \hat{\mathbf{g}}_i\, d\xi^i\ . \tag{4.3b}$$

These infinitesimal changes of the position vectors lead to the points Q and $\hat{\mathrm{Q}}$ adjacent to P and $\hat{\mathrm{P}}$ (see Fig. 4.1).

In accordance with the rules for the transformation behaviour of tensors (Sections 2.3, 2.4) the base vectors of the deformed body can be expressed by those of the undeformed body and vice versa:

$$\hat{\mathbf{g}}_i = \beta^j_{\ i}\, \mathbf{g}_j\ , \tag{4.4a}$$

$$\mathbf{g}_i = \hat{\beta}^j_{\ i}\, \hat{\mathbf{g}}_j\ . \tag{4.4b}$$

The displacement vector can be written as follows

$$\mathbf{v} = \mathrm{v}^{\mathrm{i}}\mathbf{g}_{\mathrm{i}} = \hat{\mathrm{v}}^{\mathrm{i}}\hat{\mathbf{g}}_{\mathrm{i}} = \hat{\mathbf{r}} - \mathbf{r} \,. \tag{4.5a}$$

Differentiation of *(4.5a)* with *(2.33)* leads to

$$\mathbf{v}_{,\mathrm{i}} = \mathrm{v}^{\mathrm{j}}|_{\mathrm{i}}\,\mathbf{g}_{\mathrm{j}} = \hat{\mathrm{v}}^{\mathrm{j}}||_{\mathrm{i}}\,\hat{\mathbf{g}}_{\mathrm{j}} = \hat{\mathbf{r}}_{,\mathrm{i}} - \mathbf{r}_{,\mathrm{i}} = \hat{\mathbf{g}}_{\mathrm{i}} - \mathbf{g}_{\mathrm{i}} \,. \tag{4.5b}$$

Since the base of the undeformed body or the base of the deformed body can be used alternately in order to illustrate the displacement vector $\mathbf{v}$, we have two different covariant derivatives of the displacement components as a result (according to *(4.5b)*). Here, one line stands for the covariant derivative applied to the base of the undeformed body, a double line stands for the covariant derivative related to the base of the deformed body. With the KRONECKER symbol we obtain the relation between the base vectors of the deformed body and the undeformed body [A.7]:

$$\hat{\mathbf{g}}_{\mathrm{i}} = (\,\delta^{\mathrm{j}}_{\mathrm{i}} + \mathrm{v}^{\mathrm{j}}|_{\mathrm{i}}\,)\,\mathbf{g}_{\mathrm{j}} \,, \tag{4.6a}$$

$$\mathbf{g}_{\mathrm{i}} = (\,\delta^{\mathrm{j}}_{\mathrm{i}} - \hat{\mathrm{v}}^{\mathrm{j}}||_{\mathrm{i}}\,)\,\hat{\mathbf{g}}_{\mathrm{j}} \,. \tag{4.6b}$$

The elements of the transformation tensor then read

$$\beta^{\mathrm{j}}_{\mathrm{i}} = \delta^{\mathrm{j}}_{\mathrm{i}} + \mathrm{v}^{\mathrm{j}}|_{\mathrm{i}} \,, \tag{4.7a}$$

$$\hat{\beta}^{\mathrm{j}}_{\mathrm{i}} = \delta^{\mathrm{j}}_{\mathrm{i}} - \hat{\mathrm{v}}^{\mathrm{j}}||_{\mathrm{i}} \,. \tag{4.7b}$$

Corresponding transformation relations are valid for the line elements $\mathrm{d}\mathbf{r}$ and $\mathrm{d}\hat{\mathbf{r}}$ in analogy with the base vectors. If we define a mixed tensor of order two according to *(2.12)*

$$\mathbf{F} = (\,\delta^{\mathrm{j}}_{\mathrm{i}} + \mathrm{v}^{\mathrm{j}}|_{\mathrm{i}}\,)\,\mathbf{g}_{\mathrm{j}}\mathbf{g}^{\mathrm{i}} = \mathbf{I} + \mathbf{V} \,, \tag{4.8a}$$

$$\hat{\mathbf{F}} = (\,\delta^{\mathrm{j}}_{\mathrm{i}} - \hat{\mathrm{v}}^{\mathrm{j}}||_{\mathrm{i}}\,)\,\hat{\mathbf{g}}_{\mathrm{j}}\hat{\mathbf{g}}^{\mathrm{i}} = \mathbf{I} - \hat{\mathbf{V}} \,, \tag{4.8b}$$

for the line elements this results in

$$\mathrm{d}\hat{\mathbf{r}} = \mathbf{F}\,\mathrm{d}\mathbf{r} \,, \tag{4.9a}$$

$$\mathrm{d}\mathbf{r} = \hat{\mathbf{F}}\,\mathrm{d}\hat{\mathbf{r}} \,. \tag{4.9b}$$

By means of *(4.2)* and due to *(4.5a)* we obtain the following total differential of the displacement vector $\mathrm{d}\mathbf{v}$:

$$\mathrm{d}\mathbf{v} = \mathrm{d}\hat{\mathbf{r}} - \mathrm{d}\mathbf{r} = (\,\mathbf{F} - \mathbf{I}\,)\,\mathrm{d}\mathbf{r} = \mathbf{V}\,\mathrm{d}\mathbf{r} \,. \tag{4.10a}$$

According to *(4.2)*, $\mathbf{V}$ is called the *tensor of the displacement derivatives* or the *deformation gradient*.

Due to *(4.10a)*, the total differential $\mathbf{dv}$ can be written in other notation by means of *(2.36b)*

$$\mathbf{dv} = \mathrm{Grad}\,\mathbf{v} \cdot \mathbf{dr} \,. \tag{4.10b}$$

In Cartesian coordinates the relation *(4.10b)* for a time independent displacement vector

$$\mathbf{v}(x,y,z) = \begin{bmatrix} u(x,y,z) \\ v(x,y,z) \\ w(x,y,z) \end{bmatrix}$$

reads as follows in matrix notation

$$\mathbf{dv} = \begin{bmatrix} du \\ dv \\ dw \end{bmatrix} = \begin{bmatrix} \frac{\partial u}{\partial x} & \frac{\partial u}{\partial y} & \frac{\partial u}{\partial z} \\ \frac{\partial v}{\partial x} & \frac{\partial v}{\partial y} & \frac{\partial v}{\partial z} \\ \frac{\partial w}{\partial x} & \frac{\partial w}{\partial y} & \frac{\partial w}{\partial z} \end{bmatrix} \begin{bmatrix} dx \\ dy \\ dz \end{bmatrix} . \tag{4.10c}$$

4.2 Strain tensor

The state of strain of an elastic body is obtained by subtraction of the squared line elements of a deformed and an undeformed body. Thus, we obtain a measure of how the distances of single points have changed due to a load [A.7 , A.8].
We write

$$d\hat{\mathbf{r}} \cdot d\hat{\mathbf{r}} - \mathbf{dr} \cdot \mathbf{dr} = d\hat{s}^2 - ds^2 = (\hat{\mathbf{g}}_i \cdot \hat{\mathbf{g}}_j - \mathbf{g}_i \cdot \mathbf{g}_j)\, d\xi^i d\xi^j =$$

$$= (\hat{g}_{ij} - g_{ij})\, d\xi^i d\xi^j = 2\,\gamma_{ij}\, d\xi^i d\xi^j \,, \tag{4.11a}$$

where γ_{ij} are the components of the *strain tensor*.
Accordingly, they can be determined as follows

$$\gamma_{ij} = \frac{1}{2}(\hat{g}_{ij} - g_{ij}) \,. \tag{4.11b}$$

Expressing the metric components of the deformed body by those of the undeformed body, we obtain the GREEN – LAGRANGE's components of strain [A.6]:

$$\gamma_{ij} = \frac{1}{2}\left(v_i|_j + v_j|_i + v^k|_i\, v_k|_j \right) . \tag{4.12a}$$

Linearized components of strain by neglecting the quadratic terms in *(4.12a)*:

$$\boxed{\gamma_{ij} = \frac{1}{2}(v_i|_j + v_j|_i)} \;. \tag{4.12b}$$

4.3 Strain–displacement relations

– Cartesian coordinates

From the tensor of the displacement derivatives **V** follows as the symmetric part the *linear* strain tensor. According to the rules *(2.18)*, it becomes in Cartesian coordinates:

$$\mathbf{V}_s = \frac{1}{2}(\mathbf{V} + \mathbf{V}^T) = \begin{bmatrix} \varepsilon_{xx} & \frac{1}{2}\gamma_{xy} & \frac{1}{2}\gamma_{xz} \\ \cdot & \varepsilon_{yy} & \frac{1}{2}\gamma_{yz} \\ \cdot & \cdot & \varepsilon_{zz} \end{bmatrix}, \qquad (4.13)$$

where due to *(4.12b)*

$$\left.\begin{aligned} \varepsilon_{xx} &= \frac{\partial u}{\partial x}, & \gamma_{xy} &= \frac{\partial u}{\partial y} + \frac{\partial v}{\partial x}, \\ \varepsilon_{yy} &= \frac{\partial v}{\partial y}, & \gamma_{xz} &= \frac{\partial u}{\partial z} + \frac{\partial w}{\partial x}, \\ \varepsilon_{zz} &= \frac{\partial w}{\partial z}, & \gamma_{yz} &= \frac{\partial v}{\partial z} + \frac{\partial w}{\partial y}. \end{aligned}\right\} \qquad (4.14)$$

γ_{ij} (i,j = x,y,z) are the so–called *technical* shears, and ε_{ii} (i = x,y,z) are the normal strains.

Special cases:

– Cylindrical coordinates (Fig. 2.5) (r,u; φ,v; z,w)

$$\left.\begin{aligned} \varepsilon_{rr} &= \frac{\partial u}{\partial r}, & \gamma_{r\varphi} &= \frac{1}{r}\frac{\partial u}{\partial \varphi} + \frac{\partial v}{\partial r} - \frac{v}{r}, \\ \varepsilon_{\varphi\varphi} &= \frac{1}{r}\frac{\partial v}{\partial \varphi} + \frac{u}{r}, & \gamma_{rz} &= \frac{\partial u}{\partial z} + \frac{\partial w}{\partial r}, \\ \varepsilon_{zz} &= \frac{\partial w}{\partial z}, & \gamma_{\varphi z} &= \frac{\partial v}{\partial z} + \frac{1}{r}\frac{\partial w}{\partial \varphi}. \end{aligned}\right\} \qquad (4.15)$$

– Spherical coordinates (Fig. 3.9) (r,u; ϑ,v; φ,w)

$$\left.\begin{aligned} \varepsilon_{rr} &= \frac{\partial u}{\partial r}, \\ \varepsilon_{\varphi\varphi} &= \frac{1}{r\sin\vartheta}\left(\frac{\partial w}{\partial \varphi} + u\sin\vartheta + v\cos\vartheta\right), \\ \varepsilon_{\vartheta\vartheta} &= \frac{1}{r}\frac{\partial v}{\partial \vartheta} + \frac{u}{r}, \\ \gamma_{r\varphi} &= \frac{1}{r\sin\vartheta}\frac{\partial u}{\partial \varphi} + \frac{\partial w}{\partial r} - \frac{w}{r}, \\ \gamma_{r\vartheta} &= \frac{1}{r}\frac{\partial u}{\partial \vartheta} + \frac{\partial v}{\partial r} - \frac{v}{r}, \\ \gamma_{\varphi\vartheta} &= \frac{1}{r\sin\vartheta}\left(\frac{\partial v}{\partial \varphi} + \frac{\partial w}{\partial \vartheta}\sin\vartheta - w\cos\vartheta\right). \end{aligned}\right\} \qquad (4.16)$$

4.4 Transformation of principal axis

The principal strains are determined in analogy to the principal stresses.

Characteristic equation according to *(2.22c)*:

$$\lambda^3 - I_1 \lambda^2 + I_2 \lambda - I_3 = 0 \,. \tag{4.17a}$$

The first invariant corresponds to the so-called *volume dilatation* e:

$$I_1 = e = \operatorname{div} \mathbf{v} = v^j|_j = \gamma^j_j \,. \tag{4.17b}$$

4.5 Compatibility conditions

The linear strain-displacement relations *(4.12b)* form a system of six coupled, partial differential equations for the three components v_i of the displacements for given values of the strain tensor. Thus, the system is kinematically redundant. In order that there will exist a displacement vector $\mathbf{v}_i$ subject to given values of the six mutually independent components of the strain tensor, it is necessary that the three components of the displacement vector satisfy the following compatibility conditions (DE SAINT VENANT):

$$\gamma_{ij}|_{kl} + \gamma_{kl}|_{ij} - \gamma_{il}|_{kj} - \gamma_{kj}|_{il} = \gamma_{ij}|_{kl}\, \varepsilon^{jln}\, \varepsilon^{ikm} = 0 \,. \tag{4.18}$$

Mechanical interpretation:

The interior coherence of the body has to be preserved after the deformation, i.e. material gaps or overlaps must not occur.

For a two-dimensional state of stress or strain the compatibility condition in Cartesian coordinates reads as follows:

$$\boxed{\frac{\partial^2 \varepsilon_{xx}}{\partial y^2} + \frac{\partial^2 \varepsilon_{yy}}{\partial x^2} - \frac{\partial^2 \gamma_{xy}}{\partial x \, \partial y} = 0} \,. \tag{4.19}$$

5 Constitutive laws of linearly elastic bodies

5.1 Basic concepts

In the following we are going to deal with bodies for which there exist reversibly unique relations between the components of the strain tensor and the stress tensor, and we furthermore assume that these relations are *time independent*. The behaviour of the bodies is denoted as elastic, i.e. there are *no* permanent strains ε_{pl} after removing the load of the body (Fig. 5.1). The bodies considered shall furthermore, as it is usual in the classical elasticity theory, be made of a *linearly elastic* material such that their constitutive law expresses linear relationship between the components of the stress tensor and the strain tensor (range 0 - A - in Fig. 5.1). Such bodies are usually called *HOOKEAN bodies*.

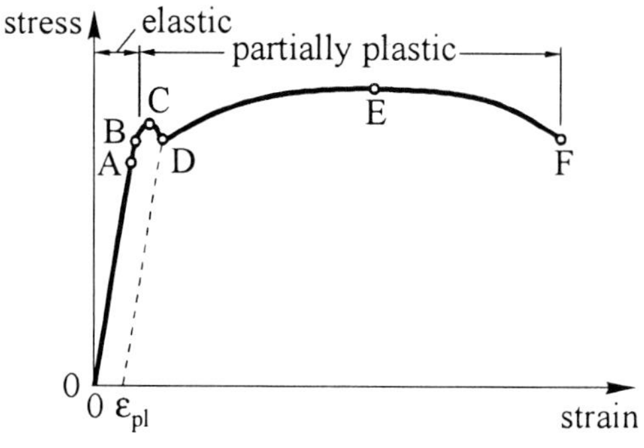

Fig. 5.1: σ, ε – diagram of a real material with a linear – elastic range

A	≙ limit of proportionality
B	≙ elastic limit
C	≙ upper yield point
D	≙ lower yield point
D – E	≙ elastic – plastic state
F	≙ ultimate stress limit
ε_{pl}	≙ plastic strain

For a great number of problems in practice this assumption is feasible, even if we have to consider non – linear strain – displacement relations (e.g. geometrical non – linearities for the post – buckling of plates and shells).

5.2 Generalized HOOKE–DUHAMEL's law for thermo–elastic, isotropic materials

– *Cartesian coordinates*

$$\left.\begin{aligned}
\varepsilon_{xx} &= \frac{1}{E}[\sigma_{xx} - \nu(\sigma_{yy} + \sigma_{zz})] + \alpha_T\Theta \quad , \quad \gamma_{xy} = \frac{1}{G}\tau_{xy} , \\
\varepsilon_{yy} &= \frac{1}{E}[\sigma_{yy} - \nu(\sigma_{zz} + \sigma_{xx})] + \alpha_T\Theta \quad , \quad \gamma_{xz} = \frac{1}{G}\tau_{xz} , \\
\varepsilon_{zz} &= \frac{1}{E}[\sigma_{zz} - \nu(\sigma_{xx} + \sigma_{yy})] + \alpha_T\Theta \quad , \quad \gamma_{yz} = \frac{1}{G}\tau_{yz}
\end{aligned}\right\} \tag{5.1}$$

with

E	YOUNG's modulus ,
ν	POISSON's ratio ,
$G = \dfrac{E}{2(1+\nu)}$	shear modulus ,
α_T	one – dimensional thermal expansion coefficient ,
$\Theta = T_1 - T_0$	difference between final and initial temperature .

Symbolic notation

$$\boxed{\mathbf{V}_s = \frac{1+\nu}{E}\mathbf{S} - \frac{\nu}{E}s\mathbf{I} + \alpha_T\Theta\mathbf{I}} \tag{5.2}$$

with s = sum of normal stresses .

Solving *(5.1)* with respect to stresses yields

$$\left.\begin{aligned}
\sigma_{xx} &= \frac{E}{1+\nu}\left(\varepsilon_{xx} + \frac{\nu}{1-2\nu}\,e\right) - \frac{E}{1-2\nu}\,\alpha_T\,\Theta \;, \quad \tau_{xy} = G\,\gamma_{xy} \;, \\
\sigma_{yy} &= \frac{E}{1+\nu}\left(\varepsilon_{yy} + \frac{\nu}{1-2\nu}\,e\right) - \frac{E}{1-2\nu}\,\alpha_T\,\Theta \;, \quad \tau_{xz} = G\,\gamma_{xz} \;, \\
\sigma_{zz} &= \frac{E}{1+\nu}\left(\varepsilon_{zz} + \frac{\nu}{1-2\nu}\,e\right) - \frac{E}{1-2\nu}\,\alpha_T\,\Theta \;, \quad \tau_{yz} = G\,\gamma_{yz} \;.
\end{aligned}\right\} \qquad (5.3)$$

Symbolic notation

$$\boxed{\mathbf{S} = 2\,G\left[\mathbf{V}_s + \frac{\nu}{1-2\nu}\,e\,\mathbf{I} - \frac{1+\nu}{1-2\nu}\,\alpha_T\,\Theta\,\mathbf{I}\right]} \qquad (5.4)$$

with $\quad e$ = volume dilatation .

– *Curvilinear coordinates in index notation*

According to *(5.1)* it follows that

$$\gamma_{kl} = \frac{1+\nu}{E}\,\tau_{kl} - \frac{\nu}{E}\,g_{kl}\,\tau^{j}_{\ j} + \alpha_T\,g_{kl}\,\Theta = D_{ijkl}\,\tau^{ij} + \alpha_T\,g_{kl}\,\Theta \qquad (5.5a)$$

with the flexibility tensor of fourth order

$$D_{ijkl} = \frac{1+\nu}{2\,E}\,(g_{ik}\,g_{jl} + g_{il}\,g_{jk}) - \frac{\nu}{E}\,g_{ij}\,g_{kl} \;. \qquad (5.5b)$$

Solving *(5.5a)* with regard to stresses leads to

$$\tau^{ij} = C^{ijkl}\,\gamma_{kl} - \beta^{ij}\,\Theta \qquad (5.6a)$$

with the elasticity tensor of fourth order

$$C^{ijkl} = G\,(g^{ik}\,g^{jl} + g^{il}\,g^{jk} + \frac{2\,\nu}{1-2\nu}\,g^{ij}\,g^{kl}) \qquad (5.6b)$$

and the thermo – elastic tensor of second order

$$\beta^{ij} = \beta\,g^{ij} = \frac{E\,\alpha_T}{1-2\nu}\,g^{ij} \;. \qquad (5.6c)$$

Other notation of *(5.6a)*:

$$\tau^{ij} = C^{*ijkl}\,(\gamma_{kl} - \alpha_T\,g_{kl}\,\Theta) \qquad (5.7a)$$

with $\quad C^{*ijkl} = \lambda\,g^{ij}\,g^{kl} + \mu\,(g^{ik}\,g^{jl} + g^{il}\,g^{jk}) \qquad (5.7b)$

and the LAMÉ constants

$$\mu \mathrel{\widehat{=}} G \ , \quad \lambda \mathrel{\widehat{=}} G \frac{2\nu}{1-2\nu} \ .$$

The relations between the different specific elasticity constants can be drawn from the following table:

	$\lambda =$	$\mu = G =$	$E =$	$\nu =$
λ, μ	λ	μ	$\frac{\mu(3\lambda + 2\mu)}{\lambda + \mu}$	$\frac{\lambda}{2(\lambda + \mu)}$
λ, ν	λ	$\frac{\lambda(1-2\nu)}{2\nu}$	$\frac{(1+\nu)(1-2\nu)\lambda}{\nu}$	ν
μ, E	$\frac{\mu(E-2\mu)}{3\mu - E}$	μ	E	$\frac{E-2\mu}{2\mu}$
E, ν	$\frac{E\nu}{(1+\nu)(1-2\nu)}$	$\frac{E}{2(1+\nu)}$	E	ν

Table 5.1: Elasticity properties

The linearly elastic constitutive equations shall be augmented by another system of equations which allows a physical interpretation, and which is applied in elastoplastic structures. Therefore, we split both the strain and the stress tensor in a spherical–symmetrical and a deviatoric part according to the following relations:

$$\left.\begin{aligned} \gamma_k^m &= \frac{1}{3} e \, \delta_k^m + {}'\gamma_k^m \ , \\ \tau_k^m &= \frac{1}{3} s \, \delta_k^m + {}'\tau_k^m \ . \end{aligned}\right\} \qquad (5.8)$$

In *(5.8)* the known expressions for the sum of strains $e = \gamma_k^k$ or the sum of stresses $s = \tau_k^k$ occur which are the first invariants of the strain tensor or the stress tensor according to *(4.17b)*. Substituting *(5.8)* in the generalized HOOKE's law leads to the following two equations

$$m \neq k \quad \longrightarrow \quad {}'\gamma_k^m = \frac{1}{2G} \, {}'\tau_k^m \ , \qquad (5.9a)$$

$$m = k \quad \longrightarrow \quad e = \frac{1}{K} \sigma_M + 3 \alpha_T \Theta \qquad (5.9b)$$

with

$K = \dfrac{E}{3(1-2\nu)}$	compression modulus ,
$\sigma_M = s/3$	mean value of the normal stresses ,
$3\alpha_T$	volumetric thermal expansion coefficient .

With *(5.9a)* a change of the shape without a change of volume $'\gamma_k^k = 0$ is described physically according to *(5.8)*, whereas *(5.9b)* describes a change of volume without a change in shape ($'\gamma_k^m = 0 \longrightarrow \gamma_k^m = 1/3\, e\, \delta_k^m$). Both relations *(5.9)* give the proportionality between strains and stresses for linearly elastic materials. It has to be emphasized once again that an isotropic, linearly elastic body only possesses two mutually independent material properties, and most often E and ν are chosen.
POISSON's ratio ν can be more closely limited from *(5.9b)* neglecting all effects of thermal stresses

$$e = \frac{3(1 - 2\nu)}{E}\,\sigma_M\,. \tag{5.10}$$

Since e and σ_M always have the same sign, ν must be smaller than 1/2. According to *(5.10)*, e = 0 for $\nu = 1/2$, which corresponds to an incompressible medium (constant volume). $\nu = 0{,}3 \div 0{,}33$ is valid for steel and light metals.

5.3 Material law for plane states

a) *State of plane stress*

– Cartesian coordinates

Definition: $\sigma_{zz} = \tau_{xz} = \tau_{yz} = 0\,.$ *(5.11)*

Strain – stress relations

$$\left.\begin{aligned}
\varepsilon_{xx} &= \frac{1}{E}(\sigma_{xx} - \nu\,\sigma_{yy}) + \alpha_T\,{}^0\Theta \quad , \quad \gamma_{xy} = \frac{1}{G}\tau_{xy}\,,\\
\varepsilon_{yy} &= \frac{1}{E}(\sigma_{yy} - \nu\,\sigma_{xx}) + \alpha_T\,{}^0\Theta \quad , \quad \gamma_{xz} = 0\,,\\
\varepsilon_{zz} &= -\frac{\nu}{E}(\sigma_{xx} + \sigma_{yy}) + \alpha_T\,{}^0\Theta \quad , \quad \gamma_{yz} = 0\,.
\end{aligned}\right\} \tag{5.12}$$

Stress – strain relations

$$\left.\begin{aligned}
\sigma_{xx} &= \frac{E}{1-\nu^2}\left[\varepsilon_{xx} + \nu\,\varepsilon_{yy} - (1+\nu)\,\alpha_T\,{}^0\Theta\right] \; , \quad \tau_{xy} = G\,\gamma_{xy}\,,\\
\sigma_{yy} &= \frac{E}{1-\nu^2}\left[\varepsilon_{yy} + \nu\,\varepsilon_{xx} - (1+\nu)\,\alpha_T\,{}^0\Theta\right] \; , \quad \tau_{xz} = 0\,,\\
\sigma_{zz} &= 0\,, \qquad\qquad\qquad\qquad\qquad\qquad\qquad\qquad \tau_{yz} = 0
\end{aligned}\right\} \tag{5.13}$$

with $\quad {}^0\Theta(x,y) = T_1(x,y) - T_0\,.$

Symbolic notation of *(5.13)*

$$\boldsymbol{\sigma} = \mathbf{E}\left[\boldsymbol{\varepsilon} - \boldsymbol{\varepsilon}_\Theta\right] \tag{5.14}$$

with

$$\boldsymbol{\sigma} = \begin{bmatrix} \sigma_{xx} \\ \sigma_{yy} \\ \tau_{xy} \end{bmatrix} , \quad \mathbf{E} = \frac{E}{1-\nu^2} \begin{bmatrix} 1 & \nu & 0 \\ \nu & 1 & 0 \\ 0 & 0 & \frac{1-\nu}{2} \end{bmatrix} ,$$

$$\boldsymbol{\varepsilon} = \begin{bmatrix} \varepsilon_{xx} \\ \varepsilon_{yy} \\ \gamma_{xy} \end{bmatrix} , \quad \boldsymbol{\varepsilon}_{\Theta} = \alpha_T \, {}^0\Theta \begin{bmatrix} 1 \\ 1 \\ 0 \end{bmatrix} .$$

– Curvilinear coordinates

The equations *(5.12)* and *(5.13)* read in index notation

$$\gamma_{\alpha\beta} = D_{\alpha\beta\gamma\delta} \, \tau^{\gamma\delta} + \alpha_T \, g_{\alpha\beta} \, {}^0\Theta \tag{5.15a}$$

$$\tau^{\alpha\beta} = E^{\alpha\beta\gamma\delta} (\gamma_{\gamma\delta} - \alpha_T \, g_{\gamma\delta} \, {}^0\Theta) \tag{5.15b}$$

with the plane elasticity and the plane flexibility tensors of fourth order

$$D_{\alpha\beta\gamma\delta} = \frac{1+\nu}{2E} (g_{\alpha\gamma} g_{\beta\delta} + g_{\alpha\delta} g_{\beta\gamma}) - \frac{\nu}{E} g_{\alpha\beta} g_{\gamma\delta} ,$$

$$E^{\alpha\beta\gamma\delta} = \frac{E}{2(1+\nu)} (g^{\alpha\gamma} g^{\beta\delta} + g^{\alpha\delta} g^{\beta\gamma} + \frac{2\nu}{1-\nu} g^{\alpha\beta} g^{\gamma\delta}) .$$

b) *State of plane strain* in Cartesian coordinates

Definitions: $\varepsilon_{zz} = \gamma_{xz} = \gamma_{yz} = 0 \, .$ (5.16)

Strain – stress relations

$$\left.\begin{aligned} \varepsilon_{xx} &= \frac{1+\nu}{E} [(1-\nu)\sigma_{xx} - \nu \sigma_{yy}] + (1+\nu) \alpha_T \, {}^0\Theta , & \gamma_{xy} &= \frac{1}{G} \tau_{xy} , \\ \varepsilon_{yy} &= \frac{1+\nu}{E} [(1-\nu)\sigma_{yy} - \nu \sigma_{xx}] + (1+\nu) \alpha_T \, {}^0\Theta , & \gamma_{xz} &= 0 , \\ \varepsilon_{zz} &= 0 , & \gamma_{yz} &= 0 . \end{aligned}\right\} \tag{5.17}$$

Stress – strain relations

$$\begin{aligned} \sigma_{xx} &= \frac{E}{(1+\nu)(1-2\nu)} [(1-\nu)\varepsilon_{xx} + \nu \varepsilon_{yy} - (1+\nu) \alpha_T \, {}^0\Theta] , \\ \sigma_{yy} &= \frac{E}{(1+\nu)(1-2\nu)} [(1-\nu)\varepsilon_{yy} + \nu \varepsilon_{xx} - (1+\nu) \alpha_T \, {}^0\Theta] , \\ \sigma_{zz} &= \frac{E\nu}{(1+\nu)(1-2\nu)} [\varepsilon_{xx} + \varepsilon_{yy} - \frac{1+\nu}{\nu} \alpha_T \, {}^0\Theta] , \\ \tau_{xy} &= G \gamma_{xy} , \quad \tau_{xz} = \tau_{yz} = 0 . \end{aligned} \tag{5.18}$$

5.4 Material law for a unidirectional layer (UD – layer) of a fibre reinforced composite

The material law for a UD – layer reads as follows according to *(5.15)* without temperature terms

$$\tau^{\alpha'\beta'} = E^{\alpha'\beta'\gamma'\delta'} \, \gamma_{\gamma'\delta'} \, . \qquad (5.19)$$

Here, indices equipped with a dash refer to the material coordinate system in the UD – layer.

Plane elasticity tensor of fourth order for a UD–layer in the $\xi^{1'}, \xi^{2'}$ – coordinate system (see Fig. 5.2):

$$\left(E^{\alpha'\beta'\gamma'\delta'} \right) = \begin{bmatrix} E^{1'1'1'1'} & E^{1'1'2'2'} & E^{1'1'1'2'} \\ E^{2'2'1'1'} & E^{2'2'2'2'} & E^{2'2'1'2'} \\ E^{1'2'1'1'} & E^{1'2'2'2'} & E^{1'2'1'2'} \end{bmatrix} =$$

$$= \begin{bmatrix} \dfrac{E_{1'}}{1 - \nu_{1'2'}\,\nu_{2'1'}} & \dfrac{\nu_{2'1'}\,E_{1'}}{1 - \nu_{1'2'}\,\nu_{2'1'}} & 0 \\ \dfrac{\nu_{1'2'}\,E_{2'}}{1 - \nu_{1'2'}\,\nu_{2'1'}} & \dfrac{E_{2'}}{1 - \nu_{1'2'}\,\nu_{2'1'}} & 0 \\ 0 & 0 & G_{1'2'} \end{bmatrix} , \qquad (5.20)$$

where the material properties have the following meaning [B.10]:

$E_{1'}$ YOUNG's modulus in $\xi^{1'}$ – direction parallel to the fibres ,

$E_{2'}$ YOUNG's modulus in $\xi^{2'}$ – direction perpendicular to the fibres ,

$\nu_{1'2'}$ POISSON's ratio perpendicular to the fibres in case of a loading parallel to the fibres ,

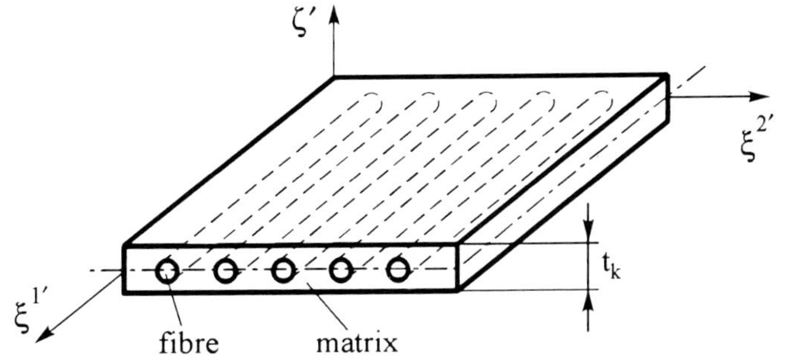

Fig. 5.2: Material coordinate system for a UD –layer

$\nu_{2'1'}$ POISSON's ratio parallel to the fibres in case of a loading perpendicular to the fibres; it holds that

$$\nu_{2'1'} = \nu_{1'2'}\, E_{2'}/E_{1'} \,,$$

$G_{1'2'}$ Shear modulus parallel and perpendicular to the fibres .

Rotation of the UD – layer by an angle α (see Fig. 5.3) is obtained by application of the transformation formula for a tensor of fourth order (generalization of *(2.14)*):

$$E^{\alpha\beta\mu\nu} = \beta^{\alpha}_{\gamma'}\, \beta^{\beta}_{\delta'}\, \beta^{\mu}_{\varrho'}\, \beta^{\nu}_{\sigma'}\, E^{\gamma'\delta'\varrho'\sigma'} \,. \qquad (5.21)$$

Transformation coefficients according to *(3.16)* in matrix notation:

$$\left(\beta^{\alpha}_{\gamma'}\right) = \begin{bmatrix} \beta^{1}_{1'} & \beta^{1}_{2'} \\ \beta^{2}_{1'} & \beta^{2}_{2'} \end{bmatrix} = \begin{bmatrix} \cos\alpha & \sin\alpha \\ -\sin\alpha & \cos\alpha \end{bmatrix} . \qquad (5.22)$$

Substitution of the components of the elasticity tensor by simplifying the notation yields :

$$\begin{bmatrix} E^{1'1'1'1'} & E^{1'1'2'2'} & E^{1'1'1'2'} \\ E^{2'2'1'1'} & E^{2'2'2'2'} & E^{2'2'1'2'} \\ E^{1'2'1'1'} & E^{1'2'2'2'} & E^{1'2'1'2'} \end{bmatrix} \longrightarrow \begin{bmatrix} E^{1'1'} & E^{1'2'} & E^{1'3'} \\ E^{2'1'} & E^{2'2'} & E^{2'3'} \\ E^{3'1'} & E^{3'2'} & E^{3'3'} \end{bmatrix} . \qquad (5.23)$$

Components of the elasticity tensor for the rotated vector base read then as follows :

$$\left.\begin{aligned} E^{11} &= E^{1'1'} \cos^4\alpha + E^{2'2'} \sin^4\alpha + \frac{1}{2} A^{1'} \sin^2 2\alpha \,, \\ E^{22} &= E^{2'2'} \cos^4\alpha + E^{1'1'} \sin^4\alpha + \frac{1}{2} A^{1'} \sin^2 2\alpha \,, \\ E^{33} &= E^{3'3'} + \frac{1}{4} A^{2'} \sin^2 2\alpha \,, \end{aligned}\right\} \qquad (5.24a)$$

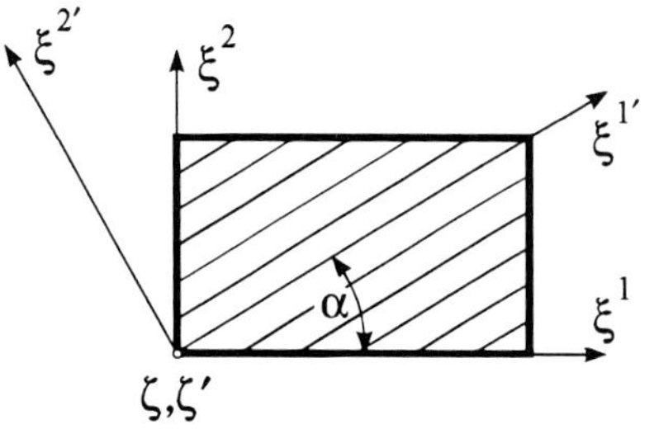

Fig. 5.3 : Rotation of a UD – layer by an angle α

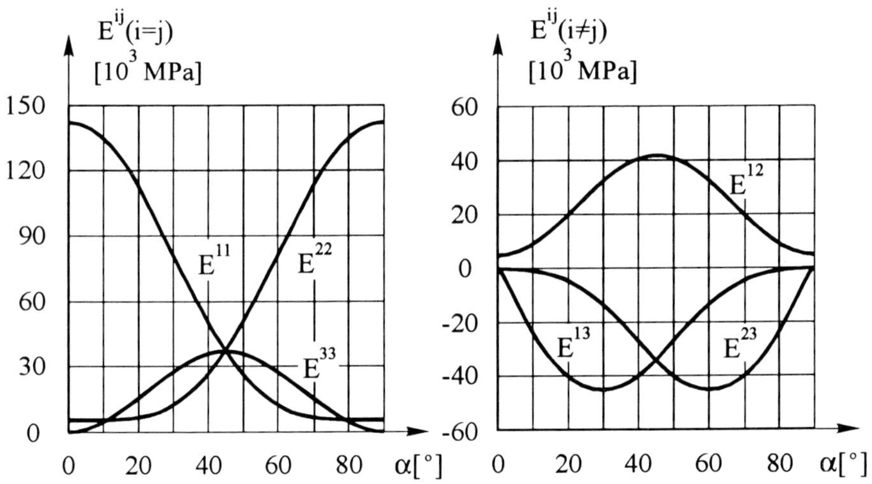

Fig. 5.4: Components of the elasticity tensors vs. the fibre – orientation α of a High Tensile fibre (HT – fibre) [B.10]

Material values: $E_{1'} = 143\,000\,MPa$, $E_{2'} = 5\,140\,MPa$,
$G_{1'2'} = 5\,280\,MPa$, $\nu_{1'2'} = 0.28$, $\nu_{2'1'} = 0.01$.

$$\left.\begin{aligned} E^{12} &= E^{21} = E^{1'2'} + \frac{1}{4} A^{2'} \sin^2 2\alpha \, , \\ E^{13} &= E^{31} = \frac{1}{2} \left[- E^{1'1'} + A^{1'} + A^{2'} \sin^2 \alpha \right] \sin 2\alpha \, , \\ E^{23} &= E^{32} = \frac{1}{2} \left[\ E^{2'2'} - A^{1'} - A^{2'} \sin^2 \alpha \right] \sin 2\alpha \end{aligned} \right\} \qquad (5.24b)$$

with $\quad A^{1'} = E^{2'1'} + 2\,E^{3'3'} \quad , \quad A^{2'} = E^{1'1'} + E^{2'2'} - 2\,A^{1'} \; .$

6 Energy principles

6.1 Basic terminology and assumptions

Our consideration of solid bodies in this section is based on the following assumptions [A.9, A.15, A.16, A.18]:

a) The processes produced in a stressed body are reversible, i.e. no dissipative effects (e.g. plastic deformations) occur. We limit ourselves to the scope of the *classical* elasticity theory.

b) The deformation process takes an *isothermal* course, i.e. there is no interaction between deformation and temperature.

c) The load process is quasi-static, i.e. the kinetic energy or the forces of inertia can be neglected.

d) The state of displacement of a solid body is described according to a LAGRANGEAN approach.

e) The theorem of mass conservation ($d\hat{V} = dV$) and the volume forces in the deformed and undeformed bodies ($\hat{f} \equiv f$) are equal.

6.2 Energy expressions

First, we consider the uniaxial state of stress of a rod subjected to a single force F. The relation between force and displacement can be assumed to be nonlinear as well as linear (Fig. 6.1). The external work done by the normal force F against the displacement δu is given by

$$\delta W = \bar{F}\,\delta\bar{u} \,. \tag{6.1}$$

Here, we use the differential δ for the changes of state, e.g. deformation differentials, strain differentials. For these quantities it is assumed that they are virtual (not existing in reality), infinitesimally small and geometrically compatible. Eq. *(6.1)* illustrates the area of a thin strip with the width $\delta\bar{u}$ and the height $\bar{F}$ in a force-deformation diagram (Fig. 6.1), where terms of higher order have been neglected. The total work of the single force results from an integration over the deformation differentials

$$W = \int\limits_{\bar{u}=0}^{u} \bar{F}\,\delta\bar{u} \,. \tag{6.2}$$

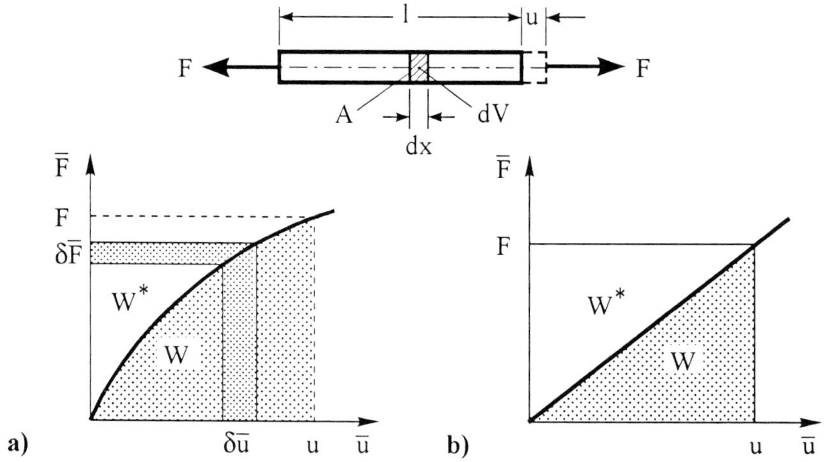

Fig. 6.1: Nonlinear and linear force-deformation curve of a rod subject to a single load

In Fig. 6.1, the area W* represents the complementary work, because W and W* *complement one another* and their sum is represented by the rectangle $F \cdot u = W + W^*$.
By analogy to *(6.1)* and *(6.2)* the following holds for the complementary work

$$\delta W^* = \bar{u}\, \delta \bar{F} \tag{6.3}$$

or

$$W^* = \int_{\bar{F}=0}^{F} \bar{u}\, \delta \bar{F} \,. \tag{6.4}$$

In the case of a linear force - deformation curve $\bar{F} = c \cdot \bar{u}$ (Fig. 6.1b) an integration over the deformation differentials can be carried out. Thus, we obtain

$$W = W^* = \frac{1}{2} F \cdot u \,. \tag{6.5}$$

The external work is stored as so - called *internal* energy or deformation energy in the rod. Substituting the increase of deformation δu by $\delta \varepsilon\, dx$ in *(6.1)*, we can write ($\bar{u}$ is denoted by u now, because it cannot be changed with the final value u):

$$\delta W = F \delta u = \frac{F}{A} A\, \delta \varepsilon\, dx = \sigma\, \delta \varepsilon\, A\, dx = \sigma\, \delta \varepsilon\, dV = \delta U \,. \tag{6.6}$$

If we divide by the volume element $dV = A\, dx$, we obtain the expression for the *specific deformation energy*

$$\delta \bar{U} = \sigma\, \delta \varepsilon \tag{6.7}$$

and by analogy, for the *specific complementary energy* we obtain

$$\delta \bar{U}^* = \varepsilon\, \delta \sigma \,. \tag{6.8}$$

The relation between the stress σ and the strain ε is given by a non - linear curve similar to the one shown in Fig. 6.1a. If a linear σ, ε - curve exists, by analogy to *(6.5)* we obtain the following for the specific deformation energy and the specific complementary energy

$$\bar{U} = \bar{U}^* = \frac{1}{2} \sigma \varepsilon \,. \tag{6.9}$$

The expression is now extended to a three - dimensional elastic body subjected to external forces (volume forces **f**, distributed surface tractions **p**, and concentrated forces $\mathbf{F}_k$) (Fig. 6.2).

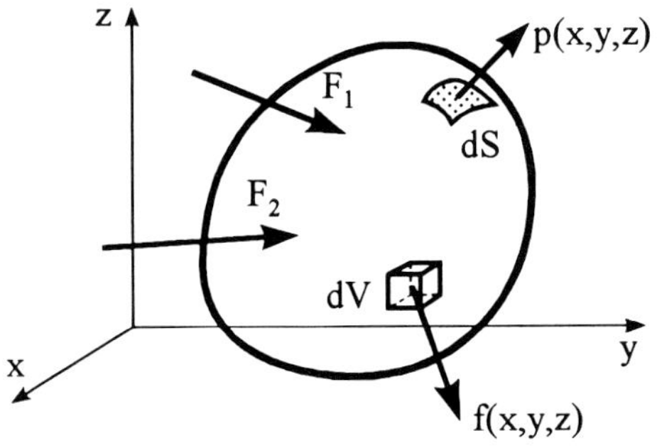

Fig. 6.2: Elastic body subjected to external forces

In vector notation, the external work can be written as follows

$$\delta W = \int_V \mathbf{f}^T \delta \mathbf{v}\, dV + \int_S \mathbf{p}^T \delta \mathbf{v}\, dS + \mathbf{F}^T \delta \mathbf{v}^0 \tag{6.10}$$

with

$\mathbf{f}^T = (f_x, f_y, f_z)$ vector of volume forces ,

$\mathbf{p}^T = (p_x, p_y, p_z)$ vector of surface tractions ,

$\mathbf{v} = \begin{bmatrix} u \\ v \\ w \end{bmatrix}$ displacement vector of an elastic body ,

$\mathbf{F}^T = (\mathbf{F}_1^T, \mathbf{F}_2^T, \ldots, \mathbf{F}_i^T)$ vector of concentrated forces

$$\mathbf{F}_i^T = (F_x, F_y, F_z)_i ,$$

$\mathbf{v}^0 = \begin{bmatrix} \mathbf{v}_1^0 \\ \mathbf{v}_2^0 \\ \vdots \\ \mathbf{v}_n^0 \end{bmatrix}$ vector of displacement vectors for points of action of concentrated forces

$$\mathbf{v}_i^0 = \begin{bmatrix} u^0 \\ v^0 \\ w^0 \end{bmatrix}_i .$$

Transition to isotropic, linearly elastic body

Specific deformation energy and complementary energy

$$\overline{U} = \overline{U}^* = \frac{1}{2} \boldsymbol{\sigma}^T \boldsymbol{\varepsilon} = \frac{1}{2} \boldsymbol{\varepsilon}^T \boldsymbol{\sigma} . \tag{6.11}$$

Introduction of HOOKE's law

$$\overline{U} = \frac{1}{2} \boldsymbol{\varepsilon}^T \mathbf{C} \boldsymbol{\varepsilon} , \tag{6.12a}$$

$$\overline{U}^* = \frac{1}{2} \boldsymbol{\sigma}^T \mathbf{D} \boldsymbol{\sigma} \tag{6.12b}$$

with vectors for the strains and stresses in Cartesian coordinates defined by

$$\boldsymbol{\varepsilon}^T = (\varepsilon_{xx}, \varepsilon_{yy}, \varepsilon_{zz}, \gamma_{xy}, \gamma_{yz}, \gamma_{zx}) \ , \quad \boldsymbol{\sigma}^T = (\sigma_{xx}, \sigma_{yy}, \sigma_{zz}, \tau_{xy}, \tau_{yz}, \tau_{zx}) \ ,$$

leads to the *elasticity matrix*

$$\mathbf{C} = \frac{E}{(1+\nu)(1-2\nu)} \left[\begin{array}{ccc|ccc} 1-\nu & \nu & \nu & & & \\ \nu & 1-\nu & \nu & & \mathbf{0} & \\ \nu & \nu & 1-\nu & & & \\ \hline & & & \frac{1-2\nu}{2} & 0 & 0 \\ & \mathbf{0} & & 0 & \frac{1-2\nu}{2} & 0 \\ & & & 0 & 0 & \frac{1-2\nu}{2} \end{array} \right] = \mathbf{C}^T \qquad (6.13a)$$

and the *flexibility matrix*

$$\mathbf{D} = \frac{1}{E} \left[\begin{array}{ccc|ccc} 1 & -\nu & -\nu & & & \\ -\nu & 1 & -\nu & & \mathbf{0} & \\ -\nu & -\nu & 1 & & & \\ \hline & & & 2(1+\nu) & 0 & 0 \\ & \mathbf{0} & & 0 & 2(1+\nu) & 0 \\ & & & 0 & 0 & 2(1+\nu) \end{array} \right] = \mathbf{D}^T . \qquad (6.13b)$$

The expressions *(6.12)* are bilinear forms which are positive definite because $\overline{U} > 0$ and $\overline{U}^* > 0$.

In usual *index notation*

$$\overline{U} = \overline{U}^* = \frac{1}{2} \tau^{ij} \gamma_{ij} \qquad (6.14)$$

or

$$\overline{U} = \frac{1}{2} C^{ijkl} \gamma_{ij} \gamma_{kl} \ , \qquad (6.15a)$$

$$\overline{U}^* = \frac{1}{2} D_{ijkl} \tau^{ij} \tau^{kl} \ . \qquad (6.15b)$$

Consideration of thermal influences

$$\overline{U} = \frac{1}{2} \tau^{ij} (\gamma_{ij} - \alpha_T g_{ij} \Theta) = \frac{1}{2} C^{ijkl} \gamma_{ij} \gamma_{kl} - \beta_T \gamma_i^i \Theta \ , \qquad (6.16a)$$

$$\overline{U}^* = \frac{1}{2} \tau^{ij} (\gamma_{ij} + \alpha_T g_{ij} \Theta) = \frac{1}{2} D_{ijkl} \tau^{ij} \tau^{kl} + \alpha_T \tau_i^i \Theta \ . \qquad (6.16b)$$

In *(6.16)*, the quadratic temperature terms are neglected [B.1].

6.3 Principle of virtual displacements (Pvd)

The virtual work δW of external forces is equal to the increase of virtual strain energy δU according to *(6.6)*:

$$\boxed{\delta W = \delta U = \int_V \tau^{ij}\, \delta \gamma_{ij}\, dV} \quad . \tag{6.17}$$

With the strain energy U $\widehat{=}$ *internal* potential Π_i

$$U = \int_V \overline{U}\, dV = \Pi_i \tag{6.18a}$$

and the work of the *external* forces (here without concentrated forces) of a conservative system equals the negative of the potential Π_e of the external forces

$$W = \int_V f^i\, v_i\, dV + \int_S p^i\, v_i\, dS = -\Pi_e \, , \tag{6.18b}$$

the total potential of the elastic body reads as follows

$$\Pi = \Pi_i + \Pi_e \, . \tag{6.19}$$

Principle of stationarity of the virtual total potential

$$\delta \Pi = \delta(\Pi_i + \Pi_e) = 0 \, . \tag{6.20}$$

This implies an extremum value of the total potential

$$\Pi = \Pi_i + \Pi_e = \text{extremum} \, . \tag{6.21a}$$

GREEN – DIRICHLET's principle of a minimum (valid for linearly elastic behaviour of material) [A.8]

$$\Pi = \Pi_i + \Pi_e = \text{minimum} \, . \tag{6.21b}$$

6.4 Principle of virtual forces (Pvf)

The complementary virtual work δW^* of the external forces is equal to the increase of the complementary virtual energy δU^*:

$$\delta W^* = \delta U^* = \int_V \gamma_{ij}\, \delta \tau^{ij}\, dV \, . \tag{6.22}$$

Total complementary potential follows by analogy to *(6.19)*

$$\Pi^* = \Pi_i^* + \Pi_e^* \tag{6.23}$$

with *internal* complementary potential $U^* = \int_V \overline{U}^* \, dV = \Pi_i^*$ *(6.24a)*

and *external* complementary potential Π_e^* defined as the negative of the external complementary work W^*

$$W^* = \int_V v_i f^i \, dV + \int_{S_v} v_i p^i \, dS = -\Pi_e^* . \tag{6.24b}$$

Principle of stationarity of the virtual total complementary potential

$$\delta \Pi^* = \delta (\Pi_i^* + \Pi_e^*) = 0 . \tag{6.25}$$

This implies an extremum value of the total complementary potential

$$\Pi^* = \Pi_i^* + \Pi_e^* = \text{extremum} . \tag{6.26a}$$

The CASTIGLIANO and MENABREA principle (valid for linear-elastic behaviour of material) [A.9]

$$\Pi^* = \Pi_i^* + \Pi_e^* = \text{minimum} . \tag{6.26b}$$

First theorem by CASTIGLIANO $v_i = \dfrac{\partial U^*(F^j)}{\partial F^i}$. *(6.27a)*

Theorem by MENABREA $\dfrac{\partial U^*(F^j)}{\partial (F^k)^R} = 0$, *(6.28)*

where the index R refers to the reaction forces.

Second theorem by CASTIGLIANO $F^i = \dfrac{\partial U(v_j)}{\partial v_i}$. *(6.27b)*

Generalized variational functional by HELLINGER and REISSNER [A.5, A.17, A.18, A.19]:

$$\Pi_R = \int_V \Big\{ \overline{U}(\gamma_{ij}) - f^i v_i + \tau^{ij} \Big[\frac{1}{2} (v_i|_j + v_j|_i) - \gamma_{ij} \Big] \Big\} dV - \\ - \int_{S_t} p_0^i v_i \, dS + \int_{S_d} (v_{i0} - v_i) p^i \, dS , \tag{6.29}$$

where p_0^i, v_{i0} are prescribed loads or displacements at the boundary.

HELLINGER - REISSNER stationary principle:

$$\delta \Pi_R = 0 . \tag{6.30}$$

6.5 Reciprocity theorems and "Unit–Load" method

Theorem by BETTI

$$\boxed{W_{21} = W_{12}} \quad . \tag{6.31a}$$

In words: If two sets of loads are acting on an elastic body, the work of loadset 1 against the displacements due to loadset 2 is equal to the work of loadset 2 against the displacements due to loadset 1.

If displacements at points 1 and 2 are expressed by MAXWELL's influence coefficients δ_{ij}, the *Theorem by MAXWELL* follows proving the symmetry of the coefficients:

$$\boxed{\delta_{ij} = \delta_{ji}} \quad . \tag{6.31b}$$

In words: The displacements at a point i due to a unit load at another point j is equal to the displacements at j due to a unit load at i. (It is assumed that the displacements of the points are measured in the directions of the applied forces.)

Unit-Load method
The *Unit–Load* method plays an important role in elasto–mechanics. By means of this method, the deformations of an elastic body at a certain point can be calculated [A.18, A.19]:

$$\sum\big(\underbrace{\overbrace{\bar{F}^i}^{\text{virtual static group}} v_i + \bar{M}^i \varphi_i}_{} \big) = \int_V \bar{\tau}^{ij} \gamma_{ij} \, dV \quad . \tag{6.32}$$

(virtual static group: $\bar{F}^i$, $\bar{M}^i$, $\bar{\tau}^{ij}$; real kinematic group: v_i, φ_i, γ_{ij})

6.6 Treatment of a variational problem

A curve $y = y(x)$ is to be determined in such a way that an integral I depending on x, $y = y(x)$, and the derivatives $y' = y'(x)$ to $y^{(n)} = y^{(n)}(x)$,

$$I = I(x, y, y', \ldots, y^{(n)}) = \int_{x_1}^{x_2} F(x, y, y', \ldots, y^{(n)})\, dx \tag{6.33}$$

attains an extremum value. This implies stationarity of I

$$\boxed{\delta I = \delta \int_{x_1}^{x_2} F(x, y, y', \ldots, y^{(n)})\, dx = 0} \quad . \tag{6.34}$$

For an integrand function F containing only derivatives up to the second order of the unknown function $y = y(x)$, follows from partial integration:

$$\delta I = \int_{x_1}^{x_2} \underbrace{\left(\frac{\partial F}{\partial y} - \frac{d}{dx}\frac{\partial F}{\partial y'} + \frac{d^2}{dx^2}\frac{\partial F}{\partial y''} \right)}_{\Longrightarrow \text{ EULER's equation}} \delta y \, dx +$$

$$+ \left[\underbrace{\left(\frac{\partial F}{\partial y'} - \frac{d}{dx}\frac{\partial F}{\partial y''} \right)}_{} \delta y \right]_{x_1}^{x_2} + \left[\frac{\partial F}{\partial y''} \delta y' \right]_{x_1}^{x_2} = 0 \,. \tag{6.35}$$

$\Longrightarrow$ residual (physical) boundary conditions

essential (geometric) boundary conditions

6.7 Continuous approximation methods

The following approaches belong to the group of continuous methods [A.3] (as opposed to discrete methods like the Finite Element Method, or the Boundary Element Method).

1) *Method by RAYLEIGH – RITZ* [A.14]
Point of reference $\longrightarrow$ variational expression *(6.33)*

$$I = \int_{x_1}^{x_2} F(x, y, y', \ldots, y^{(n)}) \, dx \quad \longrightarrow \quad \text{Extremum} \,.$$

Choice of a set of linearly independent approximation functions

$$y^*(x) = \sum_{n=0}^{N} a_n \, y_n^*(x) \,, \tag{6.36}$$

where the y_n^* must at least satisfy the geometric boundary conditions.

Demand of a minimum:

$$\frac{\partial I}{\partial a_n} = 0 \quad , \quad (n = 0, \ldots, N). \tag{6.37}$$

Assuming a quadratic form of the functional, this leads to a linear system of equations for determination of the coefficients a_n .

2) *Method by GALERKIN* [A.6, A.16]
Point of reference $\longrightarrow$ variational functional in the varied form according to *(6.35)*.

Choice of a function $y^*(x)$ in analogy to the RITZ approach *(6.36)*. Functions $y_n^*(x)$ must fulfill *all* boundary conditions.

Demand of a minimum:
Fulfilling of the GALERKIN equations

$$\int_{x_1}^{x_2} L(y^*) \, y_n^* \, dx = 0 \,, \quad (n = 1, 2, 3, \ldots, N) \tag{6.38}$$

with $L(y^*)$ as the differential equation for the problem (see *(6.35)*).

This leads to a system of linear equations with respect to the coefficients a_n if a quadratic form of the functional (≙ linear differential equation) is assumed. In the case of functionals of higher order than quadratic, we get a non – linear system of equations.

7 Problem formulations in the theory of linear elasticity

7.1 Basic equations and boundary – value problems

- *three* equilibrium conditions *(3.28a)* ,
- *six* strain – displacement relations *(4.12)* ,
- *six* equations of the material law *(5.5)* or *(5.6)* ,

i.e. altogether 15 equations for 15 unknown field quantities (6 stresses τ^{ij}, 6 strains γ_{ij} , 3 displacements v_i).

Problem of elasticity theory : solution of basic equations with given boundary conditions
$\longrightarrow$ *boundary – value problems.*

We distinguish between three kinds of *boundary – value problems*:

– *First boundary – value problem*
On the total surface S of a body B, the *tractions* t^j are given (Fig. 7.1a). The following is valid for the components of the stress vector

$$t^j_{(S)} = (\tau^{ij} n_i)_S \, . \tag{7.1a}$$

– *Second boundary – value problem*
The displacements of the total surface S of the body B are given (Fig. 7.1b). On the surface the following displacements are given:

$$v_{i(S)} = (v_i)_S \, . \tag{7.2a}$$

– *Mixed boundary – value problem*
On one part S_t of the surface S of the body B, the tractions are given, and on the remaining part S_d of the surface the displacements are given (Fig. 7.1c). The boundary conditions then read

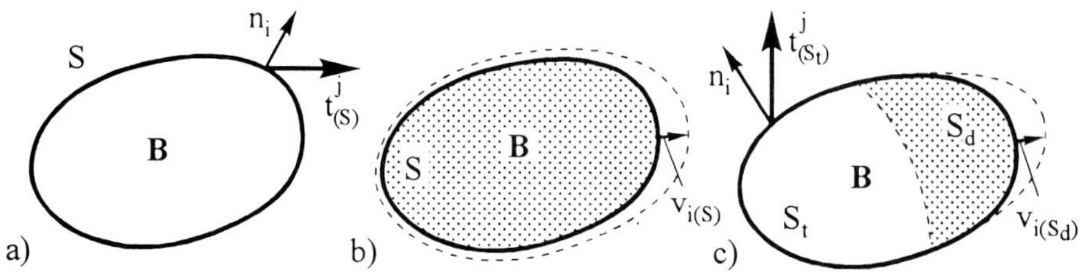

Fig. 7.1 : Illustration of boundary – value problems

$$t^j_{(S_t)} = (\tau^{ij} n_i)_{S_t} , \qquad (7.1b)$$

$$v_{i(S_d)} = (v_i)_{S_d} . \qquad (7.2b)$$

7.2 Solutions of basic equations

LAMÉ - NAVIER's equations – solving with regard to the displacements

$$\boxed{\frac{\mu}{\lambda + \mu} \Delta v^i + e|^i - \frac{3\lambda + 2\mu}{\lambda + \mu} \alpha_T \Theta|^i + \frac{1}{\lambda + \mu} f^i = 0} \qquad (7.3)$$

with the LAMÉ constants μ, λ according to *(5.7b)* and the volume dilatation e according to *(4.17b)*.
These are three coupled, partial differential equations of second order for the three unknown displacement components v^i.

BELTRAMI - MICHELL's equations – solving with regard to the stresses

$$\boxed{\Delta \tau^{ij} + \frac{1}{1 + \nu} s|^{ij} - \frac{\nu}{1 + \nu} g^{ij} \Delta s = 0} \qquad (7.4)$$

with s as the sum of the normal stresses according to *(3.13a)*.

These are six coupled, partial differential equations of second order for the six unkown stress components τ^{ij}.

7.3 Special equations for three-dimensional problems

Solved with respect to displacements [A.5, A.10, A.11, A.12]
Use of the LOVE function χ leads to

$$\Delta\Delta \chi = 0 . \qquad (7.5)$$

This bipotential equation has an infinite number of solutions, e.g. feasible solutions in cylindrical coordinates are for the axisymmetric case [A.9]

$$\chi = r^2 , \ln r , r^2 \ln r ; z , z^2 , z^3 ; z \ln r , R , \frac{1}{R} , \ln \frac{R + z}{R - z} , z \ln (z + R) \qquad (7.6)$$

with $R = \sqrt{r^2 + z^2}$.

All linear combinations of *(7.6)* with arbitrary constants are solutions of *(7.5)* as well.
Displacements in the axisymmetric case

$$u = - \frac{1}{1 - 2\nu} \frac{\partial^2 \chi}{\partial r \partial z} , \qquad (7.7a)$$

$$w = \frac{2(1 - \nu)}{1 - 2\nu} \Delta \chi - \frac{1}{1 - 2\nu} \frac{\partial^2 \chi}{\partial z^2} . \qquad (7.7b)$$

Stresses in the axisymmetrical case

$$\sigma_{rr} = 2\,G\,\frac{\nu}{1-2\nu}\,\frac{\partial}{\partial z}\left(\Delta\chi - \frac{1}{\nu}\,\frac{\partial^2\chi}{\partial r^2}\right) \quad , \tag{7.8a}$$

$$\sigma_{\varphi\varphi} = 2\,G\,\frac{\nu}{1-2\nu}\,\frac{\partial}{\partial z}\left(\Delta\chi - \frac{1}{\nu}\,\frac{1}{r}\,\frac{\partial\chi}{\partial r}\right) \quad , \tag{7.8b}$$

$$\sigma_{zz} = 2\,G\,\frac{2-\nu}{1-2\nu}\,\frac{\partial}{\partial z}\left(\Delta\chi - \frac{1}{2-\nu}\,\frac{\partial^2\chi}{\partial z^2}\right) , \tag{7.8c}$$

$$\tau_{rz} = 2\,G\,\frac{1-\nu}{1-2\nu}\,\frac{\partial}{\partial r}\left(\Delta\chi - \frac{1}{1-\nu}\,\frac{\partial^2\chi}{\partial z^2}\right) . \tag{7.8d}$$

7.4 Special equations for plane problems

a) *State of plane stress*

– Solved with regard to displacements

NAVIER's equation

$$\boxed{v^{\alpha}\big|^{\beta}_{\beta} + \frac{1+\nu}{1-\nu}\,v^{\beta}\big|^{\alpha}_{\beta} - 2\,\frac{1+\nu}{1-\nu}\,\alpha_T\,{}^0\Theta\big|^{\alpha} + \frac{1}{G}\,f^{\alpha} = 0} \quad . \tag{7.9}$$

Coupled system of two partial differential equations for the two unknown displacement components v^{α}.
Introduction of a displacement function Ψ

$$v^{\alpha} = \Psi\big|^{\alpha} . \tag{7.10}$$

For vanishing volume forces f^{α} the POISSON's equation is obtained from *(7.9)*

$$\boxed{\Psi\big|^{\beta}_{\beta} = \Delta\Psi = \nabla^2\Psi = (1+\nu)\,\alpha_T\,{}^0\Theta} \quad , \tag{7.11}$$

where Ψ is called the *thermo – elastic displacement potential* [A.13].

– Solving with regard to stresses

$$\boxed{\varepsilon^{\alpha\mu}\,\varepsilon^{\beta\nu}\,D_{\alpha\beta\gamma\delta}\,\tau^{\gamma\delta}\big|_{\mu\nu} + \alpha_T\,{}^0\Theta\big|^{\delta}_{\delta} = 0} \quad . \tag{7.12}$$

Introduction of AIRY's stress function in *(7.12)* assuming conservative volume forces ($f^{\alpha} = -V\big|^{\alpha}$) [A.4, A.8]

$$\tau^{\gamma\delta} = \varepsilon^{\gamma\sigma}\,\varepsilon^{\delta\tau}\,\Phi\big|_{\sigma\tau} + V\,g^{\gamma\delta} \tag{7.13}$$

provides the bipotential equation

$$\Phi\big|^{\sigma\tau}_{\sigma\tau} = -E\,\alpha_T\,{}^0\Theta\big|^{\delta}_{\delta} - (1-\nu)\,V\big|^{\varepsilon}_{\varepsilon}$$

or
$$\boxed{\Delta\Delta\,\Phi = -\,E\,\alpha_T\,\Delta\,{}^0\Theta - (1-\nu)\,\Delta\,V} \; . \tag{7.14}$$

b) *State of plane strain*

– Solved with regard to displacements

Analogous to *(7.11)*

$$\boxed{\Delta\,\Psi = \frac{1+\nu}{1-\nu}\,\alpha_T\,{}^0\Theta} \; . \tag{7.15}$$

– Solved with regard to the stresses

In analogy to *(7.14)*

$$\boxed{\Delta\Delta\,\Phi = -\frac{E}{1-\nu}\,\alpha_T\,\Delta\,{}^0\Theta - \frac{1-2\nu}{1-\nu}\,\Delta\,V} \; . \tag{7.16}$$

7.5 Comparison of state of plane stress and state of plane strain

Since many problems can be described in Cartesian coordinates, we would like to list the notations for both two-dimensional states in these coordinates again.

State of plane stress	State of plane strain

The stresses must fulfill the equilibrium conditions by analogy to *(3.27a)*

$$\begin{aligned} \frac{\partial\,\sigma_{xx}}{\partial\,x} + \frac{\partial\,\tau_{xy}}{\partial\,y} + f_x &= 0\,, \\ \frac{\partial\,\tau_{xy}}{\partial\,x} + \frac{\partial\,\sigma_{yy}}{\partial\,y} + f_y &= 0\,. \end{aligned} \tag{7.17}$$

The boundary conditions *(7.1b, 7.2b)* can be given in a mixed form

$$\begin{aligned} (\sigma_{xx}\cos\alpha + \tau_{xy}\sin\alpha)\big|_S &= t_{x0}\,, \\ (\sigma_{yy}\sin\alpha + \tau_{xy}\cos\alpha)\big|_S &= t_{y0} \end{aligned} \tag{7.18a}$$

or
$$u_S = u_0 \quad , \quad v_S = v_0 \, . \tag{7.18b}$$

Here, equations *(7.18a)* are valid at points on S, where external loads are acting with the components t_{x0} and t_{y0} per surface unit. For points on S at which boundary displacements are given by the components u_0 and v_0, *(7.18b)* is valid. These quantities, but also the tractions may equal zero.

For $\Theta = 0$, the material laws *(5.12)* and *(5.17)* read

$$\varepsilon_{xx} = \frac{1}{E}(\sigma_{xx} - \nu\sigma_{yy}), \qquad \varepsilon_{xx} = \frac{1}{E}[(1-\nu^2)\sigma_{xx} - \nu(1+\nu)\sigma_{yy}],$$

$$\varepsilon_{yy} = \frac{1}{E}(\sigma_{yy} - \nu\sigma_{xx}), \qquad \varepsilon_{yy} = \frac{1}{E}[(1-\nu^2)\sigma_{yy} - \nu(1+\nu)\sigma_{xx}], \tag{7.19}$$

$$\gamma_{xy} = \frac{1}{G}\tau_{xy}, \qquad \gamma_{xy} = \frac{1}{G}\tau_{xy},$$

and the corresponding compatibility condition according to *(4.19)* follows to be

$$\frac{\partial^2 \varepsilon_{xx}}{\partial y^2} + \frac{\partial^2 \varepsilon_{yy}}{\partial x^2} - \frac{\partial^2 \gamma_{xy}}{\partial x\, \partial y} = 0 . \tag{7.20}$$

One should pay attention to the fact that γ_{xy} is the *technical shear strain* which differs from the tensorial quantity $\gamma_{\alpha\beta}$ by a factor of 2!
Substituting the material law *(7.19)* into *(7.20)* leads to

$$\Delta(\sigma_{xx} + \sigma_{yy}) = -(1+\nu)\left(\frac{\partial f_x}{\partial x} + \frac{\partial f_y}{\partial y}\right) \qquad \Delta(\sigma_{xx} + \sigma_{yy}) = +\frac{1}{(1-\nu)}\left(\frac{\partial f_x}{\partial x} + \frac{\partial f_y}{\partial y}\right). \tag{7.21}$$

According to this, the sum of stresses as the first invariant of the stress state is a harmonic function. If the external load is known at the whole boundary, the stresses can be determined from *(7.17)*, *(7.18)*, and *(7.21)* without considering the displacement field.
By introducing AIRY's stress function Φ according to *(7.13)*

$$\sigma_{xx} = \frac{\partial^2 \Phi}{\partial y^2} + V , \quad \sigma_{yy} = \frac{\partial^2 \Phi}{\partial x^2} + V , \quad \tau_{xy} = -\frac{\partial^2 \Phi}{\partial x\, \partial y} \tag{7.22}$$

the equilibrium conditions *(7.17)* are identically fulfilled with

$$\frac{\partial V}{\partial x} = -f_x , \quad \frac{\partial V}{\partial y} = -f_y .$$

From *(7.21)* result

$$\Delta\Delta\Phi + (1-\nu)\Delta V = 0 \qquad \Delta\Delta\Phi + \frac{1-2\nu}{1-\nu}\Delta V = 0 . \tag{7.23}$$

If no volume forces are present, the biharmonic equations *(7.23)* take the same form for both states

$$\Delta\Delta\Phi = \frac{\partial^4 \Phi}{\partial x^4} + 2\frac{\partial^4 \Phi}{\partial x^2\, \partial y^2} + \frac{\partial^4 \Phi}{\partial y^4} = 0 . \tag{7.24}$$

A.2 Exercises

Exercise A-2-1:

An oblique base $\mathbf{g}_i$ (i = 1,2,3) expressed by the orthonormalized base vectors $\mathbf{e}_i$ ($i = 1,2,3$; $|\mathbf{e}_i| = 1$) is given:

$$\mathbf{g}_1 = \mathbf{e}_1 ,$$

$$\mathbf{g}_2 = \mathbf{e}_1 + \mathbf{e}_2 ,$$

$$\mathbf{g}_3 = \mathbf{e}_1 + \mathbf{e}_2 + \mathbf{e}_3 .$$

a) Determine for the vectors

$$\mathbf{x} = 2\,\mathbf{e}_1 + 2\,\mathbf{e}_2 + 2\,\mathbf{e}_3 \quad , \quad \mathbf{y} = -\mathbf{e}_1 - \mathbf{e}_2 + 4\,\mathbf{e}_3$$

the scalar product $\mathbf{x} \cdot \mathbf{y}$, $|\mathbf{x}|$, and the angle φ between the vectors (Fig. A-1).

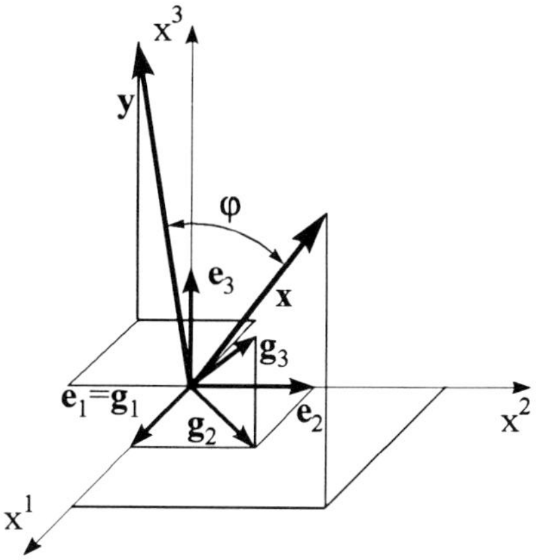

Fig. A-1: Oblique base with two vectors **x** and **y**

b) State the contravariant metric tensor and the contravariant base vectors.

c) Apply the vector $\mathbf{a} = \mathbf{g}_1 + 2\,\mathbf{g}_2 + \mathbf{g}_3$ to the contravariant base $\mathbf{g}^i$.

d) Determine the physical components of the vector **a** .

e) For the given transformation matrix between two bases

$$\beta^j_{i'} = \begin{bmatrix} 1 & -1 & -1 \\ 1 & 0 & -1 \\ 1 & 0 & 1 \end{bmatrix} \qquad (i' = 1,2,3) ,$$

express the "old" base $\mathbf{g}_j$ in terms of the *new* base $\mathbf{g}_{i'}$, and vice versa.

Solution:

a) We first determine the required quantities in the orthonormalized base:

$$\mathbf{x} = 2\,\mathbf{e}_1 + 2\,\mathbf{e}_2 + \mathbf{e}_3 = x^i\,\mathbf{e}_i \quad ; \quad x^1 = 2 \ , \quad x^2 = 2 \ , \quad x^3 = 1 \ ,$$

$$\mathbf{y} = -\,\mathbf{e}_1 - \mathbf{e}_2 + 4\,\mathbf{e}_3 = y^j\,\mathbf{e}_j \quad ; \quad y^1 = -1 \ , \quad y^2 = -1 \ , \quad y^3 = 4 \ .$$

Scalar product in orthonormalized base *(2.2a)* ($g_{ij} \longrightarrow \delta_{ij}$)

$$\mathbf{x} \cdot \mathbf{y} = x^i y^j \underbrace{\mathbf{e}_i \cdot \mathbf{e}_j}_{\delta_{ij}} = x^1 y^1 + x^2 y^2 + x^3 y^3 = 2(-1) + 2(-1) + 1 \cdot 4 = 0 \ .$$

Length of vector $\mathbf{x}$ *(2.2b)*

$$|\mathbf{x}| = \sqrt{\mathbf{x} \cdot \mathbf{x}} = \sqrt{x^i x^j \mathbf{e}_i \cdot \mathbf{e}_j} = \sqrt{x^i x^j \delta_{ij}} =$$

$$= \sqrt{(x^1)^2 + (x^2)^2 + (x^3)^2} = \sqrt{4 + 4 + 1} = \sqrt{9} = 3 \ .$$

Angle between vectors $\mathbf{x}$ and $\mathbf{y}$ *(2.2c)*

$$\cos\varphi = \frac{x^i y^j \delta_{ij}}{\sqrt{x^i x^j \delta_{ij}}\,\sqrt{y^k y^l \delta_{kl}}} = \frac{x^1 y^1 + x^2 y^2 + x^3 y^3}{\sqrt{(x^1)^2 + (x^2)^2 + (x^3)^2}\,\sqrt{(y^1)^2 + (y^2)^2 + (y^3)^2}} =$$

$$= \frac{0}{3\sqrt{18}} = 0 \longrightarrow \varphi = \frac{\pi}{2} \ , \quad \text{i.e.} \quad \mathbf{y} \perp \mathbf{x} \ .$$

Transformation into the oblique base then yields :

$$\mathbf{e}_1 = \mathbf{g}_1 \ ,$$
$$\mathbf{e}_2 = \mathbf{g}_2 - \mathbf{g}_1 \ ,$$
$$\mathbf{e}_3 = \mathbf{g}_3 - \mathbf{g}_2 \ .$$

Vector in covariant base

$$\mathbf{x} = \mathbf{g}_2 + \mathbf{g}_3 = \bar{x}^i\,\mathbf{g}_i \quad ; \quad \bar{x}^1 = 0 \ , \quad \bar{x}^2 = 1 \ , \quad \bar{x}^3 = 1 \ ,$$

$$\mathbf{y} = -5\,\mathbf{g}_2 + 4\,\mathbf{g}_3 = \bar{y}^j\,\mathbf{g}_j \quad ; \quad \bar{y}^1 = 0 \ , \quad \bar{y}^2 = -5 \ , \quad \bar{y}^3 = 4 \ .$$

Covariant metric components according to *(2.1a,b)*

$$g_i \cdot g_j = g_{ij} \longrightarrow (g_{ij}) = \begin{bmatrix} g_{11} & g_{12} & g_{13} \\ g_{12} & g_{22} & g_{23} \\ g_{13} & g_{23} & g_{33} \end{bmatrix} ,$$

$$g_{11} = \mathbf{g}_1 \cdot \mathbf{g}_1 = \mathbf{e}_1 \cdot \mathbf{e}_1 = 1 \cdot 1 = 1 \ ,$$
$$g_{12} = \mathbf{g}_1 \cdot \mathbf{g}_2 = \mathbf{e}_1(\mathbf{e}_1 + \mathbf{e}_2) = 1 \ ,$$
$$g_{13} = \mathbf{g}_1 \cdot \mathbf{g}_3 = \mathbf{e}_1(\mathbf{e}_1 + \mathbf{e}_2 + \mathbf{e}_3) = 1 \ ,$$

$$g_{22} = \mathbf{g}_2 \cdot \mathbf{g}_2 = (\mathbf{e}_1 + \mathbf{e}_2)(\mathbf{e}_1 + \mathbf{e}_2) = 1 + 1 = 2\,,$$
$$g_{23} = \mathbf{g}_2 \cdot \mathbf{g}_3 = (\mathbf{e}_1 + \mathbf{e}_2)(\mathbf{e}_1 + \mathbf{e}_2 + \mathbf{e}_3) = 1 + 1 = 2\,,$$
$$g_{33} = \mathbf{g}_3 \cdot \mathbf{g}_3 = (\mathbf{e}_1 + \mathbf{e}_2 + \mathbf{e}_3)(\mathbf{e}_1 + \mathbf{e}_2 + \mathbf{e}_3) = 1 + 1 + 1 = 3\,.$$

The metric tensor then reads

$$(g_{ij}) = \begin{bmatrix} 1 & 1 & 1 \\ 1 & 2 & 2 \\ 1 & 2 & 3 \end{bmatrix}. \tag{1}$$

Formulation of the *scalar product* for the covariant base yields due to *(2.2a)*:

$$\begin{aligned}
\mathbf{x} \cdot \mathbf{y} &= g_{ij}\,\bar{x}^i\,\bar{y}^j = \\
&= g_{1j}\,\bar{x}^1\,\bar{y}^j + g_{2j}\,\bar{x}^2\,\bar{y}^j + g_{3j}\,\bar{x}^3\,\bar{y}^j = \\
&= g_{11}\,\bar{x}^1\,\bar{y}^1 + g_{12}\,\bar{x}^1\,\bar{y}^2 + g_{13}\,\bar{x}^1\,\bar{y}^3 + \\
&\quad + g_{21}\,\bar{x}^2\,\bar{y}^1 + g_{22}\,\bar{x}^2\,\bar{y}^2 + g_{23}\,\bar{x}^2\,\bar{y}^3 + \\
&\quad + g_{31}\,\bar{x}^3\,\bar{y}^1 + g_{32}\,\bar{x}^3\,\bar{y}^2 + g_{33}\,\bar{x}^3\,\bar{y}^3 = \\
&= 1 \cdot 0 \cdot 0 + 1 \cdot 0 \cdot (-5) + 1 \cdot 0 \cdot 4 + \\
&\quad + 1 \cdot 1 \cdot 0 + 2 \cdot 1 \cdot (-5) + 2 \cdot 1 \cdot 4 + \\
&\quad + 1 \cdot 1 \cdot 0 + 2 \cdot 1 \cdot (-5) + 3 \cdot 1 \cdot 4 = \\
&= -10 + 8 - 10 + 12 = 0\,.
\end{aligned}$$

Length of vector $\mathbf{x}$ in the covariant base

$$\begin{aligned}
|\mathbf{x}|^2 &= g_{11}\,\bar{x}^1\,\bar{x}^1 + g_{12}\,\bar{x}^1\,\bar{x}^2 + g_{13}\,\bar{x}^1\,\bar{x}^3 + g_{21}\,\bar{x}^2\,\bar{x}^1 + g_{22}\,\bar{x}^2\,\bar{x}^2 + g_{23}\,\bar{x}^2\,\bar{x}^3 + \\
&\quad + g_{31}\,\bar{x}^3\,\bar{x}^1 + g_{32}\,\bar{x}^3\,\bar{x}^2 + g_{33}\,\bar{x}^3\,\bar{x}^3 = \\
&= 0 + 0 + 0 + 0 + 2 \cdot 1 \cdot 1 + 2 \cdot 1 \cdot 1 + 0 + 2 \cdot 1 \cdot 1 + 3 \cdot 1 \cdot 1 = 9\,;
\end{aligned}$$

$|\mathbf{x}| = \sqrt{9} = 3 \quad \longrightarrow$ herewith, the length is proved to be an invariant .

In oblique base the *angle* φ follows from *(2.2c)*

$$\cos\varphi = \frac{g_{ij}\,\bar{x}^i\,\bar{y}^j}{\sqrt{g_{kl}\,\bar{x}^k\,\bar{x}^l}\,\sqrt{g_{mn}\,\bar{y}^m\,\bar{y}^n}} = \frac{0}{3\sqrt{18}} = 0 \quad \longrightarrow \quad \varphi = \frac{\pi}{2}\,.$$

b) The contravariant base can be calculated by two methods :

First way :

According to *(2.5a)* $\qquad \mathbf{g}^i = g^{ij}\,\mathbf{g}_j$

and to *(2.5b)* $\qquad \mathbf{g}_i = g_{ij}\,\mathbf{g}^j\,.$

The scalar product of the base vectors is determined by *(2.5c)*

$$\delta^i_k = g^{ij} g_{kj} = g^{i1} g_{k1} + g^{i2} g_{k2} + g^{i3} g_{k3}$$

and further

$$\delta^1_k = g^{11} g_{k1} + g^{12} g_{k2} + g^{13} g_{k3} \,,$$

$$\delta^2_k = g^{21} g_{k1} + g^{22} g_{k2} + g^{23} g_{k3} \,,$$

$$\delta^3_k = g^{31} g_{k1} + g^{32} g_{k2} + g^{33} g_{k3} \,;$$

$$\Longrightarrow \begin{bmatrix} g^{11} & g^{12} & g^{13} \\ g^{12} & g^{22} & g^{23} \\ g^{13} & g^{23} & g^{33} \end{bmatrix} \begin{bmatrix} g_{11} & g_{12} & g_{13} \\ g_{12} & g_{22} & g_{23} \\ g_{13} & g_{23} & g_{33} \end{bmatrix} = \begin{bmatrix} 1 & 0 & 0 \\ 0 & 1 & 0 \\ 0 & 0 & 1 \end{bmatrix} .$$

Forming the inverse of (g_{ij}) yields the contravariant matrix (g^{ij}):

$$(g^{ij}) = (g_{ij})^{-1} = \begin{bmatrix} 1 & 1 & 1 \\ 1 & 2 & 2 \\ 1 & 2 & 3 \end{bmatrix}^{-1} .$$

From the rules of matrix inversion follows

$$(g_{ij})^{-1} = \frac{1}{g} \begin{bmatrix} \begin{vmatrix} 2 & 2 \\ 2 & 3 \end{vmatrix} & -\begin{vmatrix} 1 & 1 \\ 2 & 3 \end{vmatrix} & \begin{vmatrix} 1 & 1 \\ 2 & 2 \end{vmatrix} \\ -\begin{vmatrix} 1 & 2 \\ 1 & 3 \end{vmatrix} & \begin{vmatrix} 1 & 1 \\ 1 & 3 \end{vmatrix} & -\begin{vmatrix} 1 & 1 \\ 1 & 2 \end{vmatrix} \\ \begin{vmatrix} 1 & 2 \\ 1 & 2 \end{vmatrix} & -\begin{vmatrix} 1 & 1 \\ 1 & 2 \end{vmatrix} & \begin{vmatrix} 1 & 1 \\ 1 & 2 \end{vmatrix} \end{bmatrix} .$$

The determinant is

$$g = |g_{ij}| = \begin{vmatrix} 1 & 1 & 1 \\ 1 & 2 & 2 \\ 1 & 2 & 3 \end{vmatrix} = 6 + 2 + 2 - 2 - 4 - 3 = 1 .$$

The contravariant metric tensor now becomes

$$(g^{ij}) = (g_{ij})^{-1} = \frac{1}{1} \begin{bmatrix} 2 & -1 & 0 \\ -1 & 2 & -1 \\ 0 & -1 & 1 \end{bmatrix} . \qquad (2)$$

From *(2.5a)* follows herewith

$$\mathbf{g}^i = g^{ij} \mathbf{g}_j \longrightarrow$$

$$\mathbf{g}^1 = g^{11} \mathbf{g}_1 + g^{12} \mathbf{g}_2 + g^{13} \mathbf{g}_3 = 2\mathbf{g}_1 - \mathbf{g}_2 \,,$$

$$\mathbf{g}^2 = g^{12} \mathbf{g}_1 + g^{22} \mathbf{g}_2 + g^{23} \mathbf{g}_3 = -\mathbf{g}_1 + 2\mathbf{g}_2 - \mathbf{g}_3 \,, \qquad (3)$$

$$\mathbf{g}^3 = g^{13} \mathbf{g}_1 + g^{23} \mathbf{g}_2 + g^{33} \mathbf{g}_3 = -\mathbf{g}_2 + \mathbf{g}_3 \,.$$

The contravariant base can be expressed by the orthonormalized base as

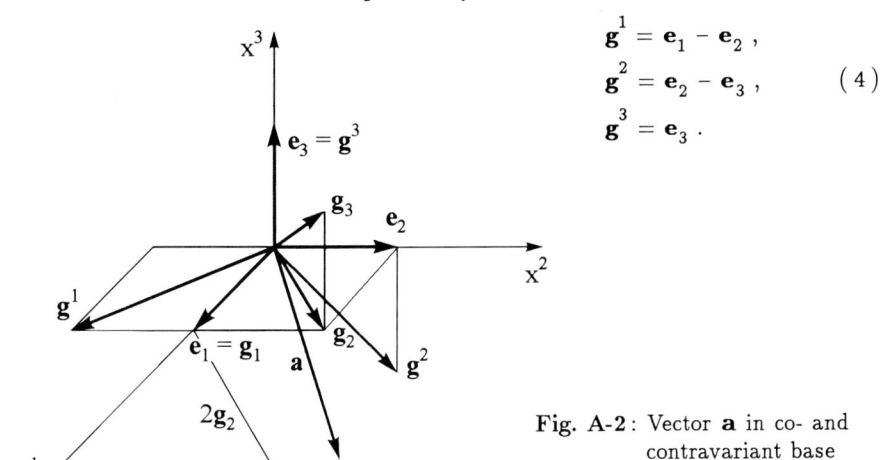

$$\mathbf{g}^1 = \mathbf{e}_1 - \mathbf{e}_2 ,\quad \mathbf{g}^2 = \mathbf{e}_2 - \mathbf{e}_3 ,\quad \mathbf{g}^3 = \mathbf{e}_3 . \tag{4}$$

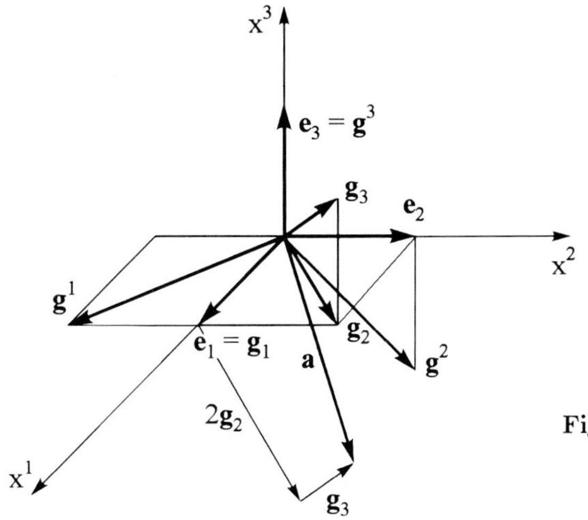

Fig. A-2: Vector **a** in co- and contravariant base

Check:

According to *(2.3a)* $\quad \mathbf{g}_i \cdot \mathbf{g}^j = \delta_i^j$

with (4) yields

$$\mathbf{g}_1 \cdot \mathbf{g}^1 = \mathbf{e}_1 (\mathbf{e}_1 - \mathbf{e}_2) = 1$$

or with (3) and *(2.5b)* with (1)

$$\mathbf{g}_1 \cdot \mathbf{g}^1 = (\mathbf{g}^1 + \mathbf{g}^2 + \mathbf{g}^3)(2\mathbf{g}_1 - \mathbf{g}_2) = 2\,\delta_1^1 - \delta_2^2 = 2 - 1 = 1 ,$$

$$\mathbf{g}_1 \cdot \mathbf{g}^2 = \mathbf{e}_1 (\mathbf{e}_2 - \mathbf{e}_3) = (\mathbf{g}^1 + \mathbf{g}^2 + \mathbf{g}^3)(-\mathbf{g}_1 + 2\mathbf{g}_2 - \mathbf{g}_3) =$$
$$= -\,\delta_1^1 + 2\,\delta_2^2 - \delta_3^3 = -1 + 2 - 1 = 0 .$$

Second way:

The contravariant base vectors can be calculated by forming the vector products according to *(2.6)*:

$$\mathbf{g}^1 = \frac{\mathbf{g}_2 \times \mathbf{g}_3}{[\mathbf{g}_1, \mathbf{g}_2, \mathbf{g}_3]} , \quad \mathbf{g}^2 = \frac{\mathbf{g}_3 \times \mathbf{g}_1}{[\mathbf{g}_1, \mathbf{g}_2, \mathbf{g}_3]} , \quad \mathbf{g}^3 = \frac{\mathbf{g}_1 \times \mathbf{g}_2}{[\mathbf{g}_1, \mathbf{g}_2, \mathbf{g}_3]} ,$$

$$[\mathbf{g}_1, \mathbf{g}_2, \mathbf{g}_3] = \begin{vmatrix} 1 & 0 & 0 \\ 1 & 1 & 0 \\ 1 & 1 & 1 \end{vmatrix} = 1 \quad \longrightarrow$$

$$\mathbf{g}^1 = \mathbf{g}_2 \times \mathbf{g}_3 = \begin{vmatrix} \mathbf{e}_1 & \mathbf{e}_2 & \mathbf{e}_3 \\ 1 & 1 & 0 \\ 1 & 1 & 1 \end{vmatrix} = \mathbf{e}_1 - \mathbf{e}_2 = \mathbf{g}_1 - (\mathbf{g}_2 - \mathbf{g}_1) = 2\mathbf{g}_1 - \mathbf{g}_2 ,$$

$$\mathbf{g}^2 = \mathbf{g}_3 \times \mathbf{g}_1 = \begin{vmatrix} \mathbf{e}_1 & \mathbf{e}_2 & \mathbf{e}_3 \\ 1 & 1 & 1 \\ 1 & 0 & 0 \end{vmatrix} = \mathbf{e}_2 - \mathbf{e}_3 = \mathbf{g}_2 - \mathbf{g}_1 - (\mathbf{g}_3 - \mathbf{g}_2) =$$

$$= -\mathbf{g}_1 + 2\mathbf{g}_2 - \mathbf{g}_3 ,$$

$$\mathbf{g}^3 = \mathbf{g}_1 \times \mathbf{g}_2 = \begin{vmatrix} \mathbf{e}_1 & \mathbf{e}_2 & \mathbf{e}_3 \\ 1 & 0 & 0 \\ 1 & 1 & 0 \end{vmatrix} = \mathbf{e}_3 = \mathbf{g}_3 - \mathbf{g}_2 .$$

According to *(2.4b)*, the contravariant metric components become

$$g^{ij} = \mathbf{g}^i \cdot \mathbf{g}^j \longrightarrow$$

$$g^{11} = \mathbf{g}^1 \cdot \mathbf{g}^1 = (2\mathbf{g}_1 - \mathbf{g}_2)(2\mathbf{g}_1 - \mathbf{g}_2) = 4g_{11} - 4g_{12} + g_{22} = 4 - 4 + 2 = 2 ,$$

$$g^{12} = \mathbf{g}^1 \cdot \mathbf{g}^2 = (2\mathbf{g}_1 - \mathbf{g}_2)(-\mathbf{g}_1 + 2\mathbf{g}_2 - \mathbf{g}_3) =$$

$$= -2g_{11} + 4g_{12} - 2g_{13} + g_{12} - 2g_{22} + g_{23} =$$

$$= -2 \cdot 1 + 4 \cdot 1 - 2 \cdot 1 + 1 - 2 \cdot 2 + 2 = -1 ,$$

etc.

c) The vector **a** in the covariant base

$$\mathbf{a} = \mathbf{g}_1 + 2\mathbf{g}_2 + \mathbf{g}_3 = a^i \mathbf{g}_i \qquad (a^1 = 1 , a^2 = 2 , a^3 = 1)$$

shall be applied to the contravariant base.

Analogously to *(2.5b)* it holds that

$$a_i = g_{ij} a^j \longrightarrow$$

$$a_1 = g_{11} a^1 + g_{12} a^2 + g_{13} a^3 \longrightarrow a_1 = 1 \cdot 1 + 1 \cdot 2 + 1 \cdot 1 = 4 ,$$

$$a_2 = g_{12} a^1 + g_{22} a^2 + g_{23} a^3 \longrightarrow a_2 = 1 \cdot 1 + 2 \cdot 2 + 2 \cdot 1 = 7 ,$$

$$a_3 = g_{13} a^1 + g_{23} a^2 + g_{33} a^3 \longrightarrow a_3 = 1 \cdot 1 + 2 \cdot 2 + 3 \cdot 1 = 8 .$$

Therefore, the vector in the contravariant base is

$$\mathbf{a} = a_i \mathbf{g}^i = 4\mathbf{g}^1 + 7\mathbf{g}^2 + 8\mathbf{g}^3 .$$

d) The physical components of the vector according to *(2.10)* become

$$a^{*i} = a^i \sqrt{g_{(ii)}} \qquad \text{or} \qquad a^*_i = a_i \sqrt{g^{(ii)}} .$$

With the vector in co- and/or contravariant base

$$\mathbf{a} = \mathbf{g}_1 + 2\mathbf{g}_2 + \mathbf{g}_3 = 4\mathbf{g}^1 + 7\mathbf{g}^2 + 8\mathbf{g}^3 ,$$

it then follows that

$$a^{*1} = 1\sqrt{g_{11}} = 1\sqrt{1} = 1 \quad ; \quad a_1^* = 4\sqrt{g^{11}} = 4\sqrt{2} \,,$$

$$a^{*2} = 2\sqrt{g_{22}} = 2\sqrt{2} \quad ; \quad a_2^* = 7\sqrt{g^{22}} = 7\sqrt{2} \,,$$

$$a^{*3} = 1\sqrt{g_{33}} = \sqrt{3} \quad ; \quad a_3^* = 8\sqrt{g^{33}} = 8\sqrt{1} = 8 \,.$$

e) The "new" base $\mathbf{g}_{i'}$ is expressed by the "old" base $\mathbf{g}_j$ by means of the transformation matrix

$$\beta^j_{i'} = \begin{bmatrix} 1 & -1 & -1 \\ 1 & 0 & -1 \\ 1 & 0 & 1 \end{bmatrix}, \quad i' = 1,2,3 \quad \text{new base} \,.$$

For this, we use *(2.7a)*

$$\mathbf{g}_{i'} = \beta^j_{i'}\,\mathbf{g}_j \quad \longrightarrow$$

$$\mathbf{g}_{1'} = \beta^1_{1'}\,\mathbf{g}_1 + \beta^2_{1'}\,\mathbf{g}_2 + \beta^3_{1'}\,\mathbf{g}_3 = \mathbf{g}_1 - \mathbf{g}_2 - \mathbf{g}_3 = -\mathbf{e}_1 - 2\,\mathbf{e}_2 - \mathbf{e}_3 \,,$$

$$\mathbf{g}_{2'} = \beta^1_{2'}\,\mathbf{g}_1 + \beta^2_{2'}\,\mathbf{g}_2 + \beta^3_{2'}\,\mathbf{g}_3 = \mathbf{g}_1 - \mathbf{g}_3 = -\mathbf{e}_2 - \mathbf{e}_3 \,,$$

$$\mathbf{g}_{3'} = \beta^1_{3'}\,\mathbf{g}_1 + \beta^2_{3'}\,\mathbf{g}_2 + \beta^3_{3'}\,\mathbf{g}_3 = \mathbf{g}_1 + \mathbf{g}_3 = 2\,\mathbf{e}_1 + \mathbf{e}_2 + \mathbf{e}_3 \,.$$

The "old" base can be expressed by the "new" base as

$$\mathbf{g}_1 = \frac{1}{2}\mathbf{g}_{2'} + \frac{1}{2}\mathbf{g}_{3'} \,,$$

$$\mathbf{g}_2 = -\mathbf{g}_{1'} + \mathbf{g}_{2'} \,,$$

$$\mathbf{g}_3 = -\frac{1}{2}\mathbf{g}_{2'} + \frac{1}{2}\mathbf{g}_{3'} \,.$$

The transformation matrix then follows to be

$$\beta^{k'}_j = \begin{bmatrix} 0 & 1/2 & 1/2 \\ -1 & 1 & 0 \\ 0 & -1/2 & 1/2 \end{bmatrix}.$$

In order to check this matrix, we form according to *(2.8)*

$$(\beta^j_{i'})(\beta^{k'}_j) = \begin{bmatrix} 1 & -1 & -1 \\ 1 & 0 & -1 \\ 1 & 0 & 1 \end{bmatrix} \begin{bmatrix} 0 & 1/2 & 1/2 \\ -1 & 1 & 0 \\ 0 & -1/2 & 1/2 \end{bmatrix} = \begin{bmatrix} 1 & 0 & 0 \\ 0 & 1 & 0 \\ 0 & 0 & 1 \end{bmatrix} = \delta^{k'}_{i'} \,.$$

Exercise A-2-2:

In order to determine the state of stress in a thin parallelogram disk (Fig. A-3), a transformation in an oblique coordinate system ξ^{α} ($\alpha = 1, 2$) is advantageously carried out for establishing the required equations.

a) Determine the base vectors, the co- and the contravariant metric tensor, and the CHRISTOFFEL symbols.

b) How does the bipotential equation $\Delta\Delta\,\Phi = 0$ (Φ = AIRY's stress function) read in the oblique coordinate system, and how do we determine the associated stresses from it?

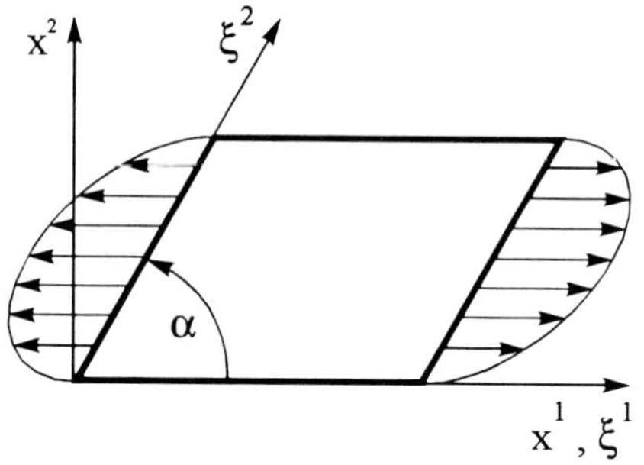

Fig. A-3: Parallelogram disk

Solution:

a) According to Fig. A-4, we express the relation between Cartesian and oblique coordinates as follows:

$$
\begin{aligned}
x^1 &= \xi^1 + \xi^2 \cos\alpha \,, \\
x^2 &= \xi^2 \sin\alpha \quad , \quad (\alpha = \text{const}) \,.
\end{aligned}
\tag{1}
$$

Base vectors

According to *(2.26)*, for the coordinates in the plane the relation between a base $\mathbf{g}_{\alpha}(\xi^{\alpha})$ and the orthonormalized base $\mathbf{e}_{\alpha}$ is

$$\mathbf{g}_{\beta} = \frac{\partial x^{\alpha}}{\partial \xi^{\beta}}\, \mathbf{e}_{\alpha} \,.$$

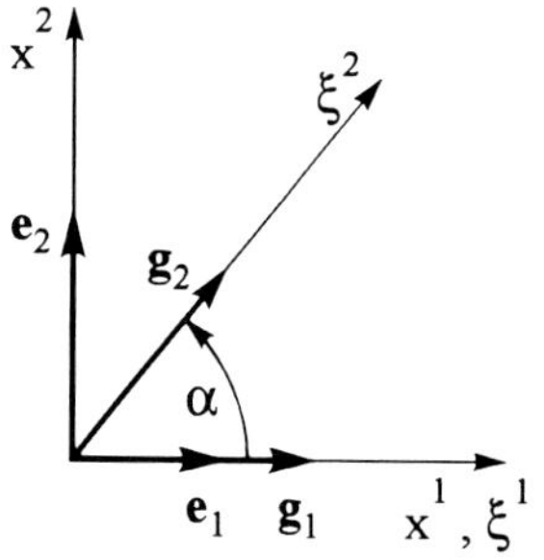

Fig. A-4: Cartesian and oblique coordinate system

Herewith, the base vectors are calculated as follows:

$$\mathbf{g}_1 = \frac{\partial x^1}{\partial \xi^1}\,\mathbf{e}_1 + \frac{\partial x^2}{\partial \xi^1}\,\mathbf{e}_2 = \mathbf{e}_1 \ ,$$

$$\mathbf{g}_2 = \frac{\partial x^1}{\partial \xi^2}\,\mathbf{e}_1 + \frac{\partial x^2}{\partial \xi^2}\,\mathbf{e}_2 = \cos\alpha\,\mathbf{e}_1 + \sin\alpha\,\mathbf{e}_2 \ .$$

Covariant metric tensor

According to *(2.1a)*, the covariant metric components become

$$g_{\alpha\beta} = \mathbf{g}_\alpha \cdot \mathbf{g}_\beta$$

$$\Longrightarrow \quad g_{11} = \mathbf{g}_1 \cdot \mathbf{g}_1 = \mathbf{e}_1 \cdot \mathbf{e}_1 = 1 \ ,$$

$$g_{12} = g_{21} = \mathbf{g}_1 \cdot \mathbf{g}_2 = \mathbf{e}_1 \cdot (\cos\alpha\,\mathbf{e}_1 + \sin\alpha\,\mathbf{e}_2) = \cos\alpha \ ,$$

$$g_{22} = \mathbf{g}_2 \cdot \mathbf{g}_2 = \cos^2\alpha\,(\mathbf{e}_1 \cdot \mathbf{e}_1) + \sin^2\alpha\,(\mathbf{e}_2 \cdot \mathbf{e}_2) = 1 \ ,$$

regarding $\mathbf{e}_1 \cdot \mathbf{e}_2 = 0$.

Thus, the covariant metric tensor is

$$g_{\alpha\beta} = \begin{bmatrix} 1 & \cos\alpha \\ \cos\alpha & 1 \end{bmatrix} . \tag{2}$$

As (2) shows, the metric tensor of a non-orthogonal base is completely occupied.

Determinant of the metric tensor according to *(2.1c)*

$$g = |g_{\alpha\beta}| = 1 - \cos^2\alpha = \sin^2\alpha = \text{const} \ . \tag{3}$$

Contravariant metric tensor

The contravariant metric component $g^{\alpha\beta}$ can be determined according to *(2.5c)* by inversion of the covariant metric tensor $(g_{\alpha\beta})$.

This yields

$$(g^{\alpha\beta}) = (g_{\alpha\beta})^{-1} = \frac{1}{\sin^2\alpha}\begin{bmatrix} 1 & -\cos\alpha \\ -\cos\alpha & 1 \end{bmatrix} . \tag{4}$$

CHRISTOFFEL symbols of the first and second kind from *(2.30)* and *(2.31)*

$$\Gamma_{\alpha\beta\gamma} = \Gamma^{\delta}_{\alpha\beta} = 0 \quad , \quad \text{since} \quad g_{\alpha\beta} = \text{const} \ . \tag{5}$$

This is always valid for rectilinear coordinates.

b) According to *(2.40)*, the following expression is valid for the *bipotential operator* $\Delta\Delta\,\Phi$:

$$\Delta\Delta\,\Phi = \frac{1}{\sqrt{g}}\left\{\sqrt{g}\,g^{\alpha\beta}\left[\frac{1}{\sqrt{g}}\left(\sqrt{g}\,g^{\gamma\delta}\,\Phi_{,\delta}\right)_{,\gamma}\right]_{,\alpha}\right\}_{,\beta} \ .$$

With (3) the bipotential operator is simplified to

$$\Delta\Delta\,\Phi = g^{\alpha\beta}\,g^{\gamma\delta}\,\Phi_{,\alpha\beta\gamma\delta} \quad \Longrightarrow \tag{6}$$

$$\begin{aligned}\Delta\Delta\,\Phi = {} & g^{11}g^{11}\,\Phi_{,1111} + g^{11}g^{12}\,\Phi_{,1112} + g^{11}g^{21}\,\Phi_{,1121} + g^{11}g^{22}\,\Phi_{,1122} + \\ & + g^{12}g^{11}\,\Phi_{,1211} + g^{12}g^{12}\,\Phi_{,1212} + g^{12}g^{21}\,\Phi_{,1221} + g^{12}g^{22}\,\Phi_{,1222} + \\ & + g^{21}g^{11}\,\Phi_{,2111} + g^{21}g^{12}\,\Phi_{,2112} + g^{21}g^{21}\,\Phi_{,2121} + g^{21}g^{22}\,\Phi_{,2122} + \\ & + g^{22}g^{11}\,\Phi_{,2211} + g^{22}g^{12}\,\Phi_{,2212} + g^{22}g^{21}\,\Phi_{,2221} + g^{22}g^{22}\,\Phi_{,2222}\,.\end{aligned}$$

By using the symmetry $g^{\alpha\beta} = g^{\beta\alpha}$ and SCHWARZ's permutation equation we can further simplify the fully expanded expression in (6):

$$\begin{aligned}\Delta\Delta\,\Phi = {} & g^{11}g^{11}\,\Phi_{,1111} + 4\,g^{11}g^{12}\,\Phi_{,1112} + 4\,g^{12}g^{12}\,\Phi_{,1212} + \\ & + 2\,g^{11}g^{22}\,\Phi_{,1122} + 4\,g^{12}g^{22}\,\Phi_{,1222} + g^{22}g^{22}\,\Phi_{,2222}\,.\end{aligned}$$

Application of (4) yields

$$\begin{aligned}\Delta\Delta\,\Phi = {} & \big[\,\Phi_{,1111} + 4(-\cos\alpha)\,\Phi_{,1112} + 4\cos^2\alpha\,\Phi_{,1212} + \\ & + 2\,\Phi_{,1122} + 4(-\cos\alpha)\,\Phi_{,1222} + \Phi_{,2222}\big]\frac{1}{\sin^4\alpha}\end{aligned}$$

$$\longrightarrow \quad \begin{aligned}\Delta\Delta\,\Phi = {} & \big[\,\Phi_{,1111} - 4\cos\alpha\,\Phi_{,1112} + (2 + 4\cos^2\alpha)\,\Phi_{,1122} - \\ & - 4\cos\alpha\,\Phi_{,1222} + \Phi_{,2222}\big]\frac{1}{\sin^4\alpha}\,.\end{aligned}$$

The bipotential equation then reads

$$\Phi_{,1111} - 4\cos\alpha\,\Phi_{,1112} + 2(1 + 2\cos^2\alpha)\,\Phi_{,1122} - 4\cos\alpha\,\Phi_{,1222} + \Phi_{,2222} = 0\,. \tag{7}$$

With Φ as AIRY's stress function, the stresses follow from *(7.13)* by means of the two-dimensional permutation tensor:

$$\tau^{\gamma\delta} = \varepsilon^{\gamma\sigma}\,\varepsilon^{\delta\tau}\,\Phi\big|_{\sigma\tau}\,.$$

Analogous to *(2.19)* and *(2.20)* the two-dimensional permutation tensor $\varepsilon^{\gamma\delta}$ corresponds to

$$\varepsilon^{\gamma\sigma} = \frac{1}{\sqrt{g}}\,e^{\gamma\sigma} = \frac{1}{\sqrt{g}}\begin{bmatrix} 0 & +1 \\ -1 & 0 \end{bmatrix}$$

and with (3) $$\varepsilon^{\gamma\sigma} = \begin{bmatrix} 0 & \dfrac{1}{\sin\alpha} \\ -\dfrac{1}{\sin\alpha} & 0 \end{bmatrix}. \tag{8}$$

With (5), the covariant derivative *(2.34b)* changes into a partial derivative

$$\Phi\big|_{\sigma\tau} = (\,\Phi_{,\sigma})\big|_{\tau} = \Phi_{,\sigma\tau} - \Phi_{,\beta}\,\Gamma^{\beta}_{\sigma\tau} = \Phi_{,\sigma\tau}\,.$$

Each component of the stress tensor can then be obtained from *(7.13)*

$$\tau^{11} = \varepsilon^{1\delta}\varepsilon^{1\tau}\,\Phi\big|_{\delta\tau} = \underbrace{\varepsilon^{11}\varepsilon^{11}\,\Phi\big|_{11}}_{0} + \underbrace{\varepsilon^{12}\varepsilon^{11}\,\Phi\big|_{21}}_{0} + \underbrace{\varepsilon^{11}\varepsilon^{12}\,\Phi\big|_{12}}_{0} + \varepsilon^{12}\varepsilon^{12}\,\Phi\big|_{22}$$

$$\Longrightarrow \quad \tau^{11} = \varepsilon^{12}\varepsilon^{12}\,\Phi_{,22} = \frac{1}{\sin^2\alpha}\,\Phi_{,22}\,,$$

$$\tau^{12} = \varepsilon^{1\delta}\varepsilon^{2\tau}\,\Phi_{,\delta\tau} = \varepsilon^{11}\varepsilon^{21}\,\Phi_{,11} + \varepsilon^{12}\varepsilon^{21}\,\Phi_{,21} + \varepsilon^{11}\varepsilon^{22}\,\Phi_{,12} + \varepsilon^{12}\varepsilon^{22}\,\Phi_{,22}$$

$$\Longrightarrow \quad \tau^{12} = \varepsilon^{12}\varepsilon^{21}\,\Phi_{,21} = -\frac{1}{\sin^2\alpha}\,\Phi_{,21}\,,$$

$$\tau^{22} = \varepsilon^{2\delta}\varepsilon^{2\tau}\,\Phi_{,\delta\tau} = \varepsilon^{21}\varepsilon^{21}\,\Phi_{,11} + \varepsilon^{22}\varepsilon^{21}\,\Phi_{,21} + \varepsilon^{21}\varepsilon^{22}\,\Phi_{,12} + \varepsilon^{22}\varepsilon^{22}\,\Phi_{,22}$$

$$\Longrightarrow \quad \tau^{22} = \varepsilon^{21}\varepsilon^{21}\,\Phi_{,11} = \frac{1}{\sin^2\alpha}\,\Phi_{,11}\,.$$

Exercise A-2-3:

An infinite strip under constant tension σ_x has a crack which can be presented as an elliptical hole (see Fig. A-5). The geometry is described by an elliptical-hyperbolical coordinate system ξ^1, ξ^2. The relationship between Cartesian and elliptical-hyperbolical coordinates is given by:

$$\left.\begin{aligned} x^1 &= c\cosh\xi^1\cos\xi^2 \\ x^2 &= c\sinh\xi^1\sin\xi^2 \end{aligned}\right\} \quad \begin{aligned} &\text{with } c = \text{focal distance to the origin} \\ &\text{and } \; 0 < \xi^1 \le \infty \;\; ; \;\; 0 \le \xi^2 \le 2\pi\,. \end{aligned}$$

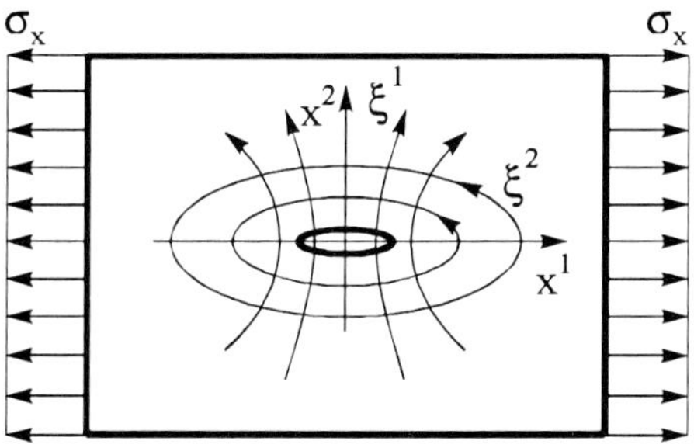

Fig. A-5: Elliptical hole in an infinite strip

Determine

a) the covariant base vectors, the covariant metric tensor, its determinant, and the contravariant metric tensor ;

b) the CHRISTOFFEL symbols of the first and second kind ;

c) the physical components of a vector $\mathbf{v}$.

Solution:

a) For determining the covariant metric components we establish the base vectors according to *(2.26)*

$$\mathbf{g}_\alpha = \frac{\partial x^\beta}{\partial \xi^\alpha}\, \mathbf{e}_\beta \quad \longrightarrow$$

$$\mathbf{g}_1 = \frac{\partial x^1}{\partial \xi^1}\, \mathbf{e}_1 + \frac{\partial x^2}{\partial \xi^1}\, \mathbf{e}_2 = \ c \sinh \xi^1 \cos \xi^2\, \mathbf{e}_1 + c \cosh \xi^1 \sin \xi^2\, \mathbf{e}_2 \ ,$$

$$\mathbf{g}_2 = \frac{\partial x^1}{\partial \xi^2}\, \mathbf{e}_1 + \frac{\partial x^2}{\partial \xi^2}\, \mathbf{e}_2 = -c \cosh \xi^1 \sin \xi^2\, \mathbf{e}_1 + c \sinh \xi^1 \cos \xi^2\, \mathbf{e}_2 \ . \qquad (1)$$

The metric components then follow from *(2.1a)*

$$g_{\alpha\beta} = \mathbf{g}_\alpha \cdot \mathbf{g}_\beta = \sum_{\nu=1}^{2} \frac{\partial x^\nu}{\partial \xi^\alpha} \frac{\partial x^\nu}{\partial \xi^\beta} \quad \longrightarrow$$

$$g_{11} = \frac{\partial x^1}{\partial \xi^1} \frac{\partial x^1}{\partial \xi^1} + \frac{\partial x^2}{\partial \xi^1} \frac{\partial x^2}{\partial \xi^1} = c^2 (\sinh^2 \xi^1 \cos^2 \xi^2 + \cosh^2 \xi^1 \sin^2 \xi^2) =$$

$$= c^2 (\sinh^2 \xi^1 (1 - \sin^2 \xi^2) + (1 + \sinh^2 \xi^1) \sin^2 \xi^2) = c^2 (\sinh^2 \xi^1 + \sin^2 \xi^2) \ ,$$

$$g_{22} = \frac{\partial x^1}{\partial \xi^2} \frac{\partial x^1}{\partial \xi^2} + \frac{\partial x^2}{\partial \xi^2} \frac{\partial x^2}{\partial \xi^2} = c^2 (\cosh^2 \xi^1 \sin^2 \xi^2 + \sinh^2 \xi^1 \cos^2 \xi^2) =$$

$$= c^2 (\sinh^2 \xi^1 + \sin^2 \xi^2) = g_{11} \ ,$$

$$g_{12} = \frac{\partial x^1}{\partial \xi^1} \frac{\partial x^1}{\partial \xi^2} + \frac{\partial x^2}{\partial \xi^1} \frac{\partial x^2}{\partial \xi^2} =$$

$$= c^2 (\sinh \xi^1 \cos \xi^2 (-\cosh \xi^1 \sin \xi^2) + \cosh \xi^1 \sin \xi^2 \sinh \xi^1 \cos \xi^2) = 0 \ .$$

Since $g_{12} = g_{21} = 0$, the elliptical – hyperbolical coordinate system is orthogonal, i.e. the two families of curve parameter lines meet each other at right angles.

The determinant of the metric tensor follows from *(2.1c)*

$$g = \det (g_{\alpha\beta}) = g_{11}\, g_{11} = c^4 (\sinh^2 \xi^1 + \sin^2 \xi^2)^2 \ . \qquad (2)$$

The covariant metric tensor then is

$$(g_{\alpha\beta}) = \begin{bmatrix} g_{11} & 0 \\ 0 & g_{11} \end{bmatrix} = \begin{bmatrix} \sqrt{g} & 0 \\ 0 & \sqrt{g} \end{bmatrix} \qquad (3)$$

with $\sqrt{g} = c^2 (\sinh^2 \xi^1 + \sin^2 \xi^2)$.

Since $g_{\alpha\beta}$ is an orthogonal basis (diagonal matrix) according to

$$g^{\alpha\beta}\, g_{\beta\gamma} = \delta^\alpha_\gamma \ ,$$

the non – vanishing contravariant metric components are inverses of the elements of the covariant metric components

$$g^{(\alpha\alpha)} = \frac{1}{g_{(\alpha\alpha)}} \quad \text{for} \quad \alpha = 1,2\ ,$$

i.e. the contravariant metric tensor is

$$(g^{\alpha\beta}) = \begin{bmatrix} 1/\sqrt{g} & 0 \\ 0 & 1/\sqrt{g} \end{bmatrix} . \tag{4}$$

b) The CHRISTOFFEL symbols of the first kind can be calculated from *(2.30)*

$$\Gamma_{\alpha\beta\gamma} = \frac{1}{2}(g_{\beta\gamma,\alpha} + g_{\gamma\alpha,\beta} - g_{\alpha\beta,\gamma}) \quad \Longrightarrow$$

$$\gamma = 1: \quad \Gamma_{111} = \frac{1}{2}(g_{11,1} + g_{11,1} - g_{11,1}) = \frac{1}{2}\frac{\partial g_{11}}{\partial \xi^1} = c^2 \cosh \xi^1 \sinh \xi^1\ ,$$

$$\Gamma_{121} = \frac{1}{2}(g_{21,1} + g_{11,2} - g_{12,1}) = \frac{1}{2}\frac{\partial g_{11}}{\partial \xi^2} = c^2 \sin \xi^2 \cos \xi^2\ ,$$

$$\Gamma_{211} = \Gamma_{121}\ ,$$

$$\Gamma_{221} = \frac{1}{2}(g_{21,2} + g_{12,2} - g_{22,1}) = -\frac{1}{2}\frac{\partial g_{22}}{\partial \xi^1} = -c^2 \cosh \xi^1 \sinh \xi^1\ .$$

$$\gamma = 2: \quad \Gamma_{112} = \frac{1}{2}(g_{12,1} + g_{21,1} - g_{11,2}) = -\frac{1}{2}\frac{\partial g_{11}}{\partial \xi^2} = -c^2 \sin \xi^2 \cos \xi^2\ ,$$

$$\Gamma_{122} = \frac{1}{2}(g_{22,1} + g_{21,2} - g_{12,2}) = \frac{1}{2}\frac{\partial g_{22}}{\partial \xi^1} = c^2 \cosh \xi^1 \sinh \xi^1\ ,$$

$$\Gamma_{212} = \Gamma_{122}\ ,$$

$$\Gamma_{222} = \frac{1}{2}(g_{22,2} + g_{22,2} - g_{22,2}) = \frac{1}{2}\frac{\partial g_{22}}{\partial \xi^2} = c^2 \sin \xi^2 \cos \xi^2\ .$$

The CHRISTOFFEL symbols of the second kind follow from *(2.31)* with (4)

$$\Gamma^{\delta}_{\alpha\beta} = g^{\delta\gamma}\,\Gamma_{\alpha\beta\gamma} \quad \Longrightarrow$$

$$\delta = 1: \quad \Gamma^1_{11} = g^{1\gamma}\,\Gamma_{11\gamma} = g^{11}\,\Gamma_{111} + g^{12}\,\Gamma_{112} = \frac{1}{\sqrt{g}}\,c^2 \sinh \xi^1 \cosh \xi^1\ ,$$

$$\Gamma^1_{12} = g^{1\gamma}\,\Gamma_{12\gamma} = g^{11}\,\Gamma_{121} + g^{12}\,\Gamma_{122} = \frac{1}{\sqrt{g}}\,c^2 \sin \xi^2 \cos \xi^2\ ,$$

$$\Gamma^1_{21} = \Gamma^2_{12}\ ,$$

$$\Gamma^1_{22} = g^{1\gamma}\,\Gamma_{22\gamma} = g^{11}\,\Gamma_{221} + g^{12}\,\Gamma_{222} = -\frac{1}{\sqrt{g}}\,c^2 \sinh \xi^1 \cosh \xi^1\ .$$

$$\delta = 2: \quad \Gamma_{11}^{2} = g^{2\gamma}\,\Gamma_{11\gamma} = g^{21}\,\Gamma_{111} + g^{22}\,\Gamma_{112} = -\frac{1}{\sqrt{g}}\,c^{2}\sin\xi^{2}\cos\xi^{2}\,,$$

$$\Gamma_{12}^{2} = g^{2\gamma}\,\Gamma_{12\gamma} = g^{21}\,\Gamma_{121} + g^{22}\,\Gamma_{122} = \frac{1}{\sqrt{g}}\,c^{2}\sinh\xi^{1}\cosh\xi^{1}\,,$$

$$\Gamma_{21}^{2} = \Gamma_{12}^{2}\,,$$

$$\Gamma_{22}^{2} = g^{2\gamma}\,\Gamma_{22\gamma} = g^{21}\,\Gamma_{221} + g^{22}\,\Gamma_{222} = \frac{1}{\sqrt{g}}\,c^{2}\sin\xi^{2}\cos\xi^{2}\,.$$

With the abbreviations

$$\frac{1}{\sqrt{g}}\,c^{2}\sinh\xi^{1}\cosh\xi^{1} = \frac{1}{2}\,\frac{\sinh 2\,\xi^{1}}{\sinh^{2}\xi^{1} + \sin^{2}\xi^{2}} = A\,,$$

$$\frac{1}{\sqrt{g}}\,c^{2}\sin\xi^{2}\cos\xi^{2} = \frac{1}{2}\,\frac{\sin 2\,\xi^{2}}{\sinh^{2}\xi^{1} + \sin^{2}\xi^{2}} = B\,,$$

the CHRISTOFFEL symbols of the second kind read in matrix form

$$\Gamma_{\alpha\beta}^{1} = \begin{bmatrix} A & B \\ B & -A \end{bmatrix}\,, \quad \Gamma_{\alpha\beta}^{2} = \begin{bmatrix} -B & A \\ A & B \end{bmatrix}. \tag{5}$$

c) The physical components u, v of a vector **v** are calculated from *(2.10)*:

$$v_{\alpha}^{*} = v_{\alpha}\sqrt{g^{(\alpha\alpha)}}$$

$$\left.\begin{aligned} \Longrightarrow\quad v_{1}^{*} &\mathrel{\hat=} u = v_{1}\sqrt{g^{11}} = v_{1}\,g^{-1/4} \longrightarrow v_{1} = g^{1/4}\,u\,, \\ v_{2}^{*} &\mathrel{\hat=} v = v_{2}\sqrt{g^{22}} = v_{2}\,g^{-1/4} \longrightarrow v_{2} = g^{1/4}\,v\,. \end{aligned}\right\} \tag{6}$$

Exercise A-3-1:

For the state of plane stress

$$\sigma_{xx} = 60\,MPa\,,\ \sigma_{yy} = -20\,MPa\,,\ \tau_{xy} = 30\,MPa\,,$$

determine by means of MOHR's circle the magnitude and orientation of the principal stresses, and the magnitude of the principal shear stress.

Solution:

For the given σ_{xx}, σ_{yy} und τ_{xy} we can determine the stresses on arbitrary sections from MOHR's circle.

For drawing the circle shown in Fig. A-6 we first plot the given stresses on the σ-axis due to the signs. At these points we then plot the shear stresses τ_{xy} and/ or τ_{yx} as ordinates. Here, as opposed to the usual rule of signs, a shear stress is considered positive, if it rotates clockwise. This special sign convention, which is only valid for MOHR's circle, is necessary, if we want to display the rotation direction of an arbitrary angle α between the coordinate systems in the circle. Thus, τ_{xy} is to apply down, and τ_{yx} to apply up in the sketch. With A and A', two points of the circle are fixed. Their connection line is the diameter of length 2 r of MOHR's circle, and provides the centre point M. Now, the circle can be drawn.

The principal stresses σ_I and/or σ_{II} following from the intersection points of the circle with the σ-axis, lie under the angles α^* and/or $\alpha^* + \pi/2$. The highest shear stress τ_{max} can be read as the highest ordinate of the circle, i.e., $\tau_{max} = r$ (cf. *(3.22b)*). The corresponding section has the angle

$$\alpha^{**} = \alpha^* + \frac{\pi}{4}$$

with respect to the x-axis. From the drawn circle the results can be read for the given task:

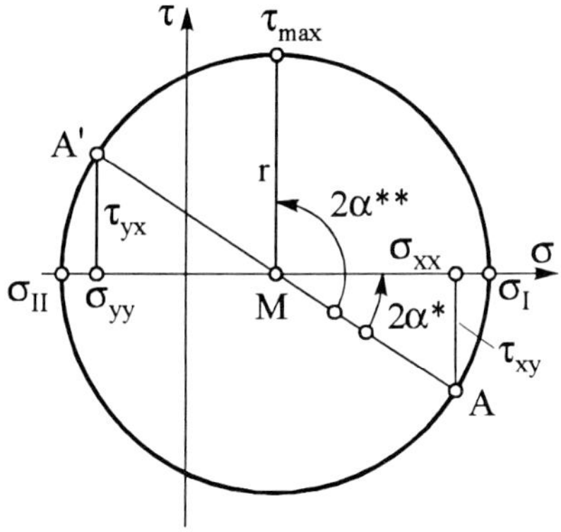

Scale: ⊢—⊣ $\widehat{=}$ $20\,MPa$.

$\sigma_I = 70\,MPa$,

$\sigma_{II} = -\,30\,MPa$,

$2\alpha^* \approx 37^\circ$,

$\tau_{max} = 50\,MPa$,

$2\alpha^{**} \approx 127^\circ$.

Fig. A-6: MOHR's circle

Exercise A-3-2:

The stress components of a three-dimensional state of stress

$$\sigma_{xx} = 100\,MPa\ , \qquad \sigma_{yy} = 60\,MPa\ , \quad \sigma_{zz} = 10\,MPa\ ,$$

$$\tau_{xy} = 20\sqrt{3}\ MPa\ , \quad \tau_{yz} = \tau_{zx} = 0$$

are given.

Determine:
- a) Principal stresses ,
- b) Principal axes ,
- c) Stress deviator ,
- d) Invariants of the stress deviator .

Solution:

a) With the invariants *(3.13)*

$$I_1 = 100 + 60 + 10 = 170\,[\,MPa\,]\,,$$

$$I_2 = \begin{vmatrix} 100 & 20\sqrt{3} \\ 20\sqrt{3} & 60 \end{vmatrix} + \begin{vmatrix} 60 & 0 \\ 0 & 10 \end{vmatrix} + \begin{vmatrix} 10 & 0 \\ 0 & 100 \end{vmatrix} = 4800 + 600 + 1000 =$$

$$= 6400\,[\,MPa\,]^2\,,$$

$$I_3 = \begin{vmatrix} 100 & 20\sqrt{3} & 0 \\ 20\sqrt{3} & 60 & 0 \\ 0 & 0 & 10 \end{vmatrix} = 48000\,[\,MPa\,]^3\,.$$

According to *(3.12)*, the principal stresses follow from

$$\sigma^3 - 170\,\sigma^2 + 6400\,\sigma - 48000 = 0\,.$$

As no shear stresses are acting on the section z = const in the example, the component $\sigma_{III} = \sigma_{zz} = 10\,MPa$ is a first root of this cubic equation. Dividing by $(\sigma - \sigma_{III})$, we come to the quadratic equation

$$\sigma^2 - 160\,\sigma + 4800 = 0$$

and with that to the following eigenvalues

$$\left.\begin{matrix} \sigma_I \\ \\ \sigma_{II} \end{matrix}\right\} = 80 \pm \sqrt{6400 - 4800} = \left\{\begin{matrix} 120\,MPa\,, \\ \\ 40\,MPa\,. \end{matrix}\right.$$

b) By substituting the eigenvalues into the homogeneous equation system *(3.11)* the orientations of the principal stresses follow. With $\sigma_I = 120\,MPa$ this yields:

$$\begin{aligned} (100 - 120)\,n_x^* + 20\sqrt{3}\,n_y^* &= 0\,, \\ 20\sqrt{3}\,n_x^* + (60 - 120)\,n_y^* &= 0\,, \\ (10 - 120)\,n_z^* &= 0\,. \end{aligned}$$

From this, it follows that $\quad n_z^* = 0 \quad , \quad n_x^* = \sqrt{3}\,n_y^*$

and herewith the eigenvector $\quad \mathbf{n}_I^* = C \begin{bmatrix} \sqrt{3} \\ 1 \\ 0 \end{bmatrix}.$

Normalizing the eigenvector to the length 1

$$|\mathbf{n}_I^*| = 1 = C^2\left[(\sqrt{3}\,)^2 + 1^2 + 0^2\right] \quad \Longrightarrow \quad C = \frac{1}{2}\,,$$

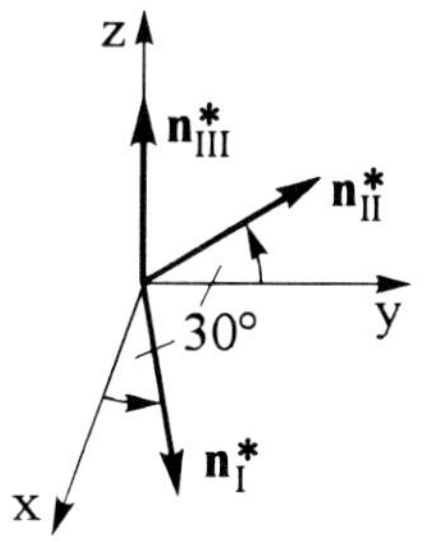

Fig. A-7: The initial system and the system of principal axes

yields

$$\mathbf{n}^*_I = \begin{bmatrix} \sqrt{3}/2 \\ 1/2 \\ 0 \end{bmatrix}, \text{ and in analogy one obtains } \mathbf{n}^*_{II} = \begin{bmatrix} -1/2 \\ \sqrt{3}/2 \\ 0 \end{bmatrix}, \quad \mathbf{n}^*_{III} = \begin{bmatrix} 0 \\ 0 \\ 1 \end{bmatrix}.$$

As the components of the normal vectors correspond to the cosine directions relative to the initial system, we can draw the principal axes in the x, y, z−system (Fig. A-7).

In order to check, we can carry out the transformation from the initial system to the system of principal axes. With the rotation matrix according to *(3.10)*

$$\mathbf{B} = \begin{bmatrix} \frac{1}{2}\sqrt{3} & \frac{1}{2} & 0 \\ -\frac{1}{2} & \frac{1}{2}\sqrt{3} & 0 \\ 0 & 0 & 1 \end{bmatrix}$$

follows from the matrix product *(3.9)*

$$\mathbf{S}' = \mathbf{B}\,\mathbf{S}\,\mathbf{B}^T = \begin{bmatrix} \frac{1}{2}\sqrt{3} & \frac{1}{2} & 0 \\ -\frac{1}{2} & \frac{1}{2}\sqrt{3} & 0 \\ 0 & 0 & 1 \end{bmatrix} \begin{bmatrix} 100 & 20\sqrt{3} & 0 \\ 20\sqrt{3} & 60 & 0 \\ 0 & 0 & 10 \end{bmatrix} \begin{bmatrix} \frac{1}{2}\sqrt{3} & -\frac{1}{2} & 0 \\ \frac{1}{2} & \frac{1}{2}\sqrt{3} & 0 \\ 0 & 0 & 1 \end{bmatrix} =$$

$$= \begin{bmatrix} 120 & 0 & 0 \\ 0 & 40 & 0 \\ 0 & 0 & 10 \end{bmatrix} = \begin{bmatrix} \sigma_I & 0 & 0 \\ 0 & \sigma_{II} & 0 \\ 0 & 0 & \sigma_{III} \end{bmatrix}$$

the diagonal matrix of the principal stresses.

c) With *(3.24)*

$$\sigma_M = \frac{1}{3}(100 + 60 + 10) = \frac{170}{3}\ MPa\,,$$

we calculate according to *(3.23)* the stress deviator

$$'\mathbf{S} = \begin{bmatrix} 100 - \frac{170}{3} & 20\sqrt{3} & 0 \\ 20\sqrt{3} & 60 - \frac{170}{3} & 0 \\ 0 & 0 & 10 - \frac{170}{3} \end{bmatrix} = \begin{bmatrix} \frac{130}{3} & 20\sqrt{3} & 0 \\ 20\sqrt{3} & \frac{10}{3} & 0 \\ 0 & 0 & -\frac{140}{3} \end{bmatrix}.$$

d) According to *(3.26)*, we obtain the invariants of the deviator as

$$'I_1 = \frac{130}{3} + \frac{10}{3} - \frac{140}{3} = 0 ,$$

$$'I_2 = 6400 - 3\left(\frac{170}{3}\right)^2 = -\frac{9700}{3}\left[MPa\right]^2 ,$$

$$'I_3 = 48000 - 6400\,\frac{170}{3} + 2\left(\frac{170}{3}\right)^3 = \frac{1330000}{27}\left[MPa\right]^3 .$$

In order to check, we directly calculate the invariants of the deviator from $'\mathbf{S}$. Thus, we get for example

$$'I_3 = \left|\,'\mathbf{S}\right| = -\frac{140}{3}\begin{vmatrix} \frac{130}{3} & 20\sqrt{3} \\ 20\sqrt{3} & \frac{10}{3} \end{vmatrix} = \frac{1330000}{27}\left[MPa\right]^3 .$$

Exercise A-3-3:

For the infinite strip with an elliptical hole (see **A-2-3**), establish the equilibrium conditions in elliptical–hyperbolical coordinates.

Solution:

The equilibrium equations follow from *(3.28a)*:

$$\tau^{\beta\alpha}\big|_\beta + f^\alpha = 0 .$$

The two equilibrium equations then read:

$$\alpha = 1: \qquad \tau^{\beta 1}\big|_\beta + f^1 = 0 ,$$

$$\alpha = 2: \qquad \tau^{\beta 2}\big|_\beta + f^2 = 0 .$$

The covariant derivatives may be determined from *(2.35b)*:

$$\tau^{\beta\alpha}\big|_\beta = \tau^{\beta\alpha}{}_{,\beta} + \Gamma^\alpha_{\beta\gamma}\,\tau^{\gamma\beta} + \Gamma^\beta_{\beta\gamma}\,\tau^{\alpha\gamma} .$$

Fully written, this yields for

$$\alpha = 1: \quad \tau^{11}\big|_1 + \tau^{21}\big|_2 + f^1 = 0 \quad \longrightarrow$$

$$\tau^{11}{}_{,1} + \Gamma^1_{11}\tau^{11} + \Gamma^1_{12}\tau^{21} + \Gamma^1_{11}\tau^{11} + \Gamma^1_{12}\tau^{12} +$$

$$+ \tau^{21}{}_{,2} + \Gamma^1_{21}\tau^{12} + \Gamma^1_{22}\tau^{22} + \Gamma^2_{21}\tau^{11} + \Gamma^2_{22}\tau^{12} + f^1 = 0$$

$$\longrightarrow \qquad \tau^{11}{}_{,1} + \tau^{21}{}_{,2} + 3\,A\,\tau^{11} + 4\,B\,\tau^{12} - A\,\tau^{22} + f^1 = 0 .$$

Using the abbreviations A and B for the components of the matrix of the CHRISTOFFEL symbols from **A-2-3**, one obtains

$$\frac{\partial \tau^{11}}{\partial \xi^1} + \frac{\partial \tau^{21}}{\partial \xi^2} + \frac{1}{2\,(\sinh^2 \xi^1 + \sin^2 \xi^2)}\,(\,3 \sinh 2\,\xi^1\,\tau^{11} +$$

$$+\ 4 \sin 2\,\xi^2\,\tau^{12} - \sinh 2\,\xi^1\,\tau^{22}) + f^1 = 0\ .$$

In analogy, one obtains for $\alpha = 2$:

$$\tau^{12}{}_{,1} + \tau^{22}{}_{,2} - B\,\tau^{11} + 4\,A\,\tau^{12} + 3\,B\,\tau^{22} + f^2 = 0$$

$$\Longrightarrow \quad \frac{\partial \tau^{12}}{\partial \xi^1} + \frac{\partial \tau^{22}}{\partial \xi^2} + \frac{1}{2\,(\sinh^2 \xi^1 + \sin^2 \xi^2)}\,(4 \sinh 2\,\xi^1\,\tau^{12} +$$

$$+\ 3 \sin 2\,\xi^2\,\tau^{22} - \sin 2\,\xi^2\,\tau^{11}) + f^2 = 0\ .$$

Finally, the physical components according to *(2.17)* should be introduced into these equations.

Exercise A-4-1:

For a disk, which is clamped along the boundaries $x = 0$, $y = 0$ as shown in Fig. A-8, and has the opposite edges free, the strains have been determined from strain gauge measurements during the action of an external load. The results can be approximately described by

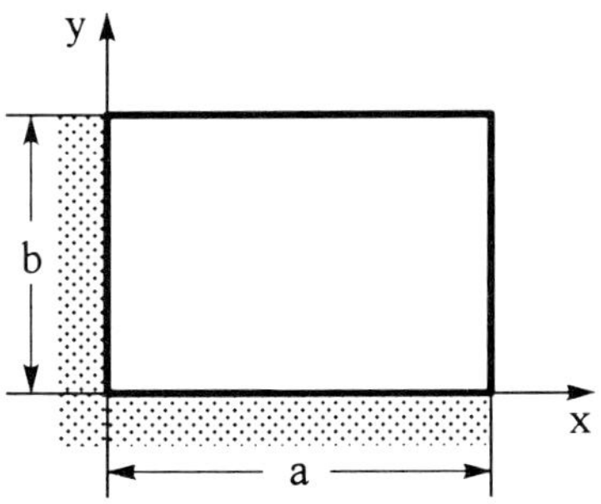

$$\varepsilon_{xx} = k\left(\frac{b}{a}\,x^2 y + a\,y^2\right),$$

$$\varepsilon_{yy} = k\,\frac{a}{b}\,y^2 x$$

$(k = \text{constant factor})$.

Fig. A-8: Disk with two clamped and two free edges

a) Calculate the displacements u and v and the shear γ_{xy} over the domain of the disk.

b) Is this a compatible state of strain?

Solution:

a) For the given strains the following strain-displacement relations *(4.14)* are valid:

$$\left.\begin{aligned} \varepsilon_{xx} &= \frac{\partial u}{\partial x} = k\left(\frac{b}{a}x^2 y + a y^2\right), \\ \varepsilon_{yy} &= \frac{\partial v}{\partial y} = k\,\frac{a}{b}y^2 x . \end{aligned}\right\} \quad (1)$$

Integration of (1) yields

$$\left.\begin{aligned} u(x,y) &= k\left(\frac{b}{a}\frac{1}{3}x^3 y + a y^2 x\right) + f(y), \\ v(x,y) &= k\frac{a}{b}\frac{1}{3}y^3 x + g(x) . \end{aligned}\right\} \quad (2)$$

The unknown functions f(y) and g(x) can be calculated from the following *boundary conditions*:

$$\left.\begin{aligned} &1) \quad u(0,y) = 0 \;, \quad u(x,0) = 0 \quad \longrightarrow \quad f(y) = 0 , \\ &2) \quad v(0,y) = 0 \;, \quad v(x,0) = 0 \quad \longrightarrow \quad g(x) = 0 . \end{aligned}\right\} \quad (3)$$

Thus, we obtain for the displacements

$$\left.\begin{aligned} u(x,y) &= k\left(\frac{b}{a}\frac{1}{3}x^3 y + a y^2 x\right), \\ v(x,y) &= k\frac{a}{b}\frac{1}{3}y^3 x . \end{aligned}\right\} \quad (4a)$$

In dimensionless form (4a) may be written as

$$u\left(\frac{x}{a},\frac{y}{b}\right) = u_0\left(\frac{1}{3}\frac{x^3}{a^3}\frac{y}{b} + \frac{y^2}{b^2}\frac{x}{a}\right),$$

$$v\left(\frac{x}{a},\frac{y}{b}\right) = v_0\frac{1}{3}\frac{y^3}{b^3}\frac{x}{a} \quad (4b)$$

with $u_0 = v_0 = k\,a^2 b^2$.

The displacements u and v at the free boundaries are presented in Fig. A-9.

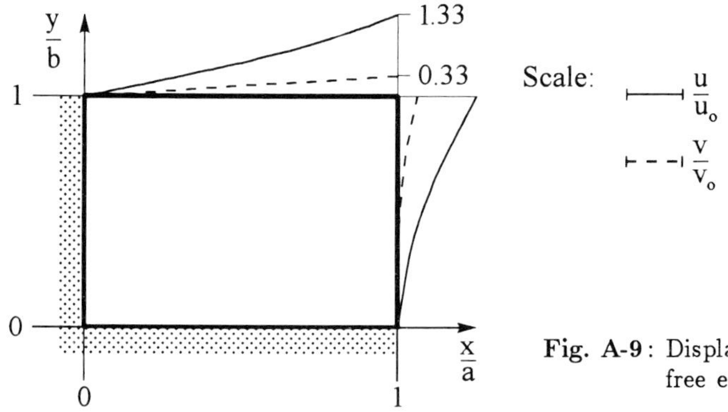

Fig. A-9: Displacements of the free edges of the disk

Finally, with (4a) and according to *(4.14)* the shear strain becomes

$$\gamma_{xy} = \frac{\partial u}{\partial y} + \frac{\partial v}{\partial x} = \frac{1}{3} k \left(\frac{b}{a} x^3 + 6 a x y + \frac{a}{b} y^3 \right) = \frac{1}{3} \frac{u_0}{b} \left(\frac{x^3}{a^3} + 6 \frac{x y}{a b} + \frac{b}{a} \frac{y^3}{b^3} \right) . \qquad (5)$$

b) According to *(4.19)*, the compatibility condition is

$$\frac{\partial^2 \varepsilon_{xx}}{\partial y^2} + \frac{\partial^2 \varepsilon_{yy}}{\partial x^2} - \frac{\partial^2 \gamma_{xy}}{\partial x \, \partial y} = 0 .$$

Inserting of (1) and (5) yields

$$\frac{\partial^2}{\partial y^2} \left[k \left(\frac{b}{a} x^2 y + a y^2 \right) \right] + \frac{\partial^2}{\partial x^2} \left(k \frac{a}{b} y^2 x \right) - \frac{\partial^2}{\partial x \, \partial y} \left[\frac{1}{3} k \left(\frac{b}{a} x^3 + 6 a x y + \frac{a}{b} y^3 \right) \right] = 2 a k - 2 a k = 0 .$$

It is hereby proved that the state of strain is compatible.

Exercise A-4-2:

A strain gauge rosette as shown in Fig. A-10 has measured the following strains in the directions 1 to 3

$$\varepsilon_1 = \varepsilon_0 ,$$
$$\varepsilon_2 = - \sqrt{3} \, \varepsilon_0 ,$$
$$\varepsilon_3 = - \varepsilon_0 .$$

Determine the principal strains in terms of direction and magnitude.

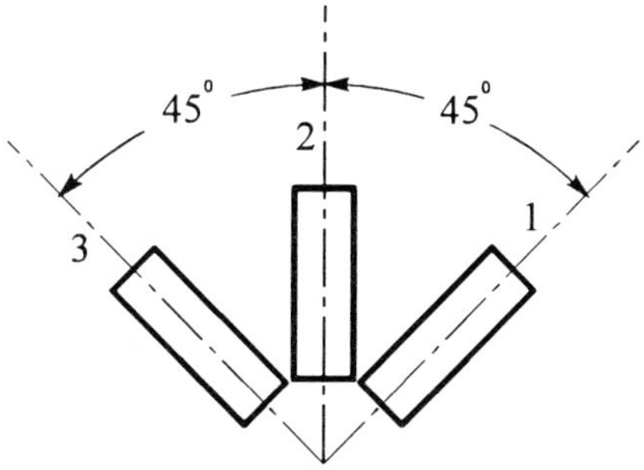

Fig. A-10: Strain gauge rosette

Solution:

Suppose that the direction 1 lies under the yet unknown angle α against the principal axis I. Applying the transformation formula for stresses *(3.17)*, which is analogous for strains, and taking into consideration that no shear strains occur between the principal directions, the following is valid for the directions 1 and 3:

$$\varepsilon_1 = \frac{1}{2}(\varepsilon_I + \varepsilon_{II}) + \frac{1}{2}(\varepsilon_I - \varepsilon_{II})\cos 2\alpha \,, \qquad (1a)$$

$$\varepsilon_3 = \frac{1}{2}(\varepsilon_I + \varepsilon_{II}) - \frac{1}{2}(\varepsilon_I - \varepsilon_{II})\cos 2\alpha \,. \qquad (1b)$$

From this follows

$$\varepsilon_I + \varepsilon_{II} = \varepsilon_1 + \varepsilon_3 \quad , \quad (\varepsilon_I - \varepsilon_{II})\cos 2\alpha = \varepsilon_1 - \varepsilon_3 \quad . \qquad (2)$$

Inserting (2) into the relation for the direction 2, we obtain

$$\varepsilon_2 = \frac{\varepsilon_I + \varepsilon_{II}}{2} + \frac{\varepsilon_I - \varepsilon_{II}}{2}\cos[2(\alpha + 45^\circ)] = \frac{\varepsilon_1 + \varepsilon_3}{2} - \frac{\varepsilon_1 - \varepsilon_3}{2}\tan 2\alpha$$

and from this $\qquad \tan 2\alpha = -\dfrac{2\varepsilon_2 - \varepsilon_1 - \varepsilon_3}{\varepsilon_1 - \varepsilon_3}$.

From the given values it becomes

$$\tan 2\alpha = +\sqrt{3} \quad \longrightarrow \quad \alpha = 30^\circ \,.$$

From

$$\varepsilon_I + \varepsilon_{II} = 0 \quad , \quad (\varepsilon_I - \varepsilon_{II})\cos 60^\circ = 2\,\varepsilon_0 \,,$$

we then obtain the principal strains to be

$$\varepsilon_I = 2\,\varepsilon_0 \quad , \quad \varepsilon_{II} = -2\,\varepsilon_0 \,.$$

Exercise A-4-3:

In Cartesian coordinates a displacement vector is given as

$$\mathbf{v}(x,y,z) = \begin{bmatrix} u \\ v \\ w \end{bmatrix} = k \begin{bmatrix} x \\ y + 4z \\ 4\sqrt{2}\,x + 3z \end{bmatrix} .$$

Determine

a) the strain tensor $\mathbf{V}_s$,

b) the principal strains with regard to magnitude and direction ,

c) the volume dilatation .

Solution:

a) The strain tensor $\mathbf{V_s}$ follows from *(4.13)*

$$\mathbf{V_s} = k \begin{bmatrix} \frac{\partial u}{\partial x} & \frac{1}{2}\left(\frac{\partial u}{\partial y} + \frac{\partial v}{\partial x}\right) & \frac{1}{2}\left(\frac{\partial u}{\partial z} + \frac{\partial w}{\partial x}\right) \\ & \frac{\partial v}{\partial y} & \frac{1}{2}\left(\frac{\partial v}{\partial z} + \frac{\partial w}{\partial y}\right) \\ \text{sym.} & & \frac{\partial w}{\partial z} \end{bmatrix} ,$$

i.e.,
$$\mathbf{V_s} = k \begin{bmatrix} 1 & 0 & 2\sqrt{2} \\ 0 & 1 & 2 \\ 2\sqrt{2} & 2 & 3 \end{bmatrix} .$$

From this we obtain the normal and the shear strains

$$\varepsilon_{xx} = k \quad , \quad \gamma_{xy} = 0 \, ,$$
$$\varepsilon_{yy} = k \quad , \quad \gamma_{xz} = 4\sqrt{2}\, k \, ,$$
$$\varepsilon_{zz} = 3\,k \quad , \quad \gamma_{yz} = 4\,k \, .$$

b) The principal strains are determined from the corresponding eigenvalue problem *(2.22)*:

$$\det(\mathbf{V_s} - \lambda\,\mathbf{I}) = \begin{vmatrix} \varepsilon_{xx}-\lambda & \frac{1}{2}\gamma_{xy} & \frac{1}{2}\gamma_{xz} \\ \frac{1}{2}\gamma_{xy} & \varepsilon_{yy}-\lambda & \frac{1}{2}\gamma_{yz} \\ \frac{1}{2}\gamma_{xz} & \frac{1}{2}\gamma_{yz} & \varepsilon_{zz}-\lambda \end{vmatrix} =$$

$$= \begin{vmatrix} k-\lambda & 0 & 2\sqrt{2}\,k \\ 0 & k-\lambda & 2\,k \\ 2\sqrt{2}\,k & 2\,k & 3k-\lambda \end{vmatrix} = 0 \, .$$

Calculation of the determinant yields the characteristic equation

$$(k - \lambda)\,(\lambda^2 - 4\,k\lambda - 9\,k^2) = 0$$

with the roots

$$\lambda_1 = k \quad , \quad \lambda_{2,3} = (2 \pm \sqrt{13}\,)\,k \, ,$$

i.e., the principal strains

$$\varepsilon_I = k \quad , \quad \varepsilon_{II} = (2 + \sqrt{13}\,)\,k \quad , \quad \varepsilon_{III} = (2 - \sqrt{13}\,)\,k \, .$$

The *directions* of the principal strains are calculated by the corresponding eigenvectors. This will be shown for the case of $\varepsilon_I = \lambda_1 = k$, where we insert this root into the homogeneous system of linear equations *(2.22a)*:

$$\left.\begin{array}{l} 0 \cdot a_{1x} + 0 \cdot a_{1y} + 2\sqrt{2}\,k\,a_{1z} = 0 \\ 0 \cdot a_{1x} + 0 \cdot a_{1y} + \quad 2\,k\,a_{1z} = 0 \end{array}\right\} \longrightarrow \quad a_{1z} = 0 \;,$$

$$2\sqrt{2}\,k\,a_{1x} + 2\,k\,a_{1y} + \quad 2\,k\,a_{1z} = 0 \quad \longrightarrow \quad a_{1x} = C \quad \text{arbitrary} \;,$$

$$a_{1y} = -\sqrt{2}\,a_{1x} = -\sqrt{2}\,C \;.$$

Thus, it yields the eigenvector

$$\mathbf{a}_1 = C \begin{bmatrix} 1 \\ -\sqrt{2} \\ 0 \end{bmatrix} \quad \text{with arbitrary } C \in \mathbb{R} \quad .$$

By analogy, the reader may calculate the two other eigenvectors and obtain

$$\mathbf{a}_2 = C \begin{bmatrix} 1 \\ \frac{1}{2}\sqrt{2} \\ \frac{1}{4}\sqrt{2}\,(1 + \sqrt{13}\,) \end{bmatrix} \;, \quad \mathbf{a}_3 = C \begin{bmatrix} 1 \\ \frac{1}{2}\sqrt{2} \\ \frac{1}{4}\sqrt{2}\,(1 - \sqrt{13}\,) \end{bmatrix}$$

with arbitrary $C \in \mathbb{R}$.

c) According to *(4.17b)*, the volume dilatation is equal to the first invariant

$$e = I_1 = \varepsilon_{xx} + \varepsilon_{yy} + \varepsilon_{zz} = k + k + 3\,k = 5\,k \;,$$

which may be alternatively calculated as

$$e = \varepsilon_{I} + \varepsilon_{II} + \varepsilon_{III} = k + (2 + \sqrt{13}\,)\,k + (2 - \sqrt{13}\,)\,k = 5\,k \;.$$

Exercise A-4-4:

The strain-displacement relation for the normal strain ε_{22} and the material law of an isotropic, infinite disk with an elliptical hole are to be determined by using the elliptical-hyperbolical coordinates from **Exercise A-2-3**.

Solution:

The physical components of the strain tensor (tensor of second order) are first calculated according to *(2.17)*:

$$\gamma^*_{\alpha\beta} = \gamma_{\alpha\beta}\sqrt{g^{(\alpha\alpha)}\,g^{(\beta\beta)}} \;.$$

Inserting (5) and (6) from **A-2-3** yields the following for ε_{22}:

$$\gamma^*_{22} \mathrel{\widehat{=}} \varepsilon_{22} = \gamma_{22}\sqrt{g^{22}\,g^{22}} = \gamma_{22}\frac{1}{\sqrt{g}} = v_2|_2\,\frac{1}{\sqrt{g}} = \tag{1}$$

$$= \left(\frac{\partial v_2}{\partial \xi^2} + A\,v_1 - B\,v_2\right)\frac{1}{\sqrt{g}} = \left[\frac{\partial}{\partial \xi^2}(g^{1/4}\,v) + A\,g^{1/4}\,u - B\,g^{1/4}\,v\right]\frac{1}{\sqrt{g}} =$$

$$= \left[\frac{\partial}{\partial \xi^2}\left(c(\sinh^2\xi^1 + \sin^2\xi^2)^{1/2}\,v\right) + \frac{1}{2}\sinh 2\,\xi^1\,c(\sinh^2\xi^1 + \sin^2\xi^2)^{-1/2}\,u -\right.$$

$$\left. - \frac{1}{2}\sin 2\,\xi^2\,c(\sinh^2\xi^1 + \sin^2\xi^2)^{-1/2}\,v\right]\frac{1}{c^2(\sinh^2\xi^1 + \sin^2\xi^2)} =$$

$$= \left[(\sinh^2\xi^1 + \sin^2\xi^2)^{-1/2}\,\frac{1}{2}\sin 2\,\xi^2\,v + (\sinh^2\xi^1 + \sin^2\xi^2)^{1/2}\,\frac{\partial v}{\partial \xi^2} + \right.$$

$$+ \frac{1}{2}\sinh 2\,\xi^1(\sinh^2\xi^1 + \sin^2\xi^2)^{-1/2}\,u -$$

$$\left. - \frac{1}{2}\sin 2\,\xi^2(\sinh^2\xi^1 + \sin^2\xi^2)^{-1/2}\,v\right]\frac{1}{c(\sinh^2\xi^1 + \sin^2\xi^2)} =$$

$$= \frac{1}{c(\sinh^2\xi^1 + \sin^2\xi^2)}\left[\sqrt{\sinh^2\xi^1 + \sin^2\xi^2}\,\frac{\partial v}{\partial \xi^2} + \right.$$

$$\left. + \frac{1}{2}\,\frac{\sinh 2\,\xi^1}{\sqrt{\sinh^2\xi^1 + \sin^2\xi^2}}\,u\right] \longrightarrow$$

$$\varepsilon_{22} = \frac{1}{c}\left[\frac{1}{\sqrt{\sinh^2\xi^1 + \sin^2\xi^2}}\,\frac{\partial v}{\partial \xi^2} + \frac{\sinh 2\,\xi^1}{2\sqrt{(\sinh^2\xi^1 + \sin^2\xi^2)^3}}\,u\right]. \tag{2}$$

The material law for the state of plane stress without consideration of temperature terms reads according to *(5.5)*

$$\gamma_{\alpha\beta} = D_{\alpha\beta\gamma\delta}\,\tau^{\gamma\delta} \tag{3a}$$

with
$$D_{\alpha\beta\gamma\delta} = \frac{1+\nu}{2E}(g_{\alpha\gamma}\,g_{\beta\delta} + g_{\alpha\delta}\,g_{\beta\gamma}) - \frac{\nu}{E}g_{\alpha\beta}\,g_{\gamma\delta}\,. \tag{3b}$$

Insertion yields

$$\gamma_{\alpha\beta} = \left[\frac{1+\nu}{2E}(g_{\alpha\gamma}\,g_{\beta\delta} + g_{\alpha\delta}\,g_{\beta\gamma}) - \frac{\nu}{E}g_{\alpha\beta}\,g_{\gamma\delta}\right]\tau^{\gamma\delta}\,.$$

We now obtain the material law for γ_{22} :

$$\gamma_{22} = \left[\frac{1+\nu}{2E}(g_{2\gamma}\,g_{2\delta} + g_{2\delta}\,g_{2\gamma}) - \frac{\nu}{E}g_{22}\,g_{\gamma\delta}\right]\tau^{\gamma\delta}\,.$$

With the covariant metric components

$$g_{\varrho\sigma} = 0 \qquad \text{for} \quad \rho \neq \sigma$$

and
$$g_{\varrho\sigma} = \sqrt{g} \qquad \text{for} \quad \rho = \sigma$$

finally follows

$$\gamma_{22} = \left[-\frac{\nu}{E} g_{22}\, g_{11} \right] \tau^{11} + \left[\frac{1+\nu}{2E} (g_{22}\, g_{22} + g_{22}\, g_{22}) - \frac{\nu}{E} g_{22}\, g_{22} \right] \tau^{22} =$$

$$= \left[\frac{1+\nu}{2E} (g + g) - \frac{\nu}{E} g \right] \tau^{22} - \frac{\nu}{E} g\, \tau^{11} =$$

$$\longrightarrow \quad \gamma_{22} = \frac{1}{E} c^4 (\sinh^2 \xi^1 + \sin^2 \xi^2)^2 \left[\tau^{22} - \nu\, \tau^{11} \right] . \tag{4}$$

Here, the physical components of γ_{22} , τ^{22} and τ^{11} are still to be introduced:

$$\gamma_{22} = \sqrt{g}\ \varepsilon_{22} \quad , \quad \tau^{22} = \frac{1}{\sqrt{g}} \sigma_{22} \quad \longrightarrow \quad \varepsilon_{22} = \frac{1}{E} (\sigma_{22} - \nu \sigma_{11}) .$$

Exercise A-5-1:

A square-shaped steel ingot is embedded in a rigid concrete base and both are stress-free at room temperature (Fig. A-11). The upper end of the ingot is free. The ingot only is now subjected to a constant temperature increase Θ (coefficient of thermal expansion α_T).

a) Which strain occurs in the z-direction of the ingot, provided that the vertical boundaries are frictionless ?

b) How does the strain change if, additionally, the boundaries at x = const are moved so that no contact occurs ?

Solution:

a) We solve the problem by means of HOOKE-DUHAMEL's law *(5.2)*

$$\mathbf{V}_s = \frac{1+\nu}{E} \mathbf{S} - \frac{\nu}{E} s\, \mathbf{I} + \alpha_T\, \Theta\, \mathbf{I} .$$

From the problem formulation the conditions follow

- stress-free surface

$$\sigma_{zz} = 0 , \tag{1a}$$

- impeded strains

$$\varepsilon_{xx} = \varepsilon_{yy} = 0 , \tag{1b}$$

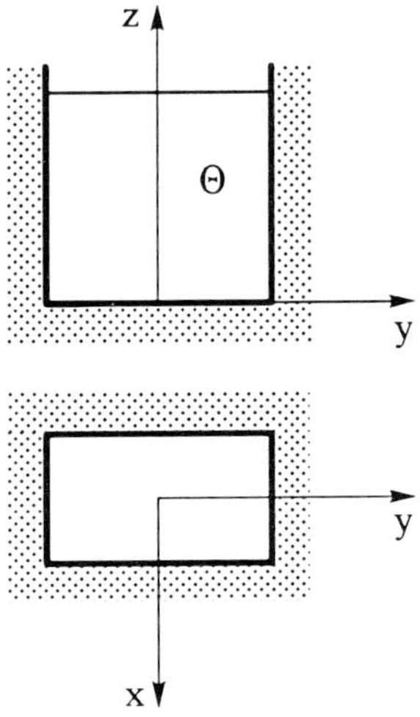

Fig. A-11: Steel ingot

– shear – free state of strain

$$\gamma_{xy} = \gamma_{xz} = \gamma_{yz} = 0 \ . \tag{1c}$$

According to *(5.1)*, the relations for the strains in component notation read

$$\varepsilon_{xx} = \frac{1}{E}[\sigma_{xx} - \nu(\sigma_{yy} + \sigma_{zz})] + \alpha_T \Theta \ , \tag{2a}$$

$$\varepsilon_{yy} = \frac{1}{E}[\sigma_{yy} - \nu(\sigma_{zz} + \sigma_{xx})] + \alpha_T \Theta \ , \tag{2b}$$

$$\varepsilon_{zz} = \frac{1}{E}[\sigma_{zz} - \nu(\sigma_{xx} + \sigma_{yy})] + \alpha_T \Theta \ . \tag{2c}$$

Inserting the conditions (1a,b) into (2a,b) then yields the two equations

$$\sigma_{xx} - \nu\sigma_{yy} = - E\,\alpha_T \Theta \ , \tag{3a}$$

$$- \nu\sigma_{xx} + \sigma_{yy} = - E\,\alpha_T \Theta \ . \tag{3b}$$

From (3a,b), σ_{xx} and σ_{yy} are calculated as

$$\sigma_{xx} = \sigma_{yy} = - \frac{E\,\alpha_T \Theta}{1-\nu} \ . \tag{4}$$

The strain in the z – direction follows from (2c) and (4):

$$\varepsilon_{zz} = - \frac{\nu}{E}(\sigma_{xx} + \sigma_{yy}) + \alpha_T \Theta \quad \Rightarrow \quad \varepsilon_{zz} = \frac{1+\nu}{1-\nu}\,\alpha_T \Theta \ .$$

b) In this case, for free boundaries at $x = \text{const}$

$$\sigma_{xx} = 0 \tag{5}$$

is valid instead of $\varepsilon_{xx} = 0$.

Then follows from the equations (2a,b,c)

$$\varepsilon_{xx} = - \frac{\nu}{E}\sigma_{yy} + \alpha_T \Theta \ , \tag{6a}$$

$$\varepsilon_{yy} = \frac{\sigma_{yy}}{E} + \alpha_T \Theta \ , \tag{6b}$$

$$\varepsilon_{zz} = - \frac{\nu}{E}\sigma_{yy} + \alpha_T \Theta \ . \tag{6c}$$

Furthermore, no strain ε_{yy} occurs in the y – direction; thus, from (6b) follows

$$\sigma_{yy} = - E\,\alpha_T \Theta$$

and from (6a) or (6c)

$$\varepsilon_{xx} = (1 + \nu)\,\alpha_T \Theta = \varepsilon_{zz} \ .$$

The strain in the z – direction is now smaller than in a).

Exercise A-6-1:

A BERNOULLI–EULER beam with one end clamped and the other simply supported, is subjected to a transverse load q (x) per unit length of the axis of the beam (plane beam bending) (Fig. A-12).

The differential equation and the boundary conditions for the plane beam are to be determined

a) by means of the principle of virtual displacements,

b) by solution of the corresponding variational principle.

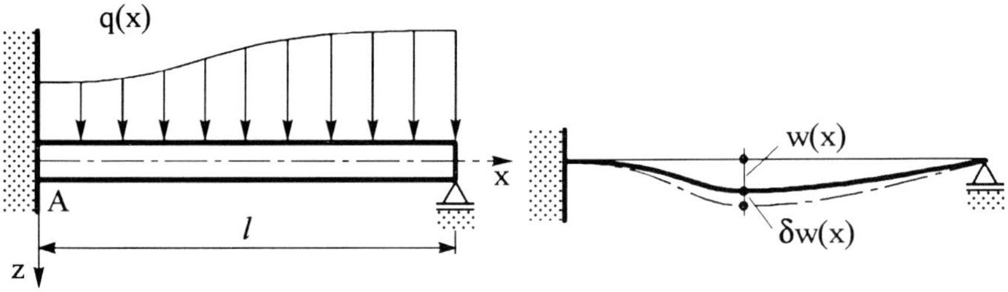

Fig. A-12: Clamped, simply supported beam under uniformly distributed load

Solution:

a) At first, we have to state the strain energy of the beam with d/dx $\widehat{=}$ ( )$_{,x}$:

$$U = \frac{1}{2}\int_0^l E\,I_y\, w_{,xx}{}^2\, dx = \Pi_i \,. \tag{1a}$$

For the work of the external forces we have

$$W = -\Pi_e = \int_0^l q(x)\, w\, dx \,. \tag{1b}$$

From (1a) and (1b) we obtain the *total potential*

$$\Pi = \Pi_i + \Pi_e = \int_0^l \left[\frac{1}{2} E\,I_y\, w_{,xx}{}^2 - q(x)\, w\right] dx \,. \tag{2}$$

The minimum principle *(6.21b)* is now applied:

$$\Pi = \Pi_i + \Pi_e = \text{minimum} \,.$$

The stationarity, see *(6.20)*, implies that the variation of the total potential must vanish for arbitrary kinematically admissible variations δ w

$$\delta \Pi = \delta \left\{ \int_0^l \left[\frac{1}{2} E I_y w_{,xx}^{\;2} - q(x) w \right] dx \right\} = 0 \tag{3}$$

$$\longrightarrow \qquad \delta \Pi = \int_0^l \left[E I_y w_{,xx} \, \delta w_{,xx} - q(x) \, \delta w \right] dx = 0 \, . \tag{4}$$

After double partial integration, (4) can be written as:

$$\delta \Pi = \int_0^l \left[(E I_y w_{,xx})_{,xx} - q(x) \right] \delta w \, dx - \left[(E I_y w_{,xx})_{,x} \, \delta w \right]_{x=0}^{l} + \\ + \left[E I_y w_{,xx} \, \delta w_{,x} \right]_{x=0}^{l} = 0 \, . \tag{5}$$

Vanishing of the integral for arbitrary variations δw leads to the differential equation for the elastic line of the beam

$$(E I_y w_{,xx})_{,xx} = q(x) \, . \tag{6}$$

The *boundary conditions* follow from the boundary terms. In order that these vanish, we must have at both $x = 0$ and $x = l$,

either $\quad \delta w = 0 \quad$ or $\quad (E I_y w_{,xx})_{,x} = - Q_z = 0 \, ,$ $\qquad$ (7a)

and

either $\quad \delta w_{,x} = 0 \quad$ or $\quad E I_y w_{,xx} = - M_y = 0 \, .$ $\qquad$ (7b)

The boundary conditions for the clamped end $x = 0$ are:

for $\quad x = 0 \longrightarrow w = 0 \; , \; w_{,x} = 0 \, .$ $\qquad$ (8a)

Hence, at $x = 0$ we have $\delta w = 0$ and $\delta w_{,x} = 0$, and according to (7a,b), $Q_z \neq 0$ and $M_y \neq 0$ in general.

The boundary conditions for the simply supported end are:

for $\quad x = l \longrightarrow w = 0 \; , \; M_y \sim w_{,xx} = 0 \, .$ $\qquad$ (8b)

Here, the condition $w = 0$ implies that $\delta w = 0$ and thus, by (7a) that $Q_z \neq 0$ in general. The second condition $M_y = 0$ is obtained from (7b) as, in general, $w_{,x} \neq 0$ and thus $\delta w_{,x} \neq 0$ at a simply supported end.

b) We naturally obtain the same result for the example if we apply the general solution *(6.35)* to the problem of variation *(6.34)*.

According to (3), the basic function reads

$$F(x, w, w_{,xx}) = \frac{1}{2} E I_y w_{,xx}^{\;2} - q(x) w \, . \tag{9}$$

From *(6.35)* now follows *EULER's differential equation*:

$$\left(\frac{\partial F}{\partial w_{,xx}} \right)_{,xx} + \frac{\partial F}{\partial w} = 0 \longrightarrow (E I_y w_{,xx})_{,xx} - q(x) = 0 \, . \tag{10}$$

This differential equation is identical with (6).

As *boundary conditions* we obtain according to *(6.35)*:

$$\left\{\left[\frac{\partial F}{\partial w_{,x}} - \left(\frac{\partial F}{\partial w_{,xx}}\right)_{,x}\right]\delta w + \frac{\partial F}{\partial w_{,xx}}\,\delta w_{,x}\right\}_{x_1}^{x_2} = 0$$

$$\Longrightarrow \quad \Big[(E I_y w_{,xx})_{,x}\,\delta w\Big]_{x=0} = 0 \quad , \quad \Big[E I_y w_{,xx}\,\delta w_{,x}\Big]_{x=0} = 0 \; ,$$

$$\Big[(E I_y w_{,xx})_{,x}\,\delta w\Big]_{x=l} = 0 \quad , \quad \Big[E I_y w_{,xx}\,\delta w_{,x}\Big]_{x=l} = 0 \; . \qquad (11)$$

In (11) the boundary conditions (7a,b) as well as (8a,b) are included (see discussion in Section 6.6).

Exercise A-6-2:

A linearly elastic body is subjected to a stationary temperature field $\Theta(\xi^i)$. Set up the variational functional by HELLINGER and REISSNER for small strains and show that one obtains the basic equations of linear thermo-elasticity as necessary conditions from the corresponding variational problem. Volume, surface and inertia forces are to be neglected.

Solution:

Substitution of the expression for strain energy $\overline{U}$ from *(6.16a)* into the variational functional *(6.29)* yields

$$\Pi_R = \int\limits_V \left\{\frac{1}{2} C^{ijkl}\gamma_{ij}\gamma_{kl} - \beta_T \gamma_i^i \Theta + \tau^{ij}\left[\frac{1}{2}(v_i|_j + v_j|_i) - \gamma_{ij}\right]\right\} dV \, .$$

This functional contains altogether 15 unknown functions γ_{ij}, τ^{ij}, v_i. From the problem of variation

$$\delta \Pi_R = \delta \int\limits_V \overline{\Pi}_R \, dV = 0$$

follow in analogy with *(6.35)* the corresponding EULER's equations as necessary conditions:

1. $$\frac{\partial \overline{\Pi}_R}{\partial \tau^{ij}} = 0 \quad \longrightarrow \quad \gamma_{ij} = \frac{1}{2}(v_i|_j + v_j|_i)$$

   $\longrightarrow$ Strain-displacement equations according to *(4.12b)* .

2. $$\frac{\partial \overline{\Pi}_R}{\partial \gamma_{ij}} = 0 \quad \longrightarrow \quad \tau^{ij} = C^{ijkl}\gamma_{kl} - \beta g^{ij}\Theta$$

   $\longrightarrow$ Material law according to *(5.6a)* .

3. $$\left(\frac{\partial \overline{\Pi}_R}{\partial v_i|_j}\right)\Big|_j = 0 \quad \longrightarrow \quad \tau^{ij}|_j = 0$$

   $\longrightarrow$ Equilibrium conditions according to *(3.28a)* .

Exercise A-6-3:

Structural components in mechanical and civil engineering often have a complex shape. Furthermore, they are subjected to multiple loads and their boundary conditions can be quite complicated. Starting from the energy principles of the theory of elasticity the *Finite Element Method (FEM)* was developed for determining strains and stresses of such components, and taking the principle of virtual displacements as a basis, displacements are introduced as unknowns and can be derived from the equilibrium conditions. We proceed in such a way that the supporting framework is divided into so-called finite elements (bar, beam, disk, plate elements etc.) with approximate expressions chosen for the displacement fields in each element (for details, see e.g. [A.1, A.2, A.21]).

With the example of a disk under temperature loads the basic relations for the method should be determined. Therefore, as shown in Fig. A-13, a triangular element is taken out of the domain of the disk and is described in a Cartesian coordinate system.

At each of the three nodes 1, 2, 3 of the triangular disk element two nodal displacements are admissible and can be assembled in a row vector as follows:

$$\mathbf{d}^T = (u_1, u_2, u_3, v_1, v_2, v_3)^T .$$

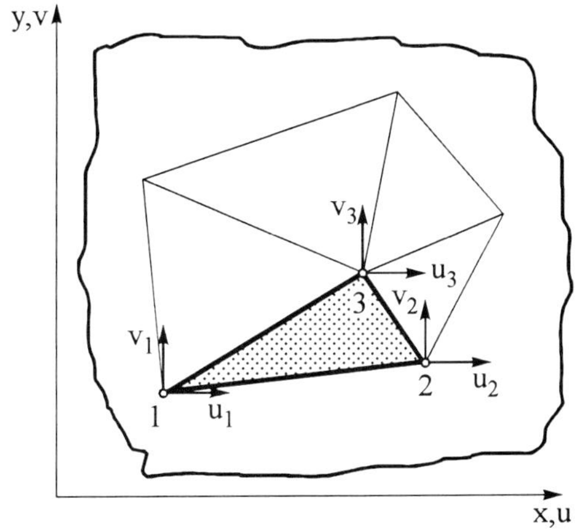

Fig. A-13: Triangular element with 3 x 2 nodal displacements of the corners

a) The strain-displacement equations are to be stated in matrix notation in the element-inherent system. Therefore, a linear displacement approximation (constant strains) of the following form will be assumed:

$$u(x,y) = a_1 + a_2 x + a_3 y ,$$

$$v(x,y) = a_4 + a_5 x + a_6 y .$$

The relation between strains and nodal displacements is to be given with $a_1, \ldots, a_6$ as free values.

b) The equilibrium of the element is to be formulated with the help of the principle of virtual displacements.

Solution:

a) The strains in the element-inherent system (upper index e) are given for the state of plane stress according to *(4.12b)*:

$$(\gamma_{\alpha\beta})^e = \frac{1}{2}(v_\alpha|_\beta + v_\beta|_\alpha)^e .$$

In matrix form this relation reads

$$\boldsymbol{\varepsilon}^e = (\mathbf{D}\,\mathbf{v})^e , \qquad (1a)$$

where

$$\boldsymbol{\varepsilon}^e = \begin{bmatrix} \varepsilon_{xx} \\ \varepsilon_{yy} \\ \gamma_{xy} \end{bmatrix}^e , \quad \mathbf{D}^e = \begin{bmatrix} \dfrac{\partial}{\partial x} & 0 \\ 0 & \dfrac{\partial}{\partial y} \\ \dfrac{\partial}{\partial y} & \dfrac{\partial}{\partial x} \end{bmatrix}^e , \quad \mathbf{v}^e = \begin{bmatrix} u \\ v \end{bmatrix}^e . \qquad (1b)$$

If we introduce the assumed linear displacement field with the six free values $a_1, \ldots, a_6$, we can write the following vector in column notation for the displacements:

$$\mathbf{v}^e = \mathbf{H}^e \cdot \mathbf{a}^e \qquad (2a)$$

with $$\mathbf{H}^e = \begin{bmatrix} 1 & x & y & 0 & 0 & 0 \\ 0 & 0 & 0 & 1 & x & y \end{bmatrix}^e , \quad \mathbf{a} = \begin{bmatrix} a_1 \\ a_2 \\ a_3 \\ a_4 \\ a_5 \\ a_6 \end{bmatrix}^e .$$

Further, we obtain the vector of the nodal displacements as

$$\mathbf{d}^e = (\boldsymbol{\Omega}\,\mathbf{a})^e \qquad (3a)$$

with $$\boldsymbol{\Omega}^e = \left[\begin{array}{ccc|ccc} 1 & x_1 & y_1 & & & \\ 1 & x_2 & y_2 & & \mathbf{0} & \\ 1 & x_3 & y_3 & & & \\ \hline & & & 1 & x_1 & y_1 \\ & \mathbf{0} & & 1 & x_2 & y_2 \\ & & & 1 & x_3 & y_3 \end{array}\right] . \qquad (3b)$$

We now insert the relations (2a) and (3a) into (1a) and obtain the following relation between strains and nodal displacements:

$$\boldsymbol{\varepsilon}^e = (\mathbf{D}\,\mathbf{H}\,\boldsymbol{\Omega}^{-1})^e\,\mathbf{d}^e =: \mathbf{V}^e\,\mathbf{d}^e\,. \tag{4}$$

b) We start from the *DUHAMEL–NEUMANN form* of HOOKE's law for disks in *(5.15b)*:

$$\tau^{\alpha\beta} = E^{\alpha\beta\gamma\delta}\left[\gamma_{\gamma\delta} - \alpha_T\, a_{\gamma\delta}\, {}^0\Theta\right].$$

In matrix notation this law reads, see *(5.14)*

$$\boldsymbol{\sigma}^e = \mathbf{E}\left[\boldsymbol{\varepsilon}^e - \boldsymbol{\varepsilon}_\Theta\right] \tag{5}$$

with

$$\boldsymbol{\sigma}^e = \begin{bmatrix}\sigma_{xx}\\ \sigma_{yy}\\ \tau_{xy}\end{bmatrix}\,,\quad \mathbf{E} = \frac{E}{1-\nu^2}\begin{bmatrix}1 & \nu & 0\\ \nu & 1 & 0\\ 0 & 0 & \dfrac{1-\nu}{2}\end{bmatrix} = \mathbf{E}^T\,,\quad \boldsymbol{\varepsilon}_\Theta = \alpha_T\,{}^0\Theta\begin{bmatrix}1\\1\\0\end{bmatrix}.$$

The equilibrium conditions can now be derived from the principle of virtual displacements. Thus, according to *(6.20)*, a body is in static equilibrium, if the virtual total potential is equal to zero:

$$\delta\,\Pi_i + \delta\,\Pi_e = 0\,.$$

Here, Π_i and Π_e are the internal and external potential, respectively.

With the material law (5) it follows from *(6.12a)* that the internal potential for a thermally loaded disk is given by

$$\Pi_i = \frac{1}{2}\,t\int_A\left[\boldsymbol{\varepsilon}^T\,\mathbf{E}\,\boldsymbol{\varepsilon} - 2\,\boldsymbol{\varepsilon}^T\,\mathbf{E}\,\boldsymbol{\varepsilon}_\Theta\right]dA\,. \tag{6}$$

We now insert (4) into (6) and obtain

$$\Pi_i^e = \frac{1}{2}(\mathbf{d}^T)^e\,\mathbf{K}^e\,\mathbf{d}^e + (\mathbf{d}^T)^e\,\mathbf{k}_\Theta^e \tag{7}$$

with
$$\mathbf{K}^e = t\left(\int_A \mathbf{V}^T\,\mathbf{E}\,\mathbf{V}\,dA\right)^e \tag{8}$$

as element stiffness matrix and

$$\mathbf{k}_\Theta^e = -t\left(\int_A \mathbf{V}^T\,\mathbf{E}\,\boldsymbol{\varepsilon}_\Theta\,dA\right)^e$$

as a vector of the *temperature forces*.

With the principle of virtual displacements we obtain after introducing the virtual work of nodal loads (vector $\mathbf{p}$) from (7):

$$\delta\,\Pi_i^e + \delta\,\Pi_a^e = \delta\,(\mathbf{d}^T)^e(\mathbf{K}\,\mathbf{d} + \mathbf{k}_\Theta - \mathbf{p})^e = \mathbf{0}$$

$$\longrightarrow \quad \mathbf{p}^e = \mathbf{K}^e\,\mathbf{d}^e + \mathbf{k}_\Theta^e \,.$$

After determination of the nodal forces, we can now establish the equilibrium for the entire domain. Therefore, the variation of the virtual energy for the entire structure has to vanish. This energy is obtained by summation over the elements (see [A.1, A.2, A.21, C.25]).

Remark: It should be emphasized that this example only serves to illustrate the way of developing a triangular finite element. Such an element with constant strain is only a very simple one. Most software systems contain more accurate and improved elements.

Exercise A-7-1:

Determine the stresses in a hollow sphere (outer radius a, inner radius b) under constant internal pressure p from the basic equations of the theory of elasticity.

Solution:

A spherical – symmetrical state of stress is given, in which

$$\frac{\partial}{\partial\varphi} = \frac{\partial}{\partial\vartheta} = 0 \quad , \quad \frac{\partial}{\partial r} = \frac{d}{dr} \mathrel{\widehat{=}} (\ )_{,r} \,,$$

$$\tau_{r\varphi} = \tau_{r\vartheta} = \tau_{\varphi\vartheta} = \gamma_{r\varphi} = \gamma_{r\vartheta} = \gamma_{\varphi\vartheta} = v = w = 0 \,, \qquad (1)$$

$$\sigma_{\varphi\varphi} = \sigma_{\vartheta\vartheta} \quad , \quad \varepsilon_{\varphi\varphi} = \varepsilon_{\vartheta\vartheta} \,.$$

In view of (1), the strain – displacement equations *(4.16)* can be reduced to

$$\varepsilon_{rr} = u_{,r} \quad , \quad \varepsilon_{\varphi\varphi} = \varepsilon_{\vartheta\vartheta} = \frac{u}{r} \,. \qquad (2)$$

Since $\gamma_{ij} = 0$ the current strains are principal strains.

Combination of the two equations (2) yields the compatibility condition

$$\varepsilon_{rr} = (r\,\varepsilon_{\varphi\varphi})_{,r} \,. \qquad (3)$$

Equation (3) contains the two unknown strains ε_{rr} und $\varepsilon_{\varphi\varphi}$. We now replace by means of (1) the principal strains by the material law according to *(5.1)* (spherical coordinates are orthogonal, therefore the same structure as in Cartesian coordinates):

$$\begin{aligned} \varepsilon_{rr} &= \frac{1}{E}[\sigma_{rr} - 2\nu\sigma_{\varphi\varphi}] , \\ \varepsilon_{\varphi\varphi} &= \frac{1}{E}[\sigma_{\varphi\varphi} - \nu(\sigma_{rr} + \sigma_{\varphi\varphi})] = \varepsilon_{\vartheta\vartheta} . \end{aligned} \tag{4}$$

Inserting (4) into (3) yields

$$\sigma_{rr} - 2\nu\sigma_{\varphi\varphi} = \left\{ r[(1-\nu)\sigma_{\varphi\varphi} - \nu\sigma_{rr}] \right\}_{,r} . \tag{5}$$

Furthermore, we need the equilibrium conditions *(3.30)*, which by means of (1) can be reduced to

$$\sigma_{rr,r} + \frac{2}{r}(\sigma_{rr} - \sigma_{\varphi\varphi}) = 0$$

or

$$(r^2\sigma_{rr})_{,r} = 2r\sigma_{\varphi\varphi} . \tag{6}$$

With (5) and (6) we now have two differential equations for σ_{rr} and $\sigma_{\varphi\varphi}$. If we replace $\sigma_{\varphi\varphi}$ in (5) by means of (6), we obtain one differential equation for σ_{rr}:

$$(r^2\sigma_{rr})_{,rr} - 2\sigma_{rr} = 0 . \tag{7a}$$

This differential equation corresponds to the BELTRAMI differential equation *(7.4)* in spherical coordinates for an axisymmetrical state of stress. This ordinary, second order differential equation leads to the following EULER differential equation

$$r^2\sigma_{rr,rr} + 4r\sigma_{rr,r} = 0 , \tag{7b}$$

the general solution of which can be determined with the aid of a power approach

$$\sigma_{rr} = A r^n . \tag{8}$$

The characteristic equation is

$$n(n-1) + 4n = 0$$

with the roots

$$n_1 = 0 \quad , \quad n_2 = -3 .$$

The solution of (7b) then becomes

$$\sigma_{rr} = A_1 + \frac{A_2}{r^3} \tag{9}$$

with the two constants A_1 and A_2 to be determined by the *boundary conditions*. These read as follows:

$$\sigma_{rr}(r=b) = -p \quad , \quad \sigma_{rr}(r=a) = 0 . \tag{10}$$

From (10) we obtain with (9)

$$\left. \begin{aligned} A_1 + \frac{A_2}{b^3} &= -p \\ A_1 + \frac{A_2}{a^3} &= 0 \end{aligned} \right\} \longrightarrow \quad \begin{aligned} A_1 &= p\frac{b^3}{a^3 - b^3} , \\ A_2 &= -p\frac{a^3 b^3}{a^3 - b^3} . \end{aligned}$$

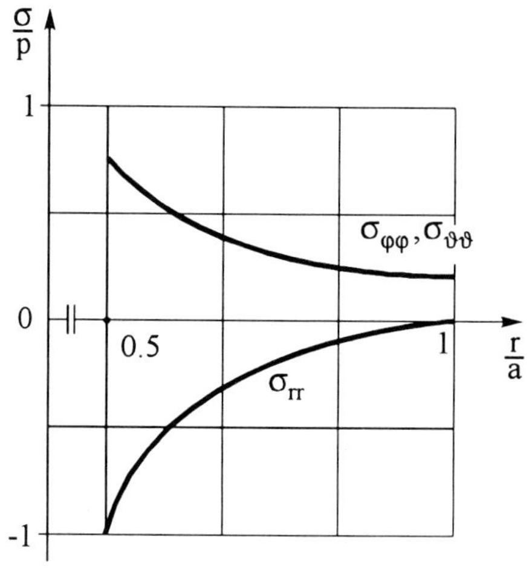

Fig. A-14: Stresses in a hollow sphere with a/b = 2

The stresses are thus found to be

$$\sigma_{rr} = -p\,\frac{\left(\frac{a}{r}\right)^3 - 1}{\left(\frac{a}{b}\right)^3 - 1} \quad , \quad \sigma_{\varphi\varphi} = \sigma_{\vartheta\vartheta} = \frac{p}{2}\,\frac{\left(\frac{a}{r}\right)^3 + 2}{\left(\frac{a}{b}\right)^3 - 1} \,. \tag{11}$$

They are presented in Fig. A-14 for the case of $a/b = 2$. The maximum tensile stress occurs at the inner surface and is given by

$$\sigma_{\varphi\varphi\max} = \frac{p}{2}\,\frac{\left(\frac{a}{b}\right)^3 + 2}{\left(\frac{a}{b}\right)^3 - 1} \,. \tag{12}$$

If the stresses in a *thin-walled* sphere are to be determined, one starts from a medium radius

$$r_m = \frac{1}{2}\,(a + b) \,.$$

With the wall thickness $t = a - b$ we obtain

$$a = r_m + \frac{t}{2} \quad , \quad b = r_m - \frac{t}{2} \,,$$

and thus for $t/r_m \ll 1$ the approximations

$$r \approx r_m \quad , \quad \left(\frac{a}{b}\right)^3 \approx 1 + 3\frac{t}{r_m} \,.$$

It now follows from (11) that for a thin-walled sphere the tangential stresses are constant over the thickness, and from (12) we obtain

$$\sigma_{\varphi\varphi} = \sigma_{\vartheta\vartheta} \approx \frac{p\,r_m}{2\,t} \,.$$

Because of its importance in boiler design this relation is called the *boiler formula*.

Exercise A-7-2:

A concentrated force F acts upon an elastic half-space as shown in Fig. A-15.

Determine the strains and stresses with the help of LOVE's displacement function $\chi(r,z)$.

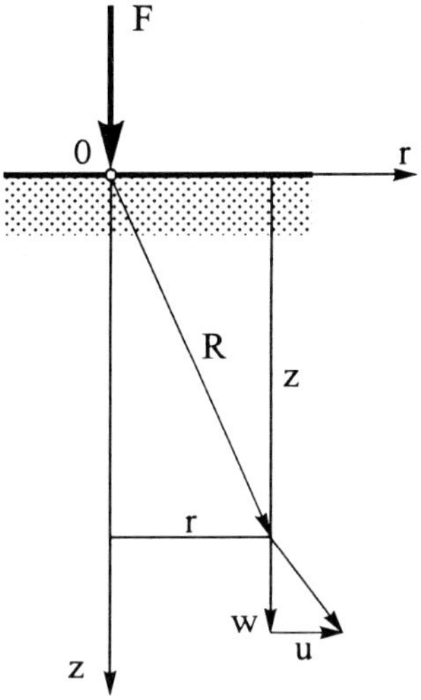

Fig. A-15: Elastic half-space subjected to a concentrated force

Solution:

We first seek a LOVE's displacement function $\chi(r,z)$, which

a) satisfies the bipotential equation *(7.5)*

$$\Delta\Delta\,\chi = 0\ , \tag{1}$$

b) fulfills the boundary conditions

$$\sigma_{zz}(r, z = 0) = 0\ , \tag{2a}$$

$$\tau_{rz}(r, z = 0) = 0\ . \tag{2b}$$

We choose a linear combination of two solutions according to *(7.6)*

$$\chi = C_1\sqrt{r^2 + z^2} + C_2\, z\,\ln\left(z + \sqrt{r^2 + z^2}\right). \tag{3}$$

With this approach (1) is satisfied. C_1 and C_2 are arbitrary constants to be determined from proper *boundary conditions*. We first determine the displacements from *(7.7a,b)* with $R = \sqrt{r^2 + z^2}$

$$u = \frac{1}{1-2\nu}\left[(C_1 + C_2)\frac{rz}{R^3} - C_2\frac{r}{R(z+R)}\right],$$
$$w = \frac{1}{1-2\nu}\left[(3-4\nu)C_1 + 2(1-2\nu)C_2\right]\frac{1}{R} + \frac{(C_1+C_2)}{1-2\nu}\frac{z^2}{R^3}, \qquad (4)$$

and the stresses from *(7.8a–d)*

$$\sigma_{rr} = 2G\left[\frac{1}{1-2\nu}\frac{C_2}{R(z+R)} + \left(C_1 - \frac{2\nu C_2}{1-2\nu}\right)\frac{z}{R^3} - \frac{3}{1-2\nu}(C_1 + C_2)\frac{zr^2}{R^5}\right],$$
$$\sigma_{\varphi\varphi} = 2G\left[(C_1 + C_2)\frac{z}{R^3} - \frac{1}{1-2\nu}\frac{C_2}{R(z+R)}\right],$$
$$\sigma_{zz} = -2G\left[\left(C_1 - \frac{2\nu C_2}{1-2\nu}\right)\frac{z}{R^3} + \frac{3}{1-2\nu}(C_1 + C_2)\frac{z^3}{R^5}\right], \qquad (5)$$
$$\tau_{rz} = -2G\left[\left(C_1 - \frac{2\nu C_2}{1-2\nu}\right)\frac{r}{R^3} + \frac{3}{1-2\nu}(C_1 + C_2)\frac{rz^2}{R^5}\right].$$

Substitution of the shear stress τ_{rz} into (2b) yields

$$\tau_{rz}(r, z = 0) = 0 \longrightarrow C_2 = \frac{1-2\nu}{2\nu}C_1 .$$

Thus, it follows that

$$\sigma_{zz} = -C_1\frac{3G}{\nu(1-2\nu)}\frac{z^3}{R^5} . \qquad (6)$$

With (6) the boundary condition (2a) is fulfilled. In addition, the integral of the normal stresses σ_{zz} acting on arbitrary sections $z = \text{const}$ $(z > 0)$ must be equal to the magnitude of the concentrated force F, i.e.

$$F = -\int_A \sigma_{zz}\, dA = -\int_{r=0}^{\infty} \sigma_{zz}\, 2\pi r\, dr . \qquad (7)$$

Inserting (6) into (7) yields the integral for the concentrated force

$$F = \frac{6C_1 G z^3 \pi}{\nu(1-2\nu)}\int_{r=0}^{\infty}\frac{r\, dr}{(r^2 + z^2)^{5/2}} .$$

By using the substitution $r^2 + z^2 = v^2$ the integral can be solved. The constant is then found to be

$$C_1 = \frac{\nu(1-2\nu)}{2\pi G}F$$

which with the use of (4) and (5) leads to the final expressions for the displacements and stresses :

$$u = \frac{F}{4\pi G}\left[\frac{rz}{R^3} - (1-2\nu)\frac{r}{R(z+R)}\right], \qquad (8a)$$

$$w = \frac{F}{4\pi G}\left[2(1-\nu)\frac{1}{R} + \frac{z^2}{R^3}\right], \qquad (8b)$$

$$\sigma_{rr} = \frac{F}{2\pi}\left[(1-2\nu)\frac{1}{R(z+R)} - \frac{3zr^2}{R^5}\right], \qquad (9a)$$

$$\sigma_{\varphi\varphi} = \frac{F}{2\pi}(1-2\nu)\left[\frac{z}{R^3} - \frac{1}{R(z+R)}\right], \qquad (9b)$$

$$\sigma_{zz} = -\frac{3F}{2\pi}\frac{z^3}{R^5}, \qquad (9c)$$

$$\tau_{rz} = -\frac{3F}{2\pi}\frac{rz^2}{R^5}. \qquad (9d)$$

The relations (9a-d) are called BOUSSINESQ's formulas.

Finally, let us discuss the results:

– Resultant stress

From (9c,d) we find the simple relationship

$$\frac{\sigma_{zz}}{\tau_{rz}} = \frac{z}{r}. \qquad (10)$$

Thus, according to Fig. A-16, the *resultant stress* at points of cut planes $z = \text{const}$ has the magnitude

$$\sigma_{res} = \sqrt{\sigma_{zz}^2 + \tau_{rz}^2} = \frac{3F}{2\pi}\frac{z^2}{(r^2+z^2)^2} = \frac{3F}{2\pi}\frac{\cos^2\beta}{r^2+z^2}, \qquad (11)$$

and points towards the origin 0.

If we lay a spherical surface of diameter d tangentially to the boundary plane $z = 0$ at the origin 0, the following is valid for every point on this surface

$$R^2 = r^2 + z^2 = d^2\cos^2\beta .$$

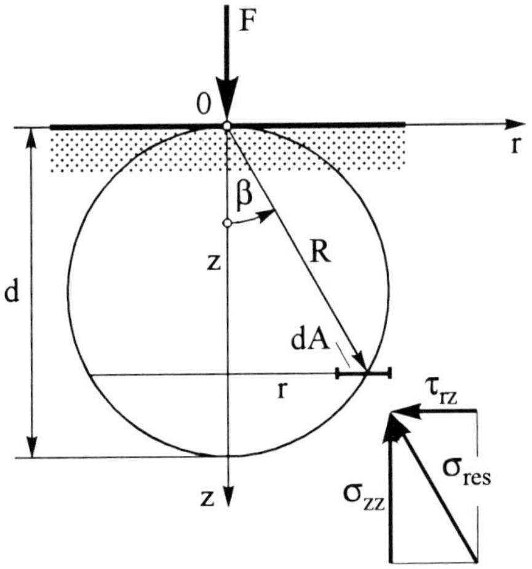

Fig. A-16: Resultant stress from a concentrated force acting on a half-space

Thus, according to (11), in the intersection points between any plane z = const and the spherical surface, the magnitude of the resultant stress is the same and given by [A.16]

$$\sigma_{res} = \frac{3F}{2\pi d^2} .$$

– Displacements of the boundary plane

For the displacements of the boundary plane z = 0, we obtain from (8a,b)

$$u(r, z = 0) = -\frac{(1-2\nu)(1+\nu)}{2\pi E r} F , \qquad (12a)$$

$$w(r, z = 0) = \frac{1-\nu^2}{\pi E r} F . \qquad (12b)$$

Thus, the theory yields singular values for the displacements u and w at the point of force application, and according to (9) similarly for the stresses. These singularities vanish if the force is distributed over a small area of the surface.

B Plane load–bearing structures

B.1 Definitions – Formulas –Concepts

8 Disks

8.1 Definitions – Assumptions – Basic Equations

Disks are *plane load–bearing structures* the thicknesses t of which are small in comparison with the other dimensions (Fig. 8.1) and which are subjected to loads acting in the mid–plane. All stresses are assumed uniformly distributed over the thickness, i.e. they do not depend on z. We therefore have a State of Plane Stress for which the most important basic equations in Cartesian and in polar coordinates are summarized in the following, where the thickness of the disk is assumed to be constant.

a) *Isotropic disk in Cartesian coordinates*

Bipotential equation (*disk equation*) according to *(7.14)*:

$$\Delta\Delta\,\Phi = -E\,\alpha_T\,\Delta\,{}^0\Theta - (1-\nu)\,\Delta\,V \tag{8.1}$$

with $\partial/\partial x \mathrel{\widehat{=}} (\ )_{,x}$, $\partial/\partial y \mathrel{\widehat{=}} (\ )_{,y}$.

$\Delta = (\ )_{,xx} + (\ )_{,yy}$ LAPLACE operator due to *(2.39)* ,

$\Phi = \Phi(x,y)$ AIRY's stress function due to *(7.22)* ,

${}^0\Theta = {}^0\Theta(x,y)$ Temperature difference relative to the initial stress–free state ,

$V = V(x,y)$ Potential of volume forces .

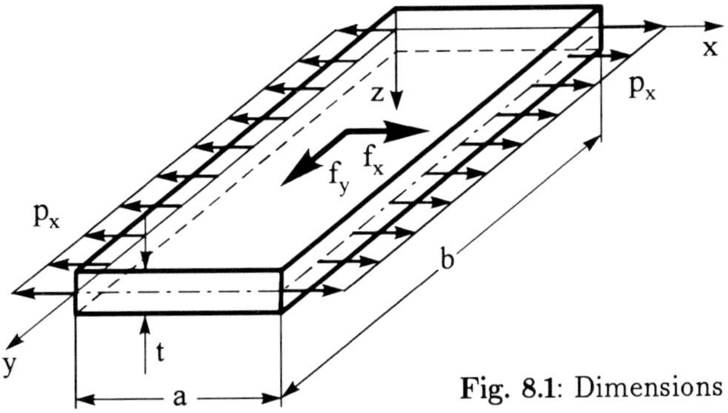

Fig. 8.1: Dimensions and loads on a disk

Stresses from derivatives of AIRY's stress function due to *(7.22)*

$$\sigma_{xx} = \Phi_{,yy} + V \quad , \quad \sigma_{yy} = \Phi_{,xx} + V \quad , \quad \tau_{xy} = -\Phi_{,xy} \,. \tag{8.2}$$

Strain-displacement equations

$$\varepsilon_{xx} = u_{,x} \quad , \quad \varepsilon_{yy} = v_{,y} \quad , \quad \gamma_{xy} = u_{,y} + v_{,x} \,. \tag{8.3}$$

Material law due to *(5.12)*

$$\left.\begin{aligned} \varepsilon_{xx} &= \frac{1}{E}(\sigma_{xx} - \nu\,\sigma_{yy}) + \alpha_T\,{}^0\Theta \,, \\ \varepsilon_{yy} &= \frac{1}{E}(\sigma_{yy} - \nu\,\sigma_{xx}) + \alpha_T\,{}^0\Theta \,, \\ \gamma_{xy} &= \frac{1}{G}\tau_{xy} \end{aligned}\right\} \tag{8.4}$$

or due to *(5.13)*

$$\left.\begin{aligned} \sigma_{xx} &= \frac{E}{1-\nu^2}\left[\varepsilon_{xx} + \nu\,\varepsilon_{yy} - (1+\nu)\alpha_T\,{}^0\Theta\right] \,, \\ \sigma_{yy} &= \frac{E}{1-\nu^2}\left[\varepsilon_{yy} + \nu\,\varepsilon_{xx} - (1+\nu)\alpha_T\,{}^0\Theta\right] \,, \\ \tau_{xy} &= G\,\gamma_{xy} \end{aligned}\right\} \tag{8.5}$$

with the YOUNG's modulus E, the shear modulus G, and the POISSON's ratio ν.

b) *Isotropic disk in polar coordinates*

Bipotential equation using *(2.49)* with $\partial/\partial r \,\widehat{=}\, (\;)_{,r}$, $\partial/\partial\varphi \,\widehat{=}\, (\;)_{,\varphi}$

$$\begin{aligned} \Delta\Delta\,\Phi &= \Phi_{,rrrr} + \frac{2}{r}\Phi_{,rrr} - \frac{1}{r^2}(\Phi_{,rr} - 2\Phi_{,rr\varphi\varphi}) + \\ &\quad + \frac{1}{r^3}(\Phi_{,r} - 2\Phi_{,r\varphi\varphi}) + \frac{1}{r^4}(4\Phi_{,\varphi\varphi} + \Phi_{,\varphi\varphi\varphi\varphi}) = \\ &= -E\,\alpha_T\left({}^0\Theta_{,rr} + \frac{1}{r}{}^0\Theta_{,r} + \frac{1}{r^2}{}^0\Theta_{,\varphi\varphi}\right) - \\ &\quad - (1-\nu)\left(V_{,rr} + \frac{1}{r}V_{,r} + \frac{1}{r^2}V_{,\varphi\varphi}\right) . \end{aligned} \tag{8.6}$$

Stresses by *(7.13)*

$$\left.\begin{aligned} \sigma_{rr} &= \frac{1}{r^2}\Phi_{,\varphi\varphi} + \frac{1}{r}\Phi_{,r} + V(r,\varphi) \,, \\ \sigma_{\varphi\varphi} &= \Phi_{,rr} + V(r,\varphi) \,, \\ \tau_{r\varphi} &= \frac{1}{r^2}\Phi_{,\varphi} - \frac{1}{r}\Phi_{,r\varphi} = -\left(\frac{1}{r}\Phi_{,\varphi}\right)_{,r} \,. \end{aligned}\right\} \tag{8.7}$$

Strain – displacement equations due to *(4.15)*

$$\left.\begin{aligned} \varepsilon_{rr} &= u_{,r} , \\ \varepsilon_{\varphi\varphi} &= \frac{1}{r} v_{,\varphi} + \frac{u}{r} , \qquad \gamma_{r\varphi} = \frac{1}{r} u_{,\varphi} + v_{,r} - \frac{v}{r} . \end{aligned}\right\} \quad (8.8)$$

Material law due to *(5.12)* and *(5.13)*

$$\left.\begin{aligned} \varepsilon_{rr} &= \frac{1}{E}(\sigma_{rr} - \nu\,\sigma_{\varphi\varphi}) + \alpha_T {}^0\Theta , \\ \varepsilon_{\varphi\varphi} &= \frac{1}{E}(\sigma_{\varphi\varphi} - \nu\,\sigma_{rr}) + \alpha_T {}^0\Theta , \\ \gamma_{r\varphi} &= \frac{1}{G}\tau_{r\varphi} \end{aligned}\right\} \quad (8.9)$$

or

$$\left.\begin{aligned} \sigma_{rr} &= \frac{E}{1-\nu^2}[\varepsilon_{rr} + \nu\,\varepsilon_{\varphi\varphi} - (1+\nu)\alpha_T {}^0\Theta] , \\ \sigma_{\varphi\varphi} &= \frac{E}{1-\nu^2}[\varepsilon_{\varphi\varphi} + \nu\,\varepsilon_{rr} - (1+\nu)\alpha_T {}^0\Theta] , \\ \tau_{r\varphi} &= G\,\gamma_{r\varphi} . \end{aligned}\right\} \quad (8.10)$$

8.2 Analytical solutions to the homogeneous bipotential equation

a) *Cartesian coordinates*

– *Approach with power series expansion*

$$\Phi = \sum_i \sum_k a_{ik}\, x^i y^k \qquad (8.11)$$

with the free coefficients a_{ik} and arbitrary integer exponents i and k (including zero).
Since according to *(8.2)* the stresses result from second derivatives of the stress function, terms with

$$i + k < 2$$

do not contribute. Power series with

$$i + k < 4$$

fulfill the basic equation for arbitrary constants a_{ik} , because only derivatives of fourth order occur in the bipotential equation. The most important special cases are listed in Table 8.1.

$\Phi(x,y)$	σ_{xx}	σ_{yy}	τ_{xy}	comment
$a_{02}\,y^2$	$2\,a_{02}$	0	0	constant tension in x-direction
$a_{11}\,x\,y$	0	0	$-a_{11}$	constant shear
$a_{20}\,x^2$	0	$2\,a_{20}$	0	constant tension in y-direction
$a_{03}\,y^3$	$6\,a_{03}\,y$	0	0	pure bending moment M_x
$a_{30}\,x^3$	0	$6\,a_{30}\,x$	0	pure bending moment M_y

Table 8.1: States of stress in power approaches

For $\quad i + k \geq 4$

$\Delta\Delta\Phi = 0$ is only fulfilled if single constants a_{ik} satisfy the necessary coupling conditions.

– *Approach with FOURIER series expansion*

Periodic functions
A load is given as a periodic function along a boundary, or it varies periodically.

FOURIER expansion of a boundary load $q(x)$, $0 \leq x \leq l$

$$q(x) = a_0 + \sum_n a_n \cos\alpha_n x + \sum_n b_n \sin\alpha_n x \tag{8.12a}$$

with $\quad \alpha_n = \dfrac{2n\pi}{l} \qquad (n = 1,2,3,\dots)\,,$

$$a_0 = \frac{1}{l}\int_0^l q(x)\,dx \quad , \quad a_n = \frac{2}{l}\int_0^l q(x)\cos\alpha_n x\,dx\,, \tag{8.12b}$$

$$b_n = \frac{2}{l}\int_0^l q(x)\sin\alpha_n x\,dx\,.$$

Expansion of a stress function in FOURIER series in case of an odd function

$$\Phi(x,y) = \sum_n \Phi_n(y)\sin\alpha_n x\,. \tag{8.13}$$

Transformation of the homogeneous bipotential equation leads to an ordinary differential equation with constant coefficients for every $\Phi_n(y)$:

$$\Phi_{n,yyyy} - 2\alpha_n^2\Phi_{n,yy} + \alpha_n^4\Phi_n = 0 \quad , \quad (d/dy \;\widehat{=}\; ,y)\,. \tag{8.14}$$

Solutions to *(8.14)*:

$$\Phi_n(y) = \sum_{m=1}^{4} C_{mn}\, e^{\lambda_{mn} y}$$

or

$$\Phi_n(y) = \frac{1}{\alpha_n^2}(A_n \cosh\alpha_n y + B_n \alpha_n y \cosh\alpha_n y + C_n \sinh\alpha_n y + D_n \alpha_n y \sinh\alpha_n y)\,. \tag{8.15}$$

Non-periodic functions

Load described by the FOURIER integral formula

$$q(x) = \frac{1}{\pi}\int_0^{\infty}\Big[\cos\alpha x \int_{-\infty}^{+\infty} q(\xi)\cos\alpha\xi\, d\xi\Big]d\alpha + \frac{1}{\pi}\int_0^{\infty}\Big[\sin\alpha x \int_{-\infty}^{+\infty} q(\xi)\sin\alpha\xi\, d\xi\Big]d\alpha\,. \tag{8.16}$$

– *Approach with complex stress functions*

Instead of the real variables x and y, the complex variable $z = x + iy$ and its complex conjugate $\overline{z} = x - iy$ are introduced. Because of

$$x = \frac{1}{2}(z + \overline{z})\ , \quad y = -\frac{1}{2}i(z - \overline{z})\ , \tag{8.17}$$

follow the derivatives

$$\frac{\partial}{\partial x} = \frac{\partial}{\partial z} + \frac{\partial}{\partial \overline{z}}\ , \qquad \frac{\partial^2}{\partial x^2} = \frac{\partial^2}{\partial z^2} + 2\frac{\partial^2}{\partial z\,\partial \overline{z}} + \frac{\partial^2}{\partial \overline{z}^2}\ ,$$

$$\frac{\partial}{\partial y} = i\Big(\frac{\partial}{\partial z} - \frac{\partial}{\partial \overline{z}}\Big)\ , \qquad \frac{\partial^2}{\partial y^2} = -\frac{\partial^2}{\partial z^2} + 2\frac{\partial^2}{\partial z\,\partial \overline{z}} - \frac{\partial^2}{\partial \overline{z}^2} \tag{8.18}$$

and the LAPLACE operator

$$\Delta = \frac{\partial^2}{\partial x^2} + \frac{\partial^2}{\partial y^2} = 4\frac{\partial^2}{\partial z\,\partial \overline{z}}\ .$$

Approach for a stress function in complex notation (cf. [B.2, B.5]):

$$\Phi(z,\overline{z}) = \frac{1}{2}\Big[\overline{z}\,\varphi(z) + z\overline{\varphi}(\overline{z}) + \int \psi(z)\,dz + \int \overline{\psi}(\overline{z})\,d\overline{z}\Big]\,.$$

Equations for determining states of plane stress and the displacements:

$$\left.\begin{aligned} \sigma_{xx} + \sigma_{yy} &= 2(\varphi' + \overline{\varphi}') = 4\,\mathrm{Re}\,\varphi'(z)\,,\\ \sigma_{yy} - \sigma_{xx} + 2i\tau_{xy} &= 2(\overline{z}\varphi'' + \psi')\,,\\ 2G(u + iv) &= -z\overline{\varphi}' - \overline{\psi} + \kappa\varphi \end{aligned}\right\} \tag{8.19}$$

with $\kappa = \dfrac{3 - \nu}{1 + \nu}$ for state of plane stress ,

$\kappa = 3 - 4\nu$ for state of plane strain .

The superscript prime ′ denotes derivatives with respect to z or $\overline{z}$. φ and ψ are two arbitrary analytical functions by means of which all stresses and displacements can be calculated.

GOURSAT already presented this solution at the turn of the century, and KOLOSOV improved this procedure which was augmented and presented in detail by MUSKHELISHVILI [B.6].

b) ***Polar coordinates***

– *Axisymmetrical states of stress* $\Phi = \Phi(r)$

Differential equation from *(8.6)* with $d/dr \mathrel{\hat{=}} (\,)_{,r}$, $V = 0$, ${}^0\Theta = 0$

$$\Phi_{,rrrr} + \frac{2}{r}\Phi_{,rrr} - \frac{1}{r^2}\Phi_{,rr} + \frac{1}{r^3}\Phi_{,r} = 0 . \qquad (8.20)$$

Solution: $\Phi = C_0 + C_1 r^2 + C_2 \ln\dfrac{r}{a} + C_3 r^2 \ln\dfrac{r}{a}$, *(8.21)*

where a denotes the reference length.

Stresses

$$\sigma_{rr} = \frac{1}{r}\Phi_{,r} \;, \quad \sigma_{\varphi\varphi} = \Phi_{,rr} \;, \quad \tau_{r\varphi} = 0 . \qquad (8.22)$$

– *Radius-independent states of stress* $\Phi = \Phi(\varphi)$

Differential equation from *(8.6)* with $d/d\varphi \mathrel{\hat{=}} (\,)_{,\varphi}$

$$\Phi_{,\varphi\varphi\varphi\varphi} + 4\Phi_{,\varphi\varphi} = 0 . \qquad (8.23)$$

Solution: $\Phi = C_1\varphi + C_2 + C_3\cos 2\varphi + C_4 \sin 2\varphi$. *(8.24)*

Stresses

$$\sigma_{rr} = \frac{1}{r^2}\Phi_{,\varphi\varphi} \;, \quad \sigma_{\varphi\varphi} = 0 \;, \quad \tau_{r\varphi} = -\left(\frac{1}{r}\Phi_{,\varphi}\right)_{,r} . \qquad (8.25)$$

– *Radiating states of stress* $\tau_{r\varphi} = 0$

Stress function $\Phi = r\,g(\varphi) + h(r)$, *(8.26a)*

where h(r) denotes an axisymmetric state of stress according to *(8.20)*.

Differential equation for $g(\varphi)$

$$g_{,\varphi\varphi\varphi\varphi} + 2g_{,\varphi\varphi} + g = 0 . \qquad (8.26b)$$

Solution: $\Phi = C_1 r\cos\varphi + C_2 r\sin\varphi + C_3 r\varphi\cos\varphi + C_4 r\varphi\sin\varphi$. *(8.27)*

Stresses due to *(8.7)*

$$\sigma_{rr} = \frac{1}{r}\Phi_{,r} + \frac{1}{r^2}\Phi_{,\varphi\varphi} \quad , \quad \sigma_{\varphi\varphi} = 0 \quad , \quad \tau_{r\varphi} = 0 \ . \tag{8.28}$$

– *Non-axisymmetric states of stress*

Further solutions are obtained from *(8.6)* by means of separation approaches of the form $r^n \cos n\varphi$ $(n \geq 2)$ [see Exercise B-8-4].

– *Complex stress function*

Transformation of *(8.19)* into polar coordinates with $z = x + iy = r e^{i\vartheta}$ yields

$$\left.\begin{aligned} \sigma_{rr} + \sigma_{\vartheta\vartheta} &= 4\,\mathrm{Re}\,\varphi'(z)\,, \\ \sigma_{\vartheta\vartheta} - \sigma_{rr} + 2i\tau_{r\vartheta} &= 2(\bar{z}\varphi'' + \psi')\,e^{2i\vartheta}\,, \\ 2G(u + iv) &= (-z\overline{\varphi}' - \overline{\psi} + \kappa\varphi)\,e^{-i\vartheta} \end{aligned}\right\} \tag{8.29}$$

where u, v are the components of the displacements in the r– and ϑ–direction, respectively (see [B.5]).

9 Plates

9.1 Definitions – Assumptions – Basic equations

A plate is a structure like a disk with small thickness t in comparison with other dimensions. The plane which halves the plate thickness is called the *mid-plane*. As shown in Fig. 9.1 a), the plate is subjected to surface loads p perpendicular to the mid-plane. An arbitrary load is resolved vertically and parallel to the surface. The in-plane forces can then be dealt with by means of the disk theory (Ch. 8). The interest in this chapter is restricted to the influence of the transverse loading on the plate. The thickness of the plate is assumed constant in the following.

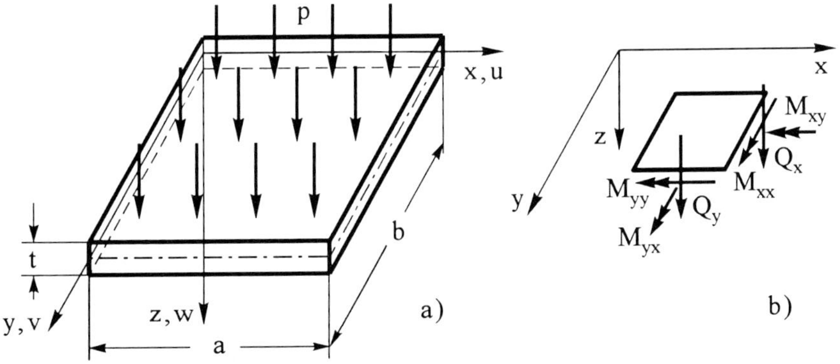

Fig. 9.1: a) Dimensions and loads of a plate
b) Sign convention for stress resultants of a plate element

a) *Plates in Cartesian coordinates*

– *Shear-elastic, isotropic plate*

Displacements of an arbitrary point P at a distance z from the mid-plane (cross-sections remain plane, see Fig. 9.2):

$$u(x,y,z) = z\,\psi_x(x,y)\,,$$
$$v(x,y,z) = z\,\psi_y(x,y)\,, \qquad (9.1)$$
$$w(x,y,z) \approx w(x,y,z=0)$$

with the bending angles ψ_x and ψ_y.

Strain-displacement relations from *(4.14)* with (9.1) ($\partial/\partial x \mathrel{\widehat{=}} (\ )_{,x}$, $\partial/\partial y \mathrel{\widehat{=}} (\ )_{,y}$)

$$\left.\begin{aligned} \varepsilon_{xx} &= z\,\psi_{x,x}\,, & \gamma_{xy} &= (\psi_{x,y} + \psi_{y,x})\,z\,,\\ \varepsilon_{yy} &= z\,\psi_{y,y}\,, & \gamma_{yz} &= \psi_y + w_{,y}\,,\\ \varepsilon_{zz} &= 0\,, & \gamma_{zx} &= w_{,x} + \psi_x\,. \end{aligned}\right\} \qquad (9.2)$$

Stress resultants (defined per unit length of a line y = const or x = const in the plate mid-plane):

$$\left.\begin{aligned} M_{xx} &= \int_{-t/2}^{+t/2} \sigma_{xx}\, z\, dz\,, \quad M_{yy} = \int_{-t/2}^{+t/2} \sigma_{yy}\, z\, dz && \text{bending moments}\\ M_{xy} &= M_{yx} = \int_{-t/2}^{+t/2} \tau_{xy}\, z\, dz && \text{torsional moments}\\ Q_x &= \int_{-t/2}^{+t/2} \tau_{xz}\, dz\,, \quad Q_y = \int_{-t/2}^{+t/2} \tau_{yz}\, dz && \text{transverse shear forces} \end{aligned}\right\} \qquad (9.3a)$$

The sign convention consistent with *(9.3a)* for the stress resultants is shown in Fig. 9.1 b).

Definitions *(9.3a)* with the material law *(5.12)* and *(9.2)* lead to the stress resultant-deformation relations

$$\left.\begin{aligned} M_{xx} &= K(\psi_{x,x} + \nu\,\psi_{y,y})\,,\\ M_{yy} &= K(\psi_{y,y} + \nu\,\psi_{x,x})\,,\\ M_{xy} &= M_{yx} = \frac{1-\nu}{2} K(\psi_{y,x} + \psi_{x,y})\,,\\ Q_x &= G\,t_s(\psi_x + w_{,x})\,,\\ Q_y &= G\,t_s(\psi_y + w_{,y}) \end{aligned}\right\} \qquad (9.3b)$$

with the plate stiffness $K = \dfrac{E\,t^3}{12(1-\nu^2)}$, the shear modulus G, and the shear thickness $t_s < t$.

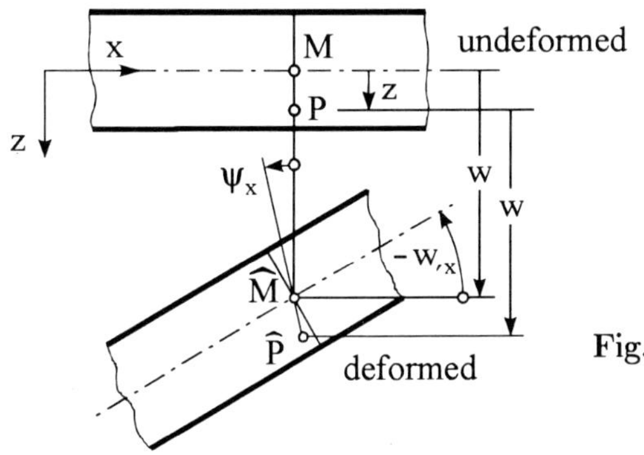

Fig. 9.2: Deformation of a plate in a cut y = const

Equilibrium conditions

$$\left.\begin{aligned} Q_{x,x} + Q_{y,y} + p &= 0\,, \\ M_{xx,x} + M_{xy,y} - Q_x &= 0\,, \\ M_{yx,x} + M_{yy,y} - Q_y &= 0\,. \end{aligned}\right\} \qquad (9.4)$$

The relations *(9.3a)* and *(9.4)* result in eight equations for the eight unknowns (five stress resultants, three deformation quantities).

Reduction of the equations

The combinations

$$\Phi = \psi_{x,x} + \psi_{y,y} \quad , \quad \Psi = \psi_{y,x} - \psi_{x,y} \qquad (9.5)$$

are the basis for the derivatives (see [B.5]):

$$\begin{aligned} w_{,x} &= -\psi_x + \frac{K}{G\,t_s}\left(\Phi_{,x} - \frac{1-\nu}{2}\Psi_{,y}\right), \\ w_{,y} &= -\psi_y + \frac{K}{G\,t_s}\left(\Phi_{,y} + \frac{1-\nu}{2}\Psi_{,x}\right). \end{aligned} \qquad (9.6)$$

This yields three equations for the three unknown functions w, Φ and Ψ:

$$\left.\boxed{\begin{aligned} K\,\Delta\Phi &= -p \\ \frac{K}{G\,t_s}\Delta\Phi - \Phi - \Delta w &= 0 \\ \Delta\Psi - \kappa_s\Psi &= 0 \end{aligned}}\right\} \qquad (9.7)$$

with the shear influence factor $\quad \frac{1}{\kappa_s} = \frac{1-\nu}{2}\,\frac{K}{G\,t_s}\,.$

By means of an additional auxiliary function

$$w_s = w - \frac{K}{G\,t_s}\Phi \tag{9.8}$$

two uncoupled equations are derived

$$\left.\boxed{\begin{aligned} K\Delta\Delta\, w_s &= p \\ \Psi - \frac{1}{\kappa_s}\Delta\Psi &= 0 \end{aligned}}\ .\right\} \tag{9.9}$$

Hereby, a partial differential problem of the sixth order is generated. Three quantities can be prescribed at each boundary [B.5].

– *Shear–rigid, isotropic plates with temperature gradient*

For such plates the shear stiffness $G\,t_s \longrightarrow \infty$, i.e. the terms multiplied by $K/G\,t_s$ can be neglected. From *(9.6)* follows:

$$\psi_x = -w_{,x}\,, \qquad \psi_y = -w_{,y}\,. \tag{9.10}$$

This means that after deformation a normal to the mid-plane remains a normal. Thus, no shear deformation occurs in cross direction ($\gamma_{yz} = \gamma_{xz} = 0$) $\Longrightarrow$ *KIRCHHOFF´s Plate Theory.*

Material law – stress resultant-displacement equations due to *(9.3)*

$$\begin{aligned} M_{xx} &= -K\left[\,w_{,xx} + \nu\, w_{,yy} + (1+\nu)\,\alpha_T\,{}^1\Theta\,\right], \\ M_{yy} &= -K\left[\,w_{,yy} + \nu\, w_{,xx} + (1+\nu)\,\alpha_T\,{}^1\Theta\,\right], \\ M_{xy} &= -K(1-\nu)\,w_{,xy} \end{aligned} \tag{9.11}$$

with the constant temperature gradient ${}^1\Theta(x,y)$ through the thickness of the plate.

Transverse shear forces from *(9.4)*

$$\left.\begin{aligned} Q_x &= -K(\Delta w)_{,x} - (1+\nu)\,\alpha_T K\,{}^1\Theta_{,x}\,, \\ Q_y &= -K(\Delta w)_{,y} - (1+\nu)\,\alpha_T K\,{}^1\Theta_{,y}\,. \end{aligned}\right\} \tag{9.12}$$

Note: As the shear deformation vanishes, *no* law of elasticity for Q_x and Q_y as in *(9.3b)* exists.

Basic equation of *KIRCHHOFF´s plate theory*

$$\boxed{K\Delta\Delta\, w = p - \alpha_T(1+\nu)\,K\Delta\,{}^1\Theta}\ . \tag{9.13}$$

The above equation is a partial differential equation of fourth order. At each boundary only two boundary conditions can be fulfilled.

Boundary conditions at a boundary x = const:

– Free boundary

$$M_{xx} = 0 \quad \text{or} \quad w_{,xx} + \nu\, w_{,yy} + (1+\nu)\,\alpha_T\, {}^1\Theta = 0\,, \tag{9.14a}$$

$$\left.\begin{aligned} &\overline{Q}_x = Q_x + M_{xy,y} = 0 \\ \text{or}\quad &w_{,xxx} + (2-\nu)\, w_{,yyx} + (1+\nu)\,\alpha_T\, {}^1\Theta_{,x} = 0\,, \end{aligned}\right\} \tag{9.14b}$$

$$\text{where}\quad \overline{Q}_x = -K\left[\, w_{,xxx} + (2-\nu)\, w_{,yyx} + (1+\nu)\,\alpha_T\, {}^1\Theta_{,x}\right] \tag{9.14c}$$

is one of the KIRCHHOFF´s *effective transverse shear forces.*

– Simply supported boundary

$$w = 0 \quad , \quad M_{xx} = 0 \tag{9.15a}$$

$$\text{or}\quad w = 0 \quad , \quad \Delta w = 0\,. \tag{9.15b}$$

Eqs. *(9.15b)* are called NAVIER´s boundary conditions.

– Clamped boundary

$$w = 0 \quad , \quad w_{,x} = 0\,. \tag{9.16}$$

Analogous boundary conditions can be formulated for a boundary y = const.

– Corner force

$$A = 2\, M_{xy} = -2\, K\, (1-\nu)\, w_{,xy}\,. \tag{9.17}$$

Determination of maximum stresses

$$\sigma_{xx\,max} = \pm\, 6\,\frac{M_{xx}}{t^2} \quad , \quad \sigma_{yy\,max} = \pm\, 6\,\frac{M_{yy}}{t^2} \quad , \quad \tau_{xy\,max} = \pm\, 6\,\frac{M_{xy}}{t^2}\,. \tag{9.18}$$

– *Transversely vibrating isotropic plate*

Differential equation for free transverse vibrations

$$\boxed{K\,\Delta\Delta\, w = -\mu\,\frac{\partial^2 w}{\partial \tau^2}}\;, \tag{9.19}$$

where τ denotes time and $\mu = \varrho\, t$ denotes the mass per unit plate area (ϱ = mass density of the plate material).

Product approach due to D. BERNOULLI for the calculation of natural vibrations:

$$w = \overline{w}(x,y) \cdot T(\tau)\,. \tag{9.20}$$

Differential equation for the time-independent vibration mode $\overline{w}(x,y)$:

$$\Delta\Delta\,\overline{w} = \lambda^4\,\overline{w}\,. \tag{9.21}$$

For an overall simply supported plate as example, the natural angular frequencies ω_{mn} are calculated from

$$\lambda^4_{mn} = \frac{\mu\,\omega^2_{mn}}{K}\,.$$

Refer to the shear-rigid plate *(9.14)* to *(9.16)* for the *boundary conditions* to *(9.21)*.

- *Shear-rigid, orthotropic plate*

Material law - stress resultant-deformation equations according to *(9.11)*

$$M_{xx} = -K_x\,(w_{,xx} + \nu_y\,w_{,yy})\;,$$

$$M_{yy} = -K_y\,(w_{,yy} + \nu_x\,w_{,xx})\;,\qquad M_{xy} = -2\,K_{xy}w_{,xy} \tag{9.22a}$$

with the stiffnesses

$$K_x = \frac{E_x\,t^3}{12\,(1-\nu_x\nu_y)}\;,\quad K_y = \frac{E_y\,t^3}{12\,(1-\nu_x\nu_y)}\;,\quad K_{xy} = \frac{G_{xy}\,t^3}{12}\,. \tag{9.22b}$$

The equilibrium conditions are the same as in the case of the isotropic plate (see *(9.4b)*).

HUBER's differential equation

$$\boxed{K_x\,w_{,xxxx} + 2\,H\,w_{,xxyy} + K_y\,w_{,yyyy} = p} \tag{9.23}$$

with the *effective torsional stiffness*

$$2\,H = 4\,K_{xy} + \nu_x\,K_y + \nu_y\,K_x\,. \tag{9.24}$$

b) ***Plates in polar coordinates***

- *Shear-rigid, isotropic circular plates*

Differential equation due to *(2.40)* and *(2.49)* with $\partial/\partial r \mathrel{\widehat{=}} {}_{,r}$, $\partial/\partial\varphi \mathrel{\widehat{=}} {}_{,\varphi}$

$$\Delta\Delta\,w = \left(w_{,rr} + \frac{1}{r}\,w_{,r} + \frac{1}{r^2}\,w_{,\varphi\varphi}\right)^2 =$$

$$= \frac{p}{K} - \alpha_T\,(1+\nu)\left({}^1\Theta_{,rr} + \frac{1}{r}\,{}^1\Theta_{,r} + \frac{1}{r^2}\,{}^1\Theta_{,\varphi\varphi}\right). \tag{9.25}$$

Material law – stress resultant–displacement relations

$$\left.\begin{aligned} M_{rr} &= -K\Big[w_{,rr} + \nu\Big(\frac{1}{r}w_{,r} + \frac{1}{r^2}w_{,\varphi\varphi}\Big) + (1+\nu)\,\alpha_T\,{}^1\Theta\Big], \\ M_{\varphi\varphi} &= -K\Big[\frac{1}{r}w_{,r} + \frac{1}{r^2}w_{,\varphi\varphi} + \nu\,w_{,rr} + (1+\nu)\,\alpha_T\,{}^1\Theta\Big], \\ M_{r\varphi} &= -(1-\nu)\,K\Big(\frac{1}{r}w_{,r\varphi} - \frac{1}{r^2}w_{,\varphi}\Big). \end{aligned}\right\} \quad (9.26)$$

Effective transverse shear forces

$$\begin{aligned} \overline{Q}_r = -K\Big[(\Delta w)_{,r} + \frac{1-\nu}{r}\Big(\frac{1}{r}w_{,r\varphi} - \frac{1}{r^2}w_{,\varphi}\Big)_{,\varphi} + \\ + (1+\nu)\,\alpha_T\Big({}^1\Theta_{,r}\cos\varphi - {}^1\Theta_{,\varphi}\frac{\sin\varphi}{r}\Big)\Big], \end{aligned} \quad (9.27a)$$

$$\begin{aligned} \overline{Q}_\varphi = -K\Big[\frac{1}{r}(\Delta w)_{,\varphi} + (1-\nu)\Big(\frac{1}{r}w_{,r\varphi} - \frac{1}{r^2}w_{,\varphi}\Big)_{,r} + \\ + (1+\nu)\,\alpha_T\Big({}^1\Theta_{,r}\sin\varphi - {}^1\Theta_{,\varphi}\frac{\cos\varphi}{r}\Big)\Big]. \end{aligned} \quad (9.27b)$$

– *Transversely vibrating circular plates*

Differential equation for the time-independent vibration mode $\overline{w}(r,\varphi)$ according to *(9.21)* [B.8, B.9]

$$\Big(\overline{w}_{,rr} + \frac{1}{r}\overline{w}_{,r} + \frac{1}{r^2}\overline{w}_{,\varphi\varphi}\Big)^2 = \lambda^4\,\overline{w} \quad \text{with} \quad \lambda^4 = \frac{\mu\,\omega^2}{K} \;. \quad (9.28)$$

Separation of the vibration mode

$$\overline{w}(r,\varphi) = R(r)\cdot\Phi(\varphi) = \sum_{n=0}^{\infty} R_n(r)\cos n\varphi \quad (9.29)$$

⟶ BESSEL's differential equations:

$$\frac{d^2R_n}{dr^2} + \frac{1}{r}\frac{dR_n}{dr} + \Big(\lambda^2 - \frac{n^2}{r^2}\Big)R_n = 0 \;, \quad (9.30a)$$

$$\frac{d^2R_n}{dr^2} + \frac{1}{r}\frac{dR_n}{dr} - \Big(\lambda^2 + \frac{n^2}{r^2}\Big)R_n = 0 \;. \quad (9.30b)$$

This type of differential equation is dealt with in [B.3].

c) ***Plates in curvilinear coordinates***

Equilibrium conditions

$$\left.\begin{aligned} Q^\alpha\big|_\alpha + p &= 0\,, \\ M^{\alpha\beta}\big|_\beta - Q^\alpha &= 0\,. \end{aligned}\right\} \quad (9.31)$$

Material law – stress resultant–displacement relations

$$M^{\alpha\beta} = -K\,E^{*\alpha\beta\gamma\delta}\,(\,w|_{\gamma\delta} + \alpha_T\, a_{\gamma\delta}\,{}^1\Theta\,) \tag{9.32}$$

with

K plate stiffness,

$a_{\alpha\beta}$, $a^{\alpha\beta}$ components of the metric tensor in the mid–plane of the plate,

${}^1\Theta(\xi^\alpha)$ temperature gradient,

$E^{*\alpha\beta\gamma\delta} = \frac{1-\nu}{2}\,(\,a^{\alpha\gamma}a^{\beta\delta} + a^{\alpha\delta}a^{\beta\gamma}\,) + \nu\, a^{\alpha\beta}a^{\gamma\delta}$ plane elasticity tensor.

Differential equation

$$K\,\Delta\Delta\, w = p - \alpha_T\,(1+\nu)\,K\,\Delta\,{}^1\Theta \tag{9.13}$$

with the LAPLACE operator Δ given by *(2.39)* in terms of the applied curvilinear coordinates.

Energy expressions

– *Cartesian coordinates* ($\partial/\partial x \mathrel{\hat=} (\;)_{,x}$, $\partial/\partial y \mathrel{\hat=} (\;)_{,y}$) from *(6.16a)* [A.9]

$$\left.\begin{aligned}
\Pi_i &= \iint\Big\{\frac{1}{2}K[\,w_{,xx}^2 + w_{,yy}^2 + 2(1-\nu)\,w_{,xy}^2 + 2\nu\, w_{,xx}w_{,yy}\,] + \\
&\qquad + K\alpha_T(1+\nu)(\,w_{,xx} + w_{,yy}\,)\,{}^1\Theta\Big\}\,dx\,dy \\
&\text{or} \\
\Pi_i &= \iint\Big\{\frac{1}{2}K[\,(\,w_{,xx} + w_{,yy}\,)^2 - 2(1-\nu)(\,w_{,xx}w_{,yy} - w_{,xy}^2\,)] + \\
&\qquad + K\alpha_T(1+\nu)(\,w_{,xx} + w_{,yy}\,)\,{}^1\Theta\Big\}\,dx\,dy\,.
\end{aligned}\right\} \tag{9.33}$$

– *Polar coordinates* ($\partial/\partial r \mathrel{\hat=} (\;)_{,r}$, $\partial/\partial\varphi \mathrel{\hat=} (\;)_{,\varphi}$)

$$\left.\begin{aligned}
\Pi_i &= \iint\Big\{\frac{1}{2}K\Big[\,w_{,rr}^2 + 2\nu\,(\,w_{,rr}w_{,\varphi\varphi} + \frac{1}{r}w_{,rr}w_{,r}\,) + \\
&\qquad + (1-\nu)\Big(w_{,r\varphi}^2 - \frac{2}{r}w_{,r\varphi}w_{,\varphi} + \frac{1}{r^2}w_{,\varphi}^2\Big) + \\
&\qquad + \Big(w_{,\varphi\varphi}^2 + \frac{2}{r}w_{,\varphi\varphi}w_{,r} + \frac{1}{r^2}w_{,r}^2\Big)\Big]\Big\}\;r\,d\varphi\,dr\,.
\end{aligned}\right\} \tag{9.34}$$

– *Curvilinear coordinates* $\Big($ with $a = |\,a_{\alpha\beta}\,|\Big)$

$$\Pi_i = \iint\Big(\frac{1}{2}K\,E^{*\alpha\beta\gamma\delta}\,w|_{\alpha\beta}\,w|_{\gamma\delta} + \alpha_T\,K\,a_{\alpha\beta}\,E^{*\alpha\beta\gamma\delta}\,w|_{\gamma\delta}\,{}^1\Theta\Big)\sqrt{a}\;d\xi^1\,d\xi^2\,. \tag{9.35}$$

9.2 Analytical solutions for shear–rigid plates

a) *Cartesian coordinates*

– *Simply supported plate strip* ($\partial/\partial y \equiv 0$, $\partial/\partial x \mathrel{\hat{=}} (\,)_{,x}$)

Differential equation from *(9.13)* $\quad \Delta\Delta w \mathrel{\hat{=}} w_{,xxxx} = \dfrac{p(x)}{K}$.

Solution: $$w = \frac{a^4}{K}\sum_n \frac{p_n}{(n\pi)^4}\sin\frac{n\pi x}{a} \qquad (9.36)$$

for $$p(x) = \sum_n p_n \sin\frac{n\pi x}{a}$$

with $$p_n = \frac{2}{a}\int_0^a p(x)\sin\frac{n\pi x}{a}\,dx \quad , \qquad (n = 1,3,5,\dots)\,.$$

– *Rectangular plate with simply supported boundaries* (dimensions a, b; Fig. 9.3)

Differential equation from *(9.13)* ($\partial/\partial x \mathrel{\hat{=}} (\,)_{,x}$, $\partial/\partial y \mathrel{\hat{=}} w_{,y}$)

$$\Delta\Delta w = w_{,xxxx} + 2\,w_{,xxyy} + w_{,yyyy} = \frac{p(x,y)}{K}\,.$$

Solution : Double series expansion according to NAVIER

$$w(x,y) = \sum_m\sum_n w_{mn}\sin\frac{m\pi x}{a}\sin\frac{n\pi y}{b} \quad , \quad (m,n = 1,2,3,\dots)\,.$$

Load $$p(x,y) = \sum_m\sum_n p_{mn}\sin\frac{m\pi x}{a}\sin\frac{n\pi y}{b}\,. \qquad (9.37)$$

Expansion coefficients $$w_{mn} = \frac{p_{mn}}{K\pi^4\left[\left(\frac{m}{a}\right)^2 + \left(\frac{n}{b}\right)^2\right]^2}\,. \qquad (9.38a)$$

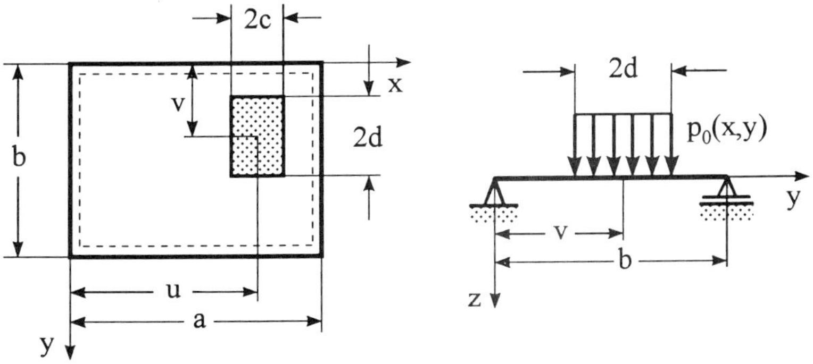

Fig. 9.3: Plate under a uniformly distributed load over a rectangular sub-domain

A plate subjected to a uniformly distributed load p_0 over a rectangular subdomain as shown in Fig. 9.3 will be considered as an example of application.

We first expand the constant load p_0 in the y-direction

$$p(y) = \sum_n p_n \sin\frac{n\pi y}{b}$$

with

$$p_n = \frac{2}{b}\int_0^b p_0 \sin\frac{n\pi y}{b}\,dy = \frac{2}{b}\int_{v-d}^{v+d} p_0 \sin\frac{n\pi y}{b}\,dy =$$

$$= \frac{2}{b}\frac{b}{n\pi}p_0\left(-\cos\frac{n\pi y}{b}\right)\Bigg|_{v-d}^{v+d} =$$

$$= 2\frac{p_0}{n\pi}2\sin\frac{n\pi v}{b}\sin\frac{n\pi d}{b} \qquad (n = 1,2,3,\ldots)\,.$$

Ensuing, this $p(y)$ is expanded in the x-direction

$$p(x,y) = \sum_m\sum_n p_{mn}\sin\frac{m\pi x}{a}\sin\frac{n\pi y}{b} \tag{9.37}$$

with $\quad p_{mn} = \dfrac{2}{a}\displaystyle\int_0^a p_n \sin\frac{m\pi x}{a}\,dx\,.$

The calculation yields

$$p_{mn} = 16\frac{p_0}{m n \pi^2}\sin\frac{m\pi u}{a}\sin\frac{m\pi c}{a}\sin\frac{n\pi v}{b}\sin\frac{n\pi d}{b} \tag{9.38b}$$

$$(m,n = 1,2,3,\ldots)\,.$$

Herewith, we obtain with *(9.38b)*

$$w = \sum_m\sum_n \frac{p_{mn}}{K\pi^4\left[\left(\frac{m}{a}\right)^2+\left(\frac{n}{b}\right)^2\right]^2}\sin\frac{m\pi x}{a}\sin\frac{n\pi y}{b}\,. \tag{9.39}$$

Two special cases:

- Full load $\longrightarrow \quad c = u = a/2 \quad , \quad d = v = b/2$

It follows that: $\quad p_{mn} = 16\dfrac{p_0}{m n \pi^2} \quad$ and

$$w = \frac{16\,p_0}{K\pi^6}\sum_m\sum_n \frac{1}{mn\left[\left(\frac{m}{a}\right)^2+\left(\frac{n}{b}\right)^2\right]^2}\sin\frac{m\pi x}{a}\sin\frac{n\pi y}{b}$$

$$(m,n = 1,3,5,\ldots)\,.$$

Herewith, for a quadratric plate (b = a) the maximum deflection (found at the centre) becomes

$$w_{max} = w\left(\frac{a}{2},\frac{a}{2}\right) = \frac{16\, p_0\, a^4}{K\,\pi^6}\left(\frac{1}{4} - \frac{1}{300} - \frac{1}{300} + \frac{1}{9\cdot 324} - \dots\right).$$

One can discern a fast convergence, particularly as the higher terms have alternating signs.

- Single load F at the point u , v

We extend the expansion coefficients

$$p_{mn} = \frac{4}{a\,b}\,4\,c\,d\,p_0 \sin\frac{m\pi u}{a}\sin\frac{n\pi v}{b}\;\frac{\sin\dfrac{m\pi c}{a}\,\sin\dfrac{n\pi d}{b}}{\dfrac{m\pi c}{a}\;\dfrac{n\pi d}{b}}$$

in such a way that the rectangle can be reduced to a point. With the limiting value

$$\lim_{c\to 0}\frac{\sin\kappa c}{\kappa c} = 1\;,\qquad \lim_{d\to 0}\frac{\sin\mu d}{\mu d} = 1$$

and $\quad \lim\limits_{\substack{c\to 0\\ d\to 0}} 4\,c\,d\,p_0 = F\,,$

we obtain $\quad p_{mn} = \dfrac{4F}{a\,b}\sin\dfrac{m\pi u}{a}\sin\dfrac{n\pi v}{b} \qquad (m, n = 1, 2, 3, \dots)\,.$

– *Plates with two parallel, simply supported boundaries and other boundaries arbitrary*

Differential equation $\longrightarrow$ *(9.13)*.

Solution approach according to LEVY:

$$w(x,y) = \sum_n w_n(y)\sin\frac{n\pi x}{a}\,. \qquad (9.40)$$

Transformation of *(9.13)* into an ordinary differential equation with constant coefficients (d/dy $\widehat{=}$ ( ),y):

$$w_n(y)_{,yyyy} - 2\left(\frac{n\pi}{a}\right)^2 w_n(y)_{,yy} + \left(\frac{n\pi}{a}\right)^4 w_n(y) = \frac{p_n}{K}\,g(y)\,. \qquad (9.41)$$

Homogeneous solution:

$$w_h = \sum_n\Big(A_n\cosh\frac{n\pi y}{a} + B_n\sinh\frac{n\pi y}{a} + C_n\frac{n\pi y}{a}\cosh\frac{n\pi y}{a} + D_n\frac{n\pi y}{a}\sinh\frac{n\pi y}{a}\Big)\sin\frac{n\pi x}{a}\,. \qquad (9.42)$$

– *Plates with arbitrarily supported boundaries*

If the surface load can be considered as a product similar to *(9.37)*, closed form solutions for plates with mixed supports can still be found. We are going to explain the solution approach by the example of an overall clamped plate according to Fig. 9.4.

Since no solution is known for the overall clamped plate which fulfills both the differential equation and the boundary conditions, we separate the problem into the following three subproblems according to usual methods of structural engineering, and obtain the solution using the superposition principle:

"0" The overall simply supported plate under uniformly distributed load with the solution w_0 following from *(9.39)*.

"1" An overall simply supported plate with a yet unknown moment distribution $M_{xx_1} \longrightarrow$ solution w_1.

"2" An overall simply supported plate with a yet unknown moment distribution $M_{yy_2} \longrightarrow$ solution w_2.

From the geometric boundary conditions – the bending angles have to vanish at the supported boundaries in the superposition, i.e.,

$$\left.\begin{aligned} \pm\frac{a}{2},\ y &: \quad w_{0,x} + w_{1,x} + w_{2,x} = 0\,, \\ x,\pm\frac{b}{2} &: \quad w_{0,y} + w_{1,y} + w_{2,y} = 0\,. \end{aligned}\right\} \qquad (9.43)$$

From *(9.43)* follow the previously unknown moment distributions, and from

$$w = w_0 + w_1 + w_2$$

we obtain the general solution.

In case of non-symmetrical support the partial solutions become more complicated and the number of geometric boundary conditions increases. Herewith, a solution in closed form can be only theoretically established at the expense of more work. Thus, the use of an energy method would be more effective [B.8].

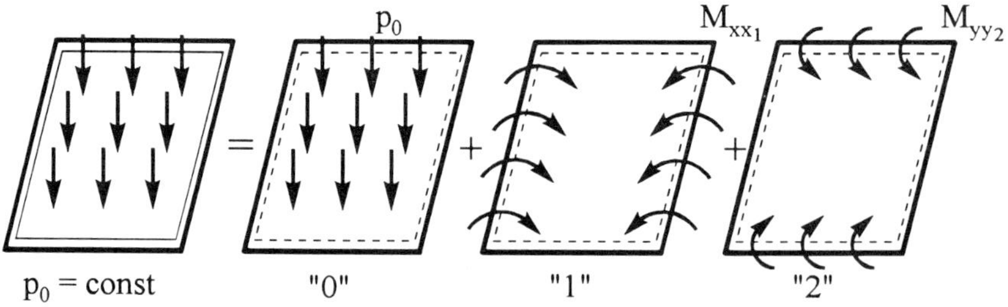

Fig. 9.4: Plate with all edges clamped

- *Orthotropic plates* [B.5, B.9]

From the series expansion $w_h = \sum_m C_m e^{\lambda_m y} \sin \frac{m \pi x}{a}$ follows the characteristic equation

$$K_x \left(\frac{m \pi}{a} \right)^4 - 2 H \lambda_m^2 \left(\frac{m \pi}{a} \right)^2 + K_y \lambda_m^4 = 0 . \tag{9.44}$$

With the four roots

$$\lambda_{m_{1,2,3,4}} = \pm \frac{m \pi}{a} \sqrt{\frac{1}{K_y} \left(H \pm \sqrt{H^2 - K_x K_y} \right)} \tag{9.45}$$

the solution procedure depends on the radicand. We distinguish between three types of plates:

1. Type: $H^2 > K_x K_y$ $\widehat{=}$ plate of high stiffness against torsion.

Since the bending stiffnesses are always positive, the inner root is less than H. Hence, all four roots are real. The solution is valid for K_x (or K_y) ≈ 0. This corresponds to a plate with a negligible bending stiffness in the x-(or y-) direction. This assumption is valid if the plate is of very high stiffness, which may be achieved by means of box-type ribs in the y-(or x-) direction (Fig. 9.5a).

2. Type: $H^2 = K_x K_y$ $\widehat{=}$ approximation according to *(9.35)*.

We find the double roots

$$\lambda_{m_1} = \lambda_{m_2} = \frac{m \pi}{a} \sqrt{\frac{H}{K_y}} = \frac{m \pi}{a} \sqrt[4]{\frac{K_x}{K_y}} \quad , \quad \lambda_{m_3} = \lambda_{m_4} = - \frac{m \pi}{a} \sqrt[4]{\frac{K_x}{K_y}} .$$

This type occurs with a crosswise reinforced concrete plate as shown in Fig. 9.5b .

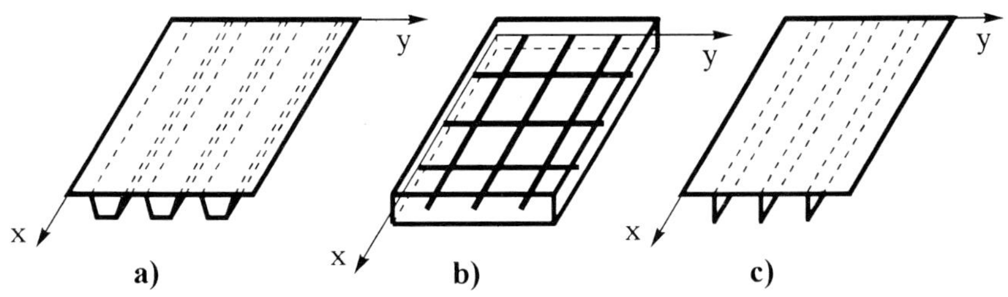

Fig. 9.5: Orthotropic plate profiles

3. Type: $H^2 < K_x K_y \mathrel{\hat{=}}$ plate of low stiffness against torsion.

In this case, the roots are complex conjugate

$$\lambda_{m_{1,2,3,4}} = \pm \frac{m\pi}{a} \sqrt{\frac{1}{K_y}\left(H \pm i \sqrt{K_x K_y - H^2} \right)} .$$

For negligibly low stiffness against torsion $H \approx 0$ we obtain

$$\lambda_{m_{1,2,3,4}} = \pm \frac{m\pi}{a} \sqrt{\pm i \sqrt{\frac{K_x}{K_y}}} = \pm \frac{m\pi}{a} \frac{1}{2} \sqrt{2}\, (1 \pm i) \sqrt[4]{\frac{K_x}{K_y}} .$$

This solution occurs, e.g. in cases of plates stiffened by bending profiles that have very low torsional stiffness (Fig. 9.5c).

b) ***Polar coordinates***

– *Axisymmetrical load case*

The loads and boundary conditions are independent of $\varphi \longrightarrow p = p(r)$, $w = w(r)$.

From *(9.25)* we obtain EULER´s differential equation

$$w_{,rrrr} + \frac{2}{r} w_{,rrr} - \frac{1}{r^2} w_{,rr} + \frac{1}{r^3} w_{,r} = \frac{p(r)}{K} . \tag{9.46}$$

Homogeneous solution:

$$w_h = C_0 + C_1 r^2 + C_2 \ln \frac{r}{a} + C_3 r^2 \ln \frac{r}{a} \tag{9.47}$$

with a suitable reference length a.

– *Non–symmetrical load case*

Load $\quad p(r,\varphi) = g(r) \sum_n p_n \cos n\varphi \quad , \quad (n \text{ integer}) .$ *(9.48)*

Expansion approach for deflection $\quad w(r,\varphi) = \sum_n w_n(r) \cos n\varphi .$ *(9.49)*

Transformation of *(9.25)* into an ordinary differential equation:

$$\left(\frac{d^2}{dr^2} + \frac{1}{r}\frac{d}{dr} - \frac{n^2}{r^2} \right)^2 w_n = \frac{1}{K} p_n g(r) . \tag{9.50}$$

$n = 0$: solution *(9.47)* for the axisymmetrical load case ,

$n = 1$: $\quad w_{h1} = C_1 r + \frac{C_2}{r} + C_3 r^3 + C_4 r \ln \frac{r}{a} ,$ *(9.51a)*

$n \geq 2$: $\quad w_{hn} = C_{1n} r^n + C_{2n} r^{-n} + C_{3n} r^{2+n} + C_{4n} r^{2-n} .$ *(9.51b)*

10 Coupled disk–plate problems

10.1 Isotropic, plane structures with large displacements

– *Basic equations*

In the previous chapters we considered elastic structures with small displacements. This simplifying assumption is not always fulfilled; especially in cases of thin-walled structures subjected to larger compressive loads, the deformations may become large compared with the thickness. The equilibrium conditions must then be formulated for the deformed state of the structure and terms of higher order must be taken into account in the strain-deformation relations. This corresponds to the *geometrical non-linearity*. Here, the material law is considered to be linear. Furthermore, the lemma of mass conservation ($\hat{\varrho}\, d\hat{V} = \varrho\, dV$) is assumed to remain valid as well as equality of the volume forces in the deformed and undeformed state ($\hat{\mathbf{f}} = \mathbf{f}$). The stress-free initial state (LAGRANGE formulation) is taken as a basis. With these assumptions, the equilibrium conditions read as follows [B.1, B.2, B.4]:

$$[(\delta^i_k + v^i|_k)\tau^{jk}]|_j + f^i = 0 . \qquad (10.1)$$

The strain-displacement relations have been introduced in Chapter 4, and we obtain according to *(4.12a)*

$$\gamma_{ij} = \frac{1}{2}(v_i|_j + v_j|_i + v^k|_i v_k|_j) . \qquad (10.2)$$

As the strains are assumed to be very small in comparison with the deformations, non-linear terms can be neglected in the compatibility conditions. The six equations of the material law are adopted in their usual form *(5.5a)* or *(5.6a)*.

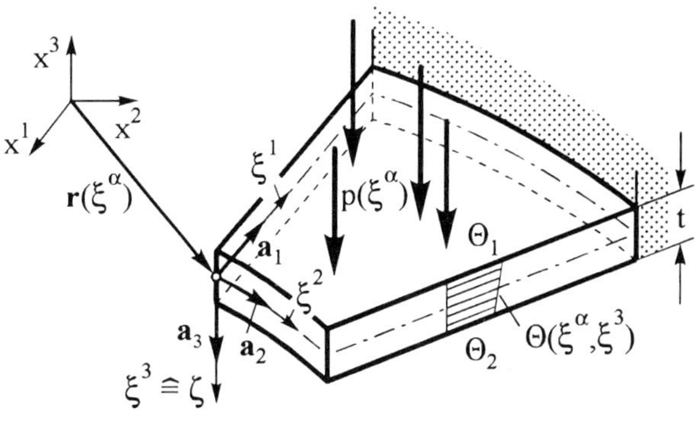

Fig. 10.1: Plane load-bearing structure under temperature and surface loads

In Chapters 8 and 9, disks and plates were considered separately because their loads were assumed to act either in the mid-plane or perpendicular to the mid-plane of the structure. Now, we extend our consideration to the coupled disk-plate problem. Besides, it will be assumed that the plane structure is subjected to an arbitrary temperature field (Fig. 10.1)

$$\Theta(\xi^\alpha,\zeta) = {}^0\Theta(\xi^\alpha) + \zeta\,\Theta^1(\xi^\alpha) \tag{10.3}$$

with $${}^0\Theta(\xi^\alpha) = \frac{1}{2}\left[\Theta_1(\xi^\alpha) + \Theta_2(\xi^\alpha)\right],$$

$${}^1\Theta(\xi^\alpha) = \frac{1}{t}\left[\Theta_1(\xi^\alpha) - \Theta_2(\xi^\alpha)\right]$$

and an external surface load $p(\xi^\alpha)$. At its boundaries, it must satisfy prescribed boundary conditions. In the following, we will restrict ourselves to the shear-rigid plane structure, but assume large deformations.

Treatment of the problem by means of the HELLINGER-REISSNER energy functional:

– Stress resultants – tensors

In addition to the moment tensor $M^{\alpha\beta}$, we introduce the tensor of in-plane forces $N^{\alpha\beta}$ (membrane tensor). According to Fig. 10.1 with $\xi^3 \mathrel{\hat=} \zeta$, they read:

$$N^{\alpha\beta} = \int_{-\frac{t}{2}}^{+\frac{t}{2}} \tau^{\alpha\beta}\,d\zeta \quad , \quad M^{\alpha\beta} = \int_{-\frac{t}{2}}^{+\frac{t}{2}} \tau^{\alpha\beta}\,\zeta\,d\zeta\,. \tag{10.4}$$

– Strain-displacement equations

A plane load-bearing structure is subjected to strains ${}^0\gamma_{\alpha\beta}$ and distortions ${}^1\gamma_{\alpha\beta}$ of the mid-plane. The total strains at an arbitrary point can be superposed from these two parts. According to *(10.2)*, the following strain-displacements relations are valid

$${}^0\gamma_{\alpha\beta} = \frac{1}{2}\left(v_\alpha|_\beta + v_\beta|_\alpha + w_{,\alpha}w_{,\beta}\right). \tag{10.5a}$$

Assuming $v_\alpha \ll w$, only the non-linear term $w_{,\alpha}w_{,\beta}$ is taken into consideration in *(10.5a)*.

The distortion of the mid-plane ${}^1\gamma_{\alpha\beta}$ is obtained from the relations of the shear-rigid plate, where the cross sectional rotations are expressed by the angles of the bending surface $w_{,\alpha}$ (KIRCHHOFF's normal hypothesis). The following relation is valid for the distortion of the shear-rigid plate

$${}^1\gamma_{\alpha\beta} = -\,w|_{\alpha\beta}\,. \tag{10.5b}$$

With ${}^0\gamma_{\alpha\beta}$ and ${}^1\gamma_{\alpha\beta}$ we form the total strain $\gamma_{\alpha\beta}$

$$\gamma_{\alpha\beta} = {}^0\gamma_{\alpha\beta} + \zeta\,{}^1\gamma_{\alpha\beta}\,. \tag{10.5c}$$

– Material law

The material law is used in the form for a *state of plane stress* of the body ($\tau_{33} \approx 0$). With the temperature field *(10.3)* and the strain tensor components *(10.5)* we obtain the following relations for the stress resultants

$$\mathrm{N}^{\alpha\beta} = \mathrm{t}\,\mathrm{E}^{\alpha\beta\gamma\delta}\,(\,{}^0\gamma_{\gamma\delta} - \alpha_{\mathrm{T}}\,\mathrm{a}_{\gamma\delta}\,{}^0\Theta\,)\,, \tag{10.6a}$$

$$\mathrm{M}^{\alpha\beta} = \frac{\mathrm{t}^3}{12}\,\mathrm{E}^{\alpha\beta\gamma\delta}\,(\,{}^1\gamma_{\gamma\delta} - \alpha_{\mathrm{T}}\,\mathrm{a}_{\gamma\delta}\,{}^1\Theta\,)\,, \tag{10.6b}$$

where, after substitution of *(10.5b)*, the equation *(10.6b)* becomes identical with *(9.32)* and with the elasticity tensor $\mathrm{E}^{\alpha\beta\gamma\delta}$ defined in connection with *(9.32)*.

Variational functional

We shall now derive the differential equations and boundary conditions for the coupled disk–plate problem by means of an energy functional without work contributions from volume forces, boundary loads, and boundary displacements [ET 2]. If we substitute the deformation energy $\overline{\mathrm{U}}$ of *(6.16a)* into *(6.29)* for the three–dimensional body, then first follows

$$\Pi_{\mathrm{R}} = \int_{\mathrm{V}} [\,\tau^{\mathrm{ij}}\,\gamma_{\mathrm{ij}} - \frac{1}{2}\,\tau^{\mathrm{ij}}\,(\,\gamma_{\mathrm{ij}} + \alpha_{\mathrm{T}}\,\mathrm{g}_{\mathrm{ij}}\,\Theta\,)\,]\,\mathrm{dV} \tag{10.7a}$$

and when regarding the plane structure as a two–dimensional body

$$\Pi_{\mathrm{R}} = \int_{\mathrm{V}} \Big\{\tau^{\alpha\beta}\,\gamma_{\alpha\beta} - \frac{1}{2}\,\tau^{\alpha\beta}\,(\,\gamma_{\alpha\beta} + \alpha_{\mathrm{T}}\,\mathrm{g}_{\alpha\beta}\,\Theta\,)\Big\}\,\mathrm{dV}\,. \tag{10.7b}$$

In *(10.7b)* we have omitted all terms with τ^{33} because of the thin–walled structure ($\tau^{33} \approx 0$) and with $\gamma_{\alpha 3}$ because of the shear–rigid behaviour ($\gamma_{\alpha 3} \approx 0$) of the plane structure. By introducing into *(10.7b)* the stress resultants according to *(10.4)* and the temperature field according to *(10.3)*, the functional becomes

$$\Pi_{\mathrm{R}} = \int_{\mathrm{A}} \Big\{\mathrm{N}^{\alpha\beta}\,{}^0\gamma_{\alpha\beta} + \mathrm{M}^{\alpha\beta}\,{}^1\gamma_{\alpha\beta} - \frac{1}{2}\Big[\frac{1}{\mathrm{t}}\,\mathrm{D}_{\alpha\beta\gamma\delta}\,\mathrm{N}^{\alpha\beta}\,\mathrm{N}^{\gamma\delta} + \\ + \frac{12}{\mathrm{t}^3}\,\mathrm{D}_{\alpha\beta\gamma\delta}\,\mathrm{M}^{\alpha\beta}\,\mathrm{M}^{\gamma\delta} + 2\,\alpha_{\mathrm{T}}\,\mathrm{N}^{\alpha}_{\alpha}\,{}^0\Theta + 2\,\alpha_{\mathrm{T}}\,\mathrm{M}^{\alpha}_{\alpha}\,{}^1\Theta\Big]\Big\}\,\mathrm{dA}\,. \tag{10.8}$$

With

$$D_{\alpha\beta\gamma\delta} = \frac{1}{E} D^*_{\alpha\beta\gamma\delta} = \frac{1}{E}\Big[\frac{(1+\nu)}{2}(a_{\alpha\gamma} a_{\beta\delta} + a_{\alpha\delta} a_{\beta\gamma}) - \nu\, a_{\alpha\beta} a_{\gamma\delta}\Big]$$

and the plate stiffness K defined by *(9.3)*, the functional takes the form

$$\begin{aligned} \Pi_R = \iint_{\xi^1 \xi^2} \Big\{ & \frac{t}{2} \varepsilon^{\alpha\sigma} \varepsilon^{\beta\tau} \Phi\big|_{\sigma\tau} (v_\alpha\big|_\beta + v_\beta\big|_\alpha + w_{,\alpha} w_{,\beta}) + \\ & + \frac{K}{2} E^{*\alpha\beta\gamma\delta} w\big|_{\alpha\beta} w\big|_{\gamma\delta} + K \alpha_T a_{\alpha\beta} E^{*\alpha\beta\gamma\delta} w\big|_{\gamma\delta} {}^1\Theta - \\ & - \frac{1}{2}\Big[\frac{t}{E} D^{*\alpha\beta\gamma\delta} \Phi\big|_{\alpha\beta} \Phi\big|_{\gamma\delta} + 2 t \alpha_T \Phi\big|_\alpha^{\alpha\, 0} {}^0\Theta \Big]\Big\} \sqrt{a}\, d\xi^1 d\xi^2 \end{aligned} \qquad (10.9a)$$

Abbreviated the functional *(10.9a)* has the form

$$\Pi_R = \int_A F(\xi^\alpha; v_\alpha, v_\alpha\big|_\beta; w, w_{,\alpha}, w\big|_{\alpha\beta}; \Phi\big|_{\alpha\beta})\, dA \ , \qquad (10.9b)$$

where the three functions v_α, w, Φ are unknown.

The external potential Π_e in *(6.18b)* for the work of the surface loads $p(\xi^\alpha)$ is given by

$$\Pi_e = -W = -\int_A p\, w\, dA \ . \qquad (10.10)$$

The total potential is now superposed from *(10.9a)* and *(10.10)*

$$\Pi = \Pi_R + \Pi_e = \Pi_R - W \ . \qquad (10.11)$$

In Cartesian coordinates the total potential of the coupled disk-plate problem is expressed as

$$\begin{aligned} \Pi = \iint \Big\{ & t\Big[\Phi_{,xx}(v_{,y} + \frac{1}{2} w_{,y}^2) + \Phi_{,yy}(u_{,x} + \frac{1}{2} w_{,x}^2) - \\ & - \Phi_{,xy}(u_{,y} + v_{,x} + w_{,x} w_{,y})\Big] + \\ & + \frac{K}{2}\Big[w_{,xx}^2 + w_{,yy}^2 + 2(1-\nu) w_{,xy}^2 + 2\nu\, w_{,xx} w_{,yy}\Big] + \\ & + K\alpha_T(1+\nu)(w_{,xx} + w_{,yy})\, {}^1\Theta - \\ & - \frac{t}{2E}\Big[\Phi_{,xx}^2 + \Phi_{,yy}^2 - 2\nu\, \Phi_{,xx} \Phi_{,yy} - 2(1+\nu)\Phi_{,xy}^2\Big] - \\ & - t\alpha_T(\Phi_{,xx} + \Phi_{,yy})\, {}^0\Theta \Big\}\, dx\, dy - \iint p\, w\, dx\, dy \ . \end{aligned} \qquad (10.12)$$

From the stationarity condition $\delta\Pi = 0$ (see *(6.20)*) in anology with *(6.35)* now, the equilibrium condition *(10.13a)*, the compatibility condition *(10.13b)* and the boundary conditions follow as:

VON KÁRMÁN's differential equations

$$K\Delta\Delta w = p - K\alpha_T(1+\nu)\Delta\,{}^1\Theta + t\,\Diamond^4(w,\Phi)\ , \qquad (10.13a)$$

$$\Delta\Delta\Phi = -E\alpha_T\Delta\,{}^0\Theta - \frac{E}{2}\Diamond^4(w,w) \qquad (10.13b)$$

or in index notation (see [B1, B.2, B.8])

$$K\,w\big|_{\gamma\delta}^{\gamma\delta} = p - K\alpha_T(1+\nu)\,{}^1\Theta\big|_\gamma^\gamma + t\,\varepsilon^{\alpha\mu}\varepsilon^{\beta\nu}\,w\big|_{\mu\nu}\,\Phi\big|_{\alpha\beta}\ , \qquad (10.14a)$$

$$\Phi\big|_{\gamma\delta}^{\gamma\delta} = -E\alpha_T\,{}^0\Theta\big|_\gamma^\gamma - \frac{E}{2}\varepsilon^{\alpha\mu}\varepsilon^{\beta\nu}\,w\big|_{\mu\nu}\,w\big|_{\alpha\beta}\ . \qquad (10.14b)$$

(10.13a) $\widehat{=}$ equilibrium condition of the forces in the z-direction in cases of large deformations,

(10.13b) $\widehat{=}$ compatibility condition of the coupled disk-plate problem.

The operator $\Diamond$ in *(10.13)* is defined in Cartesian coordinates by

$$\Diamond^4(f,g) = f_{,xx}\,g_{,yy} - 2f_{,xy}\,g_{,xy} + f_{,yy}\,g_{,xx}\ . \qquad (10.15)$$

Boundary conditions for boundaries y = const or x = const:

- Simply supported boundary

$$w = 0\ ,\quad M_{xx} = 0 \quad \text{or} \quad w = 0\ ,\quad M_{yy} = 0\ . \qquad (10.16a)$$

- Clamped boundary

$$w = 0\ ,\quad w_{,x} = 0 \quad \text{or} \quad w = 0\ ,\quad w_{,y} = 0\ . \qquad (10.16b)$$

- Free boundary

$$M_{xx} = 0\ ,\quad N_{xx}w_{,x} + N_{xy}w_{,y} + \overline{Q}_x = 0$$

$$\text{or}\quad M_{yy} = 0\ ,\quad N_{xy}w_{,x} + N_{yy}w_{,y} + \overline{Q}_y = 0\ . \qquad (10.16c)$$

Note: Due to the equilibrium considerations for the deformed element, the transverse force conditions contain additional contributions from in-plane compressive and shear forces in *(10.16c)*.

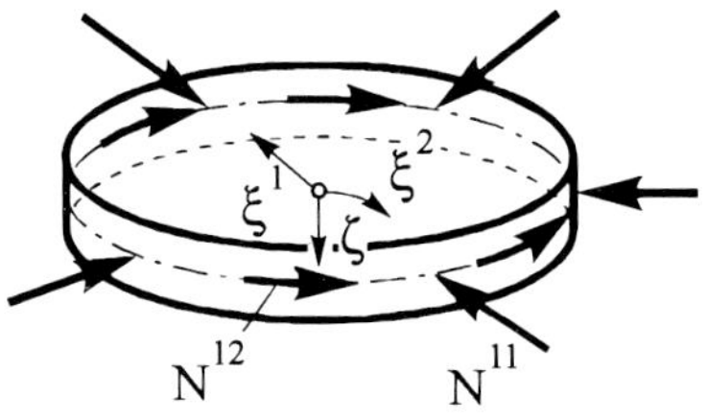

Fig. 10.2: Plate under in – plane compressive and shear forces

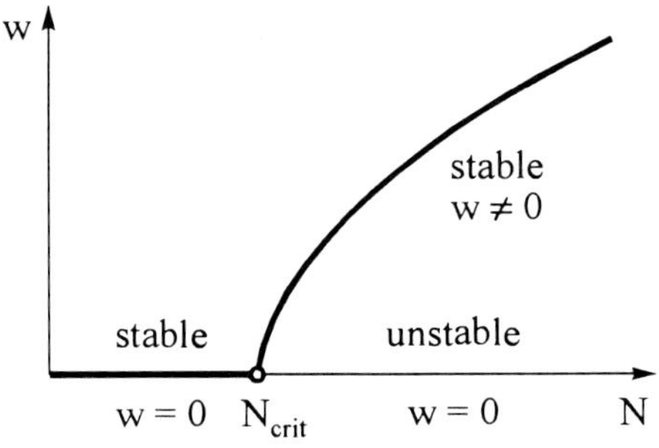

Fig. 10.3: Characteristics of a bifurcation problem

Special case: Basic equation for plate buckling

For $p = 0$, ${}^1\Theta = 0$ and the operator fully written, the differential equation *(10.13a)* reads

$$K \Delta\Delta w = t (\Phi_{,yy} w_{,xx} + \Phi_{,xx} w_{,yy} - 2 \Phi_{,xy} w_{,xy}) \quad . \tag{10.17}$$

If we introduce, by means of *(8.2)*, the in – plane forces $N_x = t\,\sigma_{xx}$ and $N_y = t\,\sigma_{yy}$ as well as the shearing force $N_{xy} = t\,\tau_{xy}$ for the derivatives of the stress function, and if we take the compressive forces to be positive, we obtain the following differential equation for plate buckling:

$$\boxed{K \Delta\Delta w + N_x (x,y)\, w_{,xx} + N_y (x,y)\, w_{,yy} + 2 N_{xy} (x,y)\, w_{,xy} = 0} \quad . \tag{10.18}$$

The solution of this equation leads to a bifurcation at a critical load (Fig. 10.3)

10.2 Load–bearing structures made of composite materials

The use of structures made of composite materials will be steadily increasing because of the possibility of *tailoring* their characteristics. Thus, very demanding technical requirements can be specified for composites, which cannot be achieved with conventional single – component materials.

Our main interest here will be directed towards composite materials with glass and carbon fibres. It is characteristic for a composite material (Fig. 10.4) that the fibre components of a single layer (lamina) are all oriented in the same direction and embedded in a matrix material. We call such a layer a unidirectional layer or, in short, a UD – layer. Characteristic parameters of a UD–layer (Fig. 10.4) are

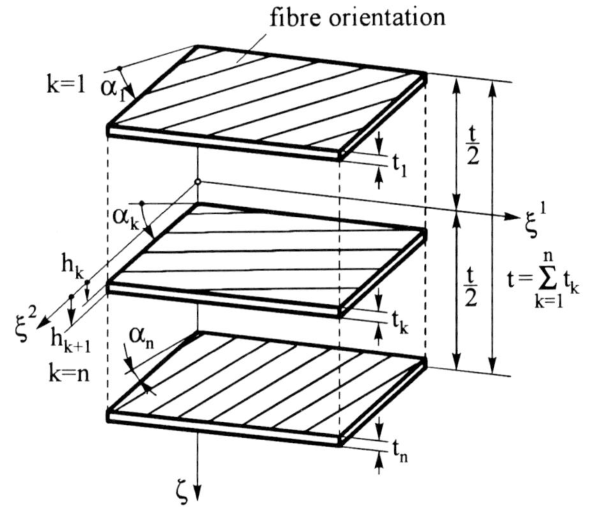

Fig. 10.4: Multilayer composite consisting of stacked single layers

- layer thicknesses t_k ,
- fibre angles α_k ,
- volume percentage of the fibres φ_F

and all material data for matrix and fibre materials.

Further to the orthotropic plate (Ch. 9.1) we shall now consider an anisotropic, plane structure made up as a laminate consisting of several layers (Fig. 10.5). Here, strains ${}^0\gamma_{\alpha\beta}$ and distortions ${}^1\gamma_{\alpha\beta}$ are treated together as discussed in 10.1.

Material law – stress resultant–strain relations

For a plane structure made up of several layers, we assume a linear stress–strain behaviour as was done for the previous isotropic plane structures. In case of a composite structure, however, the stress curve exhibits certain discontinuities at the boundaries between the single layers; here, the stress re-

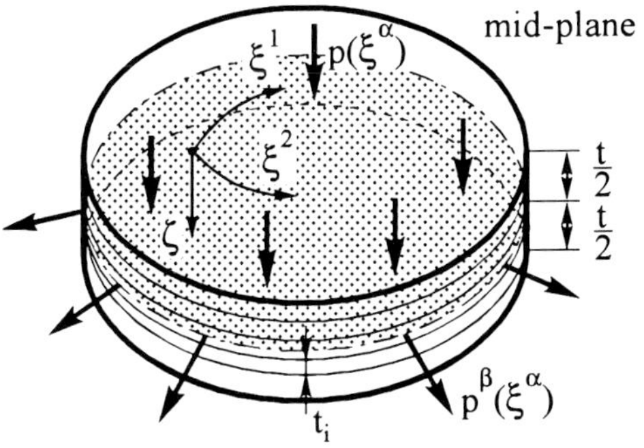

Fig. 10.5: Plate made of several layers

sultants in the single layers remain constant. For the laminate itself they depend on the thickness coordinate ζ, because the components of the tensor of elasticity differ from one layer to another. The stress resultants of the laminate follow from the equilibrium conditions by means of summation of the stress resultants of all single layers [B.11]:

$$\left.\begin{aligned} N^{\alpha\beta} &= \sum_k {}_kN^{\alpha\beta} = A^{\alpha\beta\mu\nu}\,{}^0\gamma_{\mu\nu} + B^{\alpha\beta\mu\nu}\,{}^1\gamma_{\mu\nu} - N_\Theta^{\alpha\beta}\,, \\ M^{\alpha\beta} &= \sum_k {}_kM^{\alpha\beta} = B^{\alpha\beta\mu\nu}\,{}^0\gamma_{\mu\nu} + K^{\alpha\beta\mu\nu}\,{}^1\gamma_{\mu\nu} - M_\Theta^{\alpha\beta}\,, \\ Q^{\alpha} &= \sum_k {}_kQ^{\alpha} = S^{\alpha3\beta3}\,\gamma_\beta \end{aligned}\right\} \qquad (10.19)$$

with

$$\left.\begin{aligned} A^{\alpha\beta\mu\nu} &= \sum_k {}_kA^{\alpha\beta\mu\nu}\,, \quad B^{\alpha\beta\mu\nu} = \sum_k {}_kB^{\alpha\beta\mu\nu}\,, \\ K^{\alpha\beta\mu\nu} &= \sum_k {}_kK^{\alpha\beta\mu\nu}\,, \quad S^{\alpha3\beta3} = \sum_k {}_kS^{\alpha3\beta3}\,, \\ N_\Theta^{\alpha\beta} &= \sum_k \left({}_kA^{\alpha\beta\mu\nu}\,{}_k^0\Theta_{\mu\nu} + {}_kB^{\alpha\beta\mu\nu}\,{}_k^1\Theta_{\mu\nu}\right), \\ M_\Theta^{\alpha\beta} &= \sum_k \left({}_kB^{\alpha\beta\mu\nu}\,{}_k^0\Theta_{\mu\nu} + {}_kK^{\alpha\beta\mu\nu}\,{}_k^1\Theta_{\mu\nu}\right). \end{aligned}\right\} \qquad (10.20)$$

For a physical interpretation of the relations between stress resultants and strains we write *(10.19)* with *(10.20)* in an appropriate symbolic notation:

$$\begin{bmatrix} \mathbf{N} \\ \mathbf{M} \\ \mathbf{Q} \end{bmatrix} = \begin{bmatrix} \mathbf{A} & \mathbf{B} & \mathbf{0} \\ \mathbf{B} & \mathbf{K} & \mathbf{0} \\ \mathbf{0} & \mathbf{0} & \mathbf{S} \end{bmatrix} \begin{bmatrix} {}^0\boldsymbol{\gamma} \\ {}^1\boldsymbol{\gamma} \\ \boldsymbol{\gamma} \end{bmatrix} - \begin{bmatrix} \mathbf{N}_\Theta \\ \mathbf{M}_\Theta \\ \mathbf{0} \end{bmatrix} \qquad (10.21)$$

with

$$\mathbf{A} = \begin{bmatrix} A_{11} & A_{12} & A_{13} \\ & A_{22} & A_{23} \\ \text{sym.} & & A_{33} \end{bmatrix} \qquad \text{matrix of membrane stiffnesses } (A_{ij} = A_{ji}),$$

$$\mathbf{K} = \begin{bmatrix} K_{11} & K_{12} & K_{13} \\ & K_{22} & K_{23} \\ \text{sym.} & & K_{33} \end{bmatrix} \qquad \text{matrix of bending stiffnesses } (K_{ij} = K_{ji}),$$

$$\mathbf{B} = \begin{bmatrix} B_{11} & B_{12} & B_{13} \\ & B_{22} & B_{23} \\ \text{sym.} & & B_{33} \end{bmatrix} \qquad \text{matrix of couple stiffnesses } (B_{ij} = B_{ji}),$$

$$\mathbf{S} = \begin{bmatrix} S_{11} & S_{12} \\ \text{sym.} & S_{22} \end{bmatrix}$$ matrix of shear stiffnesses $(S_{ij} = S_{ji})$,

$\mathbf{N}^T = [N_{11}, N_{22}, N_{12}]$ vector of in-plane disk (membrane) forces,

$\mathbf{M}^T = [M_{11}, M_{22}, M_{12}]$ vector of plate moments,

$\mathbf{Q}^T = [Q_1, Q_2]$ vector of shear forces,

${}^0\boldsymbol{\gamma}^T = [\varepsilon_{11}, \varepsilon_{22}, \varepsilon_{12}]$ vector of strains,

${}^1\boldsymbol{\gamma}^T = [\kappa_{11}, \kappa_{22}, \kappa_{12}]$ vector of distortions,

$\boldsymbol{\gamma}^T = [\gamma_1, \gamma_2]$ vector of shear deformations.

For calculation of the matrix components we need the components of the elasticity matrix $\mathbf{E}$ for a single layer, which we obtain from a transformation according to *(5.21)* presented in matrix notation

$$_k\mathbf{E} = {_k\mathbf{T}}\,{_k\mathbf{E}'}\,{_k\mathbf{T}^T} \qquad (10.22a)$$

with

$$_k\mathbf{T} = \begin{bmatrix} \cos^2\alpha_k & \sin^2\alpha_k & \sin 2\alpha_k \\ \sin^2\alpha_k & \cos^2\alpha_k & -\sin 2\alpha_k \\ -\frac{1}{2}\sin 2\alpha_k & \frac{1}{2}\sin 2\alpha_k & \cos 2\alpha_k \end{bmatrix}, \qquad (10.22b)$$

$$_k\mathbf{E}' = \begin{bmatrix} \dfrac{E_{1'}}{1-\nu_{1'2'}\nu_{2'1'}} & \dfrac{\nu_{2'1'}E_{1'}}{1-\nu_{1'2'}\nu_{2'1'}} & 0 \\ \dfrac{\nu_{1'2'}E_{2'}}{1-\nu_{1'2'}\nu_{2'1'}} & \dfrac{E_{2'}}{1-\nu_{1'2'}\nu_{2'1'}} & 0 \\ 0 & 0 & G_{1'2'} \end{bmatrix}$$ Elasticity matrix of a UD-layer *(5.20)*.

The material parameters can be determined by means of the relations by TSAI and HAHN [B.10, B.11].

As *(10.21)* shows, disk and plate actions occur coupled in a plane structure made of composite material. In addition, as a result of the transformation, the single components of the stiffness matrix depend on the fibre angle α_k. The components of the submatrices in *(10.21)* therefore are based on the laminate design and the fibre orientation.

Strain–displacement equations according to *(10.5c)*

$$\gamma_{\alpha\beta} = \overset{0}{\gamma}_{\alpha\beta} + \zeta\,\overset{1}{\gamma}_{\alpha\beta} \tag{10.23}$$

$$\left.\begin{aligned} &\text{with the strains} && \overset{0}{\gamma}_{\alpha\beta} = \frac{1}{2}\left(\mathrm{v}_\alpha|_\beta + \mathrm{v}_\beta|_\alpha \right) \\ &\text{and the distortions} && \overset{1}{\gamma}_{\alpha\beta} = \frac{1}{2}\left(\psi_\alpha|_\beta + \psi_\beta|_\alpha \right) . \end{aligned}\right\} \tag{10.24}$$

Shear deformation

$$\left.\begin{aligned} &\overset{0}{\gamma}_{\alpha 3} = \frac{1}{2}\gamma_\alpha \\ &\text{with} \quad \gamma_\alpha = \psi_\alpha + \mathrm{w}|_\alpha \end{aligned}\right. \tag{10.25}$$

and v_α in–plane displacements of the mid–surface ,

ψ_α bending angles ,

w displacement perpendicular to the mid–surface .

Equilibrium conditions for the *undeformed* state

From *(3.28a)* $$\mathrm{N}^{\alpha\beta}|_\alpha + \mathrm{p}^\beta = 0 \;, \quad (t\,\tau^{\alpha\beta} \mathrel{\widehat{=}} \mathrm{N}^{\alpha\beta}) \tag{10.26}$$

From *(9.31)* $$\left.\begin{aligned} \mathrm{Q}^\alpha|_\alpha + \mathrm{p} &= 0 \;, \\ \mathrm{M}^{\alpha\beta}|_\beta - \mathrm{Q}^\alpha &= 0 \;. \end{aligned}\right\} \tag{10.27}$$

In *(10.21)*, *(10.23)* and *(10.25)* the strains are expressed by means of deformations. Substituting these relations into *(10.26)* and *(10.27)* then leads to a system of five coupled differential equations for the unknown deformations v_α, w and for the angles ψ_α of rotations of the cross section [B.7, B.10, B.11].

B.2 Exercises

Exercise B-8-1:

A simply supported disk (length l, height h, thickness t) is subjected to a constant load q per unit length as shown Fig. B-1.

a) Set up an approximate AIRY's stress function in form of a power series.

b) Determine the stresses in the disk from the stress function.

c) Check that the *Equilibrium at Large* is fulfilled.

d) Calculate and compare the displacements of the disk to those of a BERNOULLI - beam ($h \ll l$).

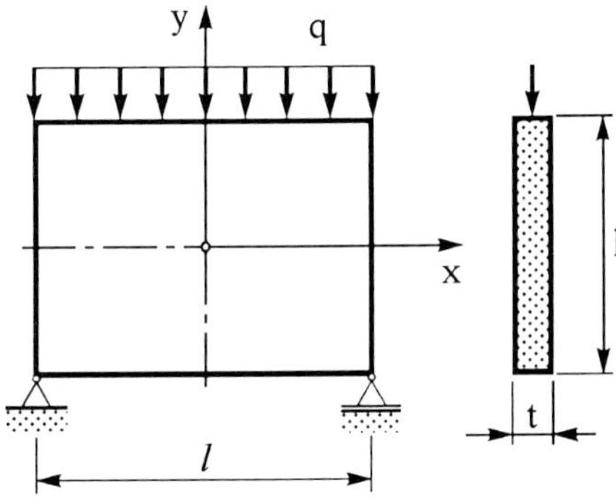

Fig. B-1: Simply supported rectangular disk under constant load q

Solution:

a) Based on *(8.11)* we consider the following power series assumption for the AIRY stress function :

$$\left.\begin{aligned}
\Phi &= a_{11}\,x\,y && \text{– constant shear}\\
&+ a_{20}\,x^2 && \text{– constant compression in } y\text{– direction}\\
&+ a_{03}\,y^3 && \text{– pure bending moment } M_z\\
&+ a_{21}\,x^2\,y && \text{– } \sigma_{yy} \text{ linear in y and } \tau_{xy} \text{ linear in x}\\
&+ a_{23}\,x^2\,y^3 && \text{– } \sigma_{xx}(x^2, y),\ \sigma_{yy}(y^3),\ \tau_{xy}(x, y^2)\\
&+ a_{05}\,y^5 && \text{– } \sigma_{xx}(y^3) \quad .
\end{aligned}\right\} \qquad (1)$$

The single a_{ik} are free coefficients.

b) The stresses are calculated by means of *(8.2)* and (1):

$$\sigma_{xx} = \Phi_{,yy} = 6\,a_{03}\,y + 6\,a_{23}\,x^2\,y + 20\,a_{05}\,y^3 , \qquad (2a)$$

$$\sigma_{yy} = \Phi_{,xx} = 2\,a_{20} + 2\,a_{21}\,y + 2\,a_{23}\,y^3 , \qquad (2b)$$

$$\tau_{xy} = -\Phi_{,xy} = -a_{11} - 2\,a_{21}\,x - 6\,a_{23}\,x\,y^2 . \qquad (2c)$$

The *boundary conditions* for the disk are formulated as follows:

$$\sigma_{yy}(x, \tfrac{h}{2}) = -\frac{q}{t}: \quad 2\,a_{20} + a_{21}\,h + \frac{1}{4}\,a_{23}\,h^3 = -\frac{q}{t} , \qquad (3a)$$

$$\sigma_{yy}(x, -\tfrac{h}{2}) = 0: \quad 2\,a_{20} - a_{21}\,h - \frac{1}{4}\,a_{23}\,h^3 = 0 , \qquad (3b)$$

$$\left.\begin{aligned} \sigma_{xx}(-\tfrac{l}{2}, y) &= 0: \\ \sigma_{xx}(\tfrac{l}{2}, y) &= 0: \end{aligned}\right\} \quad 6\,a_{03}\,y + \frac{3}{2}\,a_{23}\,l^2\,y + 20\,a_{05}\,y^3 = 0 , \qquad (3c)$$

$$\left.\begin{aligned} \tau_{xy}(x, -\tfrac{h}{2}) &= 0: \\ \tau_{xy}(x, \tfrac{h}{2}) &= 0: \end{aligned}\right\} \quad -a_{11} - 2\,a_{21}\,x - \frac{3}{2}\,a_{23}\,h^2\,x = 0 . \qquad (3d)$$

Substitution of (1) into the bipotential equation *(8.1)* yields

$$\Delta\Delta\,\Phi = \Phi_{,xxxx} + 2\,\Phi_{,xxyy} + \Phi_{,yyyy} = 24\,a_{23}\,y + 120\,a_{05}\,y \stackrel{!}{=} 0 \qquad (4)$$

$$\longrightarrow \quad a_{23} = -5\,a_{05} .$$

From (3a) + (3b) $\longrightarrow \quad a_{20} = -\dfrac{q}{4t} .$

From (3d) $\longrightarrow \quad a_{11} = 0 \quad \text{and} \quad a_{21} = -\dfrac{3}{4}\,a_{23}\,h^2 .$

Boundary condition (3d) must be fulfilled for all x.

From (3a) – (3b) + (3d) $\longrightarrow \quad a_{23} = \dfrac{q}{h^3 t} .$

From (3a) $\longrightarrow \quad a_{21} = -\dfrac{3}{4}\dfrac{q}{h\,t} .$

From (4) $\longrightarrow \quad a_{05} = -\dfrac{1}{5}\dfrac{q}{h^3 t} .$

Since (3c) cannot be fulfilled for all y, it is demanded approximately that the resultant force F_x vanishes:

$$F_x = \int_{-h/2}^{h/2} \sigma_{xx}(\pm\tfrac{l}{2}, y)\,t\,dy \stackrel{!}{=} 0 . \qquad (5)$$

This condition is fulfilled for all arbitrary a_{03}, a_{23} and a_{05}.

In addition it is therefore demanded that the moment $M_z(\sigma_{xx})$ of the stresses at the boundaries vanishes:

$$M_z = \int_{-h/2}^{h/2} \sigma_{xx}(\pm\frac{l}{2}, y)\, t\, y\, dy \stackrel{!}{=} 0 \tag{6}$$

$$\Longrightarrow \quad \int_{-h/2}^{h/2} (6\, a_{03}\, y^2 + \frac{3}{2} a_{23}\, l^2 y^2 + 20\, a_{05}\, y^4)\, t\, dy =$$

$$= t\,(2\, a_{03}\, y^3 + \frac{1}{2} a_{23}\, l^2 y^3 + 4\, a_{05}\, y^5)\Big|_{-h/2}^{h/2} =$$

$$= t\left[\frac{1}{2} a_{03}\, h^3 + \frac{1}{8} a_{23}\, l^2 h^3 + \frac{1}{4} a_{05}\, h^5\right] = 0$$

$$\Longrightarrow \quad a_{03} = \frac{q}{h^3 t}\left(\frac{h^2}{10} - \frac{l^2}{4}\right).$$

Hence, in dimensionless notation the stresses become

$$\bar{\sigma}_{xx} = \frac{\sigma_{xx}}{q/t} = 6\frac{l^2}{h^2}\left(\frac{h^2}{10\, l^2} - \frac{1}{4} + \frac{x^2}{l^2}\right)\frac{y}{h} - 4\left(\frac{y}{h}\right)^3, \tag{7a}$$

$$\bar{\sigma}_{yy} = \frac{\sigma_{yy}}{q/t} = 2\left(\frac{y}{h}\right)^3 - \frac{3}{2}\frac{y}{h} - \frac{1}{2}, \tag{7b}$$

$$\bar{\tau}_{xy} = \frac{\tau_{xy}}{q/t} = \frac{3}{2}\frac{x}{h} - 6\frac{x}{h}\left(\frac{y}{h}\right)^2 = \frac{3}{2}\frac{l}{h}\frac{x}{l} - 6\frac{l}{h}\frac{x}{l}\left(\frac{y}{h}\right)^2. \tag{7c}$$

Fig. B-2 shows the approximate stress distribution for some cuts of a disk with $l/h = 1$ and $\xi = x/l$, $\eta = y/h$.

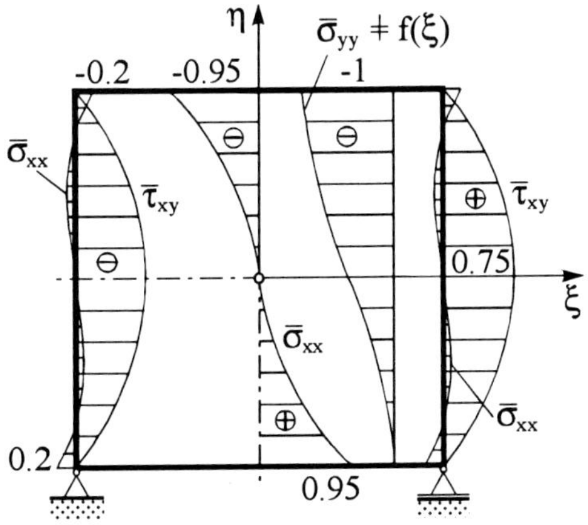

Fig. B-2: Stresses in the disk with $l/h = 1$

c) Checking the *Equilibrium at Large*

The resultants of the shear stresses at the supports must be in equilibrium with the load q.

$$t\int_{-h/2}^{h/2}\tau_{xy}\left(\frac{l}{2},y\right)dy - t\int_{-h/2}^{h/2}\tau_{xy}\left(-\frac{l}{2},y\right)dy = 2t\int_{-h/2}^{h/2}\tau_{xy}\left(\frac{l}{2},y\right)dy \overset{!}{=} q\,l$$

$$\Longrightarrow\quad 2t\int_{-h/2}^{h/2}\frac{q}{h^3 t}\left[\frac{3}{4}h^2 l - 3\,l\,y^2\right]dy = \frac{q}{h^3}\left(\frac{3}{2}h^2 l\,y - 2\,l\,y^3\right)\Bigg|_{-h/2}^{h/2} =$$

$$= \frac{q}{h^3}\left(\frac{3}{2}h^3 l - \frac{1}{2}h^3 l\right) = q\,l\,.$$

d) The strains can be calculated from the strain–displacement relations *(8.3)* and from the material law *(8.4)* for the state of plane stress:

$$\varepsilon_{xx} = u_{,x} = \frac{1}{E}\left(\sigma_{xx} - \nu\,\sigma_{yy}\right), \tag{8a}$$

$$\varepsilon_{yy} = v_{,y} = \frac{1}{E}\left(\sigma_{yy} - \nu\,\sigma_{xx}\right), \tag{8b}$$

$$\gamma_{xy} = u_{,y} + v_{,x} = \frac{1}{G}\tau_{xy}\,. \tag{8c}$$

Substitution of σ_{xx}, σ_{yy}, τ_{xy} from (7a,b,c) into (8a,b,c) leads to:

$$u_{,x} = \frac{q}{Eh^3 t}\left[6\left(\frac{h^2}{10} - \frac{l^2}{4} + x^2\right)y - 4y^3 - \nu\left(2y^3 - \frac{3}{2}h^2 y - \frac{h^3}{2}\right)\right] \longrightarrow$$

$$\begin{aligned} u = \frac{q}{Eh^3 t}\Bigg\{\Bigg[6\left(\frac{h^2}{10} - \frac{l^2}{4} + \frac{x^2}{3}\right)y - 4y^3 - \\ - \nu\left(2y^3 - \frac{3}{2}h^2 y - \frac{h^3}{2}\right)\Bigg]x + f(y)\Bigg\}, \end{aligned} \tag{9a}$$

$$v_{,y} = \frac{q}{Eh^3 t}\left\{2y^3 - \frac{3}{2}h^2 y - \frac{h^3}{2} - \nu\left[6\left(\frac{h^2}{10} - \frac{l^2}{4} + x^2\right)y - 4y^3\right]\right\} \longrightarrow$$

$$\begin{aligned} v = \frac{q}{Eh^3 t}\Bigg\{\frac{1}{2}y^4 - \frac{3}{4}h^2 y^2 - \frac{h^3}{2}y - \\ - \nu\left[3\left(\frac{h^2}{10} - \frac{l^2}{4} + x^2\right)y^2 - y^4\right] + g(x)\Bigg\}. \end{aligned} \tag{9b}$$

We now form the derivative of u with respect to y, and of v with respect to x:

$$u_{,y} = \frac{q}{Eh^3 t}\left\{\left[6\left(\frac{h^2}{10} - \frac{l^2}{4} + \frac{x^2}{3}\right) - 12y^2 - \nu\left(6y^2 - \frac{3}{2}h^2\right)\right]x + \frac{df(y)}{dy}\right\},$$

$$v_{,x} = \frac{q}{Eh^3 t}\left[-6\,\nu\,x\,y^2 + \frac{dg(x)}{dx}\right].$$

From (8c) follows:

$$\gamma_{xy} = \frac{q}{Eh^3 t}\left[6\left(\frac{h^2}{10} - \frac{l^2}{4} + \frac{x^2}{3}\right) - 12y^2 x - \nu\left(6y^2 - \frac{3}{2}h^2\right)x + \frac{df(y)}{dy} - \right.$$

$$\left. - 6\nu x y^2 + \frac{dg(x)}{dx}\right] \stackrel{!}{=} \frac{2(1+\nu)}{E}\frac{q}{h^3 t}\left[\frac{3}{2}h^2 x - 6 x y^2\right]. \qquad (10)$$

Collection of terms reduces the equation to the following:

$$2x^3 - \frac{12}{5}h^2 x - \frac{3}{2}\nu h^2 x - \frac{3}{2}l^2 x + \frac{dg(x)}{dx} + \frac{df(y)}{dy} = 0. \qquad (11)$$

Relation (11) can only be fulfilled if the term depending on y equals a constant:

$$\frac{df(y)}{dy} = C_1 \quad \Longrightarrow \quad f(y) = C_1 y + C_2 .$$

The same is valid for

$$\frac{dg(x)}{dx} = -2x^3 + \frac{12}{5}h^2 x + \frac{3}{2}\nu h^2 x + \frac{3}{2}l^2 x - C_1$$

$$\Longrightarrow \quad g(x) = -\frac{1}{2}x^4 + \frac{6}{5}h^2 x^2 + \frac{3}{4}\nu h^2 x^2 + \frac{3}{4}l^2 x^2 - C_1 x + C_3 .$$

If we substitute functions f(y) and g(x) into (9a,b), we obtain the following displacements:

$$u = \frac{q}{Eh^3 t}\left\{\left[6\left(\frac{h^2}{10} - \frac{l^2}{4} + \frac{x^2}{3}\right)y - 4y^3 - \right.\right.$$

$$\left.\left. - \nu\left(2y^3 - \frac{3}{2}h^2 y - \frac{h^3}{2}\right)\right]x + C_1 y + C_2\right\}, \qquad (12a)$$

$$v = \frac{q}{Eh^3 t}\left\{\frac{1}{2}y^4 - \frac{3}{4}h^2 y^2 - \frac{h^3}{2}y - \nu\left[3\left(\frac{h^2}{10} - \frac{l^2}{4} + x^2\right)y^2 - y^4\right] - \right.$$

$$\left. - \frac{1}{2}x^4 + \frac{6}{5}h^2 x^2 + \frac{3}{4}\nu h^2 x^2 + \frac{3}{4}l^2 x^2 - C_1 x + C_3\right\}. \qquad (12b)$$

The following boundary conditions and conditions of symmetry are employed for determining the constants C_i (i = 1, 2, 3):

$$u(0, y) = 0, \qquad (13a)$$

$$v\left(\frac{l}{2}, -\frac{h}{2}\right) = 0, \qquad (13b)$$

$$v\left(-\frac{l}{2}, -\frac{h}{2}\right) = 0. \qquad (13c)$$

From (13b,c) $\longrightarrow$ $C_1 = 0$.

From (13a) $\longrightarrow$ $C_2 = 0$.

From (13c) $\longrightarrow$ $C_3 = -h^4\left(\frac{3}{32} + \nu\frac{1}{80}\right) - \frac{5}{32}l^4 - h^2 l^2\left(\frac{3}{10} + \frac{3}{16}\nu\right).$

Thus, the displacements become

$$u(x,y) = \frac{q}{Eh^3 t}\left[6\left(\frac{h^2}{10} - \frac{l^2}{4} + x^2\right)y - 4y^3 - \nu\left(2y^3 - \frac{3}{2}h^2 y - \frac{h^3}{2}\right)x\right], \quad (14a)$$

$$\begin{aligned} v(x,y) = \frac{q}{Eh^3 t}\Bigg\{ & \frac{1}{2}y^4 - \frac{3}{4}h^2 y^2 - \frac{h^3}{2}y - \nu\left[3y^2\left(\frac{h^2}{10} - \frac{l^2}{4} + x^2\right) - y^4\right] - \\ & -\frac{1}{2}x^4 + \left(\frac{6}{5} + \frac{3}{4}\nu\right)h^2 x^2 + \frac{3}{4}l^2 x^2 - h^4\left(\frac{3}{32} + \frac{1}{80}\nu\right) - \\ & -\frac{5}{32}l^4 - h^2 l^2\left(\frac{3}{10} + \frac{3}{16}\nu\right)\Bigg\}. \end{aligned} \quad (14b)$$

For comparison with standard beam theory we transform (7a) as follows :

$$\sigma_{xx} = \frac{q}{h^3 t}\left[\underline{6\left(x^2 - \frac{l^2}{4}\right)y} + \frac{1}{20}\left(3h^2 - 20y^2\right)y\right],$$

where the underlined term corresponds to the bending stress σ_{xx} in the corresponding BERNOULLI-beam. The second term is small for $h/l \ll 1$.

The displacement function $v(x,y)$ is structured accordingly :

$$v(x,y) = \frac{q}{Eh^3 t}\left(-\frac{1}{2}x^4 + \frac{3}{4}x^2 l^2 - \frac{5}{32}l^4 + \ldots\ldots\right).$$

The above terms are identical with the deflection function of the corresponding BERNOULLI-beam; for $h/l \ll 1$, all further terms can be neglected.

Exercise B-8-2:

A circular, annular disk as shown in Fig. B-3 is assumed to be supported frictionless at the outer boundary and is subjected to a constant, axisymmetric temperature distribution over the thickness t

$$^0\Theta = \Theta_a\left(\frac{r}{a}\right)^4.$$

Determine the stresses and the radial displacement function of the disk by means of the displacement potential Ψ.

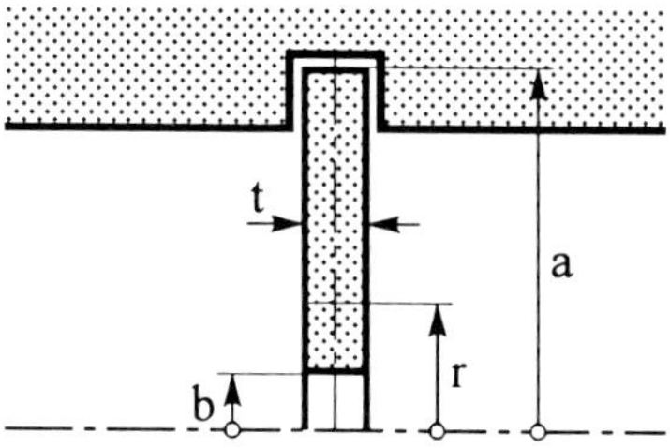

Fig. B-3 : Circular, annular disk subjected to a stationary temperature field

Solution:

We proceed from POISSON's differential equation *(7.11)*:

$$\Delta \Psi = (1+\nu)\alpha_T {}^0\Theta(r)\,.$$

Due to the axisymmetric shape, boundary conditions and temperature distribution of the disk, the derivatives with respect to φ vanish. After multiplication by r^2 the partial differential equation is transformed into an ordinary differential equation of second order of EULER type with variable coefficients,

$$r^2\Psi_{,rr} + r\Psi_{,r} = (1+\nu)\alpha_T\Theta_a\frac{r^6}{a^4}\,. \tag{1}$$

The homogeneous solution follows with the assumption $\Psi_h = r^\lambda$ as

$$\Psi_h = A_1 \ln\frac{r}{a} + A_2\,. \tag{2a}$$

After choosing an assumption for the right-hand-side type, the particular solution reads

$$\Psi_p = \frac{1+\nu}{36}\alpha_T\Theta_a\frac{r^6}{a^4}\,. \tag{2b}$$

We then have the general solution

$$\Psi = A_1\ln\frac{r}{a} + A_2 + \frac{1+\nu}{36}\alpha_T\Theta_a\frac{r^6}{a^4}\,. \tag{3}$$

By means of the material law and the strain-displacement relations for the plane case, the stresses are calculated as

$$\sigma_{rr} = -\frac{E}{1+\nu}\frac{1}{r}\frac{d\Psi}{dr} = -\frac{E}{1+\nu}\left[\frac{A_1}{r^2} + \frac{1}{6}(1+\nu)\alpha_T\Theta_a\left(\frac{r}{a}\right)^4\right], \tag{4a}$$

$$\sigma_{\varphi\varphi} = -\frac{E}{1+\nu}\frac{d^2\Psi}{dr^2} = -\frac{E}{1+\nu}\left[-\frac{A_1}{r^2} + \frac{5}{6}(1+\nu)\alpha_T\Theta_a\left(\frac{r}{a}\right)^4\right]. \tag{4b}$$

The shear stresses $\tau_{r\varphi}$ vanish identically because of the axisymmetry.

The *boundary conditions* read as follows:

$$\sigma_{rr}(a) = 0 \quad , \quad \sigma_{rr}(b) = 0\,. \tag{5a,b}$$

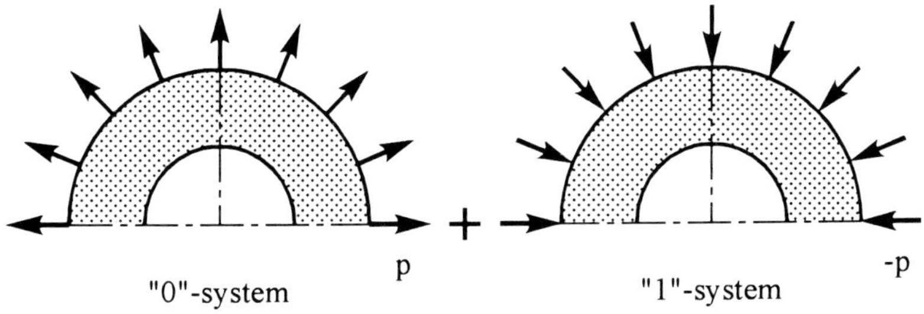

Fig. B-4: Superposition of two load cases

As (4a,b) shows, only one constant value of A_1 will occur in (5). Thus, one boundary condition can be fulfilled at the outer *or* at the inner boundary only. We obtain an additional condition if we – by analogy to statics – superpose a "1"-system onto the "0"-system (Fig. B-4). If we apply boundary condition (5a), an equal tensile stress p occurs at the outer boundary ("0"-system). We now subject the outer boundary of the disk to a compression −p ("1"-system) which has to act opposite to +p in the "0"-system so that this boundary is not subjected to a stress any longer.

According to (3a)

$$\sigma_{rr}^{(0)} = -\frac{E}{1+\nu}\left[\frac{A_1}{r^2} + \frac{(1+\nu)}{6}\alpha_T\Theta_a\left(\frac{r}{a}\right)^4\right].$$

At the boundary r = a

$$\sigma_{rr}^{(0)}(a) = \text{const} = p .$$

must be valid.

According to **A-7-1**

$$\sigma_{rr}^{(1)} = 2C_1 + \frac{C_2}{r^2} . \qquad (6a)$$

Note: The third constant C_3 equals zero for $\sigma_{rr}^{(1)}$ since the logarithmic part violates the required periodicity of v in the circumferential direction.

At the boundary r = a

$$\sigma_{rr}^{(1)}(a) = -p . \qquad (6b)$$

In order to fulfill the boundary conditions (5) we superpose the radial stresses and obtain

$$\sigma_{rr}^{(0)} + \sigma_{rr}^{(1)} = \sigma_{rr} = -\frac{E}{1+\nu}\left[\frac{A^*}{r^2} - C^* + \frac{(1+\nu)}{6}\alpha_T\Theta_a\left(\frac{r}{a}\right)^4\right] \qquad (7)$$

with $\quad A^* = A_1 - \frac{1+\nu}{E}C_2 \quad , \quad C^* = 2C_1\frac{1+\nu}{E} .$

Inserting into the boundary conditions (5) we obtain

$$\sigma_{rr}(a) = \sigma_{rr}^{(0)}(a) + \sigma_{rr}^{(1)}(a) = p - p = 0 , \qquad (8a)$$

$$\sigma_{rr}(b) = \sigma_{rr}^{(0)}(b) + \sigma_{rr}^{(1)}(b) = 0 . \qquad (8b)$$

Eqs. (8a,b) provide the equations for determining the constants A* und C*

$$\left.\begin{aligned}\frac{A^*}{a^2} - C^* &= -\frac{(1+\nu)}{6}\alpha_T\Theta_a \\ \frac{A^*}{b^2} - C^* &= -\frac{(1+\nu)}{6}\alpha_T\Theta_a\left(\frac{b}{a}\right)^4\end{aligned}\right\} \Longrightarrow \begin{aligned}A^* &= \frac{(1+\nu)}{6}\alpha_T\Theta_a(1+\rho^2)b^2 , \\ C^* &= \frac{(1+\nu)}{6}\alpha_T\Theta_a(1+\rho^2+\rho^4)\end{aligned}$$

with $\quad \rho = b/a .$

The following expressions are obtained for the radial and the tangential stresses:

$$\sigma_{rr} = \frac{1}{6}E\alpha_T\Theta_a\left\{1 + (\rho^2+\rho^4)\left[1-\left(\frac{a}{r}\right)^2\right] - \left(\frac{r}{a}\right)^4\right\} , \qquad (9a)$$

$$\sigma_{\varphi\varphi} = \frac{1}{6}E\alpha_T\Theta_a\left\{1 + (\rho^2+\rho^4)\left[1+\left(\frac{a}{r}\right)^2\right] - 5\left(\frac{r}{a}\right)^4\right\} . \qquad (9b)$$

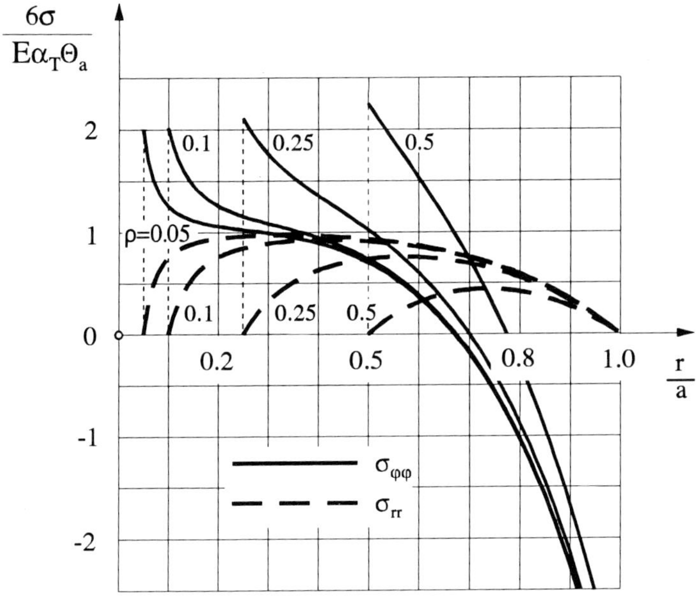

Fig. B-5: Radial and circumferential stresses in the disk

Fig. B-5 presents the non-dimensional stresses in dependence on the radius ratio ρ. Without the need for further integration, the expansion u(r) can be determined from (9a,b), the second strain-displacement equation *(8.8)* for $\varepsilon_{\varphi\varphi}$, and the material law *(8.9)*:

$$u(r) = \frac{r}{E}(\sigma_{\varphi\varphi} - \nu\sigma_{rr}) + r\alpha_T{}^0\Theta .$$

The reader is asked to check the results (9a,b) by means of the solutions to the bipotential equation *(8.6)*. For further examples of thermal stress problems refer to [A.13, B.8].

Exercise B-8-3:

A circular disk with constant thickness t rotates with a constant angular velocity ω (Fig. B-6).

Determine the location and the magnitude of the maximum stresses for a full disk ($0 \leqq r \leqq a$) and for an annular disk ($b \leqq r \leqq a$).

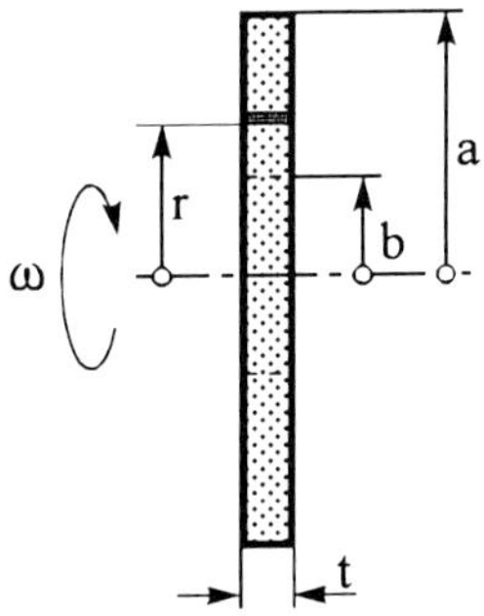

Fig. B-6: Rotating circular disk

Solution:

During rotation a centrifugal force at a distance r

$$f_r = \rho\, r\, \omega^2 \tag{1a}$$

occurs as a D´ALEMBERT inertia force at the element.

Since $$f_r = -\frac{\partial V}{\partial r}$$

this force can be derived from a potential, i.e.

$$V = -\frac{1}{2}\rho\,\omega^2 r^2 \,. \tag{1b}$$

Thus, from the bipotential equation *(8.6)*

$$\Delta\Delta\,\Phi = -(1-\nu)\,\Delta V$$

with $$\Delta V = V_{,rr} + \frac{1}{r} V_{,r} = -2\rho\,\omega^2 \tag{2a}$$

the following is obtained from *(8.20)* in view of the axisymmetry

$$r^4\,\Phi_{,rrrr} + 2\,r^3\,\Phi_{,rrr} - r^2\,\Phi_{,rr} + r\,\Phi_{,r} = 2(1-\nu)\rho\,\omega^2 r^4 \,. \tag{2b}$$

The homogeneous solution is given by *(8.21)*. The particular integral is obtained by an approximation of the right-hand-side type

$$\Phi_p = C\,r^4 \quad\longrightarrow\quad \Phi_p = \frac{1-\nu}{32}\rho\,\omega^2 r^4 \,. \tag{3}$$

If we – due to the uniqueness of the displacement v – assume the constant C_3 to be equal to zero (according to [ET1, 2]), and if we consider *(8.7)*, the total solution for the stresses reads:

$$\begin{aligned}\sigma_{rr} &= \frac{1}{r}\Phi_{,r} + V = 2\,C_1 + \frac{C_2}{r^2} + \frac{1-\nu}{8}\rho\,\omega^2 r^2 - \frac{1}{2}\rho\,\omega^2 r^2 = \\ &= 2\,C_1 + \frac{C_2}{r^2} - \frac{3+\nu}{8}\rho\,\omega^2 r^2 \,,\end{aligned} \tag{4a}$$

$$\begin{aligned}\sigma_{\varphi\varphi} &= \Phi_{,rr} + V = 2\,C_1 - \frac{C_2}{r^2} + \frac{3(1-\nu)}{8}\rho\,\omega^2 r^2 - \frac{1}{2}\rho\,\omega^2 r^2 = \\ &= 2\,C_1 - \frac{C_2}{r^2} - \frac{1+3\nu}{8}\rho\,\omega^2 r^2 \,.\end{aligned} \tag{4b}$$

a) In case of a *solid disk* the constant C_2 must equal zero because of the non-existing singularity at the origin.

From $$\sigma_{rr}(a) = 2\,C_1 - \frac{3+\nu}{8}\rho\,\omega^2 a^2 = 0$$

follows $$2\,C_1 = \frac{3+\nu}{8}\rho\,\omega^2 a^2$$

and thus $$\sigma_{rr} = \frac{3+\nu}{8}\rho\,\omega^2 (a^2 - r^2)\,, \tag{5a}$$

$$\sigma_{\varphi\varphi} = \frac{3+\nu}{8}\rho\,\omega^2 a^2 - \frac{1+3\nu}{8}\rho\,\omega^2 r^2 \,. \tag{5b}$$

The maximum values are attained at the axis r = 0 and become

$$\sigma_{rr_{max}} = \sigma_{\varphi\varphi_{max}} = \frac{3+\nu}{8}\rho\,\omega^2 a^2 . \tag{5c}$$

b) For the case of an *annular disk*, the constants C_1 und C_2 follow from the boundary conditions

$$\sigma_{rr}(b) = 0 \quad , \quad \sigma_{rr}(a) = 0 . \tag{6}$$

These conditions yield a system of two linear equations for C_1 and C_2. Having determined these, we obtain the stresses

$$\sigma_{rr} = \frac{3+\nu}{8}\rho\,\omega^2\left(a^2 + b^2 - \frac{a^2 b^2}{r^2} - r^2\right), \tag{7a}$$

$$\sigma_{\varphi\varphi} = \frac{3+\nu}{8}\rho\,\omega^2\left(a^2 + b^2 + \frac{a^2 b^2}{r^2} - \frac{1+3\nu}{3+\nu} r^2\right). \tag{7b}$$

The maximum circumferential stress is found at the inner boundary r = b

$$\sigma_{\varphi\varphi_{max}} = \frac{3+\nu}{4}\rho\,\omega^2\left(a^2 + \frac{1-\nu}{3+\nu} b^2\right). \tag{7c}$$

The maximum radial stress results from

$$\sigma_{rr,r} = \frac{3+\nu}{8}\rho\,\omega^2\left(2\frac{a^2 b^2}{r^3} - 2r\right) = 0 \quad \longrightarrow \quad r^* = \sqrt{a\,b}$$

as $$\sigma_{rr_{max}} = \sigma_{rr}(r^*) = \frac{3+\nu}{8}\rho\,\omega^2 (a-b)^2 .$$

For b = 0 the solution is equal to that obtained in a).

Exercise B-8-4:

A quarter-circle annular disk (outer radius a, inner radius b, thickness t) according to Fig. B-7 is clamped at A-A. A force F acts in the radial direction at the free end B-B of the disk.

Determine the stresses within the disk by assuming that the normal stresses in the tangential direction vary proportionally with $\sin\varphi$ (as is the case in the theory of a curved beam). Discuss the results.

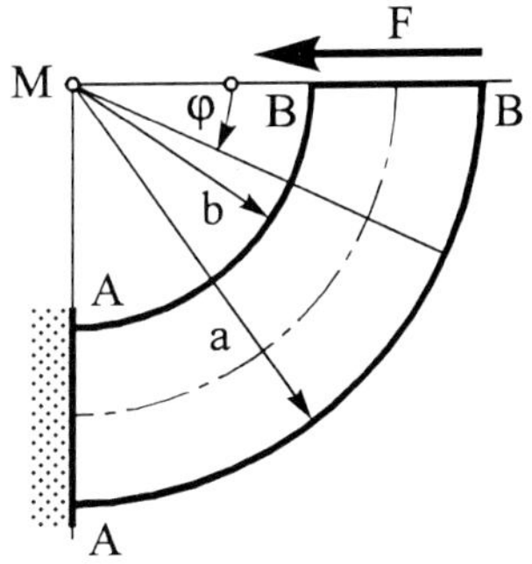

Fig. B-7: Clamped quarter-circle annular disk subjected to a single load

Solution:

The given quarter-circle annular disk is subjected to a non-axisymmetric state of stress.

The bipotential equation *(8.6)* in polar coordinates r, φ ($\overset{0}{\Theta} = 0$, $V = 0$) reads:

$$\Delta\Delta\,\Phi = \Phi_{,rrrr} + \frac{2}{r}\,\Phi_{,rrr} - \frac{1}{r^2}(\Phi_{,rr} - 2\,\Phi_{,rr\varphi\varphi}) + \\ + \frac{1}{r^3}(\Phi_{,r} - 2\,\Phi_{,r\varphi\varphi}) + \frac{1}{r^4}(4\,\Phi_{,\varphi\varphi} + \Phi_{,\varphi\varphi\varphi\varphi}) = 0\,. \tag{1}$$

Since the normal stresses are varying proportionally with $\sin\varphi$, the following assumption is chosen:

$$\Phi = f(r)\sin\varphi\,. \tag{2}$$

Substitution into (1) yields

$$\Delta\Delta\,\Phi = \left(f_{,rrrr} + \frac{2}{r}f_{,rrr} - \frac{3}{r^2}f_{,rr} + \frac{3}{r^3}f_{,r} - \frac{3}{r^4}f\right)\sin\varphi = 0\,,$$

i.e. the differential expression in the paranthesis has to vanish:

$$r^4 f_{,rrrr} + 2r^3 f_{,rrr} - 3r^2 f_{,rr} + 3rf_{,r} - 3f = 0\,. \tag{3}$$

Eq. (3) is a differential equation of the EULER type.

The solution assumption $f(r) = r^n$ leads to the characteristic equation:

$$n^4 - 4n^3 + 2n^2 + 4n - 3 = 0\,.$$

The solutions are:

$$n_1 = n_2 = 1\,, \quad n_3 = 3\,, \quad n_4 = -1\,.$$

Thus, we obtain the general solution of the differential equation:

$$f(r) = C_1 r^3 + C_2\frac{1}{r} + C_3 r + C_4 r\ln r \tag{4a}$$

or, according to (2), the AIRY stress function

$$\Phi = \left(C_1 r^3 + C_2\frac{1}{r} + C_3 r + C_4 r\ln r\right)\sin\varphi\,. \tag{4b}$$

From (4b) one obtains the stresses in polar coordinates *(8.7)* as:

$$\sigma_{rr} = \frac{1}{r^2}\Phi_{,\varphi\varphi} + \frac{1}{r}\Phi_{,r} = \left(2C_1 r - \frac{2C_2}{r^3} + \frac{C_4}{r}\right)\sin\varphi\,, \tag{5a}$$

$$\sigma_{\varphi\varphi} = \Phi_{,rr} = \left(6C_1 r + \frac{2C_2}{r^3} + \frac{C_4}{r}\right)\sin\varphi\,, \tag{5b}$$

$$\tau_{r\varphi} = -\left(\frac{1}{r}\Phi_{,\varphi}\right)_{,r} = -\left(2C_1 r - \frac{2C_2}{r^3} + \frac{C_4}{r}\right)\cos\varphi\,. \tag{5c}$$

The *boundary conditions* are:

$$\sigma_{rr}(a,\varphi) = 0\,, \quad \sigma_{rr}(b,\varphi) = 0\,, \tag{6a}$$

$$\tau_{r\varphi}(a,\varphi) = 0\,, \quad \tau_{r\varphi}(b,\varphi) = 0\,, \tag{6b}$$

$$\int_b^a \tau_{r\varphi}(r,0)\,t\,dr \stackrel{!}{=} F\,. \tag{7}$$

We can now determine the constants C_1, C_2 and C_4 by means of (6a,b) and (7). Substitution of (6a) into (5a) yields

$$2C_1 a - 2\frac{C_2}{a^3} + \frac{C_4}{a} = 0\,, \tag{8a}$$

$$2C_1 b - 2\frac{C_2}{b^3} + \frac{C_4}{b} = 0\,. \tag{8b}$$

From (8a,b) we determine

$$C_1 = -\frac{C_2}{a^2 b^2}\,. \tag{9}$$

Condition (7) with (5c) becomes:

$$\int_b^a \tau_{r\varphi}(r,0)\,t\,dr = \int_b^a -\left(2C_1 r - \frac{2C_2}{r^3} + \frac{C_4}{r}\right)\underbrace{\cos 0}_{=1}\,t\,dr =$$

$$= -t\left(C_1 r^2 + \frac{C_2}{r^2} + C_4 \ln r\right)\Bigg|_b^a \stackrel{!}{=} F$$

$$\Longrightarrow \qquad -C_1(a^2 - b^2) + C_2\frac{a^2 - b^2}{a^2 b^2} - C_4 \ln\frac{a}{b} = \frac{F}{t}\,. \tag{10}$$

Substitution of (9) into (10) leads to:

$$C_2\frac{a^2 - b^2}{a^2 b^2} + C_2\frac{a^2 - b^2}{a^2 b^2} - C_4 \ln\frac{a}{b} = \frac{F}{t}$$

$$\Longrightarrow \qquad C_4 = \frac{1}{\ln\frac{a}{b}}\left[2C_2\frac{a^2 - b^2}{a^2 b^2} - \frac{F}{t}\right]. \tag{11}$$

We then proceed with (9) and (11) substituted into (8):

$$-2C_2\frac{a}{a^2 b^2} - 2C_2\frac{1}{a^3} + \frac{1}{a}\left[\frac{2C_2(a^2 - b^2)}{a^2 b^2} - \frac{F}{t}\right]\frac{1}{\ln\frac{a}{b}} = 0$$

$$\Longrightarrow \qquad C_2 = \frac{a^2 b^2 F}{2\left[a^2 - b^2 - (a^2 + b^2)\ln\frac{a}{b}\right]t} \tag{12a}$$

and, further,

$$C_4 = \frac{(a^2 + b^2)F}{\left[a^2 - b^2 - (a^2 + b^2)\ln\frac{a}{b}\right]t}\,, \tag{12b}$$

$$C_1 = -\frac{F}{2\left[a^2 - b^2 - (a^2 + b^2)\ln\frac{a}{b}\right]t}\,. \tag{12c}$$

By substituting (12,a,b,c) into (5a,b,c) one obtains the stresses :

$$\sigma_{rr} = \frac{F}{ta} \frac{\frac{r}{a} + \frac{b^2}{a^2}\left(\frac{a}{r}\right)^3 - \left(1+\frac{b^2}{a^2}\right)\frac{a}{r}}{\frac{b^2}{a^2} - 1 - \left(1+\frac{b^2}{a^2}\right)\ln\frac{a}{b}} \sin\varphi , \qquad (13a)$$

$$\sigma_{\varphi\varphi} = \frac{F}{ta} \frac{3\frac{r}{a} - \frac{b^2}{a^2}\left(\frac{a}{r}\right)^3 - \left(1+\frac{b^2}{a^2}\right)\frac{a}{r}}{\frac{b^2}{a^2} - 1 - \left(1+\frac{b^2}{a^2}\right)\ln\frac{a}{b}} \sin\varphi , \qquad (13b)$$

$$\tau_{r\varphi} = -\frac{F}{ta} \frac{\frac{r}{a} + \frac{b^2}{a^2}\left(\frac{a}{r}\right)^3 - \left(1+\frac{b^2}{a^2}\right)\frac{a}{r}}{\frac{b^2}{a^2} - 1 - \left(1+\frac{b^2}{a^2}\right)\ln\frac{a}{b}} \cos\varphi . \qquad (13c)$$

For presenting the results, the stresses are normalized as $F/(b-a)t$. This delivers the following stress distributions where the abbreviation $\rho = b/a$ is used :

$$\bar{\sigma}_{rr} = \frac{\sigma_{rr}(a-b)\,t}{F} = (1-\rho)\frac{\frac{r}{a} + \rho^2\left(\frac{a}{r}\right)^3 - (1+\rho^2)\frac{a}{r}}{\rho^2 - 1 + (1+\rho^2)\ln\rho} \sin\varphi ,$$

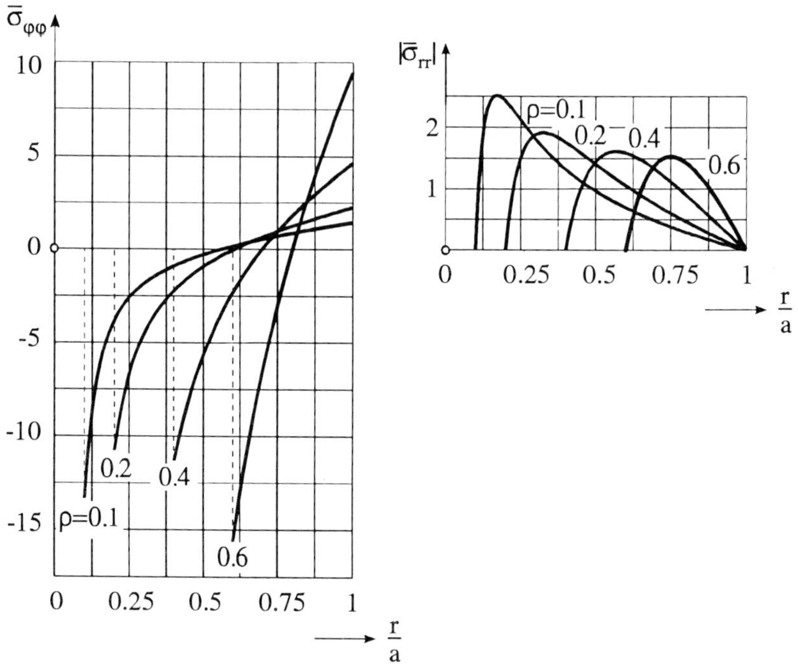

Fig. B-8 : Circumferential and radial stresses in the quarter – circle disk

$$\bar{\sigma}_{\varphi\varphi} = \frac{\sigma_{\varphi\varphi}(a-b)t}{F} = (1-\rho)\frac{3\frac{r}{a} - \rho^2\left(\frac{a}{r}\right)^3 - (1+\rho^2)\frac{a}{r}}{\rho^2 - 1 + (1+\rho^2)\ln\rho}\sin\varphi ,$$

$$\bar{\tau}_{r\varphi} = \frac{\tau_{r\varphi}(a-b)t}{F} = -(1-\rho)\frac{\frac{r}{a} + \rho^2\left(\frac{a}{r}\right)^3 - (1+\rho^2)\frac{a}{r}}{\rho^2 - 1 + (1+\rho^2)\ln\rho}\cos\varphi .$$

The maximum stresses $\bar{\sigma}_{rr_{max}}$ or $\bar{\sigma}_{\varphi\varphi_{max}}$ are obtained at $\varphi = \frac{\pi}{2}$, and $\bar{\tau}_{r\varphi}$ has its maximum at $\varphi = 0$.

Fig. B-8 presents the distribution of the normal stresses over the cross section for different ratios of the dimensionless radii ρ. If ρ increases, $\sigma_{\varphi\varphi}$ approaches a linear distribution corresponding to that of a straight beam.

Exercise B-8-5:

A semi-infinite disk $y \geq 0$, $-\infty \leq x \leq +\infty$ (thickness t) is subjected to a concentrated moment M at the origin 0 as shown in Fig. B-9a.

a) Determine the stress function for this load case by using a force couple as shown in Fig. B-9b and by applying the stress function (radiating stress state) $\Phi = (F/\pi t)\, r\varphi\cos\varphi$ for a concentrated force F as formulated in [ET2, A.16].

b) Which stresses occur in the semi-infinite disk?

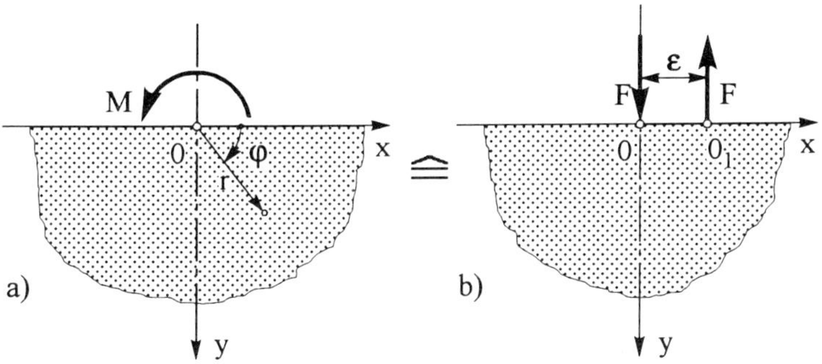

Fig. B-9: Semi-infinite disk under a concentrated moment M

Solution:

a) In order to determine the stress function Φ, we substitute the prescribed moment by a force couple as shown in Fig. B-9b. Then the stress function for F in 0_1 can be written as:

$$-\Phi(x-\varepsilon, y).$$

If we superpose the above stress function by the corresponding stress function for F in 0, i.e. $\Phi(x,y)$, we obtain

$$\Phi_1 = -\Phi(x-\varepsilon,y) + \Phi(x,y) . \tag{1}$$

According to the rules of the differential calculus the partial derivative is defined as :

$$\lim_{\varepsilon\to 0} \frac{-\Phi(x-\varepsilon,y)+\Phi(x,y)}{\varepsilon} = \frac{\partial\Phi}{\partial x} \mathrel{\widehat{=}} \Phi_{,x} . \tag{2}$$

The difference in (1) can, by applying (2), be written in the form

$$\Phi_1 = \varepsilon\,\Phi_{,x} . \tag{3a}$$

After transformation into polar coordinates $(\partial/\partial r \mathrel{\widehat{=}} (\ )_{,r}\ ,\ \partial/\partial_{\varphi} \mathrel{\widehat{=}} (\ )_{,\varphi})$ the difference reads :

$$\Phi_1 = \varepsilon\left(\Phi_{,r}\cos\varphi - \Phi_{,\varphi}\frac{\sin\varphi}{r}\right). \tag{3b}$$

In [ET2], the stress function of a single load on a semi-plane is derived as:

$$\Phi = \frac{F}{\pi t}\, r\,\varphi\cos\varphi . \tag{4}$$

The partial derivatives of (4)

$$\Phi_{,r} = \frac{F}{\pi t}\,\varphi\cos\varphi ,$$

$$\Phi_{,\varphi} - \frac{F}{\pi t}\, r(\cos\varphi - \varphi\sin\varphi)$$

yield by substitution into (3b) the following stress function :

$$\Phi_1 = \frac{F\varepsilon}{\pi t}\left[\varphi\cos^2\varphi - \sin\varphi\cos\varphi + \varphi\sin^2\varphi\right] = \frac{F\varepsilon}{\pi t}\left[\varphi - \sin\varphi\cos\varphi\right].$$

With $F\varepsilon \longrightarrow M \implies$ $$\Phi_1 = \frac{M}{\pi t}\left[\varphi - \sin\varphi\cos\varphi\right]. \tag{5}$$

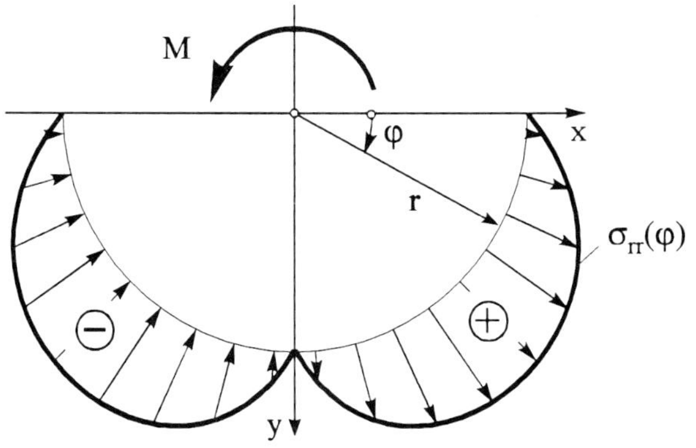

Fig. B-10 : Radial stresses in a semi-infinite disk under a concentrated moment

b) The stresses are determined according to *(8.7)*

$$\sigma_{rr} = \frac{1}{r}\Phi_{1,r} + \frac{1}{r^2}\Phi_{1,\varphi\varphi} \,. \tag{6}$$

Since $\Phi_{1,r} = 0$, it follows from (6) that: $\sigma_{rr} = \frac{1}{r^2}\Phi_{1,\varphi\varphi}$.

By

$$\Phi_{1,\varphi} = \frac{M}{\pi t}\left[1 - (\cos^2\varphi - \sin^2\varphi)\right] = \frac{M}{\pi t}\left[1 - \cos 2\varphi\right] \,, \quad \Phi_{1,\varphi\varphi} = \frac{M}{\pi t} 2 \sin 2\varphi \,,$$

we obtain $$\sigma_{rr} = \frac{2M}{\pi r^2 t}\sin 2\varphi \,, \quad \sigma_{\varphi\varphi} = \tau_{r\varphi} = 0 \,,$$

i.e., $$0 < \varphi < \frac{\pi}{2} \longrightarrow \underset{\text{tension}}{\sigma_{rr} > 0} \,, \quad \frac{\pi}{2} < \varphi < \pi \longrightarrow \underset{\text{compression}}{\sigma_{rr} < 0} \,.$$

Fig. B-10 shows the distribution of the radial stresses in this semi-infinite disk.

Exercise B-8-6:

A circular, annular disk made of carbon fibre reinforced plastic (CFRP) is subjected to an outer traction p_o, an inner pressure p_i, the centrifugal forces $f_r = \varrho r \omega^2$, and a temperature field $^0\Theta(r)$. The laminate is an angle-ply composite consisting of two layers with a fibre angle $\pm\alpha$ and the total thickness t ($t_1 = t_2 = t/2$) according to Fig. B-11.

The following is to be determined:

a) the EULER differential equation for the corresponding disk problem ;

b) the distribution of the radial and tangential stresses in dependence of the fibre angle α and the radial expansion of the disk ;

c) The results obtained shall be discussed on the basis of the following numerical values:

Geometry:
Outer radius $a = 0.25\, m$, inner radius $b = 0.0625\, m$, thickness $t = 2 \cdot 10^{-3} m$

Material characteristics of the layer (fibre T300, matrix 914C)
(Standards see Section 5.4):

$E_{1'} = 132\,680\ MPa$, $E_{2'} = 9059\ MPa$, $G_{1'2'} = 4268\ MPa$,

$\nu_{1'2'} = 0.28$, $\varrho = 1.19 \cdot 10^{-6}\ kg/mm^3$, $\varphi = 0.60$,

$\alpha_{T1'} = 0.23 \cdot 10^{-6}\ /{}^\circ C$, $\alpha_{T2'} = 53.25 \cdot 10^{-6}\ /{}^\circ C$.

Loads :

$p_i = 0\ MPa$, $p_o = 1\ MPa$, revolutions $n = 1000\ min^{-1}$.

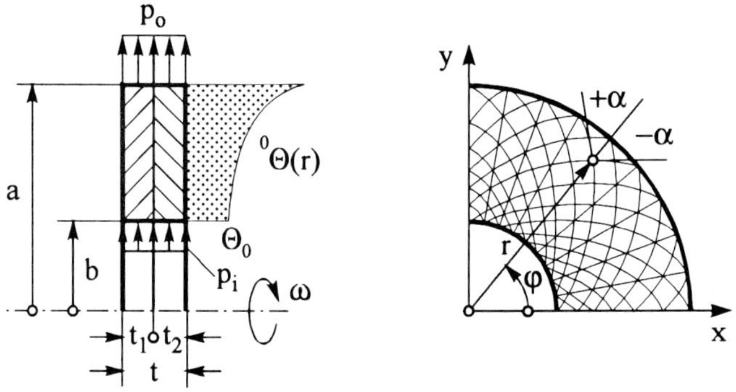

Fig. B-11: Circular annular CFRP-disk under several loads

Temperature field :

$${}^0\Theta(r) = \Theta_0\left(\frac{r}{b}\right)^4 \quad \text{with} \quad \Theta_0 = 1\ ^\circ C .$$

Solution :

a) According to *(3.29b)*, the equilibrium conditions for a circular disk can be expressed as follows in polar coordinates:

$$\sigma_{rr} + r\,\sigma_{rr,r} + \tau_{\varphi r,\varphi} - \sigma_{\varphi\varphi} + r\,f_r = 0 , \tag{1a}$$

$$r\,\tau_{\varphi r,r} + \sigma_{\varphi\varphi,\varphi} + 2\,\tau_{\varphi r} + r\,f_\varphi = 0 . \tag{1b}$$

All loads are axisymmetric so that both the stresses and the displacements are independent of the circumferential coordinate φ. Thus, all derivatives with respect to φ vanish in (1a,b). Furthermore, $\tau_{r\varphi}$, $\gamma_{r\varphi}$, and v are equal zero. The simplified equilibrium condition (1a) then reads:

$$\sigma_{rr} + r\,\sigma_{rr,r} - \sigma_{\varphi\varphi} + r\,f_r = 0 . \tag{2}$$

In a similar manner, the strain-displacement equations *(8.8)* for the axisymmetrical load case can be written as :

$$\varepsilon_{rr} = u_{,r} \quad , \quad \varepsilon_{\varphi\varphi} = \frac{u}{r} . \tag{3a,b}$$

Finally, we have to determine the material law for an *angle-ply laminate*. This is achieved by rotating a unidirectional layer (UD-layer) (see Section 5.4). For pure disk action, we have according to *(10.21)*:

$$\begin{bmatrix} N_{rr} \\ N_{\varphi\varphi} \\ N_{r\varphi} \end{bmatrix} = \begin{bmatrix} A_{11} & A_{12} & A_{13} \\ & A_{22} & A_{23} \\ \text{sym.} & & A_{33} \end{bmatrix} \left\{ \begin{bmatrix} \varepsilon_{rr} \\ \varepsilon_{\varphi\varphi} \\ \gamma_{r\varphi} \end{bmatrix} - \begin{bmatrix} \alpha_{Tr} \\ \alpha_{T\varphi} \\ \alpha_{Tr\varphi} \end{bmatrix} {}^0\Theta \right\} . \tag{4a}$$

The elasticity matrix in (4a) is fully occupied and its components depend on the angle of rotation. By special choice of the fibre orientations in a multi-layered composite (see Fig. B-11) and constant uniform layer thicknesses symmetrical with respect to the mid-plane, one obtains a simplified material law for the state of plane stress. We can prove for $\alpha_k = \pm\,\alpha$ that $A_{13} = A_{23} = 0$, and thus (4a) becomes with the assumptions made,

$$\begin{bmatrix} \sigma_{rr} \\ \sigma_{\varphi\varphi} \end{bmatrix} = \begin{bmatrix} E_{11}(\alpha) & E_{12}(\alpha) \\ E_{21}(\alpha) & E_{22}(\alpha) \end{bmatrix} \left\{ \begin{bmatrix} \varepsilon_{rr} \\ \varepsilon_{\varphi\varphi} \end{bmatrix} - \begin{bmatrix} \alpha_{Tr} \\ \alpha_{T\varphi} \end{bmatrix} {}^0\Theta \right\} . \qquad (4b)$$

In (4b) we have introduced the stresses corresponding to the stress resultants by means of $N_{rr} = \sigma_{rr}\,t$ etc. This way, the elements of the elasticity matrix correspond to those in *(5.24a,b)*. Furthermore, α_{Tr} and $\alpha_{T\varphi}$ denote the coefficients of thermal expansion in r- and φ-direction.

At this point it should be mentioned that with the chosen fibre position, the elements of the elasticity matrix are independent of the angular coordinate φ ; the assumptions concerning strains and displacements are thus confirmed.

Equations (2), (3a,b) and (4b) provide five equations for the determination of the five unknowns σ_{rr}, $\sigma_{\varphi\varphi}$, ε_{rr}, $\varepsilon_{\varphi\varphi}$ and u . In the following, this system shall be solved with respect to the stresses (see Section 7.4). In a first step we therefore transform (4b) with respect to the strains ε_{rr} und $\varepsilon_{\varphi\varphi}$ and obtain :

$$\begin{bmatrix} \varepsilon_{rr} \\ \varepsilon_{\varphi\varphi} \end{bmatrix} = \begin{bmatrix} D_{11}(\alpha) & D_{12}(\alpha) \\ D_{21}(\alpha) & D_{22}(\alpha) \end{bmatrix} \begin{bmatrix} \sigma_{rr} \\ \sigma_{\varphi\varphi} \end{bmatrix} + \begin{bmatrix} \alpha_{Tr} \\ \alpha_{T\varphi} \end{bmatrix} {}^0\Theta \qquad (4c)$$

with components of the compliance matrix denoted by $D_{\gamma\delta}$ ($\gamma, \delta = 1,2$). These are given by:

$$D_{11}(\alpha) = \frac{E_{22}(\alpha)}{\Delta(\alpha)} \quad , \quad D_{22}(\alpha) = \frac{E_{11}(\alpha)}{\Delta(\alpha)} \quad , \quad D_{12}(\alpha) = D_{21} = \frac{E_{12}(\alpha)}{\Delta(\alpha)}$$

with $\quad \Delta(\alpha) = E_{11}(\alpha)\,E_{22}(\alpha) - E_{12}^2(\alpha)$.

From (3a,b) with (4c) we obtain

$$\left.\begin{aligned} u_{,r} &= \varepsilon_{rr} \\ u_{,r} &= \frac{d}{dr}(r\,\varepsilon_{\varphi\varphi}) = \varepsilon_{\varphi\varphi} + r\,\varepsilon_{\varphi\varphi,r} \end{aligned}\right\} \quad \varepsilon_{rr} = \varepsilon_{\varphi\varphi} + r\,\varepsilon_{\varphi\varphi,r}$$

$$\Longrightarrow \quad \begin{aligned} &(D_{21} - D_{11})\,\sigma_{rr} + (D_{22} - D_{12})\,\sigma_{\varphi\varphi} + (\alpha_{T\varphi} - \alpha_{Tr})\,{}^0\Theta + \\ &\qquad + r\left(D_{12}\,\sigma_{rr,r} + D_{22}\,\sigma_{\varphi\varphi,r} + \alpha_{T\varphi}\,{}^0\Theta_{,r}\right) = 0 . \end{aligned} \qquad (5a)$$

We substitute from (2) into (5a)

$$\sigma_{\varphi\varphi} = \sigma_{rr} + r\,\sigma_{rr,r} + \rho\,r^2\,\omega^2 ,$$

and obtain the following differential equation :

$$D_{22}\, r^2 \sigma_{rr,rr} + 3\, D_{22}\, r\, \sigma_{rr,r} + (D_{22} - D_{11})\,\sigma_{rr} = \qquad (5b)$$

$$= (\alpha_{Tr} - \alpha_{T\varphi})\, {}^0\Theta(r) - \alpha_{T\varphi}\, r\, {}^0\Theta_{,r} + (D_{12} - 3\, D_{22})\,\rho\, r^2 \omega^2 .$$

After substitution of the temperature field and its derivative

$${}^0\Theta(r) = \Theta_0 \left(\frac{r}{b}\right)^4 \quad , \quad {}^0\Theta(r)_{,r} = \frac{4\,\Theta_0}{b}\left(\frac{r}{b}\right)^3 .$$

Eq. (5b) takes the form

$$D_{22}\, r^2 \sigma_{rr,rr} + 3\, D_{22}\, r\, \sigma_{rr,r} + (D_{22} - D_{11})\,\sigma_{rr} = \qquad (5c)$$

$$= (\alpha_{Tr} - 5\,\alpha_{T\varphi})\,\Theta_0 \left(\frac{r}{b}\right)^4 + (D_{12} - 3\, D_{22})\,\rho\, r^2 \omega^2 .$$

b) Equation (5c) is an inhomogeneous EULER differential equation. We derive the homogeneous solution by means of the assumption

$$\sigma_{rr_h} = C\, r^\lambda , \qquad (6)$$

which yields the characteristic equation

$$\lambda^2 + 2\lambda + \left(1 - \frac{D_{11}}{D_{22}}\right) = 0$$

with the roots

$$\lambda_{1,2} = -1 \pm \sqrt{1 - \left(1 - \frac{D_{11}}{D_{22}}\right)}$$

$$\longrightarrow \qquad \lambda_1 = -1 + \sqrt{\frac{D_{11}}{D_{22}}} \quad , \quad \lambda_2 = -1 - \sqrt{\frac{D_{11}}{D_{22}}} .$$

$\sqrt{\dfrac{D_{11}}{D_{22}}}$ is the degree of anisotropy.

Thus, the homogeneous solution is

$$\sigma_{rr_h} = C_1\, r^{\lambda_1} + C_2\, r^{\lambda_2} \qquad (7)$$

with the free constants C_1 and C_2.

We now determine the particular solution by means of an assumption for the right-hand side. Thus, we obtain

$$\sigma_{rr_p} = \frac{(\alpha_{Tr} - 5\,\alpha_{T\varphi})\,\Theta_0}{(25\, D_{22} - D_{11})}\left(\frac{r}{b}\right)^4 + \frac{(D_{12} - 3\, D_{22})}{(9\, D_{22} - D_{11})}\,\rho\, r^2 \omega^2 . \qquad (8)$$

The total solution consists of parts (7) and (8):

$$\sigma_{rr} = \sigma_{rr_h} + \sigma_{rr_p} . \qquad (9)$$

The constants are obtained from the *boundary conditions*

$$\sigma_{rr}(b) = p_i \,, \tag{10a}$$

$$\sigma_{rr}(a) = p_o \,. \tag{10b}$$

After substitution of (9) with (7) and (8) into (10a,b) the constants can be determined from the linear system of two equations

$$C_1 = A\Bigg[\left(p_i - \frac{(D_{12} - 3D_{22})\rho\omega^2}{(9D_{22} - D_{11})}\,b^2 - \frac{(\alpha_{Tr} - 5\alpha_{T\varphi})\Theta_0}{(25D_{22} - D_{11})}\right)a^{\lambda_2} -$$

$$- \left(p_o - \frac{(D_{12} - 3D_{22})\rho\omega^2}{(9D_{22} - D_{11})}\,a^2 - \frac{(\alpha_{Tr} - 5\alpha_{T\varphi})\Theta_0}{(25D_{22} - D_{11})}\,\frac{a^4}{b^4}\right)a^{\lambda_2}\Bigg],$$

$$C_2 = A\Bigg\{-\left[p_i - \frac{(D_{12} - 3D_{22})\rho\omega^2}{(9D_{22} - D_{11})}\,b^2 - \frac{(\alpha_{Tr} - 5\alpha_{T\varphi})\Theta_0}{(25D_{22} - D_{11})}\right]a^{\lambda_1} +$$

$$+ \left[p_o - \frac{(D_{12} - 3D_{22})\rho\omega^2}{(9D_{22} - D_{11})}\,a^2 - \frac{(\alpha_{Tr} - 5\alpha_{T\varphi})\Theta_0}{(25D_{22} - D_{11})}\,\frac{a^4}{b^4}\right]a^{\lambda_1}\Bigg\},$$

where $A = \dfrac{1}{b^{\lambda_1} b^{\lambda_2}\left[\left(\frac{a}{b}\right)^{\lambda_1} - \left(\frac{a}{b}\right)^{\lambda_2}\right]}$.

Introducing the abbreviation $\rho_0 = a/b$, the complete solution for the radial stresses reads :

$$\sigma_{rr} = B_1\left(\frac{r}{b}\right)^{\lambda_1} + B_2\left(\frac{r}{b}\right)^{\lambda_2} + \frac{(D_{12} - 3D_{22})\rho\omega^2}{(9D_{22} - D_{11})}\left(\frac{r}{b}\right)^2 + \frac{(\alpha_{Tr} - 5\alpha_{T\varphi})\Theta_0}{(25D_{22} - D_{11})}\left(\frac{r}{b}\right)^4, \tag{11}$$

where B_1 and B_2 are given by

$$B_1 = \frac{1}{\rho_0^{\lambda_2} - \rho_0^{\lambda_1}}\left\{p_i\rho_0^{\lambda_2} - p_o + \left(1 - \rho_0^{\lambda_2}\right)\left[\frac{(D_{12} - 3D_{22})\rho\omega^2}{(9D_{22} - D_{11})}\,a^2 + \frac{(\alpha_{Tr} - 5\alpha_{T\varphi})\Theta_0}{(25D_{22} - D_{11})}\left(\frac{a}{b}\right)^4\right]\right\},$$

$$B_2 = \frac{1}{\rho_0^{\lambda_1} - \rho_0^{\lambda_2}}\left\{p_i\rho_0^{\lambda_1} - p_o + \left(1 - \rho_0^{\lambda_1}\right)\left[\frac{(D_{12} - 3D_{22})\rho\omega^2}{(9D_{22} - D_{11})}\,a^2 + \frac{(\alpha_{Tr} - 5\alpha_{T\varphi})\Theta_0}{(25D_{22} - D_{11})}\left(\frac{a}{b}\right)^4\right]\right\}.$$

After determination of σ_{rr} , the radial stresses can be calculated from (2):

$$\sigma_{\varphi\varphi} = \sigma_{rr} + r\,\sigma_{rr,r} + \rho r^2\omega^2 \,. \tag{12}$$

The radial expansion can then be calculated from (3b) and (4c):

$$u = r\,\varepsilon_{\varphi\varphi} = r\,D_{21}\sigma_{rr} + r\,D_{22}\sigma_{\varphi\varphi} + \alpha_{T\varphi}\,r\,\overset{0}{\Theta}(r) \,.$$

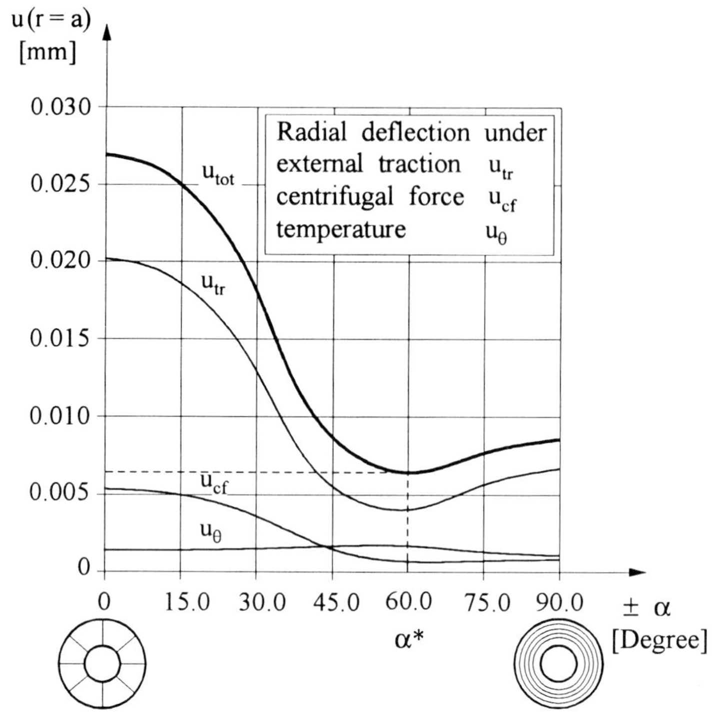

Fig. B-12 : Radial displacement of a circular hollow CFRP – disk

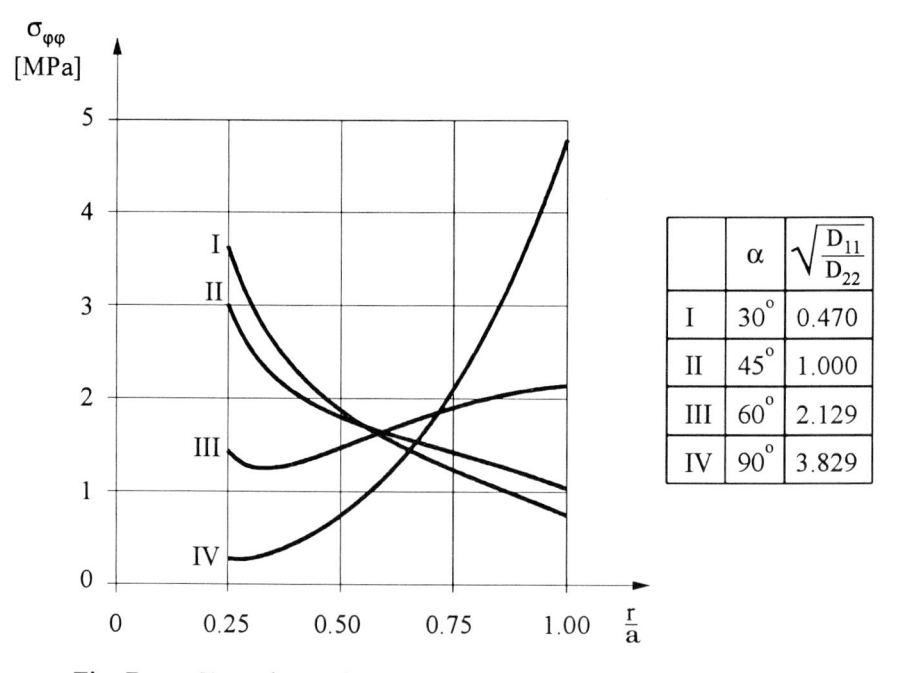

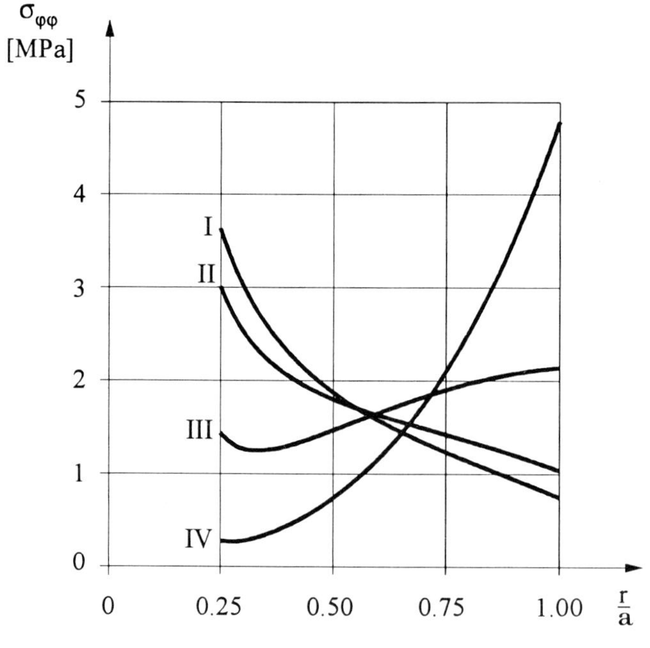

	α	$\sqrt{\frac{D_{11}}{D_{22}}}$
I	30°	0.470
II	45°	1.000
III	60°	2.129
IV	90°	3.829

Fig. B-13 : Circumferential stresses in a CFRP – disk

Fig. B-12 shows that the radial deflection depends on the angle $\pm\alpha$; the smallest value occurs at $\alpha^* = \pm 60^\circ$. The circumferential stresses in dependence of the degree of anisotropy $\sqrt{D_{11}/D_{22}}$ are presented in Fig. B-13. Here, it is obvious that the circumferential stresses are approximately constant for $\sqrt{D_{11}/D_{22}} \approx 2{,}0$ so that the high stresses at the inner and the outer boundary respectively can be reduced together.

Exercise B-8-7:

An infinite disk possesses an elliptical hole. The disk is subjected to a uniaxial tensile stress σ_0, see Fig. B-14. No external loads act at the boundary of the ellipse.

Calculate the stresses in the disk by the complex solution method. For this purpose, use the conformal mapping

$$z = f(\zeta) = c\left(\zeta + \frac{m}{\zeta}\right)$$

with $\quad c = \dfrac{a+b}{2} \quad , \quad m = \dfrac{a-b}{a+b} \quad , \quad 0 \le m \le 1$

to transform the area outside the elliptical hole in the z–plane into an area outside the unit circle ($|\zeta| = 1$) in the complex ζ–plane .

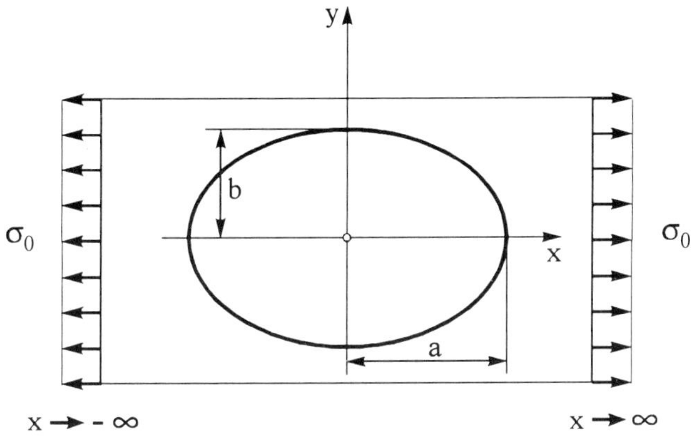

Fig. B-14: Infinite disk with an elliptical hole under uniaxial tension σ_0

Solution :

Comments on conformal mapping

The equation $z = f(\zeta)$ assigns any point $z = x + iy = r e^{i\vartheta}$ of the complex z–domain to a point $\zeta = \xi + i\eta = R e^{i\Theta}$ of the complex ζ–domain (see Fig. B-15). For further details on conformal mapping refer to, e.g., [A.8, B.6].

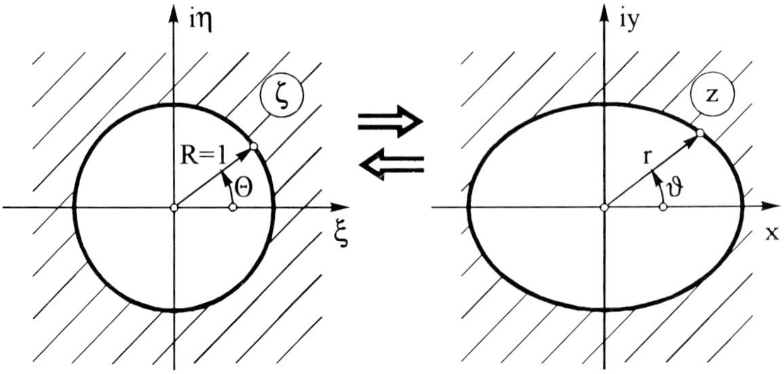

Fig. B-15: Mapping from the complex ζ – domain to the complex z – domain and vice versa

First, we will show that the given equation

$$z = f(\zeta) = c\left(\zeta + \frac{m}{\zeta}\right) \tag{1}$$

maps the area outside the unit – 1 – circle in the ζ – domain into the area outside an ellipse in the z – domain. Thus, the circle with radius 1 is transformed into an ellipse.

The equation for the circle with radius 1 in the ζ – domain reads :

$$\zeta = e^{i\Theta} \qquad (R = 1)\,. \tag{2}$$

Substitution of (2) into (1) leads to

$$z = x + iy = f(e^{i\Theta}) = c(e^{i\Theta} + me^{-i\Theta})$$

and, after some transformations by means of the MOIVRE formulas, we obtain

$$z = c\left[(1+m)\frac{e^{i\Theta}+e^{-i\Theta}}{2} + i(1-m)\frac{e^{i\Theta}-e^{-i\Theta}}{2i}\right] =$$

$$= c\left[(1+m)\cos\Theta + i(1-m)\sin\Theta\right] \tag{3}$$

or $$x = c(1+m)\cos\Theta\,, \tag{4a}$$

$$y = c(1-m)\sin\Theta\,. \tag{4b}$$

Eq. (3) and (4) are the parametric relations of the ellipse $\frac{x^2}{a^2} + \frac{y^2}{b^2} = 1.$

For Θ = 0 we obtain the large semi – axis

$$x(0) = a = c(1+m)\,.$$

The small semi – axis results from $\Theta = \frac{\pi}{2}$:

$$y\left(\frac{\pi}{2}\right) = b = c(1-m)\,.$$

Relations for the state of plane stress

The following approach holds for the stress function Φ in complex notation and fulfills the bipotential equation derived in [A.5, B.6]:

$$\Phi(z, \overline{z}) = \frac{1}{2}\left[\overline{z}\,\varphi(z) + z\,\overline{\varphi}(\overline{z}) + \int \psi(z)\,dz + \int \overline{\psi}(\overline{z})\,d\overline{z}\right]. \tag{5}$$

Here, φ and ψ are arbitrary analytical functions to be determined by means of boundary conditions. $\overline{\varphi}(\overline{z})$, $\overline{\psi}(\overline{z})$ are the corresponding complex conjugate functions with $\overline{z} = x - iy$.

Eq. (5) provides the following formulas for the stresses and deformations *(8.19)*:

$$\sigma_{xx} + \sigma_{yy} = \Phi_{,yy} + \Phi_{,xx} = 2(\varphi' + \overline{\varphi}'), \tag{6a}$$

$$\sigma_{yy} - \sigma_{xx} + 2i\tau_{xy} = \Phi_{,xx} - \Phi_{,yy} - 2i\Phi_{,xy} = 2(\overline{z}\varphi'' + \psi'), \tag{6b}$$

$$2G(u + iv) = -z\overline{\varphi}' - \overline{\psi} + \varkappa\varphi, \tag{6c}$$

where $\varkappa = \dfrac{3-\nu}{1+\nu}$ is valid for the state of plane stress.

Boundary conditions

a) Boundary of the hole

The boundary conditions at the boundary of the hole result from the fact that the load vector must equal the stress vector **t** at the boundary.

In accordance with Fig. B-16 we formulate the equilibrium conditions considering the +/− sign of dx :

x − direction : $t_x\, ds\, t = \sigma_{xx}\, dy\, t + \tau_{yx}(-dx)\, t$

$$\longrightarrow \quad t_x = \sigma_{xx}\frac{dy}{ds} - \tau_{yx}\frac{dx}{ds} = \Phi_{,yy}\frac{dy}{ds} + \Phi_{,xy}\frac{dx}{ds}.$$

y − direction : $t_y\, ds\, t = \sigma_{yy}(-dx)\, t + \tau_{xy}\, dy\, t$

$$\longrightarrow \quad t_y = -\sigma_{yy}\frac{dx}{ds} + \tau_{xy}\frac{dy}{ds} = -\Phi_{,xx}\frac{dx}{ds} - \Phi_{,xy}\frac{dy}{ds}.$$

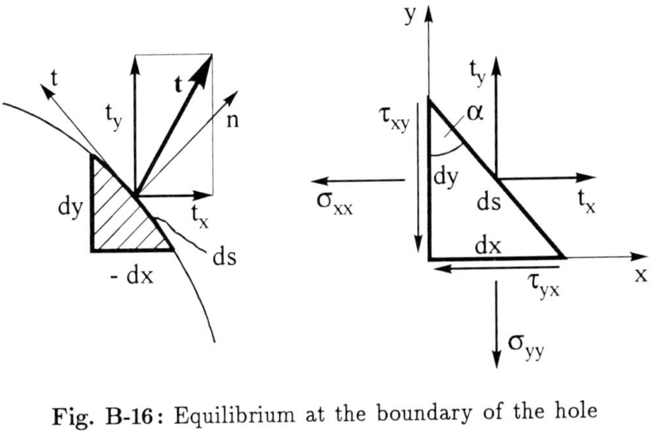

Fig. B-16: Equilibrium at the boundary of the hole

Since the boundary load is zero it holds that :

$$\mathbf{t} = t_x + i\,t_y = \frac{d}{ds}\left(\Phi_{,y} - i\,\Phi_{,x}\right) = 0$$

$$\Longrightarrow \qquad \Phi_{,y} - i\,\Phi_{,x} = \text{const} \; .$$

Since the choice of constants is not yet limited, we can set them zero :

$$\Phi_{,y} - i\,\Phi_{,x} = i\left(\frac{\partial}{\partial z} - \frac{\partial}{\partial \overline{z}} - \frac{\partial}{\partial z} - \frac{\partial}{\partial \overline{z}}\right)\Phi = -2\,i\frac{\partial \Phi}{\partial \overline{z}} = 0$$

or, with (5) at the boundary of the hole $\quad \varphi(z) + z\,\overline{\varphi}' + \overline{\psi} = 0 \; .$ (7)

Using the chain rule we transform the formulas for the stresses and displacements into functions of ζ :

$$z = f(\zeta) \Longrightarrow \varphi(z) = \varphi[f(\zeta)] = \varphi_1(\zeta) = \varphi_1 \, ,$$

$$\psi(z) = \psi[f(\zeta)] = \psi_1(\zeta) = \psi_1 \, ,$$

$$\frac{d\varphi}{dz} = \frac{d\varphi_1}{d\zeta}\frac{d\zeta}{dz} = \frac{\varphi_1'(\zeta)}{f'(\zeta)} = \frac{\varphi_1'}{f'} \, ,$$

$$\frac{d^2\varphi}{dz^2} = \frac{d}{d\zeta}\left(\frac{\varphi_1'}{f'}\right)\frac{d\zeta}{dz} = \frac{\varphi_1''f' - \varphi_1' f''}{f'^3} \, ,$$

$$\frac{d\psi}{dz} = \frac{\psi_1'}{f'} \, .$$

From (6a) through (6c) we then obtain the transformed relation :

$$\sigma_{xx} + \sigma_{yy} = 2\left(\frac{\varphi_1'}{f'} + \frac{\overline{\varphi}_1'}{\overline{f}'}\right) , \qquad (8a)$$

$$\sigma_{yy} - \sigma_{xx} + 2\,i\,\tau_{xy} = 2\left(\frac{\overline{f}}{f'^2}\varphi_1'' - \frac{\overline{f}\,f''}{f'^3}\varphi_1' + \frac{\psi_1'}{f'}\right) , \qquad (8b)$$

$$2\,G(u + i\,v) = -\frac{f}{\overline{f}'}\overline{\varphi}_1' - \overline{\psi}_1 + \varkappa\,\varphi_1 \, . \qquad (8c)$$

At the boundary, we have, according to (7):

$$\varphi_1 + \frac{f}{\overline{f}'}\overline{\varphi}_1' + \overline{\psi}_1 = 0 \, . \qquad (9a)$$

According to [B.6], (9a) can be replaced by its conjugate form. We now achieve a simplified notation in the form

$$\overline{\varphi}_1 + \frac{\overline{f}}{f'}\varphi_1' + \psi_1 = 0 \, , \qquad (9b)$$

where $\quad f = c\left(\zeta + \frac{m}{\zeta}\right) \, , \quad f' = c\left(1 - \frac{m}{\zeta^2}\right) \, , \quad f'' = 2\,c\,m\frac{1}{\zeta^3} \, .$ (10)

Determination of the stress function by means of LAURENT – series

In the following, power series shall be established for the functions φ_1 und ψ_1 [B.6]. Since the stresses in equations (8a) and (8b) must converge at infinity, no term with positive exponents are allowed to occur in the series. First, the following LAURENT – series are set up for the derivatives :

$$\varphi_1' = \sum_{n=0}^{\infty} A_n \zeta^{-n} \quad , \quad \psi_1' = \sum_{n=0}^{\infty} B_n \zeta^{-n} \tag{11}$$

with A_n and B_n as complex constants.

Integration yields:

$$\varphi_1 = A_0 \zeta + A_1 \ln \zeta + \sum_{n=2}^{\infty} \frac{A_n \zeta^{-n+1}}{-n+1} + A \, , \tag{12a}$$

$$\psi_1 = B_0 \zeta + B_1 \ln \zeta + \sum_{n=2}^{\infty} \frac{B_n \zeta^{-n+1}}{-n+1} + B \, . \tag{12b}$$

By differentiation we obtain from (11):

$$\varphi_1'' = \sum_{n=0}^{\infty} - n A_n \zeta^{-n-1} \, . \tag{12c}$$

Determination of the constants

a) *Conditions at infinity*

By means of (8a) and (11), we state for the stresses at infinity:

$$(\sigma_{xx} + \sigma_{yy})\Big|_{\zeta \to \infty} = 2\left(\frac{\varphi_1'}{f'} + \frac{\overline{\varphi_1'}}{\overline{f'}} \right)\Big|_{\zeta \to \infty} = 4 \, \mathrm{Re} \frac{A_0}{c} = \sigma_0 \, .$$

The imaginary part of A_0 only represents a rigid body displacement of the disk and is therefore set zero, i.e.

$$A_0 = \frac{c\sigma_0}{4} \, .$$

Correspondingly, it follows from (8b) that

$$(\sigma_{yy} - \sigma_{xx} + 2 i \tau_{xy})\Big|_{\zeta \to \infty} = 2\left(\frac{\overline{f}}{f'^2} \varphi_1'' - \frac{\overline{f} f''}{f'^3} \varphi_1' + \frac{\psi_1'}{f'} \right)\Big|_{\zeta \to \infty} =$$

$$= 2\left(0 - 0 + \frac{B_0}{c} \right) = - \sigma_0 \quad \Longrightarrow \quad B_0 = - \frac{c\sigma_0}{2} \, .$$

Hence, we have obtained the results for A_0 and B_0.

b) *Boundary of the hole* R = 1

For boundary condition (9b) we derive the remaining relations for the determination of the constants:

$$\overline{\varphi_1} + \frac{\overline{f}}{f'} \varphi_1' + \psi_1 = 0$$

with $$\frac{\overline{f}}{f'} = \frac{e^{-i\Theta} + m e^{i\Theta}}{1 - m e^{-2i\Theta}} = \left(e^{-i\Theta} + m e^{i\Theta} \right)\left(1 - m e^{-2i\Theta} \right)^{-1} =$$

$$= \left(e^{-i\Theta} + m e^{i\Theta} \right)\left(1 + m e^{-2i\Theta} + m^2 e^{-4i\Theta} + m^3 e^{-6i\Theta} + \ldots \right) =$$

$$= m e^{i\Theta} + (1 + m^2) \sum_{n=1}^{\infty} m^{n-1} e^{-(2n-1)i\Theta} \, ,$$

$$\varphi_1' = A_0 + A_1 e^{-i\Theta} + A_2 e^{-2i\Theta} + A_3 e^{-3i\Theta} + \dots ,$$

$$\frac{\bar{f}}{f'}\varphi_1' = A_0 m e^{i\Theta} + A_0(1+m^2)e^{-i\Theta} + A_0(1+m^2)e^{-3i\Theta} + \dots +$$

$$+ A_1 m + A_1(1+m^2)e^{-2i\Theta} + A_1(1+m^2)e^{-4i\Theta} + \dots +$$

$$+ A_2 m e^{-i\Theta} + A_2(1+m^2)e^{-3i\Theta} + A_2(1+m^2)e^{-5i\Theta} + \dots + \dots$$

and with (12a,b)

$$\bar{\varphi}_1 = \bar{A}_0 e^{-i\Theta} - i\bar{A}_1\Theta - \bar{A}_2 e^{i\Theta} - \frac{\bar{A}_3}{2}e^{2i\Theta} - \frac{\bar{A}_4}{3}e^{3i\Theta} - \dots ,$$

$$\psi_1 = B_0 e^{i\Theta} + iB_1\Theta - B_2 e^{-i\Theta} - \frac{B_3}{2}e^{-2i\Theta} - \frac{B_4}{3}e^{-3i\Theta} - \dots .$$

Now, all of the above equations are substituted into (9b) and rearranged with respect to the e-functions. After that we get

$$A_1 = 0 , \qquad B_1 = 0 ,$$

$$A_2 = \frac{c\sigma_0}{4}\left(m - 2\right) , \quad B_2 = \frac{c\sigma_0}{2}\left(1 - m + m^2\right),$$

$$A_n = 0 \quad \text{for} \quad n \geq 3 ,$$

$$B_{2n} = \frac{c\sigma_0}{2}\left(1 + m^2\right)\left(m - 1\right)(2n - 1)m^{n-2} \quad \text{for} \quad n \geq 2 ,$$

$$B_{2n+1} = 0 \quad \text{for} \quad n \geq 0 .$$

We thus obtain

$$\varphi_1' = \frac{c\sigma_0}{4}\left[1 + \left(m - 2\right)\frac{1}{\zeta^2}\right], \tag{13a}$$

$$\psi_1' = \frac{c\sigma_0}{2}\left[-1 + \left(1 - m + m^2\right)\frac{1}{\zeta^2} + \right.$$
$$\left. + (1 + m^2)\left(m - 2\right)\sum_{n=2}^{\infty}\frac{(2n-1)m^{n-2}}{\zeta^{2n}}\right]. \tag{13b}$$

Determination of the stresses

The circumferential stress can be calculated by means of equation (13a):

$$\sigma_{xx} + \sigma_{yy} = \sigma_{rr} + \sigma_{\Theta\Theta} = 4\,\mathrm{Re}\left(\frac{\varphi_1'}{f'}\right) = 4\,\mathrm{Re}\,\frac{c\sigma_0}{4}\,\frac{1+\left(m-2\right)\frac{1}{\zeta^2}}{c\left(1-\frac{m}{\zeta^2}\right)} =$$

$$= \sigma_0\,\mathrm{Re}\,\frac{\zeta^2 + m - 2}{\zeta^2 - m} = \sigma_0\,\mathrm{Re}\,\frac{R^2 e^{2i\Theta} + m - 2}{R^2 e^{2i\Theta} - m} =$$

$$= \sigma_0\,\mathrm{Re}\,\frac{(R^2 e^{2i\Theta} + m - 2)(R^2 e^{-2i\Theta} - m)}{(R^2 e^{2i\Theta} - m)(R^2 e^{-2i\Theta} - m)} =$$

$$= \sigma_0 \operatorname{Re} \frac{R^4 + m R^2 e^{-2i\Theta} - 2 R^2 e^{-2i\Theta} - m R^2 e^{2i\Theta} - m^2 + 2m}{R^4 - m R^2 e^{-2i\Theta} - m R^2 e^{2i\Theta} + m^2} =$$

$$= \sigma_0 \frac{R^4 + 2m - m^2 - 2R^2 \cos 2\Theta}{R^4 - 2m \cos 2\Theta + m^2} .$$

At the boundary of the hole $R = 1$, we have $\sigma_{rr} = 0$ and therefore the stress becomes

$$\sigma_{\Theta\Theta} = \sigma_0 \frac{1 + 2m - m^2 - 2\cos 2\Theta}{1 - 2m\cos 2\Theta + m^2} .$$

In the special case of a circular hole we have $a = b$ and, correspondingly, $m = 0$. For $\Theta = \pi / 2$ we obtain the maximum stress

$$\sigma_{max} = 3\sigma_0 .$$

The stresses at arbitrary points r, Θ are calculated from (8b). These calculations shall not be performed here. Further examples pertaining to the increase of stresses at holes and notches can be found, e.g., in [A.15, B.6].

Exercise B-8-8:

An infinite disk subjected to a uniaxial tensile stress σ_0 possesses a crack with the length 2a (Fig. B-17). The state of stress within the disk is described by the complex stress functions

$$\varphi = \frac{\sigma_0 z}{4} + \frac{\sigma_0 a}{2}\left(\sqrt{\left(\frac{z}{a}\right)^2 - 1} - \frac{z}{a}\right),$$

$$\psi = \frac{\sigma_0 z}{2} - \frac{\sigma_0 a}{2} \frac{1}{\sqrt{\left(\frac{z}{a}\right)^2 - 1}},$$

with $z = x + iy$.

Check whether the corresponding stresses fulfill all boundary conditions

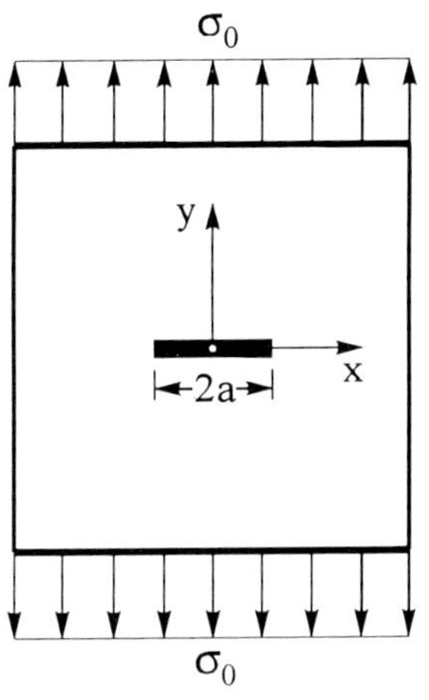

Fig. B-17: Infinite disk with a crack

Solution :

In order to apply equations *(8.19)* , we calculate the derivatives

$$\varphi' = \frac{\sigma_0}{4} + \frac{\sigma_0}{2}\left[\frac{\frac{z}{a}}{\sqrt{\left(\frac{z}{a}\right)^2 - 1}} - 1\right], \tag{1a}$$

$$\varphi'' = -\frac{\sigma_0}{2a}\frac{1}{\left[\left(\frac{z}{a}\right)^2 - 1\right]^{3/2}}, \tag{1b}$$

$$\psi' = \frac{\sigma_0}{2} + \frac{\sigma_0}{2}\frac{\frac{z}{a}}{\left[\left(\frac{z}{a}\right)^2 - 1\right]^{3/2}}. \tag{1c}$$

With (1a,b,c), the stresses follow from *(8.19)*

$$\sigma_{xx} + \sigma_{yy} = -\sigma_0 + \sigma_0\left[\frac{\frac{z}{a}}{\sqrt{\left(\frac{z}{a}\right)^2 - 1}} + \frac{\frac{\bar{z}}{a}}{\sqrt{\left(\frac{\bar{z}}{a}\right)^2 - 1}}\right], \tag{2a}$$

$$\sigma_{yy} - \sigma_{xx} + 2i\tau_{xy} = \sigma_0 + \sigma_0\left[\frac{\frac{z}{a} - \frac{\bar{z}}{a}}{\left[\left(\frac{z}{a}\right)^2 - 1\right]^{3/2}}\right]. \tag{2b}$$

In order to calculate the stresses themselves, the right – hand sides must be separated into their real and imaginary parts. For checking the boundary conditions the following considerations of the limit values are sufficient.

1) For large z und $\bar{z}$, i.e. the outer boundaries, holds

$$\sigma_{xx} + \sigma_{yy} = \sigma_0 , \tag{3a}$$

$$\sigma_{yy} - \sigma_{xx} + 2i\tau_{xy} = \sigma_0 , \tag{3b}$$

and it follows that $\sigma_{xx} = 0 \quad , \quad \sigma_{yy} = \sigma_0 \quad , \quad \tau_{xy} = 0$

in accordance with the given load.

2) Along the x – axis, i.e. $z = \bar{z} = x$, the relations *(8.19)* become

$$\sigma_{xx} + \sigma_{yy} = -\sigma_0 + 2\sigma_0\frac{\frac{x}{a}}{\sqrt{\left(\frac{x}{a}\right)^2 - 1}}, \tag{4a}$$

$$\sigma_{yy} - \sigma_{xx} + 2i\tau_{xy} = \sigma_0 . \tag{4b}$$

For reasons of symmetry, we have along the x – axis

a) $\tau_{xy} = 0$.

The above is valid both for the non – loaded crack surfaces $\left|\frac{x}{a}\right| < 1$ and along the axis $\left|\frac{x}{a}\right| > 1$.

Solution of (4a,b) now yields

b)

$$\sigma_{xx} = \begin{cases} -\sigma_0 & \text{for } \left|\frac{x}{a}\right| < 1 , \\ \sigma_0 \left[\dfrac{\frac{x}{a}}{\sqrt{\left(\frac{x}{a}\right)^2 - 1}} - 1 \right] & \text{for } \left|\frac{x}{a}\right| > 1 . \end{cases}$$

The stress σ_{xx} along the crack remains constant. At the crack tips $x = \pm a$, a singularity occurs, and for large x σ_{xx} approaches zero (load – free outer boundary).

c)

$$\sigma_{yy} = \begin{cases} 0 & \text{for } \left|\frac{x}{a}\right| < 1 , \\ \sigma_0 \dfrac{\frac{x}{a}}{\sqrt{\left(\frac{x}{a}\right)^2 - 1}} & \text{for } \left|\frac{x}{a}\right| > 1 . \end{cases}$$

The stress σ_{yy} is equal to zero at the surface of the crack $\left|\frac{x}{a}\right| < 1$, but at the crack tips $x = \pm a$, σ_{yy} (as well as σ_{xx}) exhibits a singularity.

Thus, the stresses derivable from the functions given in the problem formulation fulfill all boundary conditions.

Exercise B-9-1:

An isotropic, shear – rigid rectangular plate (dimensions a, b , thickness t) is simply supported at $x = 0$ and $x = a$, clamped at $y = 0$, and free at $y = b$ (Fig. B-18). The plate is subjected to a transverse, triangular load

$$p(y) = p_0\left(1 - \frac{y}{b}\right) .$$

Determine the complete solution to the plate equation, and formulate the boundary conditions.

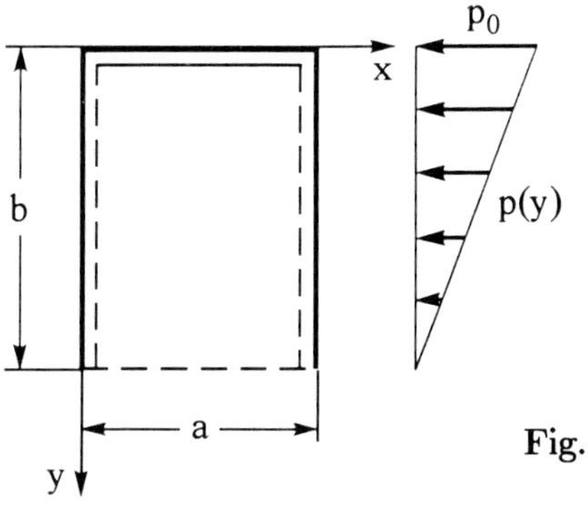

Fig. B-18: Rectangular plate with different boundary conditions

Solution:

The differential equation *(9.13)* (${}^{1}\Theta = 0$)

$$K \Delta\Delta w = p_0\left(1 - \frac{y}{b}\right) \tag{1}$$

has the homogeneous solution with the double root $\lambda_n = \pm\frac{n\pi}{a}$ (see *(9.42)*)

$$w_h = \sum_n (A^*_n \cosh\frac{n\pi y}{a} + B^*_n \frac{n\pi y}{a} \cosh\frac{n\pi y}{a} + C^*_n \sinh\frac{n\pi y}{a} + D^*_n \frac{n\pi y}{a} \sinh\frac{n\pi y}{a}) \sin\frac{n\pi x}{a} . \tag{2}$$

This solution fulfills both the homogeneous differential equation and the boundary conditions at the boundaries x = const :

$$w(0,y) = M_{xx}(0,y) = w(a,y) = M_{xx}(a,y) = 0 .$$

In order to determine the particular solution, the load in z-direction is expressed as a FOURIER series as discussed in the beginning of Section 9.2,

$$p(x,y) = p_0\left(1 - \frac{y}{b}\right) = \left(1 - \frac{y}{b}\right)\sum_n p_n \sin\frac{n\pi x}{a} \tag{3a}$$

with

$$p_n = \frac{2}{a}\int_0^a p_0 \sin\frac{n\pi x}{a}\, dx = \frac{4p_0}{n\pi} \quad , \quad n = 1,3,5\ldots . \tag{3b}$$

Using the following series expansion for the particular solution

$$w_p = \left(1 - \frac{y}{b}\right)\sum_n F_n \sin\frac{n\pi x}{a} , \tag{4}$$

it follows from the differential equation that

$$\left(\frac{n\pi}{a}\right)^4 F_n \left(1 - \frac{y}{b}\right) = \frac{4p_0}{Kn\pi}\left(1 - \frac{y}{b}\right) \quad \longrightarrow \quad F_n = \frac{4p_0}{K}\frac{a^4}{(n\pi)^5} . \tag{5}$$

The complete solution is then obtained from (2) and (4) :

$$w = w_h + w_p = \frac{4p_0}{K}\sum_n \left[\left(1 - \frac{y}{b}\right)\frac{a^4}{(n\pi)^5} + A_n \cosh\frac{n\pi y}{a} + B_n \frac{n\pi y}{a}\cosh\frac{n\pi y}{a} + C_n \sinh\frac{n\pi y}{a} + D_n \frac{n\pi y}{a}\sinh\frac{n\pi y}{a}\right] \sin\frac{n\pi x}{a} . \tag{6}$$

Owing to the sine-expansion, the complete solution fulfills the boundary conditions at x = const, too.

For the determination of the $4 \times n$ unknowns $A_n, \ldots, D_n$, we have two boundary conditions for each of the boundaries y = const., and these apply for each value of n $\Longrightarrow$

for $y = 0$ acc. to *(9.16)* $\longrightarrow$ $w(x,0) = 0 , \quad w_{,y}(x,0) = 0 ,$ (7a)

for $y = b$ acc. to *(9.14a,b)* $\longrightarrow$ $M_{yy}(x,b) = 0 , \quad \overline{Q}_y(x,b) = 0 .$ (7b)

The stress resultants are obtained from *(9.11)* according to (7b) or, alternatively, in analogy with *(9.14c)* with $\partial/\partial x \mathrel{\hat=} (\ )_{,x}$, $\partial/\partial y \mathrel{\hat=} (\ )_{,y}$:

$$M_{yy} = -K(w_{,yy} + \nu w_{,xx}) ,$$
$$\overline{Q}_y = -K(w_{,yyy} + (2-\nu) w_{,xxy}) .$$

Exercise B-9-2:

A simply supported semi-infinite plate strip (width a, thickness t) as shown in Fig. B-19 is subjected to a load in form of a uniformly distributed moment along the boundary y = 0. The moment per unit length of the boundary is denoted by M_0.

Determine the influence of the plate shear stiffness on the deflection.

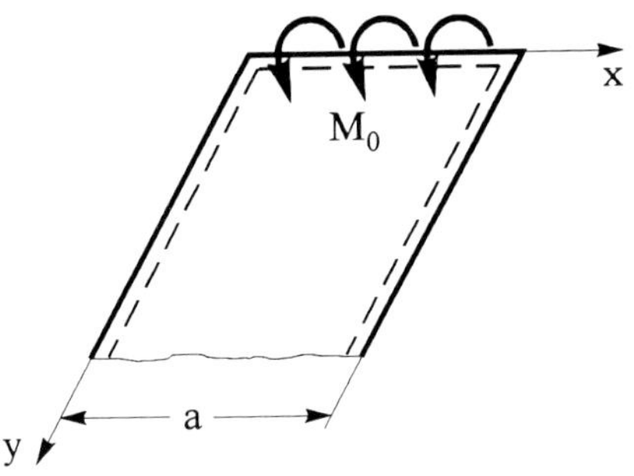

Fig. B-19: Semi-infinite plate strip subjected to boundary moment M_0

Solution:

We now deal with a *shear-elastic* plate, and refer to the basic equations *(9.7)*. Since no surface load acts, the basic equations reduce to

$$K \Delta \Phi = 0 \quad , \quad \Delta w = -\Phi \quad , \quad \Psi - \frac{1}{\varkappa_s} \Delta \Psi = 0 , \tag{1}$$

where the shear influence factor is abbreviated as

$$\frac{1}{\varkappa_s} = \frac{1-\nu}{2} \frac{K}{G t_s} .$$

In order to solve the coupled partial differential equations, we introduce into (1) the following approximation series of products with a seperation of variables,

$$\Phi = \sum_n f_n(y) \sin \alpha_n x , \tag{2a}$$

$$\Psi = \sum_n g_n(y) \cos \alpha_n x , \tag{2b}$$

$$w = \sum_n h_n(y) \sin \alpha_n x , \tag{2c}$$

where $\alpha_n = n\pi/a$. These approximations take into account the simple support at the boundaries x = const. The circular functions are now separated from (1), and for each n one obtains a coupled system of ordinary differential equations with constant coefficients, and with the abbreviations $\partial/\partial x \mathrel{\hat=} (\ )_{,x}$, $\partial/\partial y \mathrel{\hat=} (\ )_{,y}$ it follows in this case

$$f_{n,yy} - \alpha_n^2 f_n = 0 \,, \tag{3a}$$

$$h_{n,yy} - \alpha_n^2 h_n = f_n \,, \tag{3b}$$

$$g_n - \frac{1}{\varkappa_s}(g_{n,yy} - \alpha_n^2 g_n) = 0 \,. \tag{3c}$$

As solutions to these differential equations we have

$$f_n = C_n e^{-\alpha_n y} \,, \tag{4a}$$

$$g_n = D_n e^{-\lambda_n y} \,, \tag{4b}$$

$$h_n = (A_n + \frac{C_n}{2\alpha_n} y) e^{-\alpha_n y} \tag{4c}$$

with $\lambda_n^2 = \varkappa_s + \alpha_n^2$,

where only the decaying parts of the semi-infinite plate strip are of interest.

We now expand the boundary moment M_0 as a FOURIER series

$$M_0 = \sum_n M_n \sin \alpha_n x \,. \tag{5}$$

In order to state the boundary conditions, we have to express the moments in terms of the solutions (4). From *(9.3)* and *(9.6)* follows that

$$M_{yy} = K(\psi_{y,y} + \nu\psi_{x,x}) = K\Big[-w_{,yy} - \nu w_{,xx} + \frac{K}{Gt_s}(\Phi_{,yy} + \nu\Phi_{,xx}) + $$

$$+ \frac{1}{\varkappa_s}(\Psi_{,xy} - \nu\Psi_{,xy}) =$$

$$= K\Big\{\Big[-\alpha_n^2 A_n + \nu\alpha_n^2 A_n + \frac{C_n}{2\alpha_n}(2\alpha_n - \alpha_n^2 y + \nu\alpha_n^2 y)\Big] e^{-\alpha_n x} +$$

$$+ \frac{K}{Gt_s}(\alpha_n^2 - \nu\alpha_n^2) C_n e^{-\alpha_n y} + \frac{1}{\varkappa_s}\lambda_n\alpha_n(1-\nu) D_n e^{-\lambda_n y}\Big\} \sin\alpha_n x \,, \tag{6a}$$

$$M_{xy} = \frac{1-\nu}{2} K(\psi_{y,x} + \psi_{x,y}) = \frac{1-\nu}{2} K\Big\{-2w_{,xy} + \frac{K}{Gt_s}\Big[2\Phi_{,xy} + \frac{1-\nu}{2}$$

$$(\Psi_{,xx} - \Psi_{,yy})\Big]\Big\} =$$

$$= \frac{1-\nu}{2} K\Big\{\Big[2\alpha_n^2 A_n + \frac{C_n}{2\alpha_n}(\alpha_n - \alpha_n^2 y)\Big] e^{-\alpha_n y} +$$

$$+ \frac{K}{Gt_s}(-2\alpha_n^2) C_n e^{-\alpha_n y} + \frac{1-\nu}{2}(-\alpha_n^2 - \lambda_n^2) D_n e^{-\lambda_n y}\Big\} \cos\alpha_n x \,. \tag{6b}$$

The boundary conditions then yield:

$$w(x,0) = 0 \longrightarrow A_n = 0 \,, \tag{7a}$$

$$M_{yy}(x,0) = M_0 \longrightarrow \frac{M_n}{K} = C_n\Big[1 + \frac{K}{Gt_s}\alpha_n^2(1-\nu)\Big] +$$

$$+ \frac{1}{\varkappa_s}\lambda_n\alpha_n(1-\nu) D_n \,, \tag{7b}$$

$$M_{xy}(x,0)=0 \quad\longrightarrow\quad \frac{C_n}{2}+\frac{K}{Gt_s}\left[-2\alpha_n^2 C_n+\frac{1-\nu}{2}(-\alpha_n^2-\lambda_n^2)D_n\right]=0\,. \quad (7c)$$

The solution of (7b) and (7c) with the *shear area* of the rectangle

$$t_s=\frac{5}{6}t \quad\text{and}\quad \frac{1}{\varkappa_s}=\frac{1-\nu}{2}\,\frac{Et^3}{12(1-\nu^2)}\,\frac{2(1+\nu)}{E}\,\frac{6}{5t}=\frac{t^2}{10}$$

yields

$$C_n=\frac{M_n}{K}\,\frac{1}{1+2\dfrac{\alpha_n^2}{\varkappa_s}-\dfrac{1}{\alpha_n^2+\lambda_n^2}\left[(1-\nu)\alpha_n\lambda_n+\dfrac{2}{5}t^2\alpha_n^3\lambda_n\right]}=\frac{M_n}{K}\Lambda_n\,. \quad (8)$$

For the deflection in (2c) follows with the solution for h_n in (4c)

$$w=\sum_n \sin\alpha_n x\;\alpha_n y\,e^{-\alpha_n y}\,\frac{M_n}{2K\alpha_n^2}\Lambda_n\,. \quad (9)$$

In the *shear-rigid* case ($G t_s \longrightarrow \infty$, $1/\varkappa_s \longrightarrow 0$, $\lambda_n^2 \longrightarrow \infty$), only the first summand remains in the denominator of Λ_n in (8). In comparison, we now focus on the first term $n=1$. Considering the example of $a=10\,\pi\,t$, we obtain

$$\alpha_1=\frac{\pi}{a}=\frac{1}{10t}\,,\quad \lambda_1^2=\frac{10}{t^2}+\left(\frac{1}{10t}\right)^2\approx\frac{10}{t^2}\,,$$

and the factor Λ_1 is calculated as 0.98 .

Then, $$\frac{w_{el}}{w_{sh}}=\frac{1}{0.98}=1.02\,.$$

Thus, the effect of taking into account the finite shear stiffness in our plate example with $a=10\,\pi\,t$, is that the deflection increases by only 2% in comparison with that of the shear-rigid plate.

Hence, from an engineering point of view, KIRCHHOFF's plate theory generally provides sufficiently precise solutions for thin plates (in the present example $t/a = 1/10\pi \approx 1/30$).

The influence of the fast decaying part of the solution ($\lambda_n \gg \alpha_n$) on the resultant forces is discussed comprehensively in [B.5].

Exercise B-9-3:

A rectangular plate of constant thickness t (length a, width 2b, $\nu=0{,}3$) is simply supported at the boundaries $x=0,a$. As shown in Fig. B-20, two tubes with the same bending stiffness EI_y are welded to the plate along the boundaries $y=\pm b$. The plate is subjected to a temperature gradient through the plate thickness. The temperature gradient depends on y, and is given by:

$${}^1\Theta(y)=\Theta_1\left(1-m\frac{y}{b}\right),$$

where $$\Theta_1=\frac{T_1-T_2}{t}=\text{const} \quad\text{and}\quad 0\le m\le 1\,.$$

The temperatures of both tubes are equal to the temperature of the plate mid-plane which is everywhere equal to $(T_1 + T_2)/2$. (T_1 and T_2 denote the temperatures of the plate surfaces along the x-axis).

a) Derive the differential equation for the deflection of the plate under the given temperature field and boundary conditions by means of an energy principle. The tubes can be assumed to be rigid with respect to torsion.

b) Set up the general solution of the differential equations and the system of equations for determining the integration constants.

c) Discuss the influence of the elastic support from the two tubes (parameter $\lambda = (EI_y)/(aK)$, $0 \leq \lambda \leq \infty$) on the deflection and the stresses of a plate with the ratio $a/b = 1$ and ${}^1\Theta = \Theta_1 = \text{const}$ $(m = 0)$.

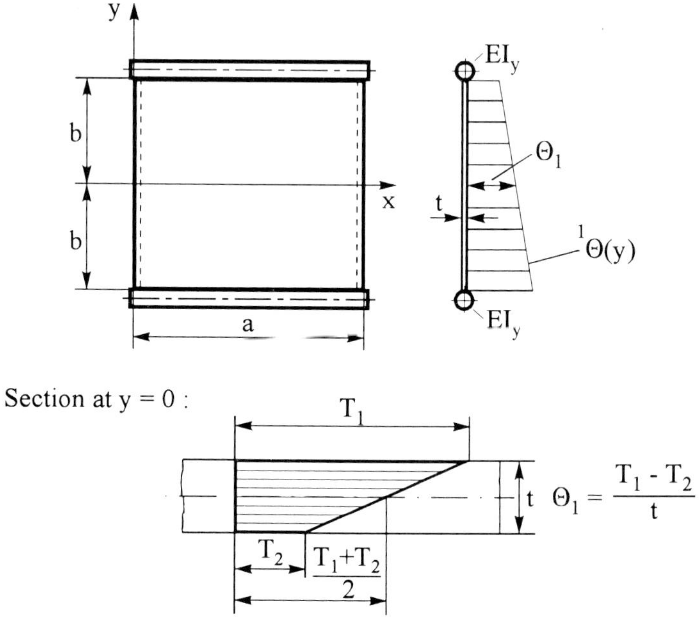

Fig. B-20: Rectangular plate with two simply supported and two elastically supported boundaries subjected to a temperature gradient ${}^1\Theta(y)$ through its thickness

Solution:

a) *Differential equation and boundary conditions*

The derivation is carried out by means of the principle of stationary (minimum) total potential energy *(6.20)*. Due to the absence of external static forces in the current example, the potential of such forces is zero. We now establish the non-vanishing energy expressions for our problem:

– *Deformation energy of the plate*

The deformation energy of the plate consists of elastic energy due to pure bending and thermal deformation energy due to the temperature gradient through the plate thickness. The total deformation energy for the plate is given by *(9.33)*

$$\Pi_{pl} = \frac{K}{2} \int\limits_{y=-b}^{+b} \int\limits_{x=0}^{a} \Big[\big(w_{,xx} + w_{,yy} \big)^2 - 2(1-\nu)\big(w_{,xx} w_{,yy} - w_{,xy}{}^2 \big) + \\ + 2(1+\nu)\alpha_T {}^1\Theta (w_{,xx} + w_{,yy}) \Big] dx\, dy \qquad (1)$$

with the notations $\partial/\partial x \mathrel{\widehat{=}} (\ )_{,x}$, $\partial/\partial y \mathrel{\widehat{=}} (\ )_{,y}$.

– *Elastic energy of the tubes*

Based on [A.8, A.18, A.19] we have for either of the tubes

$$\Pi_t = \frac{1}{2E} \int\limits_{x=0}^{a} \frac{M_y^2(x)}{I_y(x)} dx \; .$$

Considering $M_y(x) = -E I_y w(x, \pm b)_{,xx}$ yields:

$$\Pi_t = \frac{E I_y}{2} \int\limits_{x=0}^{a} w(x, \pm b)_{,xx}^2 \, dx \; . \qquad (2)$$

The total energy of the system then reads:

$\Pi = \Pi_{pl} + \Pi_t \Longrightarrow$

$$\Pi = \frac{K}{2} \int\limits_{y=-b}^{+b} \int\limits_{x=0}^{a} \Big[\big(w_{,xx} + w_{,yy} \big)^2 - 2(1-\nu)\big(w_{,xx} w_{,yy} - w_{,xy}{}^2 \big) + \\ + 2(1+\nu)\alpha_T {}^1\Theta (w_{,xx} + w_{,yy}) \Big] dx\, dy + \\ + \frac{E I_y}{2} \int\limits_{x=0}^{a} w(x, b)_{,xx}^2 \, dx + \frac{E I_y}{2} \int\limits_{x=0}^{a} w(x, -b)_{,xx}^2 \, dx$$

and in abbreviated form

$$\Pi = \iint F(x,y)\, dx\, dy + \int f(x,b)\, dx + \int f(x,-b)\, dx \qquad (3)$$

with

$$F(x,y) = \frac{K}{2} \Big[\big(w_{,xx} + w_{,yy} \big)^2 - 2(1-\nu)\big(w_{,xx} w_{,yy} - w_{,xy}{}^2 \big) + \\ + 2(1+\nu)\alpha_T {}^1\Theta (w_{,xx} + w_{,yy}) \Big] ,$$

$$f(x, \pm b) = \frac{E I_y}{2} \Big[w(x, \pm b)_{,xx}^2 \Big] .$$

These functions are independent except at the boundaries.

According to *(6.20)*, the variation of the total energy of the system has to vanish, i.e.

$$\delta \Pi = \delta \Pi_{pl} + \delta \Pi_t = 0 \; .$$

We formulate the EULER equation in analogy with *(6.35)* as the necessary condition :

$$\frac{\partial F}{\partial w} - \left(\frac{\partial F}{\partial w_{,x}}\right)_{,x} - \left(\frac{\partial F}{\partial w_{,y}}\right)_{,y} + \\ + \left(\frac{\partial F}{\partial w_{,xx}}\right)_{,xx} + \left(\frac{\partial F}{\partial w_{,xy}}\right)_{,xy} + \left(\frac{\partial F}{\partial w_{,yy}}\right)_{,yy} = 0 \,. \tag{4}$$

Note that $\delta\,\Pi_t$ only enters in the boundary integral.

With $\dfrac{\partial F}{\partial w} = 0 \;,\quad \dfrac{\partial F}{\partial w_{,x}} = 0 \;,\quad \dfrac{\partial F}{\partial w_{,y}} = 0 \;,$

we obtain

$$\frac{\partial F}{\partial w_{,xx}} = \frac{K}{2}\left[2(w_{,xx} + w_{,yy}) - 2(1-\nu)w_{,yy} + 2(1+\nu)\alpha_T\,{}^1\Theta\right],$$

$$\frac{\partial F}{\partial w_{,xy}} = \frac{K}{2}\left[4(1-\nu)w_{,xy}\right],$$

$$\frac{\partial F}{\partial w_{,yy}} = \frac{K}{2}\left[2(w_{,xx} + w_{,yy}) - 2(1-\nu)w_{,xx} + 2(1+\nu)\alpha_T\,{}^1\Theta\right].$$

Substitution into (4) yields the differential equation that describes the problem

$$w_{,xxxx} + 2\,w_{,xxyy} + w_{,yyyy} + (1+\nu)\alpha_T({}^1\Theta_{,xx} + {}^1\Theta_{,yy}) = 0$$

or, abbreviated,

$$\Delta\Delta\,w + (1+\nu)\alpha_T\Delta\,{}^1\Theta = 0 \,. \tag{5}$$

– *Boundary conditions*

The variation additionally yields the boundary conditions in analogy with *(6.35)*.

Boundary $x = 0, a$:

$$\delta\,w\left[\frac{\partial F}{\partial w_{,x}} - \left(\frac{\partial F}{\partial w_{,xx}}\right)_{,x} - \left(\frac{\partial F}{\partial w_{,xy}}\right)_{,y}\right] = 0 \,, \\ \delta\,w_{,x}\frac{\partial F}{\partial w_{,xx}} = 0 \,. \tag{6}$$

Owing to $\delta\,w = 0$; $\delta\,w_{,x} \neq 0$ (simply supported at the boundary $x = 0, a$), the boundary conditions at $x = 0, a$ are:

$$w = 0 \,, \tag{7a}$$

$$w_{,xx} + \nu\,w_{,yy} + (1+\nu)\alpha_T\,{}^1\Theta = 0 \,. \tag{7b}$$

Boundary $y = \pm b$

$$\delta\,w\left[-\left(\frac{\partial f}{\partial w_{,xx}}\right)_{,xx} + \frac{\partial F}{\partial w_{,y}} - \left(\frac{\partial F}{\partial w_{,yy}}\right)_{,y} - \left(\frac{\partial F}{\partial w_{,xy}}\right)_{,x}\right] = 0 \,, \\ \delta\,w_{,y}\frac{\partial F}{\partial w_{,yy}} = 0 \,. \tag{8}$$

Owing to $\delta w \neq 0$; $\delta w_{,y} = 0$ (tubes torsionally rigid, boundary $y = \pm b$), the boundary conditions at $y = \pm b$ are:

$$w_{,y} = 0 , \tag{9a}$$

$$-\frac{EI_y}{K} w_{,xxxx} - w_{,yyy} - (2 - \nu) w_{,xxy} - (1 + \nu)\alpha_T {}^1\Theta_{,y} = 0 . \tag{9b}$$

According to (9b), the transverse force at the boundary is transferred as a load to the beam.

b) *Solution of the differential equation*

Since $\Delta\,{}^1\Theta = 0$ in the present problem, (5) can be reduced to the bipotential equation :

$$\Delta\Delta\, w = 0 . \tag{10}$$

The solution to (10) consists of two partial solutions :

$$w = w_1 + w_2 . \tag{11}$$

Partial solution w_1: Solution of the plate strip

$$w_{1,xxxx} = 0 \quad \longrightarrow \quad w_1 = A + B\,x + C\,x^2 + D\,x^3$$

with the constants A , B , C , D.

Boundary conditions for the plate strip according to (7a,b) for $x = 0$, a:

$$w_1(0) = 0 \quad \text{and} \quad w_1(a) = 0 , \tag{12a}$$

$$w_{1,xx}(0) + (1 + \nu)\alpha_T {}^1\Theta = 0 \quad \text{and} \quad w_{1,xx}(a) + (1 + \nu)\alpha_T {}^1\Theta = 0 . \tag{12b}$$

From (12a,b) we determine the constants after which we obtain w_1 as

$$w_1 = \frac{1}{2}(1 + \nu)\alpha_T \Theta_1 (a\,x - x^2)\left(1 - m\frac{y}{b}\right) . \tag{13}$$

Partial solution w_2:

LEVY´s approximation *(9.40)*: $$w_2 = \sum_n Y_n(y) \sin\frac{n\pi x}{a} . \tag{14}$$

This approximation (14) identically fulfills the following boundary conditions:

$$w_2(0, y) = 0 \quad \text{and} \quad w_2(a, y) = 0 , \tag{15a}$$

$$w_{2,xx}(0, y) + \nu\, w_{2,yy}(0, y) = 0$$

and $$w_{2,xx}(a, y) + \nu\, w_{2,yy}(a, y) = 0. \tag{15b}$$

Thereby, the total solution $w = w_1 + w_2$ also fulfills the boundary conditions (7a,b).

Substitution of equation (14) into (10) yields for the n – th term:

$$Y_{n,xxxx} - 2\frac{n^2\pi^2}{a^2} Y_{n,xx} + \frac{n^4\pi^4}{a^4} Y_n = 0 . \tag{16}$$

The solution of (16) then reads :

$$Y_n(y) = (1+\nu)\alpha_T\Theta_1\Big(A_n\cosh\frac{n\pi y}{a} + B_n\frac{n\pi y}{a}\sinh\frac{n\pi y}{a} + C_n\sinh\frac{n\pi y}{a} + D_n\frac{n\pi y}{a}\cosh\frac{n\pi y}{a}\Big)$$

with the yet unknown constants $A_n, .., D_n$.

Now we can form the total solution $w = w_1 + w_2$ as :

$$w = (1+\nu)\alpha_T\Theta_1\Big\{\frac{1}{2}x(a-x)\Big(1-m\frac{y}{b}\Big) + \sum_{n=1,2}^{\infty}\Big(A_n\cosh\frac{n\pi y}{a} + B_n\frac{n\pi y}{a}\sinh\frac{n\pi y}{a} + C_n\sinh\frac{n\pi y}{a} + D_n\frac{n\pi y}{a}\cosh\frac{n\pi y}{a}\Big)\sin\frac{n\pi x}{a}\Big\}. \tag{17}$$

We wish to express all terms in (17) by means of *one* summation sign, i.e. we write a FOURIER series for the first term (see [ET2 | 8.2.4]):

$$f(x) = \frac{1}{2}x(a-x) = \sum_n a_n\sin\frac{n\pi x}{a}.$$

The FOURIER coefficients are then calculated as follows :

$$a_n = \frac{2}{a}\int_0^a f(x)\sin\frac{n\pi x}{a}\,dx \implies$$

$$a_n = \frac{1}{a}\int_0^a x(a-x)\sin\frac{n\pi x}{a}\,dx = \int_0^a x\sin\frac{n\pi x}{a}\,dx - \frac{1}{a}\int_0^a x^2\sin\frac{n\pi x}{a}\,dx.$$

After some calculations we obtain:

$$a_n = \frac{2a^2}{n^3\pi^3}\Big[1+(-1)^{n+1}\Big], \tag{18}$$

and the total solution thus reads:

$$w = (1+\nu)\alpha_T\Theta_1\sum_{n=1,2}^{\infty}\Big[a_n\Big(1-m\frac{y}{b}\Big) + A_n\cosh\frac{n\pi y}{a} + B_n\frac{n\pi y}{a}\sinh\frac{n\pi y}{a} + C_n\sinh\frac{n\pi y}{a} + D_n\frac{n\pi y}{a}\cosh\frac{n\pi y}{a}\Big]\sin\frac{n\pi x}{a}. \tag{19}$$

In order to determine the constants, we have to form the derivatives of (19) up to the fourth order (with $\alpha_n = (n\pi)/a$), and then to substitute into the boundary conditions (9a,b) .

First, let us consider the boundary conditions (9a):

$$y = b: \quad -\frac{a_n m}{\alpha_n b} + A_n\sinh\alpha_n b + B_n(\sinh\alpha_n b + \alpha_n b\cosh\alpha_n b) + C_n\cosh\alpha_n b + D_n(\cosh\alpha_n b + \alpha_n b\sinh\alpha_n b) = 0\,; \tag{20a}$$

$$y = -b: \quad -\frac{a_n m}{\alpha_n b} - A_n\sinh\alpha_n b - B_n(\sinh\alpha_n b + \alpha_n b\cosh\alpha_n b) + C_n\cosh\alpha_n b + D_n(\cosh\alpha_n b + \alpha_n b\sinh\alpha_n b) = 0\,. \tag{20b}$$

For establishing the transverse shear force boundary condition (9b) it is additionally required to expand

$$(1+\nu)\alpha_T\,{}^1\Theta_{,y} = (1+\nu)\alpha_T\,\Theta_1\left(-\frac{m}{b}\right)$$

in a FOURIER series:

$$-\frac{m}{b} = 2\frac{m}{b}\sum_{n=1,2}^{\infty}\frac{1}{a\alpha_n}\left[(-1)^n - 1\right]\sin\alpha_n x\,.$$

Inserting into boundary condition (9b) now yields:

$$\begin{aligned}
y = b:\quad & \frac{EI_y}{Ka}a\alpha_n\Big[a_n\Big(1-m\Big) + A_n\cosh\alpha_n b + B_n\,\alpha_n b\sinh\alpha_n b + \\
&\qquad + C_n\sinh\alpha_n b + D_n\,\alpha_n b\cosh\alpha_n b\Big] - \\
&- \Big\{\Big[A_n\sinh\alpha_n b + B_n\left(3\sinh\alpha_n b + \alpha_n b\cosh\alpha_n b\right) + \\
&\qquad + C_n\cosh\alpha_n b + D_n\left(3\cosh\alpha_n b + \alpha_n b\sinh\alpha_n b\right)\Big] - \\
&- (2-\nu)\Big[-\frac{a_n m}{\alpha_n b} + A_n\sinh\alpha_n b + B_n\left(\sinh\alpha_n b + \alpha_n b\cosh\alpha_n b\right) + \\
&\qquad + C_n\cosh\alpha_n b + D_n\left(\cosh\alpha_n b + \alpha_n b\sinh\alpha_n b\right)\Big] + \\
&+ \frac{2m}{ba\alpha_n^4}\Big[(-1)^n - 1\Big]\Big\} = 0\,;
\end{aligned} \tag{20c}$$

$$\begin{aligned}
y = -b:\quad & \frac{EI_y}{Ka}a\alpha_n\Big[a_n\Big(1+m\Big) + A_n\cosh\alpha_n b + B_n\,\alpha_n b\sinh\alpha_n b - \\
&\qquad - C_n\sinh\alpha_n b - D_n\,\alpha_n b\cosh\alpha_n b\Big] - \\
&- \Big\{\Big[-A_n\sinh\alpha_n b - B_n\left(3\sinh\alpha_n b + \alpha_n b\cosh\alpha_n b\right) + \\
&\qquad + C_n\cosh\alpha_n b + D_n\left(3\cosh\alpha_n b + \alpha_n b\sinh\alpha_n b\right)\Big] - \\
&- (2-\nu)\Big[-\frac{a_n m}{\alpha_n b} - A_n\sinh\alpha_n b - B_n\left(\sinh\alpha_n b + \alpha_n b\cosh\alpha_n b\right) + \\
&\qquad + C_n\cosh\alpha_n b + D_n\left(\cosh\alpha_n b + \alpha_n b\sinh\alpha_n b\right)\Big] + \\
&+ \frac{2m}{ba\alpha_n^4}\Big[(-1)^n - 1\Big]\Big\} = 0\,.
\end{aligned} \tag{20d}$$

Eq. (20a ÷ d) presents a linear system of four equations by means of which the four free constants can be determined.

c) *Special case of* $m = 0$

Because of symmetry we must have $C_n = D_n = 0$. Eqs. (20a,c) then yield the following reduced system of equations for the integration constants with $\lambda = EI_y / Ka$. ((20b,d) lead to the same equations) :

$$A_n\sinh\alpha_n b + B_n\left(\sinh\alpha_n b + \alpha_n b\cosh\alpha_n b\right) = 0\,, \tag{21a}$$

$$\lambda\, a\, \alpha_n \Big[a_n + A_n \cosh \alpha_n b + B_n\, \alpha_n b \sinh \alpha_n b \Big] -$$
$$- \Big\{ \Big[A_n \sinh \alpha_n b + B_n (3 \sinh \alpha_n b + \alpha_n b \cosh \alpha_n b) \Big] - \qquad (21b)$$
$$- (2 - \nu) \Big[A_n \sinh \alpha_n b + B_n (\sinh \alpha_n b + \alpha_n b \cosh \alpha_n b) \Big] \Big\} = 0 \,.$$

Transformed :

$$A_n \sinh \alpha_n b + B_n (\sinh \alpha_n b + \alpha_n b \cosh \alpha_n b) = 0 \,, \qquad (22a)$$

$$\lambda\, a\, \alpha_n \big[a_n + A_n \cosh \alpha_n b + B_n\, \alpha_n b \sinh \alpha_n b \big] - 2 B_n \sinh \alpha_n b = 0 \,. \qquad (22b)$$

Eq. (22a) leads to :

$$A_n = - B_n \frac{\sinh \alpha_n b + \alpha_n b \cosh \alpha_n b}{\sinh \alpha_n b} \,.$$

From (22b) follows :

$$\lambda\, a\, \alpha_n a_n + \lambda\, a\, \alpha_n B_n \Big[- \frac{\cosh \alpha_n b}{\sinh \alpha_n b} (\sinh \alpha_n b + \alpha_n b \cosh \alpha_n b) +$$
$$+ \alpha_n b \sinh \alpha_n b \Big] - 2 B_n \sinh \alpha_n b = 0 \,.$$

The integration constants can now be stated as:

$$B_n = \frac{a_n}{\cosh \alpha_n b + \alpha_n b + \dfrac{2}{\lambda\, a\, \alpha_n} \sinh \alpha_n b} \,,$$

$$A_n = - B_n (1 + \alpha_n b \coth \alpha_n b) \,.$$

Numerical evaluation:

– *Deflection function at* $y = 0$:

$$w = (1 + \nu)\, \alpha_T\, \Theta_1 \sum_{n=1}^{\infty} (a_n + A_n) \sin \frac{n \pi x}{a} \,.$$

The given numerical values lead to :

$$\frac{a}{b} = 1 \quad \Longrightarrow \quad \alpha_n b = \frac{n\pi}{a} b = n\pi \quad , \quad m = 0 \,,$$

$$a_1 = \frac{4 a^2}{\pi^3} = 0.1291\, a^2 \quad , \quad a_2 = 0 \quad , \quad a_3 = 0.0048\, a^2 \quad , \quad a_4 = 0 \quad , \ldots .$$

1) $\lambda = 0$ ($E I_y = 0$): $B_n = 0 \quad , \quad A_n = 0$

$$\longrightarrow \quad w \approx (1 + \nu)\, \alpha_T\, \Theta_1\, a^2 \Big(0.1291 \sin \frac{\pi x}{a} + 0.0048 \sin \frac{3 \pi x}{a} \Big) \,.$$

2) $\lambda = 1$: $B_1 = \dfrac{0.1291\, a^2}{\cosh \pi + \pi + \frac{2}{\pi} \sinh \pi} = 0.0058\, a^2 \,,$

$$A_1 = - B_1 (1 + \pi \coth \pi) = - 0.0241\, a^2 \,,$$

$$B_3 \approx 0 \quad , \quad A_3 \approx 0$$

$$\longrightarrow \quad w \approx (1+\nu)\,\alpha_T\,\Theta_1\Big[(a_1 + A_1)\sin\frac{\pi x}{a} + a_3 \sin\frac{3\pi x}{a}\Big] =$$

$$= (1+\nu)\,\alpha_T\,\Theta_1\,a^2\Big[\,0.1050 \sin\frac{\pi x}{a} + 0.0048 \sin\frac{3\pi x}{a}\Big]\,.$$

3) $\lambda = \infty$: $B_1 = \dfrac{0.1291\,a^2}{\cosh\pi + \pi} = 0.00878\,a^2$,

$$A_1 = -B_1(1 + \pi \coth\pi) = -0.0364\,a^2 \,,$$

$$B_3 \approx 0 \quad , \quad A_3 \approx 0$$

$$\longrightarrow \quad w \approx (1+\nu)\,\alpha_T\,\Theta_1\,a^2\Big[\,0.0927 \sin\frac{\pi x}{a} + 0.0048 \sin\frac{3\pi x}{a}\Big]\,.$$

Fig. B-21 presents the curves for the dimensionless deflections

$$\overline{w}(x,0) = \frac{w(x,0)}{(1+\nu)\,\alpha_T\,\Theta_1\,a^2} \qquad \text{at } y = 0\,.$$

– *Moments* M_{xx} *at* $y = 0$:

$$M_{xx} = -K\left[\,w_{,xx} + \nu\, w_{,yy} + (1+\nu)\,\alpha_T\,\Theta_1\right],$$

$$w_{,xx}(x,0) = -(1+\nu)\,\alpha_T\,\Theta_1 \sum_{n=1}^{\infty}(a_n + A_n)\Big(\frac{n\pi}{a}\Big)^2 \sin\frac{n\pi x}{a}\,.$$

The FOURIER series

$$\sum_{n=1}^{\infty} a_n \Big(\frac{n\pi}{a}\Big)^2 \sin\frac{n\pi x}{a} = \sum_{n=1}^{\infty} \frac{2a^2}{n^3\pi^3}\big(1 + (-1)^{n+1}\big)\Big(\frac{n\pi}{a}\Big)^2 \sin\frac{n\pi x}{a} =$$

$$= \frac{4}{\pi}\left(\sin\frac{\pi x}{a} + \frac{\sin\frac{3\pi x}{a}}{3} + \frac{\sin\frac{5\pi x}{a}}{5} + \dots\right)$$

presents the constant "*one*" in the interval $0 < x < a$.

This leads to

$$w_{,xx}(x,0) = -(1+\nu)\,\alpha_T\,\Theta_1\Big[1 + \sum_{n=1}^{\infty} A_n\Big(\frac{n\pi}{a}\Big)^2 \sin\frac{n\pi x}{a}\Big],$$

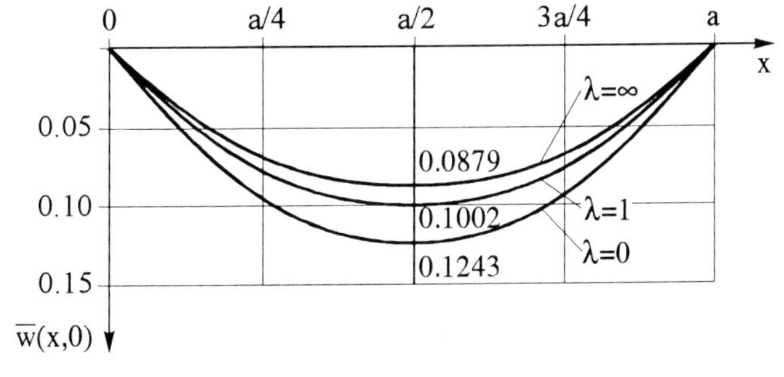

Fig. B-21: Dimensionless deflection $\overline{w}(x,0)$ of the plate

and with

$$w_{,yy}(x,0) = -(1+\nu)\alpha_T \Theta_1 \sum_{n=1}^{\infty} (A_n + B_n)\left(\frac{n\pi}{a}\right)^2 \sin\frac{n\pi x}{a}$$

the moments at $y = 0$ yield the expression:

$$M_{xx} = K(1+\nu)\alpha_T \Theta_1 \sum_{n=1}^{\infty} \Big[(1-\nu)A_n - 2\nu B_n\Big]\left(\frac{n\pi}{a}\right)^2 \sin\frac{n\pi x}{a} .$$

Here, we apply the same assumptions as in the case of the deflection functions, i.e.,

1) $\lambda = 0$: $\Big[(1-\nu)A_1 - 2\nu B_1\Big]\frac{\pi^2}{a^2} = 0$

$\longrightarrow$ $M_{xx} = 0$.

2) $\lambda = 1$: $\Big[(1-\nu)A_1 - 2\nu B_1\Big]\frac{\pi^2}{a^2} = -0.202$

$\longrightarrow$ $M_{xx} \approx -0.202 \sin\frac{\pi x}{a} K(1+\nu)\alpha_T \Theta_1$.

3) $\lambda = \infty$: $\Big[(1-\nu)A_1 - 2\nu B_1\Big]\frac{\pi^2}{a^2} = -0.314$

$\longrightarrow$ $M_{xx} \approx -0.314 \sin\frac{\pi x}{a} K(1+\nu)\alpha_T \Theta_1$.

The curves for the dimensionless bending moments

$$\overline{M}_{xx}(x,0) = \frac{-M_{xx}(x,0)}{K(1+\nu)\alpha_T \Theta_1}$$

at $y = 0$ are presented in Fig. B-22.

With vanishing stiffness of the tubes ($\lambda = 0$) at the boundaries the plate deforms without stresses.

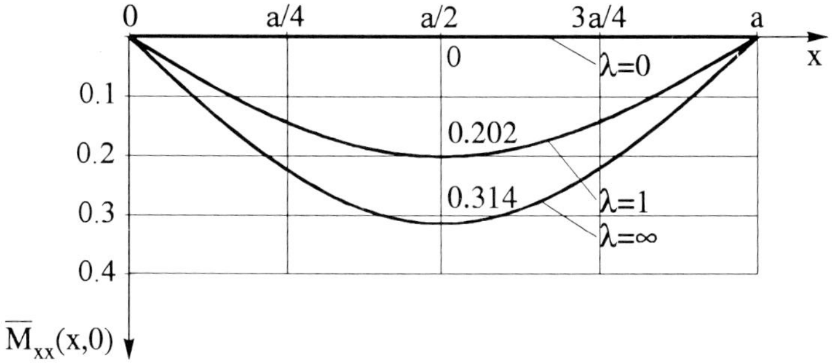

Fig. B-22: Dimensionless bending moments $\overline{M}_{xx}(x,0)$

Exercise B-9-4:

A thin, rectangular plate with all boundaries clamped (dimensions a, b, thickness t) is subjected to a uniformly distributed load with the intensity $p = p_0$ (Fig. B-23).

Determine the deflection function by means of the RITZ method, using the trigonometrical double series as an approximation function

$$w^* = \sum_{m=1}^{\infty} \sum_{n=1}^{\infty} a_{mn} \left(1 - \cos \frac{2 m \pi x}{a}\right)\left(1 - \cos \frac{2 n \pi y}{b}\right)$$

with a_{mn} as free coefficients.

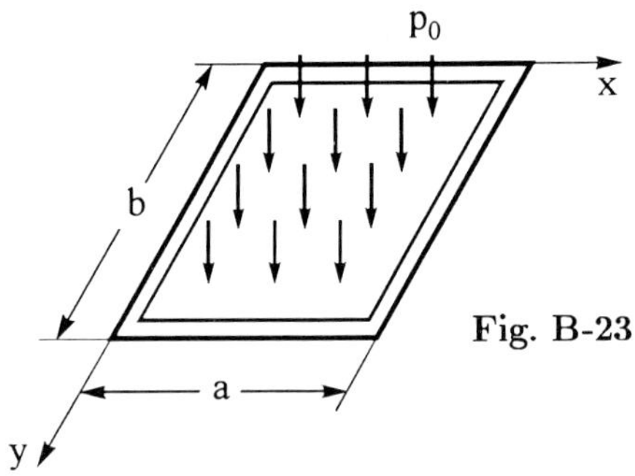

Fig. B-23: Clamped rectangular plate subjected to uniformly distributed load

Solution:

For this task, we first write the internal potential (strain energy) of a rectangular plate according to *(9.33)*, expressed in terms of the approximation function w^*:

$$\Pi_i = \frac{1}{2} K \int_0^b \int_0^a \left[\left(w^*_{,xx} + w^*_{,yy} \right)^2 - 2(1 - \nu)\left(w^*_{,xx} w^*_{,yy} - w^{*2}_{,xy} \right) \right] dx\, dy \qquad (1)$$

with the plate stiffness $K = \dfrac{Et^3}{12(1-\nu^2)}$ and the notation $\partial/\partial x \mathrel{\widehat{=}} (\ )_{,x}$, $\partial/\partial y \mathrel{\widehat{=}} (\ )_{,y}$.

For the external potential (potential of the external forces) we have, in terms of w^*,

$$\Pi_e = -\int_0^b \int_0^a p_0\, w^* \, dx\, dy \ . \qquad (2)$$

When applying the RITZ method, the approximation function has to fulfill at least the geometrical boundary conditions. These are:

$$\begin{aligned} w^*(0,y) &= w^*_{,x}(0,y) = 0 \quad , \quad w^*(a,y) = w^*_{,x}(a,y) = 0 \ ; \\ w^*(x,0) &= w^*_{,y}(x,0) = 0 \quad , \quad w^*(x,b) = w^*_{,y}(x,b) = 0 \ . \end{aligned} \qquad (3)$$

The principle of virtual displacements (see *(6.20)*) is now used as a necessary condition for determining the unknown coefficients a_{mn}

$$\frac{\partial \Pi}{\partial a_{mn}} = \frac{\partial (\Pi_i + \Pi_e)}{\partial a_{mn}} \overset{!}{=} 0 \,. \tag{4}$$

For the given approximation function

$$w^*(x,y) = \sum_{m=1}^{\infty} \sum_{n=1}^{\infty} a_{mn} \left(1 - \cos\frac{2m\pi x}{a}\right)\left(1 - \cos\frac{2n\pi y}{b}\right),$$

we now calculate the derivatives

$$w^*_{,x} = \sum_{m=1}^{\infty} \sum_{n=1}^{\infty} a_{mn} \frac{2m\pi}{a} \sin\frac{2m\pi x}{a}\left(1 - \cos\frac{2n\pi y}{b}\right),$$

$$w^*_{,y} = \sum_{m=1}^{\infty} \sum_{n=1}^{\infty} a_{mn} \left(1 - \cos\frac{2m\pi x}{a}\right)\frac{2n\pi}{b} \sin\frac{2n\pi y}{b},$$

$$w^*_{,xx} = \sum_{m=1}^{\infty} \sum_{n=1}^{\infty} a_{mn} \frac{4m^2\pi^2}{a^2} \cos\frac{2m\pi x}{a}\left(1 - \cos\frac{2n\pi y}{b}\right), \tag{5a}$$

$$w^*_{,yy} = \sum_{m=1}^{\infty} \sum_{n=1}^{\infty} a_{mn} \left(1 - \cos\frac{2m\pi x}{a}\right)\frac{4n^2\pi^2}{b^2} \cos\frac{2n\pi y}{b}, \tag{5b}$$

$$w^*_{,xy} - \sum_{m=1}^{\infty} \sum_{n=1}^{\infty} a_{mn} \frac{4mn\pi^2}{ab} \sin\frac{2m\pi x}{a} \sin\frac{2n\pi y}{b}\,. \tag{5c}$$

For simplicity, the solution based on a single term approximation ($m, n = 1$) only, will be presented in the following.

Substituting (5) into the expression $\Pi = \Pi_i + \Pi_e$ for the total potential, gives in view of (1) and (2):

$$\begin{aligned}
\Pi = \frac{1}{2} K \int_0^b \int_0^a \Bigg\{ \Bigg[& a_{11} \frac{4\pi^2}{a^2} \cos\frac{2\pi x}{a}\left(1 - \cos\frac{2\pi y}{b}\right) + \\
& + a_{11}\left(1 - \cos\frac{2\pi x}{a}\right)\frac{4\pi^2}{b^2} \cos\frac{2\pi y}{b}\Bigg]^2 - \\
- 2(1-\nu)\Bigg[& a_{11}^2 \frac{16\pi^4}{a^2 b^2} \cos\frac{2\pi x}{a}\left(1 - \cos\frac{2\pi x}{a}\right)\cos\frac{2\pi y}{b}\left(1 - \cos\frac{2\pi y}{b}\right) - \\
& - a_{11}^2 \frac{16\pi^4}{a^2 b^2} \sin^2\left(\frac{2\pi x}{a}\right)\sin^2\left(\frac{2\pi y}{b}\right)\Bigg]\Bigg\} dx\, dy - \\
- \int_0^b \int_0^a & p_0\, a_{11}\left(1 - \cos\frac{2\pi x}{a}\right)\left(1 - \cos\frac{2\pi y}{b}\right) dx\, dy\,.
\end{aligned}$$

Integration leads to the following expression :

$$\Pi = 2\,\pi^4\,K\,a\,b\,a_{11}^2\left(\frac{3}{a^4} + \frac{2}{a^2 b^2} + \frac{3}{b^4}\right) - p_0\,a_{11}\,a\,b\,. \tag{6}$$

The application of equation (4) as a necessary condition for determining the unknown coefficient a_{11} now leads to

$$\frac{\partial \Pi}{\partial a_{11}} = 4\,\pi^4\,K\,a\,b\,a_{11}\left(\frac{3}{a^4} + \frac{2}{a^2 b^2} + \frac{3}{b^4}\right) - p_0\,a\,b \overset{!}{=} 0$$

$$\Longrightarrow \qquad a_{11} = \frac{p_0}{4\,\pi^4\,K\left(\frac{3}{a^4} + \frac{2}{a^2 b^2} + \frac{3}{b^4}\right)}\,.$$

The single term approximation w^* to the deflection function is thus found to be

$$w^*(x,y) = \frac{p_0}{4\,\pi^4\,K\left(\frac{3}{a^4} + \frac{2}{a^2 b^2} + \frac{3}{b^4}\right)}\left(1 - \cos\frac{2\,\pi\,x}{a}\right)\left(1 - \cos\frac{2\,\pi\,y}{b}\right).$$

Special case of a quadratic plate ($a = b$):

$$w^*(x,y) = \frac{p_0\,a^4}{32\,\pi^4\,K}\left(1 - \cos\frac{2\,\pi\,x}{a}\right)\left(1 - \cos\frac{2\,\pi\,y}{a}\right)$$

with a maximum deflection of

$$w^*_{max} = w^*\left(\frac{a}{2},\frac{a}{2}\right) = \frac{p_0\,a^4}{8\,\pi^4\,K} = \frac{1}{8}\,w_0\,.$$

More precise values can be achieved by employing a four-term approach. Comparison of the maximum deflection obtained in the two cases with

$$w_0 = \frac{p_0 a^4}{\pi^4 K}$$

yields the following values :

approach	one-term	four-term
m	1	1 , 2
n	1	1 , 2
w^*_{max}/w_0	0.125	0.12205

This example clearly shows a relatively fast convergence of the RITZ method.

Exercise B-9-5:

Determine the maximum deflection due to a uniformly distributed load p_0 by means of the GALERKIN method for a rectangular plate with two clamped and two simply supported edges (Fig. B-24).

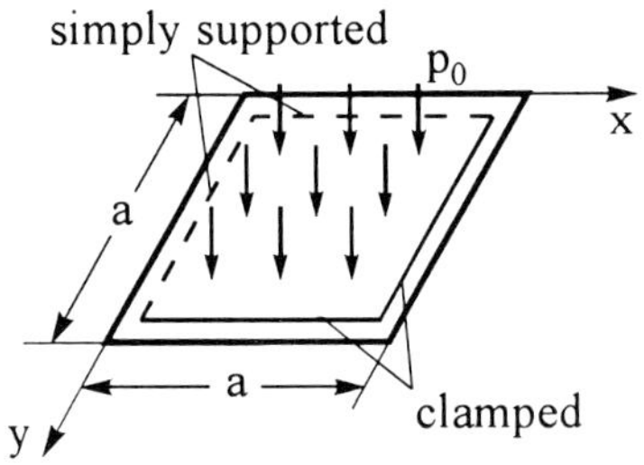

Fig. B-24: Rectangular plate with mixed boundary conditions

Solution:

The GALERKIN mèthod requires fulfillment of *all* boundary conditions. In order to comply with this demand, we formulate beam solutions for both coordinate directions. These solutions enable us to fulfill the corresponding boundary conditions of the plate. For this purpose, the eigenfunctions of the vibrating beam or of the beam subjected to buckling prove to be useful, since they also comply with desirable properties of orthogonality.

In the present task, we will proceed from the basic equations for a vibrating uniform beam (bending stiffness E I, mass density ρ, length a) [A.20, B.4]:

$$\frac{\partial^4 w}{\partial x^4} + \frac{\rho}{EI}\frac{\partial^2 w}{\partial \tau^2} = 0 . \tag{1}$$

By means of the separation approach

$$w(x,\tau) = \sum_m X_m(x) \sin \omega_m \tau \tag{2}$$

Eq. (1) is transformed with (2) into an ordinary differential equation with constant coefficients for $X_m(x)$ with $d/dx \mathrel{\widehat=} (\ )_{,x}$:

$$X_{m,xxxx}(x) - \frac{\lambda_m^4}{a^4} X_m(x) = 0 , \tag{3}$$

where $\lambda_m = a \sqrt[4]{\frac{\rho\omega_m^2}{EI}}$, $(m = 1, 2, \ldots)$ denote eigenvalues.

The solution of (3) reads:

$$X_m(x) = C_{1_m} \sin \lambda_m \frac{x}{a} + C_{2_m} \cos \lambda_m \frac{x}{a} + C_{3_m} \sinh \lambda_m \frac{x}{a} + C_{4_m} \cosh \lambda_m \frac{x}{a} . \tag{4}$$

The eigenvalues are calculated from the boundary conditions for a simply supported–clamped beam :

$$w(0)=0 \quad , \quad w_{,xx}(0)=0 \quad \longrightarrow \quad X_m(0)=0 \quad , \quad X_{m,xx}(0)=0 \, , \tag{5a}$$

$$w(a)=0 \quad , \quad w_{,x}(a)=0 \quad \longrightarrow \quad X_m(a)=0 \quad , \quad X_{m,x}(a)=0 \, . \tag{5b}$$

We substitute (4) into (5a) and (5b), and hereby obtain a homogeneous, linear system of equations for the constants. The necessary condition (for non-trivial solutions) that the determinant of the coefficients must vanish, leads to the characteristic equation for the eigenvalues,

$$\tan \lambda_m = \tanh \lambda_m \, . \tag{6}$$

Eq. (6) possesses an infinite number of real-valued solutions λ_m. Furthermore, by means of (5a,b) all constants C_{im} in (4) can be reduced by an arbitrary factor, so the *eigenfunctions* can be written

$$X_m = \sin \lambda_m \frac{x}{a} - \frac{\sin \lambda_m}{\sinh \lambda_m} \sinh \lambda_m \frac{x}{a} \, . \tag{7}$$

As mentioned above, the eigenfunctions are orthogonal, i.e.

$$\begin{aligned} \int_{x=0}^{a} X_m X_n \, dx &= \begin{cases} 0 & \text{for} \quad m \neq n \, , \\ \int_0^a X_m^2 \, dx & \text{"} \quad m = n \, ; \end{cases} \\ \int_{x=0}^{a} X_{m,xx} X_{n,xx} \, dx &= \begin{cases} 0 & \text{for} \quad m \neq n \, , \\ \int_0^a X_{m,xx}^2 \, dx & \text{"} \quad m = n \, . \end{cases} \end{aligned} \tag{8}$$

In the present example, the same boundary conditions apply for the y-direction, and we can therefore use the analogous eigenfunctions $Y_n(y)$ for this direction. Thus, the product series approximation for the plate deflection function reads as follows :

$$w^*(x,y) = \sum_m \sum_n w^*_{mn} X_m(x) Y_n(y) \, . \tag{9}$$

If the distributed load is expanded with respect to the eigenfunctions

$$p = \sum_m \sum_n p_{mn} X_m(x) Y_n(y) \tag{10a}$$

with $$p_{mn} = \frac{\iint p \, X_m \, Y_n \, dx \, dy}{\iint X_m^2 \, Y_n^2 \, dx \, dy} \tag{10b}$$

and if (9) and (10a) are substituted into *(6.38)*, we obtain

$$K \sum_m \sum_n w^*_{mn} \iint X_i \, Y_k \, (\Delta\Delta \, X_m \, Y_n) \, dx \, dy - \sum_m \sum_n p_{mn} \iint X_m \, Y_n \, X_i \, Y_k \, dx \, dy = 0 \, .$$

By considering the orthogonality this yields :

$$w^*_{mn} = \frac{p_{mn} I_3 I_6}{K(I_1 I_6 + 2 I_2 I_5 + I_3 I_4)} \, , \tag{11}$$

where

$$I_1 = \int_0^a X_{m,xxxx}\, X_m\, dx\,, \quad I_2 = \int_0^a X_{m,xx}\, X_m\, dx\,, \quad I_3 = \int_0^a X_m^2\, dx \quad (d/dx \mathrel{\hat{=}} (\ )_{,x}),$$

$$I_4 = \int_0^a Y_{n,yyyy}\, Y_n\, dy\,, \quad I_5 = \int_0^a Y_{n,yy}\, Y_n\, dy\,, \quad I_6 = \int_0^a Y_n^2\, dy \quad (d/dy \mathrel{\hat{=}} (\ )_{,y}).$$

Limitation to a single-term approximation of the form

$$w^* = w_{11}^* X_1 Y_1 = w_{11}^* \left(\sin \lambda_1 \frac{x}{a} - \frac{\sin \lambda_1}{\sinh \lambda_1} \sinh \lambda_1 \frac{x}{a}\right)\left(\sin \lambda_1 \frac{y}{a} - \frac{\sin \lambda_1}{\sinh \lambda_1} \sinh \lambda_1 \frac{y}{a}\right)$$

yields with $\lambda_1 = 3.9266$

and after substitution:

$$w_{11}^* = 0.00198 \frac{p_0 a^4}{K} .$$

Proceeding from $w^*_{,x} = w^*_{,y} = 0$, the maximum deflection is determined as

$$w^*_{max} = 0.00223 \frac{p_0 a^4}{K} \qquad \text{at } x = y = 0.383\, a\ .$$

This value differs from the exact solution by $\approx 3\%$ only.

Exercise B-9-6:

A circular plate clamped at the outer boundary $r = a$ is subjected to a constant circular line load q at a radius $r = b$ $(b < a)$ from the centre point (Fig. B-25).

a) Determine the general solutions for the inner $(r \leq b)$ and the outer part of the plate. How many free constants are obtained for the solution of this problem?

b) How do the boundary conditions and the transition conditions read?

c) Calculate the deflection functions for both the inner and the outer part of the plate. How large is the maximum deflection of the plate?

d) Prove that the radial and the tangential moments in the inner part are equal.

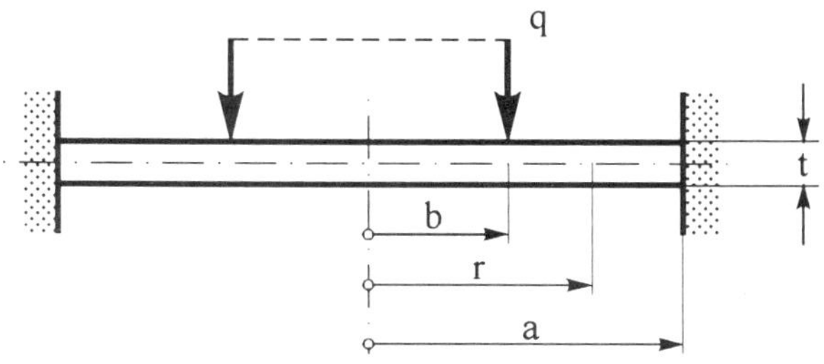

Fig. B-25: Clamped circular plate with a constant circular line load

Solution :

a) The circular plate is divided into an inner and an outer part with the following axisymmetrical loads corresponding Fig. B-26.

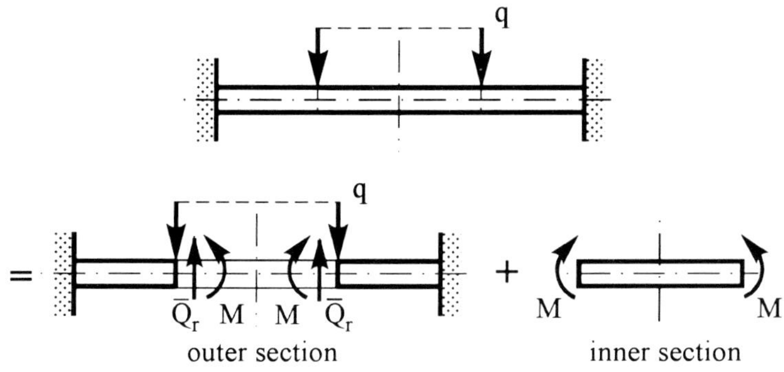

Fig. B-26: Free–body diagram for the outer and inner part of the circular plate

According to *(9.47)*, the solution for the outer part ($r \geq b$, index "o") is

$$w_o(r) = C_0 + C_1 r^2 + C_2 \ln \frac{r}{a} + C_3 r^2 \ln \frac{r}{a} , \qquad (1)$$

and for the inner part ($r \leq b$, index "i") we obtain

$$w_i(r) = C_4 + C_5 r^2 . \qquad (2)$$

Solution (2) takes into consideration that for $r \longrightarrow 0$ the deflection must be finite.

When solving this problem, we altogether obtain six free constants $C_0, \ldots, C_5$.

b) *Boundary conditions and transition conditions*

Boundary conditions for the *outer* part ($r \geq b$):

$$w_o(a) = 0 \quad , \quad w_{o,r}(a) = 0 \qquad (d/dr \mathrel{\widehat{=}} (\;)_{,r}) , \qquad (3a)$$

$$M_{rr_o}(b) = M \quad , \quad \overline{Q}_{r_o}(b) = -q . \qquad (3b)$$

Boundary conditions for the *inner* part ($r \leq b$):

$$M_{rr_i}(b) = M , \qquad (4)$$

where M denotes the yet unknown radial moment at $r = b$.

In addition, the following transition conditions have to be fulfilled for $r = b$:

$$w_i = w_o , \qquad (5a)$$

$$w_{i,r} = w_{o,r} . \qquad (5b)$$

c) *Deflection functions*

– Outer part

First, we determine the derivatives of $w_0(r)$ in (1):

$$w_{o,r} = 2C_1 r + \frac{C_2}{r} + C_3(2r\ln\frac{r}{a} + r), \tag{6a}$$

$$w_{o,rr} = 2C_1 - \frac{C_2}{r^2} + C_3(2\ln\frac{r}{a} + 3), \tag{6b}$$

$$\Delta w_o \mathrel{\widehat{=}} w_{o,rr} + \frac{1}{r} w_{o,r} = 4C_1 + 4C_3(\ln\frac{r}{a} + 1). \tag{6c}$$

Substitution of (6) and (1) into the boundary conditions (3a,b) yields:

$$w_o(a) = 0 = C_0 + C_1 a^2, \tag{7a}$$

$$w_{o,r}(a) = 0 = 2C_1 a + \frac{C_2}{a} + C_3 a, \tag{7b}$$

$$\overline{Q}_{r_o}(b) = -q = -K\frac{d}{dr}\Delta w_o = -4KC_3\frac{1}{b}, \tag{8a}$$

$$M_{rr_o}(b) = M = -K\left[w_{o,rr} + \frac{\nu}{r} w_{o,r}\right] =$$

$$= -K\left[2C_1(1+\nu) - \frac{C_2}{b^2}(1-\nu) + C_3\left(2(1+\nu)\ln\frac{b}{a} + 3 + \nu\right)\right]$$

$$\longrightarrow \quad 2C_1(1+\nu) - \frac{C_2}{b^2}(1-\nu) + C_3\left(2(1+\nu)\ln\frac{b}{a} + 3 + \nu\right) = -\frac{M}{K}. \tag{8b}$$

Eq. (8a) yields: $$C_3 = \frac{bq}{4K}. \tag{9a}$$

We divide (7b) by a, and (8b) by $(1+\nu)$ and subtract:

$$C_2\left(\frac{1}{a^2} + \frac{1}{b^2}\frac{1-\nu}{1+\nu}\right) + C_3\left[1 - \frac{1}{1+\nu}\left(2(1+\nu)\ln\frac{b}{a} + 3 + \nu\right)\right] = \frac{M}{K(1+\nu)}.$$

By substituting (9a) we obtain:

$$C_2 = \frac{1}{\frac{1}{a^2} + \frac{1}{b^2}\frac{1-\nu}{1+\nu}}\left\{\frac{M}{K(1+\nu)} - \frac{bq}{4K}\left[1 - \frac{1}{1+\nu}\left(2(1+\nu)\ln\frac{b}{a} + 3 + \nu\right)\right]\right\}. \tag{9b}$$

Solution of (7b) with respect to C_1 yields:

$$C_1 = -\frac{1}{2}\left(\frac{C_2}{a^2} + C_3\right) = -\frac{C_2}{2a^2} - \frac{bq}{8K}. \tag{9c}$$

For C_0 one obtains from (7a)

$$C_0 = -C_1 a^2 = \frac{C_2}{2} + \frac{bqa^2}{8K}. \tag{9d}$$

By that, the deflection function of the outer plate can be determined, however, still in dependence on the yet unknown radial moment M at r = b.

- Inner part:

Derivatives from (2):
$$w_{i,r} = 2\,C_5\,r\,, \tag{10a}$$
$$w_{i,rr} = 2\,C_5\,. \tag{10b}$$

Substitution of (10 a, b) into the boundary condition (4) yields :

$$M_{rr_i}(b) = M \quad\longrightarrow\quad M = -K\left(w_{i,rr} + \frac{\nu}{r}\,w_{i,r}\right) = -K\,(2\,C_5 + 2\,\nu\,C_5)$$

$$\longrightarrow \qquad C_5 = -\frac{M}{2K(1+\nu)}\,. \tag{11a}$$

The second constant C_4 can be determined by means of the transition condition (5a):

$$w_a(b) = w_i(b) = C_4 + C_5\,b^2$$

$$\longrightarrow \qquad C_4 = w_a(b) + \frac{Mb^2}{2K(1+\nu)}\,. \tag{11b}$$

Thus, the deflection of the inner plate according to (2) is

$$w_i(r) = w_a(b) - \frac{M}{2K(1+\nu)}\left(r^2 - b^2\right). \tag{12}$$

In order to write the final expressions for the deflection functions, we determine the radial moment M by means of the transition condition (5b).

According to (6a) and (9c) :

$$\begin{aligned} w_{o,r}(b) &= 2\,C_1\,b + \frac{C_2}{b} + C_3\left(2\,b\ln\frac{b}{a} + b\right) = \\ &= C_2\left(\frac{1}{b} - \frac{b}{a^2}\right) + 2\,C_3\,b\ln\frac{b}{a}\,. \end{aligned} \tag{13}$$

Herein, we substitute C_2 and C_3 from (9a,b):

$$w_{o,r}(b) = \left(\frac{1}{b} - \frac{b}{a^2}\right)\frac{1}{\dfrac{1}{a^2} + \dfrac{1}{b^2}\dfrac{1-\nu}{1+\nu}}\left\{\frac{M}{K(1+\nu)} - \right.$$
$$\left. - \frac{bq}{4K}\left[1 - \frac{1}{1+\nu}\left(2(1+\nu)\ln\frac{b}{a} + 3 + \nu\right)\right]\right\} + 2\frac{qb^2}{4K}\ln\frac{b}{a}\,,$$

and after reformulation we obtain :

$$w_{o,r}(b) = \frac{-b}{\nu + \dfrac{b^2+a^2}{b^2-a^2}}\,\frac{1}{K}\left\{M - \frac{bq}{4}\left[(1+\nu) - \right.\right.$$
$$\left.\left. - 2(1+\nu)\ln\frac{b}{a} - 3 - \nu\right]\right\} + \frac{b^2q}{2K}\ln\frac{b}{a}\,. \tag{14}$$

Equations (10a) and (11a) provide :

$$w_{i,r}(b) = -\frac{Mb}{K(1+\nu)} \; . \tag{15}$$

Substitution into the transition condition and some algebra finally lead to :

$$M = -(1+\nu)\frac{bq}{2}\left[\frac{1}{2}\left(1-\frac{b^2}{a^2}\right)+\ln\frac{b}{a}\right] . \tag{16}$$

Eqs. (9a ÷ d) substituted into (1) yields the following expressions for the outer deflection function:

$$w_o(r) = \frac{a^2 b^2 (1+\nu)}{b^2(1+\nu)+a^2(1-\nu)}\left\{\frac{M}{K(1+\nu)} - \frac{bq}{4K}\left[1 - \right.\right.$$

$$\left.\left. - \frac{1}{1+\nu}\left(2(1+\nu)\ln\frac{b}{a}+3+\nu\right)\right]\right\}\left(\frac{1}{2}-\frac{r^2}{2a^2}+\ln\frac{r}{a}\right)+$$

$$+\frac{bq}{4K}\left(\frac{a^2}{2}-\frac{r^2}{2}+r^2\ln\frac{r}{a}\right) .$$

With M given by (16) follows :

$$w_o(r) = \frac{qa^2b}{4K}\left[\left(\frac{r^2}{a^2}+\frac{b^2}{a^2}\right)\ln\frac{r}{a}+\frac{1}{2}\left(1+\frac{b^2}{a^2}\right)\left(1-\frac{r^2}{a^2}\right)\right] . \tag{17}$$

The deflection function for the inner plate then becomes :

$$w_i(r) = \frac{qa^2b}{4K}\left[\left(\frac{r^2}{a^2}+\frac{b^2}{a^2}\right)\ln\frac{b}{a}+\frac{1}{2}\left(1+\frac{r^2}{a^2}\right)\left(1-\frac{b^2}{a^2}\right)\right] . \tag{18}$$

The maximum deformation w_{max} for the circular plate is obtained from (18):

$$w_i(0) = w_{max} = \frac{qa^2b}{4K}\left[\frac{b^2}{a^2}\ln\frac{b}{a}+\frac{1}{2}\left(1-\frac{b^2}{a^2}\right)\right] . \tag{19}$$

d) *Radial and tangential moments*

The moments in the inner part are calculated by means of *(9.26)*:

$$M_{rr} = -K\left[w_{i,rr}+\frac{\nu}{r}w_{i,r}\right] ,$$

$$M_{\varphi\varphi} = -K\left[\frac{1}{r}w_{i,r}+\nu\, w_{i,rr}\right] .$$

According to (10a,b) it holds that

$$w_{i,r}(r) = 2C_5 r \quad , \quad w_{i,rr}(r) = 2C_5 \; .$$

Substitution of the above derivatives yields :

$$M_{rr} = -K\left(2C_5+2\frac{\nu}{r}C_5 r\right) = -2C_5K(1+\nu) ,$$

$$M_{\varphi\varphi} = -K\left(\frac{1}{r}2C_5 r+2\nu C_5\right) = -2C_5K(1+\nu) \implies M_{rr} \mathrel{\widehat{=}} M_{\varphi\varphi} = M \; .$$

The two moments are constant for $0 \le r \le b$ and equal to the moment M (see (11a)).

Exercise B-9-7:

A circular ring plate made of steel (E, ν) is clamped at the inner boundary $r = b$; the plate is subjected to the following harmonically varying line load at the outer boundary (Fig. B-27)

$$q(a, \varphi) = \frac{F_1}{a\pi} \cos\varphi = q_1 \cos\varphi .$$

a) Write the differential equation and the boundary conditions for the circular plate.

b) Determine the deflection function $w(r, \varphi)$ for the circular plate.

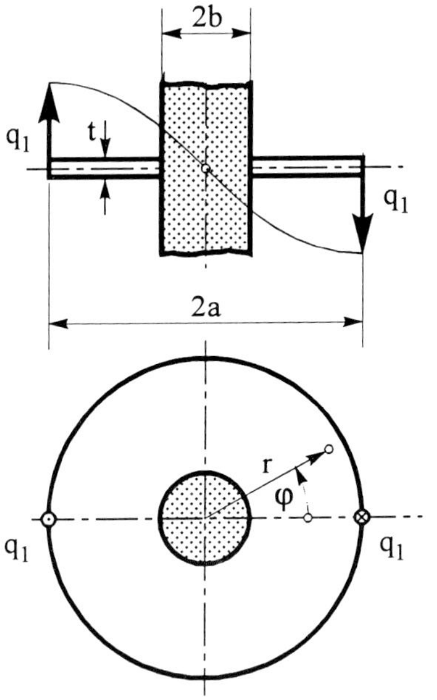

Fig. B-27: Clamped circular ring plate with a line load at the outer boundary

Solution :

a) Based upon the differential equation *(9.25)* for the shear-rigid, isotropic circular plate

$$\Delta\Delta\, w = \left(\frac{\partial^2}{\partial r^2} + \frac{1}{r}\frac{\partial}{\partial r} + \frac{1}{r^2}\frac{\partial^2}{\partial \varphi^2} \right)^2 w = \frac{p(r, \varphi)}{K}$$

and as we have $p(r, \varphi) = 0$ (no surface load) in the present case, the problem is governed by the homogeneous differential equation

$$\Delta\Delta\, w = 0 . \tag{1}$$

The *boundary conditions* for the problem read :

$$w(b, \varphi) = 0 \quad , \quad w_{,r}(b, \varphi) = 0 ; \tag{2a,b}$$

$$M_{rr}(a, \varphi) = 0 \quad , \quad \overline{Q}_r(a, \varphi) = q(a, \varphi) = q_1 \cos\varphi . \tag{2c,d}$$

b) *Deflection function* $w(r, \varphi)$

Due to the harmonic form of the line load, we assume that the deflection function $w(r, \varphi)$ can be written in the following harmonic form with a separation of variables,

$$w(r, \varphi) = w_1(r) \cos\varphi . \tag{3}$$

This form corresponds to the first term (n = 1) in *(9.49)*, which upon substitution into the plate equation *(9.25)* transforms this partial differential equation into the ordinary differential equation *(9.50)* with n = 1 for the function w_1, and the solution w_1 for the homogeneous equation is given in *(9.51a)*. As (1) is homogeneous, its solution w thus takes the form:

$$w(r,\varphi) = \left[C_1 r + \frac{C_2}{r} + C_3 r^3 + C_4 r \ln\frac{r}{b} \right] \cos\varphi \;. \tag{4}$$

In order to determine the four free constants, we calculate the derivatives

$$w_{,r} = \left[C_1 - \frac{C_2}{r^2} + 3\,C_3 r^2 + C_4 \ln\frac{r}{b} + C_4 \right] \cos\varphi \;, \tag{5a}$$

$$w_{,rr} = \left[2\,\frac{C_2}{r^3} + 6\,C_3 r + \frac{C_4}{r} \right] \cos\varphi \;, \tag{5b}$$

$$w_{,\varphi\varphi} = -\left[C_1 r + \frac{C_2}{r} + C_3 r^3 + C_4 r \ln\frac{r}{b} \right] \cos\varphi \;, \tag{5c}$$

$$\Delta w = w_{,rr} + \frac{1}{r} w_{,r} + \frac{1}{r^2} w_{,\varphi\varphi} = \left[8\,C_3 r + \frac{2C_4}{r} \right] \cos\varphi \;. \tag{5d}$$

The radial moment is obtained from *(9.26)* with (5a ÷ c)

$$M_{rr} = -K\left[w_{,rr} + \nu\left(\frac{1}{r} w_{,r} + \frac{1}{r^2} w_{,\varphi\varphi} \right) \right] =$$

$$= -K\left[2\frac{C_2}{r^3} + 6\,C_3 r + \frac{C_4}{r} + \right.$$

$$+ \nu\left(\frac{C_1}{r} - \frac{C_2}{r^3} + 3\,C_3 r + \frac{C_4}{r} \ln\frac{r}{b} + \frac{C_4}{r} - \right.$$

$$\left. \left. - \frac{C_1}{r} - \frac{C_2}{r^3} - C_3 r - \frac{C_4}{r} \ln\frac{r}{b} \right) \right] \cos\varphi$$

$$\longrightarrow \quad M_{rr} = -K\left[2\frac{C_2}{r^3}(1-\nu) + 2\,C_3 r(3+\nu) + C_4 \frac{1}{r}(1+\nu) \right] \cos\varphi \;. \tag{6}$$

By means of *(9.27a)* the effective transverse shear force is determined as:

$$\overline{Q}_r = -K\left[(\Delta w)_{,r} + \frac{1-\nu}{r} \left(\frac{1}{r} w_{,\varphi r} - \frac{1}{r^2} w_{,\varphi} \right)_{,\varphi} \right] ,$$

$$\overline{Q}_r = -K\left[+8\,C_3 - \frac{2C_4}{r^2} + \frac{1-\nu}{r}\left(-\frac{C_1}{r} + \frac{C_2}{r^3} - 3\,C_3 r - \frac{C_4}{r} \ln\frac{r}{b} - \right.\right.$$

$$\left.\left. - \frac{C_4}{r} + \frac{C_1}{r} + \frac{C_2}{r^3} + C_3 r + \frac{C_4}{r} \ln\frac{r}{b} \right) \right] \cos\varphi$$

$$\longrightarrow \quad \overline{Q}_r = -K\left[\frac{2(1-\nu)}{r^4} C_2 + 2(3+\nu) C_3 - \frac{(3-\nu)}{r^2} C_4 \right] \cos\varphi \;. \tag{7}$$

Substitution into the boundary conditions yields :

$$(4) \text{ in } (2a) \quad \longrightarrow \quad b\,C_1 + \frac{C_2}{b} + b^3 C_3 = 0 \;, \tag{8a}$$

$$(5a) \text{ in } (2b) \quad \longrightarrow \quad C_1 - \frac{C_2}{b^2} + 3\,b^2 C_3 + C_4 = 0 \;, \tag{8b}$$

$$(6) \text{ in } (2c) \quad \longrightarrow \quad \frac{2(1-\nu)}{a^3} C_2 + 2\,a(3+\nu) C_3 + (1+\nu)\frac{C_4}{a} = 0 \;, \tag{8c}$$

(7) in (2d) $\longrightarrow$

$$-K\left[\frac{2(1-\nu)}{a^4}C_2 + 2(3+\nu)C_3 - \frac{(3-\nu)}{a^2}C_4\right]\cos\varphi = q(a,\varphi) = q_1\cos\varphi$$

$$\longrightarrow \quad \frac{2(1-\nu)}{a^4}C_2 + 2(3+\nu)C_3 - \frac{(3-\nu)}{a^2}C_4 = -\frac{q_1}{K}\,. \qquad (8d)$$

C_1 can be eliminated from (8a) and (8b):

$$\left.\begin{aligned} C_1 &= -\frac{C_2}{b^2} - b^2 C_3 \\ C_1 &= \frac{C_2}{b^2} - 3b^2C_3 - C_4 \end{aligned}\right\} \longrightarrow \frac{2C_2}{b^2} - 2b^2C_3 - C_4 = 0\,. \qquad (9)$$

The relations (9), (8c) and (8d) form a system of linear equations for determination of the constants C_2, C_3 and C_4:

$$\left.\begin{aligned} \frac{2}{b^2}C_2 \quad &- 2b^2C_3 &- C_4 &= 0\,, \\ (1-\nu)\frac{2}{a^3}C_2 + &2(3+\nu)a\,C_3 + &(1+\nu)\frac{1}{a}C_4 &= 0\,, \\ (1-\nu)\frac{2}{a^4}C_2 + &2(3+\nu)\,C_3 + &(-3+\nu)\frac{1}{a^2}C_4 &= -\frac{q_1}{K}\,. \end{aligned}\right\} \qquad (10)$$

Solution of (10) yields the following constants :

$$C_2 = \frac{q_1 a^4 b^2}{8K}\,\frac{(3+\nu)a^2 - (1+\nu)b^2}{(3+\nu)a^4 + (1-\nu)b^4}\,,$$

$$C_3 = -\frac{q_1 a^2}{8K}\,\frac{(1-\nu)b^2 + (1+\nu)a^2}{(3+\nu)a^4 + (1-\nu)b^4}\,, \qquad C_4 = \frac{q_1 a^2}{4K}\,.$$

Finally, (8a) becomes :

$$C_1 = -\frac{C_2}{b^2} - b^2C_3 \longrightarrow C_1 = \frac{q_1 a^2}{8K}\,\frac{(3+\nu)a^4 - 2(1+\nu)a^2b^2 - (1-\nu)b^4}{(3+\nu)a^4 + (1-\nu)b^4}\,.$$

We can now calculate the deflection function $w(r,\varphi)$ from (4).

Assuming that $\nu = 1/3$, with $a = 4b$ we obtain a maximum deflection at the boundary given by

$$w(a,0) = \left(16\ln 4 - \frac{4440}{427}\right)\frac{q_1 b^3}{K} \approx 11.78\,\frac{q_1 b^3}{K}\,.$$

Exercise B-9-8:

A shear – rigid circular plate (radius a, thickness t) as shown in Fig. B-28 is resting on a linearly elastic foundation (foundation stiffness $k\,[N/m^3]$), and is subjected to a constant surface load p_0.

a) Express the total potential of the given system.

b) Derive the differential equation and the boundary conditions for the plate by a calculus of variations approach.

c) Calculate the maximum deflection at the centre of the plate by means of the RITZ method applying $w^* = a_0 + a_2 r^2$ (where a_0 and a_2 are free coefficients) as an approximation to the deflection function.

Numerical values:

$E = 2.1 \cdot 10^5\ MPa$, $t = 0.02\ m$, $\nu = 0.3$, $a = 0.5\ m$, $k = 0.487 \cdot 10^6\ N/m^3$.

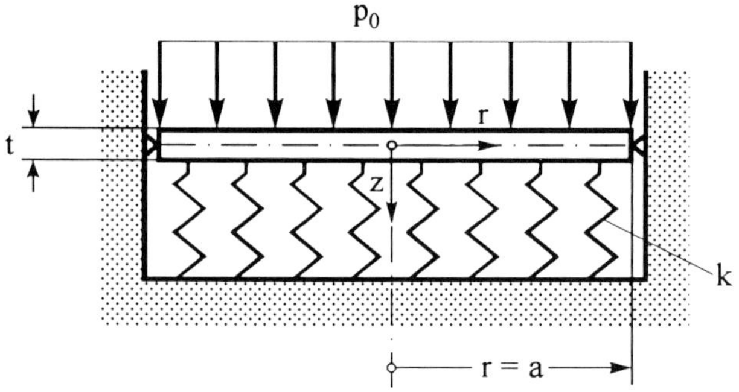

Fig. B-28: Circular plate subjected to a uniformly distributed load p_0 and resting on an elastic foundation

Solution:

a) *Total potential*

The total potential energy is the sum of the elastic energy of the plate Π_p, the elastic energy of the foundation Π_b, and the potential of the external load Π_e:

$$\Pi = \Pi_p + \Pi_b + \Pi_e \,. \tag{1}$$

The *elastic energy of the circular plate* can be obtained from *(9.35)*. Since we are dealing with an axisymmetric problem all derivatives $(\ )_{,\varphi} \longrightarrow 0$. Thus, we obtain the simplified form with $d/dr \mathrel{\widehat{=}} (\ )_{,r}$

$$\Pi_p = \frac{1}{2} \underbrace{\frac{Et^3}{12(1-\nu^2)}}_{K} \int_A \left[\left(w_{,rr} + \frac{1}{r} w_{,r} \right)^2 - \frac{2(1-\nu)}{r} w_{,r} w_{,rr} \right] dA \,. \tag{2}$$

The *elastic energy of the foundation* is

$$\Pi_b = \frac{1}{2} \int_0^{2\pi} \int_0^{a} (k\, w)\, w\, dr\, r\, d\varphi = \frac{1}{2} k \int_{\varphi=0}^{2\pi} \int_{r=0}^{a} w^2\, r\, dr\, d\varphi \,. \tag{3}$$

Finally, the *potential of the external forces* is

$$\Pi_e = -\int_0^{2\pi}\int_0^{a} p_0\, w\, r\, dr\, d\varphi\ . \tag{4}$$

After integration over φ, (1) with (2) and (4) yields the *total potential energy*

$$\Pi = \pi\int_0^{a}\left\{K\left[\left(w_{,rr} + \frac{1}{r}\,w_{,r}\right)^2 - \frac{2(1-\nu)}{r}\,w_{,r}\,w_{,rr}\right] + k\,w^2 - 2\,p_0\,w\right\} r\,dr\ . \tag{5}$$

b) *Differential equation and boundary conditions*

In the following, we will determine the differential equation and the boundary conditions of the problem. The calculus of variations is employed to achieve this goal.

The basic requirement is that the variation of the total potential energy must vanish, and in view of (5) and *(6.34)* this may be expressed as:

$$\delta\,\Pi = \pi\,\delta\int_0^{a} F(r, w, w_{,r}, w_{,rr})\,dr \overset{!}{=} 0 \tag{6a}$$

with the basic function

$$F(r, w, w_{,r}, w_{,rr}) = K\left[r\,w_{,rr}^{\;2} + 2\,w_{,r}\,w_{,rr} + \frac{1}{r}\,w_{,r}^{\;2} - 2(1-\nu)\,w_{,r}\,w_{,rr}\right] + \\ + k\,r\,w^2 - 2\,p_0\,r\,w\ . \tag{6b}$$

Based on *(6.35)*, we obtain as a necessary condition from (6a) the EULER differential equation

$$\frac{\partial F}{\partial w} - \left(\frac{\partial F}{\partial w_{,r}}\right)_{,r} + \left(\frac{\partial F}{\partial w_{,rr}}\right)_{,rr} = 0 \tag{7}$$

with the boundary conditions

$$\left[\frac{\partial F}{\partial w_{,r}} - \left(\frac{\partial F}{\partial w_{,rr}}\right)_{,r}\right]\delta w\,\Bigg|_0^a = 0\ , \tag{8a}$$

$$\frac{\partial F}{\partial w_{,rr}}\,\delta w_{,r}\,\Bigg|_0^a = 0\ . \tag{8b}$$

If we substitute the derivatives

$$\frac{\partial F}{\partial w} = 2\,k\,r\,w - 2\,p_0\,r\ ,$$

$$\left(\frac{\partial F}{\partial w_{,r}}\right)_{,r} = 2\,K\left(\nu\,w_{,rrr} + \frac{1}{r}\,w_{,rr} - \frac{1}{r^2}\,w_{,r}\right),$$

$$\left(\frac{\partial F}{\partial w_{,rr}}\right)_{,rr} = 2\,K\left[(2+\nu)\,w_{,rrr} + r\,w_{,rrrr}\right]$$

into (7), we obtain the differential equation for the present problem as

$$2krw - 2p_0 r - 2K\Big(\nu w_{,rrr} + \frac{1}{r} w_{,rr} - \frac{1}{r^2} w_{,r}\Big) + \\ + 2K\big[(2+\nu) w_{,rrr} + r w_{,rrrr}\big] = 0$$

$$\longrightarrow \quad w_{,rrrr} + \frac{2}{r} w_{,rrr} - \frac{1}{r^2} w_{,rr} + \frac{1}{r^3} w_{,r} + \frac{k}{K} w = \frac{p_0}{K} . \qquad (9)$$

Eq. (9) is a differential equation of the BESSEL type [B.3].

By substituting the derivatives into (8a,b), the boundary conditions result as

$$\underbrace{K\Big(\frac{1}{r^2} w_{,r} - \frac{1}{r} w_{,rr} - w_{,rrr}\Big)}_{\widehat{=}\, Q_r} \delta w \Big|_0^a = 0 , \qquad (10a)$$

$$\underbrace{K\big(w_{,rr} + \frac{\nu}{r} w_{,r}\big)}_{\widehat{=}\, M_{rr}} \delta w_{,r} \Big|_0^a = 0 . \qquad (10b)$$

Eq. (10a,b) allows us to establish different combinations of boundary conditions.

c) *Solution by means of the RITZ method*

The following approximation for the deflection function is given :

$$w^*(r) = a_0 + a_2 r^2 . \qquad (11a)$$

The approximation must comply with the essential (geometrical) boundary conditions, i.e.,

$$w^*(a) = 0 , \qquad (12a)$$

$$w^*_{,r}(0) = 0 . \qquad (12b)$$

Substituting (11a) into (12a) ((12b) is automatically fulfilled) gives:

$$a_0 + a_2 a^2 = 0 \quad \longrightarrow \quad a_0 = -a_2 a^2 .$$

Thus, the approximation function only depends on one free coefficient, say, a_2 :

$$w^*(r) = a_2 (r^2 - a^2) . \qquad (11b)$$

The derivatives

$$w^*_{,r} = 2a_2 r \quad , \quad w^*_{,rr} = 2a_2$$

substituted into (5) yield :

$$\Pi = \pi \int_0^a \Big\{ K\big[(2a_2 + 2a_2)^2 - \frac{2(1-\nu)}{r} 2a_2 2a_2 r\big] + \\ + k a_2^2 (r^4 - 2r^2 a^2 + a^4) - 2p_0 a_2 (r^2 - a^2)\Big\} r\,dr = $$

$$= \pi \int_0^a \Big\{ K a_2^2 8(1+\nu) r + k a_2^2 (r^5 - 2r^3 a^2 + a^4 r) - \\ - 2p_0 a_2 (r^3 - r a^2)\Big\} dr .$$

After integration we obtain

$$\Pi = \pi \left\{ a_2^2 \left[4\,K\,a^2 (1+\nu) + \frac{1}{6}\,k\,a^6 \right] + a_2\, p_0 \frac{1}{2}\, a^4 \right\} . \qquad (13)$$

From the extremum condition *(6.37)*, i.e. $\frac{\partial \Pi}{a_2} = 0$ follows that

$$a_2 \left[8\,K\,a^2 (1+\nu) + \frac{1}{3}\,k\,a^6 \right] + \frac{1}{2}\, p_0\, a^4 = 0 \quad \longrightarrow \quad a_2 = -\frac{p_0\, a^2}{16\,K(1+\nu) + \frac{2}{3} k\, a^4}$$

and thus the approximative deflection function is determined as

$$w^*(r) = \frac{p_0\, a^4}{16\,K(1+\nu) + \frac{2}{3} k\, a^4} \left(1 - \frac{r^2}{a^2}\right) .$$

For the given numerical values we obtain

$$w_0^* = w^*(0) = 1.296\, \frac{p_0}{k} .$$

The exact solution can be found in [MARKUS, G.: Theory and Calculation of axisymmetrical Structures (in German). Düsseldorf: Werner – Verlag, 4th edition 1986] as :

$$w_0 = \frac{p_0}{k} \left[1 - \frac{\mathrm{ber}\,\alpha - \frac{1-\nu}{\alpha}\,\mathrm{bei}'\alpha}{(\mathrm{ber}\,\alpha - \frac{1-\nu}{\alpha}\,\mathrm{bei}'\alpha)\,\mathrm{ber}\,\alpha + (\mathrm{bei}\,\alpha + \frac{1-\nu}{\alpha}\,\mathrm{ber}'\,\alpha)\,\mathrm{bei}\,\alpha} \right] \qquad (14)$$

with $\quad \alpha = \frac{a}{l} \quad , \quad l = \sqrt[4]{\frac{K}{k}} \; .$

In (14), " ber " and " bei " denote modified cylindrical functions with complex arguments, named after LORD KELVIN (see " Tables of BESSEL Functions $Y_0(z)$ and $Y_1(z)$ for Complex Arguments ", New York, Columbia University Press 1950). A BESSEL function with imaginary arguments consists, e.g., of the two KELVIN functions :

$$I_0\left(x \sqrt{\pm i}\, \right) = \mathrm{ber}\, x \pm \mathrm{bei}\, x \; .$$

The maximum deflection calculated by means of (14) leads to :

$$w_0 = 1.239\, \frac{p_0}{k} .$$

Hence, the ratio between the approximate and the exact value is

$$\frac{w_0^*}{w_0} = 1.046 \;, \text{ i.e. the agreement is rather good.}$$

Exercise B-9-9:

Consider a thin, circular, rotationally symmetric shear – rigid plate with radius a (Fig. B-29). The plate has nonuniform thickness described by the hyperbolic function

$$t(r) = t_0 \left(\frac{r}{a} \right)^{-x/3} ,$$

for the distance of the lower plate surface from the plane upper surface of the plate. Here, t_0 denotes the plate thickness at the outer boundary, and x is a free shape parameter. At the centre of the plate ($r = 0$) the thickness function tends to infinity, and a point support can be assumed here. This local behaviour contradicts the assumption for a thin plate, and therefore the results in close proximity to the centre point of the plate present a rough approximation only.

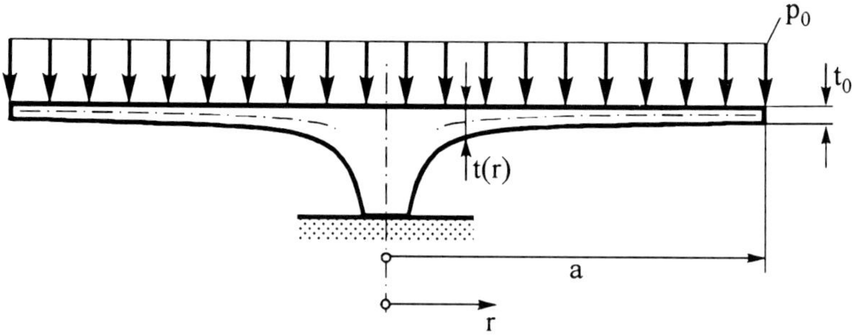

Fig. B-29: Centre–supported circular plate with nonuniform thickness

a) Derive the differential equation for a plate with variable thickness.

b) Determine the deflection functions for this type of circular plate ($a = 0.15\,m$, $t_0 = 0.005\,m$) corresponding to the three values of the shape parameter $x = 0.5$; 2.0 and 3.0. Compare the results to those for a plate with constant thickness $t_0 = 0.005\,m$.

Solution :

a) *Formulation of the differential equation*

We disregard the fact that the mid–surface of the plate is not plane, and proceed from the equilibrium conditions for a circular plate subjected to an axisymmetric load (see [ET2]):

$$(r\,Q_r)_{,r} + p(r)\,r = 0\,, \qquad (1a)$$

$$(r\,M_{rr})_{,r} - M_{\varphi\varphi} - Q_r\,r = 0\,. \qquad (1b)$$

By denoting the slope of the deflection function by ψ ($d/dr \mathrel{\widehat{=}} (\ )_{,r}$), the radial and circumferential moments read :

$$M_{rr} = K(r)\left(\psi_{,r} + \nu\frac{\psi}{r}\right), \qquad (2a)$$

$$M_{\varphi\varphi} = K(r)\left(\frac{\psi}{r} + \nu\,\psi_{,r}\right) \qquad (2b)$$

with the radius – dependent plate stiffness

$$K(r) = \frac{E\,t^3(r)}{12(1-\nu^2)} \;.$$

Substitution of (2a,b) into (1b) leads to the differential equation

$$\psi_{,rr} + \frac{1}{r}\left(1 + \frac{K(r)_{,r}}{K(r)}\,r\right)\psi_{,r} - \frac{1}{r^2}\left(1 - \frac{K(r)_{,r}}{K(r)}\,\nu\,r\right)\psi = \frac{Q(r)}{K(r)} \;. \qquad (3)$$

If we substitute the prescribed function for the plate thickness

$$t(r) = t_0\left(\frac{r}{a}\right)^{-x/3} \qquad (4)$$

into the plate stiffness $K(r)$, we obtain

$$K = K_0\left(\frac{r}{a}\right)^{-x} \qquad (5a)$$

with $\quad K_0 = \dfrac{E t_0^3}{12(1-\nu^2)} = \text{const}\;.$

The derivative of K with respect to r is

$$K(r)_{,r} = -K_0\frac{x}{r}\left(\frac{r}{a}\right)^{-x} \;. \qquad (5b)$$

The relations (5a,b) are now substituted into (3), and we herewith obtain a differential equation of the form

$$\psi_{,rr} + \frac{1-x}{r}\,\psi_{,r} - \frac{1+\nu x}{r^2}\,\psi = \frac{Q_r}{K_0}\left(\frac{r}{a}\right)^{x} \qquad (6a)$$

or, alternately, $\quad r^2\,\psi_{,rr} + (1-x)\,r\,\psi_{,r} - (1+\nu x)\,\psi = \dfrac{Q_r}{K_0}\,\dfrac{r^{x+2}}{a^x} \;. \qquad (6b)$

The above equation is an EULER differential equation for the slope ψ of the deflection. On the right – hand side, the yet unknown transverse force Q_r occurs which consists of the constant surface load p and of a concentrated load F at the centre of the plate (reaction force from the support).

To determine $Q_r(r)$ in (6b), we integrate once (1a) with $p = \text{const}$ and obtain

$$r\,Q_r(r) = C_1 - \frac{1}{2}p\,r^2 \;.$$

Here, we determine the integration constant C_1 from the boundary condition $Q_r(a) = 0$, and find $C_1 = (p\,a^2)/2$. Substituting this and solving the above equation for $Q_r(r)$ yields

$$Q_r(r) = \frac{1}{2}\left(\frac{pa^2}{r} - p\,r\right) \;. \qquad (7)$$

Eq. (7) is now substituted into (6b), and we obtain

$$r^2\,\psi_{,rr} + (1-x)\,r\,\psi_{,r} - (1+\nu x)\,\psi = \frac{1}{2K_0}\left(\frac{pa^2}{r} - p\,r\right)\frac{r^{x+2}}{a^x} =$$

$$= \frac{1}{2K_0}\left(\frac{p r^{x+1}\,a^2}{a^x} - \frac{p r^{x+3}}{a^x}\right) \qquad (8)$$

as a differential equation which describes the given problem.

b) *Deflection*

The solution to (8) consists of a homogeneous and a particular part.

Homogeneous solution:

By assuming that

$$\psi = C r^{\lambda} \tag{9}$$

the homogeneous equation yields:

$$\lambda^2 - x\lambda - (1 + \nu x) = 0 . \tag{10}$$

The roots in (10) yield

$$\lambda_{1,2} = \frac{x}{2} \pm \sqrt{\frac{x^2}{4} + 1 + \nu x} . \tag{11}$$

Particular solution:

In order to determine the particular solution, we assume it to be of the right–hand–side type. For the first particular solution we write

$$\psi_{p_1} = A r^{x+1} \quad \Longrightarrow \quad \begin{aligned} \psi_{p1,r} &= A(x+1) r^{x} , \\ \psi_{p1,rr} &= A(x+1) x r^{x-1} . \end{aligned} \tag{12}$$

Substitution of this function and its derivatives into (8) leads to:

$$r^2 A(x+1) x r^{x-1} + (1-x) r A(x+1) r^{x} - (1+\nu x) A r^{x+1} = \frac{1}{2K_0} \frac{p r^{x+1}}{a^{x-2}}$$

$$\longrightarrow \quad \underbrace{A\big[(x+1)x + (1-x)(x+1) - (1+\nu x)\big]} = \frac{1}{2K_0} \frac{p}{a^{x-2}}$$

$$A x (1-\nu) = \frac{1}{2K_0} \frac{p}{a^{x-2}}$$

$$\longrightarrow \quad A = \frac{p}{2x(1-\nu) K_0 a^{x-2}} .$$

Thus, the first particular solution is:

$$\psi_{p_1} = \frac{p r^{x+1}}{2x(1-\nu) K_0 a^{x-2}} . \tag{13}$$

For the second particular solution we assume that

$$\psi_{p2} = B r^{x+3} \quad \Longrightarrow \quad \begin{aligned} \psi_{p2,r} &= B(x+3) r^{x+2} , \\ \psi_{p2,rr} &= B(x+3)(x+2) r^{x+1} . \end{aligned} \tag{14}$$

Substitution of (14) into (8) then yields:

$$r^2 B(x+3)(x+2) r^{x+1} + (1-x) r B(x+3) r^{x+2} -$$

$$- (1+\nu x) B r^{x+3} = - \frac{p}{2K_0 a^{x}} r^{x+3}$$

$$\longrightarrow \quad B = - \frac{p}{2\big[8 + (3-\nu) x\big] K_0 a^{x}} .$$

The second particular integral is then written as

$$\psi_{p2} = -\frac{pr^{x+3}}{2\left[\,8+(3-\nu)x\right] K_0\, a^{x}}\,, \tag{15}$$

and the total solution then reads:

$$\psi = C_1 r^{\lambda_1} + C_2 r^{\lambda_2} + \frac{pr^{x+1}}{2(1-\nu)\,x\,K_0\,a^{x-2}} - \frac{pr^{x+3}}{2\left[\,8+(3-\nu)x\right] K_0\, a^{x}}\,. \tag{16}$$

The free constants C_1 and C_2 are determined from the following *boundary conditions*:

1) $\psi(0) = 0$.
According to (11), λ_2 must always be negative, i.e. the solution will always take the form $C_2 / r^{|\lambda_2|}$, and therefore we must have $C_2 = 0$.

2) $M_{rr}(a) = 0$, i.e. the radial moment at the outer boundary has to vanish.
Considering (2a) and (5a), M_{rr} can be determined as follows:

$$M_{rr} = K_0\left(\frac{r}{a}\right)^{-x}\left(\psi_{,r} + \nu\frac{\psi}{r}\right). \tag{17}$$

If we form the derivative of (16) and substitute it into (17), we obtain

$$M_{rr}(r=a) = K_0\Bigg[\lambda_1 C_1 a^{\lambda_1-1} + \frac{(x+1)pa^2}{2(1-\nu)xK_0} - \frac{(x+3)pa^2}{2\left[8+(3-\nu)x\right]K_0} + $$

$$+ \frac{\nu}{a}\left(C_1 a^{\lambda_1} + \frac{pa^3}{2(1-\nu)xK_0} - \frac{pa^3}{2\left[8+(3-\nu)x\right]K_0}\right)\Bigg] = 0$$

$$\longrightarrow \qquad C_1 = -\frac{p}{a^{\lambda_1-3}(\lambda_1+\nu)K_0}\left[\frac{x^2+2(2+\nu)x+4(1+\nu)}{(8+(3-\nu)x)(1-\nu)x}\right]. \tag{18}$$

The slope of the deflection function can now be calculated by means of (16), and the deflection itself is obtained by integration,

$$\psi = -w_{,r} \quad\longrightarrow\quad w = -\int \psi(r)\,dr + C_3\,. \tag{19}$$

The boundary condition for determining the remaining free constant C_3 reads

$$w(0) = 0 \quad\longrightarrow\quad C_3 = 0\,.$$

After integration of (19), we obtain the following function for the deflection:

$$\left.\begin{aligned} w(r) = -C_1\frac{1}{\lambda_1+1}\,r^{\lambda_1+1} - \frac{pr^{x+2}}{2(1-\nu)x(x+2)K_0\,a^{x-2}} &+ \\ + \frac{pr^{x+4}}{2\left[8+(3-\nu)x\right](x+4)K_0\,a^{x}}&\,, \end{aligned}\right\} \tag{20a}$$

and, in shorthand notation:

$$w(r) = \left(K_1 r^{\lambda_1+1} - K_2 r^{x+2} + K_3 r^{x+4}\right)p\,, \tag{20b}$$

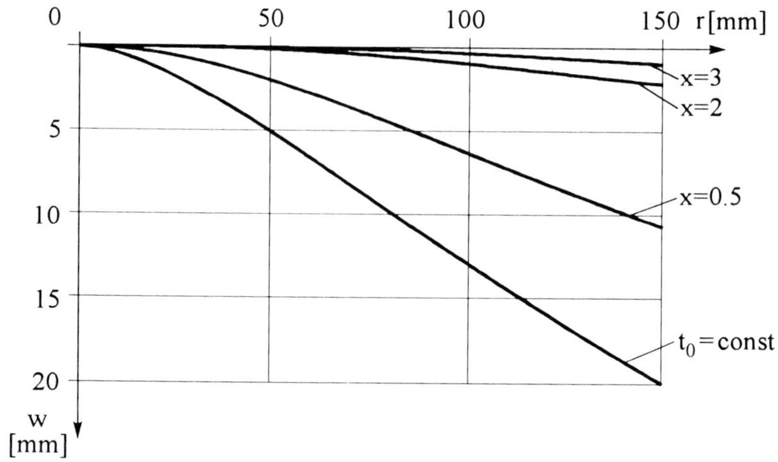

Fig. B-30: Deflection of a point–supported plate with variable thickness

where

$$K_1 = \frac{x^2 + 2(2+\nu)x + 4(1+\nu)}{a^{\lambda_1 - 3}(\lambda_1 + \nu) K_0 \left[8 + (3-\nu)x\right](1-\nu)(\lambda_1 + 1)} ,$$

$$K_2 = \frac{1}{2(1-\nu)\, x\, a^{x-2}(x+2) K_0} ,$$

$$K_3 = \frac{1}{2\left[8 + (3-\nu)\right](x+4) K_0\, a^x} .$$

Fig. B-30 shows curves for the deflection for the shape parameters $x = 0.5$; 2.0 and 3.0. These curves may be compared with the solution for a plate with central support and constant wall thickness

$$w_{const} = \frac{p_0}{64K}\left[r^4 + 2a^2 r^2\left(\frac{3+\nu}{1+\nu} + 4\ln\frac{r}{a}\right)\right]$$

which is also shown in the figure. This expression cannot be determined as a special case ($x = 0$) of (20b), since the characteristic equation (10) possesses a double root $\lambda = \pm 1$ for $x = 0$.

Exercise B-10-1:

A shear–rigid rectangular plate (dimensions a, b; thickness t) with simply supported edges and with a stiffener along its centre line $y = 0$ (cross sectional area A_s, bending stiffness $E I_{y_s}$) as shown in Fig. B-31, is subjected to an axial load $N_x = \text{const}$.

a) Establish the differential equations for buckling of the two plate parts indicated by ① and ② in the figure.

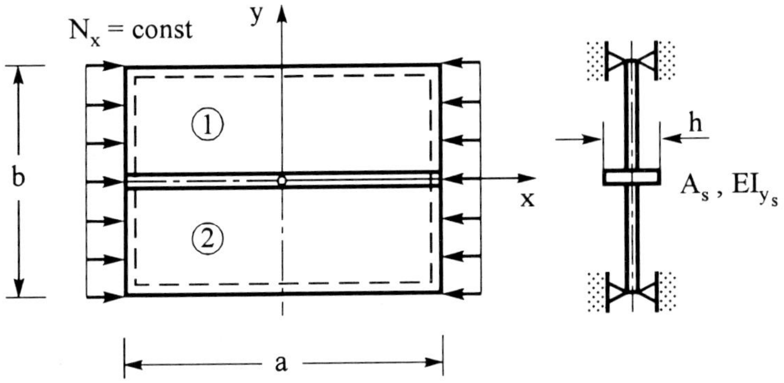

Fig. B-31: Rectangular plate with stiffener

b) Buckling modes that are symmetrical and antisymmetrical with respect to the longitudinal stiffener may occur which can be described by means of the transition conditions. Discuss the characteristic buckling equation of the stiffened plate.

Solution :

a) Each of the two plate parts ① and ② can be viewed as a rectangular plate with simply supported longitudinal boundaries at $x = \pm a/2$. The corresponding differential equation reads according to *(10.18)*

$$K \Delta\Delta w_i + N_x w_{i,xx} = 0 \qquad (i = 1, 2 \text{ number of the plate parts}) , \tag{1}$$

where N_x is considered positive if compression and $\partial/\partial x \mathrel{\hat{=}} (\,)_{,x}$.

For the solution we use LEVY's series approximation *(9.40)*:

$$w(x,y) = \sum_m w_m(y) \sin \frac{m\pi x}{a} \tag{2}$$

with $m = 2, 4, 6 \ldots$.

Substitution into (1) yields an ordinary differential equation of fourth order for $w_m(y)$ ($d/dy \mathrel{\hat{=}} (\,)_{,y}$)

$$w_m(y)_{,yyyy} - 2 w_m(y)_{,yy} \left(\frac{m\pi}{a} \right)^2 + w_m(y) \left[\left(\frac{m\pi}{a} \right)^4 - \frac{N_x}{K} \left(\frac{m\pi}{a} \right)^2 \right] = 0 . \tag{3a}$$

With the shorthand notations $\vartheta_1 = \frac{m\pi}{a}$, $n_x^2 = \frac{N_x}{K}$, (3a) reads :

$$w_m(y)_{,yyyy} - 2 \vartheta_1^2 w_m(y)_{,yy} + \vartheta_1^4 \left(1 - \frac{n_x^2}{\vartheta_1^2} \right) w_m(y) = 0 . \tag{3b}$$

Herewith, we have transformed (1) into an ordinary differential equation with constant coefficients. The assumption $w_m(y) = e^{\lambda y}$ leads to the characteristic equation

$$\lambda^4 - 2 \vartheta_1^2 \lambda^2 + \vartheta_1^4 \left(1 - \frac{n_x^2}{\vartheta_1^2} \right) = 0 \tag{4a}$$

with the roots

$$\lambda_{1,2} = \pm\sqrt{\vartheta_1(\vartheta_1 + n_x)} = \pm\sqrt{\vartheta_1(n_x + \vartheta_1)} = \pm\varkappa_1 ,$$
$$\lambda_{3,4} = \pm\sqrt{\vartheta_1(\vartheta_1 - n_x)} = \pm i\sqrt{\vartheta_1(n_x - \vartheta_1)} = \pm i\varkappa_2 . \qquad (4b)$$

The solution to (3b) then reads

$$w_m(y) = C_1 e^{\kappa_1 y} + C_2 e^{-\kappa_1 y} + C_3 e^{i\kappa_2 y} + C_4 e^{-i\kappa_2 y} . \qquad (5a)$$

If the exponential functions are substituted by trigonometrical and hyperbolical functions, (5a) is transformed into:

$$w_m(y) = A\cosh\varkappa_1 y + B\sinh\varkappa_1 y + C\cos\varkappa_2 y + D\sin\varkappa_2 y , \qquad (5b)$$

and thus the general solution of (1) including (5b) reads:

$$w(x,y) = [A\cosh\varkappa_1 y + B\sinh\varkappa_1 y + C\cos\varkappa_2 y + D\sin\varkappa_2 y]\sin\frac{m\pi x}{a} , \qquad (6)$$

where the constants A, B, C and D are to be determined by means of the boundary conditions.

Because of symmetry, solutions for the plate parts ① and ② have the same form (6).

b) The plate as a whole may exhibit buckling modes that are symmetric and antisymmetric with respect to the stiffener (see Fig. B-32).

– *Symmetrical buckling*

The corresponding boundary conditions read as follows:

$$w(x, \pm\frac{b}{2}) = 0 \quad \longrightarrow \quad w_m(\pm\frac{b}{2}) = 0 . \qquad (7a)$$

The bending moment M_{yy} has to vanish at the boundaries $y = \pm b/2$ which implies, according to *(9.11)*:

$$M_{yy}(x,\frac{b}{2}) = -K[w(x,\frac{b}{2})_{,yy} + \nu w(x,\frac{b}{2})_{,xx}] = 0 .$$

Assuming this boundary is to remain straight, we therefore demand that

$$w(x,\frac{b}{2})_{,xx} = 0 \quad , \quad \text{and thus} \quad w(x,\frac{b}{2})_{,yy} = 0 .$$

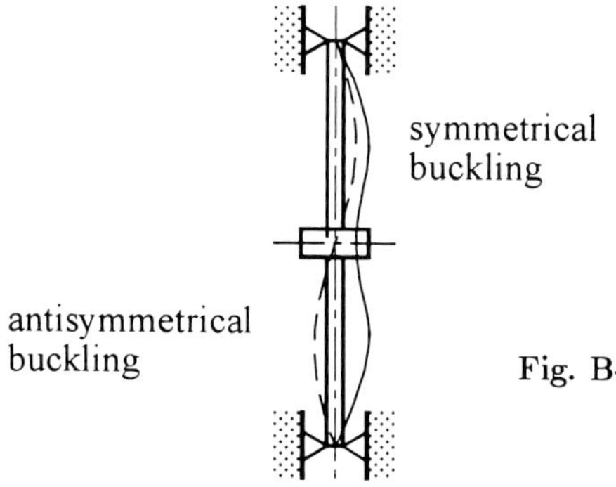

Fig. B-32: Symmetrical and antisymmetrical buckling of a plate

Considering (2) this means :

$$w_m\left(\pm\frac{b}{2}\right)_{,yy} = 0 \,. \qquad (7b)$$

With symmetrical buckling, the buckling shape at the stiffener has to have horizontal tangents in the y – direction and we therefore have

$$w(x,0)_{,y} = 0 \,,$$

i.e. according to (2):

$$w_m(0)_{,y} = 0 \,. \qquad (7c)$$

Finally, we have to formulate the transition condition between the single plate parts and the stiffener (see Fig. B-33). Assuming that $q = q(x)$ is the stiffener loading and $\overline{Q}_{y_i}$ $(i = 1,2)$ are KIRCHHOFF's effective transverse forces corresponding to *(9.14)*, we demand that:

$$q = \overline{Q}_{y_1} - \overline{Q}_{y_2} \,. \qquad (8)$$

It follows from *(9.14c)* that

$$\overline{Q}_y = -K\left[w_{,yyy} + (2-\nu)\,w_{,xxy}\right] ,$$

and from (8) follows :

$$\overline{Q}_{y_1} - \overline{Q}_{y_2} = -K\left[w_{1,yyy} - w_{2,yyy} + (2-\nu)(w_{1,xxy} - w_{2,xxy})\right]_{y=0} = q \,.$$

For reasons of symmetry we have for $y = 0$ that:

$$w_{1,y} = w_{2,y} \quad \text{and} \quad w_{1,yyy} = -w_{2,yyy} \,.$$

Thus, the following is valid for the stiffener loading at $y = 0$:

$$q(x) = -2K\,w_{1,yyy} \,. \qquad (9)$$

The stiffener loading q resulting from the difference of the effective transverse shear forces relates to the displacement w via the differential equation of the bending of a BERNOULLI – beam with an additional axial load [A.19]

$$F_x = N_x\frac{A_s}{h} \,,$$

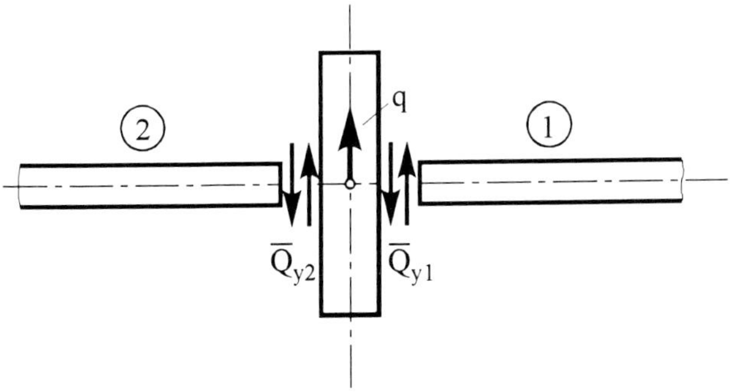

Fig. B-33 : Free – body diagram of the transition between plate and stiffener

i.e., $$q(x) = E\,I_{y_s}\, w_{,xxxx} + N_x \frac{A_s}{h} w_{,xx}$$

with the bending stiffness $E\,I_{y_s}$, the cross sectional area A_s, and the height h of the stiffener. Comparison with (9) yields:

$$\left[E\,I_{y_s}\, w_{1,xxxx} + N_x \frac{A_s}{h} w_{1,xx} + 2K\, w_{1,yyy} \right]_{y=0} = 0 \,, \tag{10}$$

and by means of (2) follows

$$\left(\frac{m\pi}{a}\right)^2 \left[E\,I_{y_s} \left(\frac{m\pi}{a}\right)^2 - N_x \frac{A_s}{h} \right] w_m(0) + 2K\, w_m(0)_{,yyy} = 0 \,.$$

By including the TIMOSHENKO parameters [B.8]

$$\gamma = \frac{EI_{y_s}}{Kb} \quad , \quad \delta = \frac{A_s}{bh} \,,$$

we obtain

$$\Phi = \left(\frac{m\pi}{a}\right)^2 \left[\frac{EI_{y_s}}{Kb} \left(\frac{m\pi}{a}\right)^2 - \frac{N_x}{K} \frac{A_s}{bh} \right] = \vartheta_1^2 \left(\gamma\, \vartheta_1^2 - n_x^2\, \delta \right) .$$

The *transition condition* (10) divided by K becomes

$$b\,\Phi\, w_m(0) + 2\, w_m(0)_{,yyy} = 0 \,. \tag{7d}$$

With $(7a \div d)$ we now have four equations for determining the four constants of (5b).

Substitution yields the four homogeneous equations:

$$\left.\begin{aligned}
\cosh \varkappa_1 \tfrac{b}{2}\, A + \sinh \varkappa_1 \tfrac{b}{2}\, B + \cos \varkappa_2 \tfrac{b}{2}\, C + \sin \varkappa_2 \tfrac{b}{2}\, D &= 0 \,, \\
\varkappa_1^2 \cosh \varkappa_1 \tfrac{b}{2}\, A + \varkappa_1^2 \sinh \varkappa_1 \tfrac{b}{2}\, B - \varkappa_2^2 \cos \varkappa_2 \tfrac{b}{2}\, C - \varkappa_2^2 \sin \varkappa_2 \tfrac{b}{2}\, D &= 0 \,, \\
\varkappa_1\, B + \varkappa_2\, D &= 0 \,, \\
b\,\Phi\, A + 2\,\varkappa_1^3\, B + b\,\Phi\, C - 2\,\varkappa_2^3\, D &= 0 \,.
\end{aligned}\right\} \tag{11a}$$

The system determinant is:

$$\Delta_N = \begin{vmatrix}
\cosh \varkappa_1 \frac{b}{2} & \sinh \varkappa_1 \frac{b}{2} & \cos \varkappa_2 \frac{b}{2} & \sin \varkappa_2 \frac{b}{2} \\
\varkappa_1^2 \cosh \varkappa_1 \frac{b}{2} & \varkappa_1^2 \sinh \varkappa_1 \frac{b}{2} & -\varkappa_2^2 \cos \varkappa_2 \frac{b}{2} & -\varkappa_2^2 \sin \varkappa_2 \frac{b}{2} \\
0 & \varkappa_1 & 0 & \varkappa_2 \\
b\,\Phi & 2\varkappa_1^3 & b\,\Phi & 2\varkappa_2^3
\end{vmatrix} = 0 \,. \tag{11b}$$

The vanishing of the *buckling determinant* delivers the so-called *buckling condition*:

$$\varkappa_1 \varkappa_2 (\varkappa_1^2 - \varkappa_2^2) + \Phi \frac{b}{2} (\varkappa_1 \tan \varkappa_2 \frac{b}{2} - \varkappa_2 \tanh \varkappa_1 \frac{b}{2}) = 0 \,. \tag{12}$$

According to (4b) and with the above introduced abbreviations we can formulate:

$$\varkappa_{1,2} = \sqrt{\vartheta_1 (n_x \pm \vartheta_1)} = \sqrt{\frac{m\pi}{a}\left(\sqrt{\frac{N_x}{K}} \pm \frac{m\pi}{a}\right)} \,.$$

Since we devote our main interest to the critical load $N_x = N_{x_{crit}} = k\,\frac{\pi^2 K}{b^2}$, we write

$$\sqrt{\frac{N_x}{K}} = \sqrt{k}\,\frac{\pi}{b}$$

with the buckling value k. Therefore, with $\alpha = a/b$ we have

$$\varkappa_{1,2} = \sqrt{\frac{m\pi}{a}\left(\sqrt{k}\,\frac{\pi}{b} \pm \frac{m\pi}{a}\right)} = \frac{\pi}{b}\sqrt{\frac{m}{\alpha}\left(\sqrt{k} \pm \frac{m}{\alpha}\right)}\,. \qquad (13)$$

In (13) the buckling value k is still contained in the values $\varkappa_1$ and $\varkappa_2$. This equation is therefore solved by trial-and-error.

Fig. B-34 presents the k-values for different stiffness ratios γ in dependence on the dimension ratio α.

For $\quad \gamma = \frac{EI_{ys}}{Kb} = 0 \quad , \quad \delta = \frac{A_s}{bh} = 0\,,$

one obtains the garland-shaped curves of the unstiffened plate as already described in [ET2] with a minimum at $k = 4$. For

$$\gamma = 0 \quad , \quad \delta = 0.2$$

the k-value drops below 4 which implies that the stiffener under compression possessing no bending stiffness ($E\,I_{ys} \approx 0$) but a certain cross-section ($A_s > 0$) has to be supported by the plate. An arrangement of this type would of course be unsuitable.

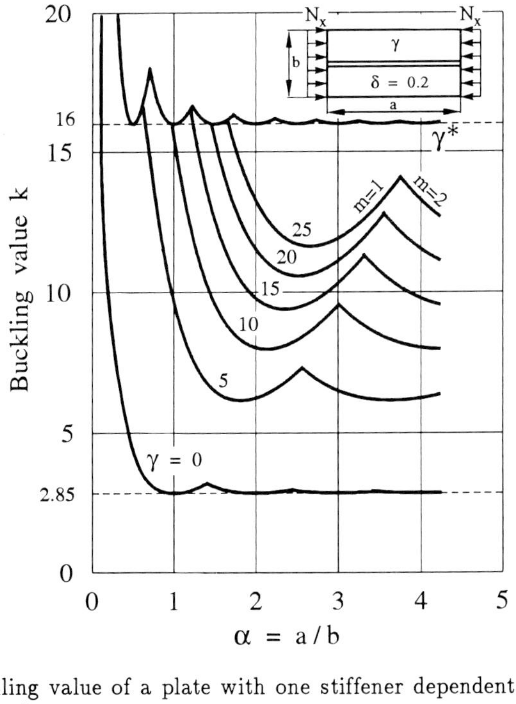

Fig. B-34: Buckling value of a plate with one stiffener dependent on the ratio $\alpha = a/b$

– *Antisymmetrical buckling*

Eq. (9) is valid only if the buckling mode is symmetric with respect to the stiffener (Fig. B-35a). However, an antisymmetrical buckling mode as shown in Fig. B-35b can also occur. If this is the case, the boundary conditions (7a,b) remain unchanged, i.e.

$$w_m\left(\pm\frac{b}{2}\right) = 0 \,, \tag{14a}$$

$$w_m\left(\pm\frac{b}{2}\right)_{,yy} = 0 \,. \tag{14b}$$

Furthermore, we have to consider the transition conditions at $y = 0$:

$$w_m(0) = 0 \,, \tag{14c}$$

$$w_m(0)_{,yy} = 0 \,. \tag{14d}$$

The above conditions correspond to those of a plate simply supported at the boundaries $y = 0$ and $y = b/2$. The present buckling case can therefore be reduced to that of a rectangular plate with the width $b/2$ and all edges simply supported. In analogy with [ET2] we obtain the buckling value:

$$k = \left(\frac{2\alpha}{m_1} + \frac{m_1}{2\alpha}\right)^2 \tag{15}$$

with m_1 as the number of longitudinal waves in the case of antisymmetrical buckling. m_1 here generally differs from the value m of the symmetrical buckling.

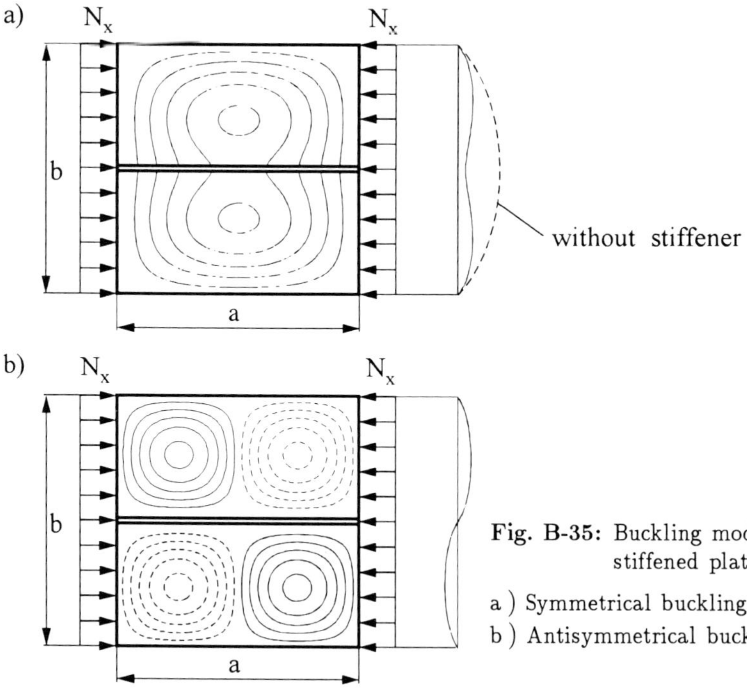

Fig. B-35: Buckling modes of the stiffened plate

a) Symmetrical buckling mode

b) Antisymmetrical buckling mode

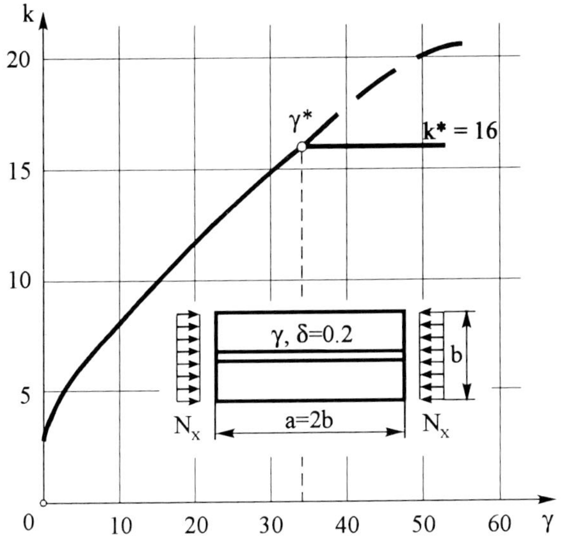

Fig. B-36 : Buckling values in dependence of bending stiffness

The buckling value belonging to the antisymmetrical buckling is independent of the dimensions and of the stiffness (if we neglect the torsional stiffness of the central stiffener), since during buckling this stiffener coincides with a nodal line of the buckling mode and is therefore not subject to deflection.

If we present the k – values of both buckling modes in dependence of the *bending stiffness* γ of the stiffener for a specific width ratio (e.g. $\alpha = a/b = 2$) and a specific cross – sectional area A_s of the stiffener (e.g. $\delta = 0.2$), we obtain the diagram in Fig. B-36 .

By increasing the stiffness of the stiffener, both the supporting force and the k – values increase; when the stiffness of the supporting stiffener reaches a specific value γ^*, either a symmetrical buckling mode with deflected stiffener, or an antisymmetrical buckling mode with an undeflected stiffener can occur. $k^* = 16$ is valid both for the former and latter buckling mode (see Fig. B-36) . An increase of the stiffness of the stiffener beyond the above value γ^* is obviously useless, since the plate would nevertheless show antimetrical buckling, and the k – values would remain unchanged. γ^* is called *least stiffness*.

Exercise B-10-2:

A thin circular plate (radius a, thickness t, YOUNG´s modulus E) is clamped along its outer boundary as shown in Fig. B-37 and is subjected to a constant surface load p that leads to large deflections.

This problem is to be treated as a coupled disk – plate problem, and the deflection is to be determined by means of the GALERKIN method with a single – term approximation. The results are to be compared with those obtained by linear theory.

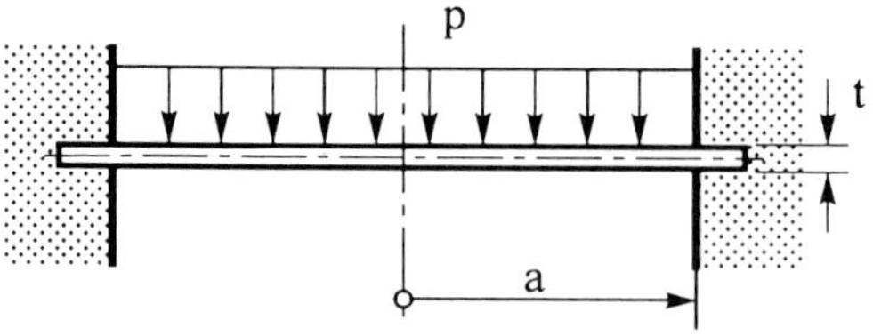

Fig. B-37: Clamped circular plate under uniform pressure

Solution :

We proceed from von KÁRMÁN's equations in operator notation *(10.13)*

$$\Delta\Delta\, w = \frac{p}{K} + \frac{t}{K} \Diamond^4 (w , \Phi) , \qquad (1a)$$

$$\Delta\Delta\, \Phi = - \frac{E}{2} \Diamond^4 (w , w) . \qquad (1b)$$

This system of nonlinear, partial differential equations, which is coupled in w and Φ, is rewritten in tensor notation according to *(10.14)* :

$$w\Big|_{\gamma\delta}^{\gamma\delta} = \frac{p}{K} + \frac{t}{K} \varepsilon^{\alpha\mu} \varepsilon^{\beta\nu} w\Big|_{\mu\nu} \Phi\Big|_{\alpha\beta} , \qquad (2a)$$

$$\Phi\Big|_{\gamma\delta}^{\gamma\delta} = - \frac{E}{2} \varepsilon^{\alpha\mu} \varepsilon^{\beta\nu} w\Big|_{\mu\nu} w\Big|_{\alpha\beta} . \qquad (2b)$$

The problem is most conveniently solved in polar coordinates for which reason we transform the covariant derivatives in the equation by means of *(2.40)*

$$\frac{1}{\sqrt{g}} \left\{ \sqrt{g}\, g^{\gamma\delta} \left[\frac{1}{\sqrt{g}} \left(\sqrt{g}\, g^{\sigma\tau} w_{,\tau} \right)_{,\sigma} \right]_{,\gamma} \right\}_{,\delta} =$$

$$= \frac{p}{K} + \frac{t}{K} \left[\varepsilon^{12} \varepsilon^{12} w\Big|_{22} \Phi\Big|_{11} + \varepsilon^{12} \varepsilon^{21} w\Big|_{21} \Phi\Big|_{12} + \right. \qquad (3a)$$

$$\left. + \varepsilon^{21} \varepsilon^{12} w\Big|_{12} \Phi\Big|_{21} + \varepsilon^{21} \varepsilon^{21} w\Big|_{11} \Phi\Big|_{22} \right] ,$$

$$\frac{1}{\sqrt{g}} \left\{ \sqrt{g}\, g^{\gamma\delta} \left[\frac{1}{\sqrt{g}} \left(\sqrt{g}\, g^{\sigma\tau} \Phi_{,\tau} \right)_{,\sigma} \right]_{,\gamma} \right\}_{,\delta} =$$

$$= - \frac{E}{2} \left[\varepsilon^{12} \varepsilon^{12} w\Big|_{22} w\Big|_{11} + \varepsilon^{12} \varepsilon^{21} w\Big|_{21} w\Big|_{12} + \right. \qquad (3b)$$

$$\left. + \varepsilon^{21} \varepsilon^{12} w\Big|_{12} w\Big|_{21} + \varepsilon^{21} \varepsilon^{21} w\Big|_{11} w\Big|_{22} \right] .$$

The metric tensors, their determinant, and the CHRISTOFFEL symbols required for the further treatment of the problem, have already been determined for cylindrical coordinates (see 2.5). If we now substitute the permutation symbols *(2.20b)* and the covariant derivatives *(2.34b)* (e.g. $w\big|_{11} = w_{,1}\big|_1$ etc.) into the right-hand sides, and if we take into consideration that the dependence on the angle coordinate $\xi^2 \mathrel{\widehat{=}} \varphi$ vanishes in the derivatives owing to the axisymmetrical shape and loads, the equations reduce with $\xi^1 \mathrel{\widehat{=}} r$ and $d/dr \mathrel{\widehat{=}} (\ )_{,r}$ to

$$\left\{ r \left[\frac{1}{r} \left(r\, w_{,r} \right)_{,r} \right]_{,r} \right\}_{,r} = \frac{pr}{K} + \frac{t}{K} (w_{,r} \Phi_{,r})_{,r} , \qquad (4a)$$

$$\left\{ r \left[\frac{1}{r} \left(r\, \Phi_{,r} \right)_{,r} \right]_{,r} \right\}_{,r} = - \frac{E}{2} \left[(w_{,r})^2 \right]_{,r} . \qquad (4b)$$

In order to solve this nonlinear system of differential equations, we first integrate (4b). This yields

$$\Phi_{,r} = -\frac{E}{2r}\int r\left(\int\frac{(w_{,r})^2}{r}\,dr\right)dr + C_1\,r\left(1 + 2\ln\frac{r}{a}\right) + 2\,C_2\,r + \frac{C_3}{r}\,. \qquad (5)$$

Eq. (5) contains terms with three constants which can be determined from the following conditions. First, we use the disk stresses according to *(8.7)*

$$\sigma_{rr} = \frac{N_{rr}}{t} = \frac{1}{r}\,\Phi_{,r}\,, \qquad (6a)$$

$$\sigma_{\varphi\varphi} = \frac{N_{\varphi\varphi}}{t} = \Phi_{,rr}\,. \qquad (6b)$$

Substitution of equation (5) into (6a,b) yields :

$$\sigma_{rr} = -\frac{E}{2r^2}\int r\left(\int\frac{(w_{,r})^2}{r}\,dr\right)dr + C_1\left(1 + 2\ln\frac{r}{a}\right) + 2\,C_2 + \frac{C_3}{r^2}\,, \qquad (7a)$$

$$\sigma_{\varphi\varphi} = -\frac{E}{2}\left[\frac{1}{r}\int r\left(\int\frac{(w_{,r})^2}{r}\,dr\right)dr\right]_{,r} + C_1\left(3 + 2\ln\frac{r}{a}\right) + 2\,C_2 - \frac{C_3}{r^2}\,. \qquad (7b)$$

Owing to the condition that the stresses in the middle of the disk ($r = 0$) cannot become infinite (*finitness condition*), the constants C_1 and C_3 must vanish.

The remaining constant can be calculated from the *boundary condition*

$$u(a) = 0\,. \qquad (8)$$

For this purpose we proceed from the material law *(8.9)* and then replace the expansion $\varepsilon_{\varphi\varphi}$ by the radial displacement u, using the corresponding linearized strain – displacement equation of the axisymmetrical stress state in *(8.8)*. (Higher order terms with respect to the deflection w in the strain – displacement equation vanish in this case.) It then follows that

$$u = r\,\varepsilon_{\varphi\varphi} = \frac{r}{E}\left(\sigma_{\varphi\varphi} - \nu\,\sigma_{rr}\right).$$

By introducing (7a,b) we obtain :

$$u = r\left\{-\frac{1}{2}\left[\frac{1}{r}\int r\left(\int\frac{(w_{,r})^2}{r}\,dr\right)dr\right]_{,r} + \right.$$
$$\left. + \frac{\nu}{2r^2}\int r\left(\int\frac{(w_{,r})^2}{r}\,dr\right)dr + \frac{2C_2(1-\nu)}{E}\right\}. \qquad (9)$$

In order to evaluate (9), we require an approximation for the deflection w. For this purpose, we choose the solution of a clamped plate with small deflection

$$w^* = c\left(1 - \frac{r^2}{a^2}\right)^2. \qquad (10)$$

Here, c is a free coefficient describing the maximum deflection at the centre. The above approximation function fulfills all boundary conditions. After calculation we obtain

$$u = r\left[-\frac{c^2}{a^2}\left(3\,\frac{r^2}{a^2} - \frac{10}{3}\,\frac{r^4}{a^4} + \frac{7}{6}\,\frac{r^6}{a^6}\right) + \right.$$
$$\left. + \nu\,\frac{c^2}{a^2}\left(\frac{r^2}{a^2} - \frac{2}{3}\,\frac{r^4}{a^4} + \frac{1}{6}\,\frac{r^6}{a^6}\right) + \frac{2C_2(1-\nu)}{E}\right]. \qquad (11)$$

From (8) then follows

$$u(r=a) = a\left[-\frac{c^2}{a^2}\frac{5-3\nu}{6} + \frac{2C_2(1-\nu)}{E}\right] = 0$$

$$\longrightarrow \quad C_2 = E\frac{c^2}{a^2}\frac{5-3\nu}{12(1-\nu)} \, .$$

Using the approximation w* according to (10), we now state with (5) an approximation for Φ :

$$\Phi^*_{,r} = -\frac{E}{r}c^2\left(\frac{r^4}{a^4} - \frac{2}{3}\frac{r^6}{a^6} + \frac{1}{6}\frac{r^8}{a^8}\right) + E\,r\frac{c^2}{a^2}\frac{5-3\nu}{6(1-\nu)} \, . \tag{12}$$

Further calculation is then carried out by means of GALERKIN's equations, stated according to *(6.38)*. By means of (4a) we write the operator

$$L(w^*,\Phi^*) = K\left\{r\left[\frac{1}{r}(r\,w^*_{,r})_{,r}\right]_{,r}\right\}_{,r} - p\,r - t(w^*_{,r}\,\Phi^*_{,r})_{,r} \, , \tag{13}$$

which we substitute into

$$\int_{\varphi=0}^{2\pi}\int_{r=0}^{a} L(w^*,\Phi^*)\,w^*\,r\,dr\,d\varphi = 0 \, .$$

After integration we finally obtain a cubic equation in terms of $\frac{c}{t} \mathrel{\widehat{=}} \frac{w_{max}}{t}$:

$$\frac{2}{21}\frac{23-9\nu}{1-\nu}\left(\frac{c}{t}\right)^3 + \frac{16}{3(1-\nu^2)}\frac{c}{t} = \frac{p}{E}\left(\frac{a}{t}\right)^4 = \bar{p} \, .$$

Fig. B-38 presents the numerical evaluation of the calculation with $\nu = 1/3$. In the case of higher loads ($\bar{p} > 5$), the nonlinear theory leads to smaller deflections than the linear theory, owing to the membrane forces in the plate which increase the stiffness against transverse deflection and thus facilitate a higher load bearing capacity [B.8].

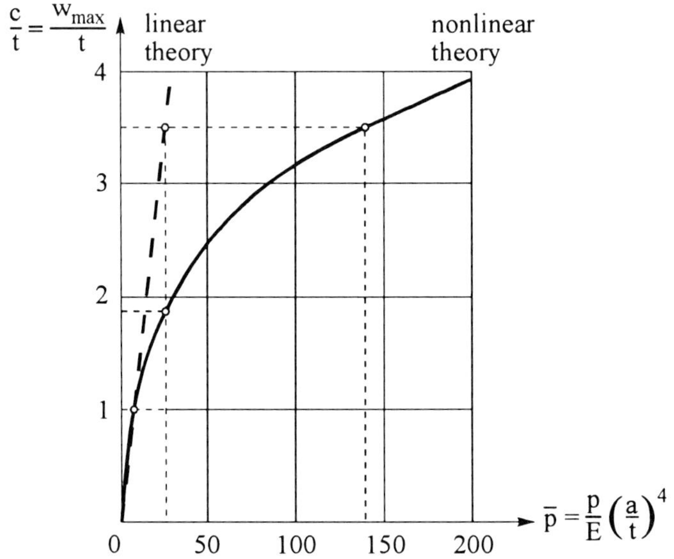

Fig. B-38: Comparison of maximum deflections obtained by linear and by nonlinear theory (dotted lines denote deviation of the pressure for special deflection values in the scope of the linear and nonlinear theory)

C Curved load-bearing structures

C.1 Definitions – Formulas – Concepts

11 General fundamentals of shells

11.1 Surface theory – description of shells

11.1.1 Representation of surfaces

We assume that there exist one-to-one relationships between the curvilinear coordinates (GAUSSIAN surface parameters) ξ^α ($\alpha = 1,2$) and the Cartesian coordinates x^i ($i = 1,2,3$) of the points P of a surface (Fig. 11.1). This correlation is expressed by

$$x^i = x^i(\xi^\alpha) \tag{11.1a}$$

or, in vector notation with the position vector $\mathbf{r}$ of a point on the surface [C.5, C.11] as

$$\mathbf{r} = \mathbf{r}(\xi^\alpha) = x^i(\xi^\alpha)\,\mathbf{e}_i \quad . \tag{11.1b}$$

Differentiability with continuous first derivatives is assumed along with non-singularity of the JACOBIAN matrix (functional matrix):

$$\mathbf{J} = \frac{\partial x^i}{\partial \xi^\alpha} = \begin{bmatrix} \dfrac{\partial x^1}{\partial \xi^1} & \dfrac{\partial x^2}{\partial \xi^1} & \dfrac{\partial x^3}{\partial \xi^1} \\ \dfrac{\partial x^1}{\partial \xi^2} & \dfrac{\partial x^2}{\partial \xi^2} & \dfrac{\partial x^3}{\partial \xi^2} \end{bmatrix} \quad . \tag{11.2}$$

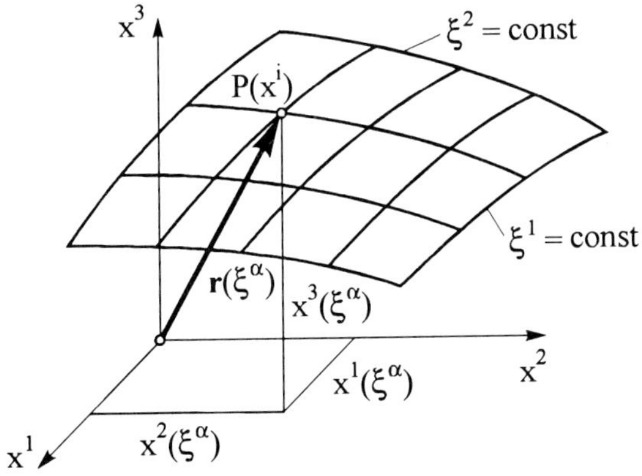

Fig. 11.1: Definition of the parametric representation of a surface

If ξ^2 = const for variable ξ^1, (*11.1b*) describes space curves embedded on the surface, and these curves are called ξ^1–lines. In analogy, with ξ^1 = const, one obtains ξ^2–lines. These ξ^1– and ξ^2–lines constitute a curvilinear coordinate mesh on the surface (Fig. 11.1).

a) Surfaces of revolution

Definition 1: One obtains a *surface of revolution* (Fig. 11.2), if a two-dimensional curve m positioned in the x^1, x^3 - plane ($x^2 = 0$)

$$x^1 = r \ , \quad x^2 = 0 \ , \quad x^3 = x^3(r)$$

is rotated around the x^3- axis as axis of revolution.

Using the polar coordinates $\xi^1 \mathrel{\widehat{=}} r$, $\xi^2 \mathrel{\widehat{=}} \vartheta$ as GAUSSIAN parameters, the vector $\mathbf{r}$ reads:

$$\mathbf{r} = \mathbf{r}(r, \vartheta) = r\cos\vartheta\,\mathbf{e}_1 + r\sin\vartheta\,\mathbf{e}_2 + x^3(r)\,\mathbf{e}_3 \tag{11.3a}$$

or according to (*11.1b*)

$$\left.\begin{aligned} x^1 &= r\cos\vartheta \ , \\ x^2 &= r\sin\vartheta \ , \\ x^3 &= x^3(r) \ . \end{aligned}\right\} \tag{11.3b}$$

Note: In (*11.3 a,b*), r describes the projection of $\mathbf{r}$ and not its value. The r-lines (ϑ = const) are called meridional curves; for $\vartheta = 0$ we obtain the zero-meridian m. The ϑ- lines (r = const) are called parallels of latitude. Both types of lines cover the surface with an orthogonal parametrical mesh (Fig.11.2).

Special case:

– Spherical surface

If the meridional curve $\vartheta = 0$ of a surface of revolution is chosen as a circle with the radius a and centre point in the origin (Fig. 11.3), a sphere is described. After introducing an angle φ, one obtains

$$x^1 = a\sin\varphi \ , \quad x^2 = 0 \ , \quad x^3 = a\cos\varphi \ . \tag{11.4 a}$$

The position vector $\mathbf{r}$ of the spherical surface then reads:

$$\mathbf{r} = \mathbf{r}(\varphi, \vartheta) = (a\sin\varphi)\cos\vartheta\,\mathbf{e}_1 + (a\sin\varphi)\sin\vartheta\,\mathbf{e}_2 + a\cos\varphi\,\mathbf{e}_3 \tag{11.4 b}$$

with the components

$$x^1 = (a\sin\varphi)\cos\vartheta \ , \quad x^2 = (a\sin\varphi)\sin\vartheta \ , \quad x^3 = a\cos\varphi \ . \tag{11.4 c}$$

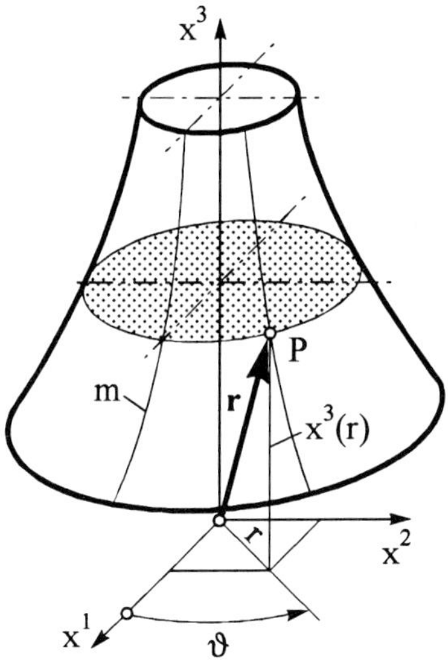

Fig. 11.2: Generation of a surface of revolution

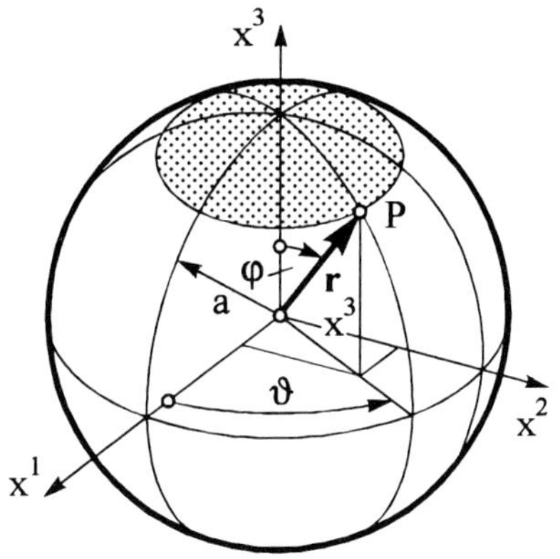

Fig. 11.3: Parameters of a spherical surface

On the surface of the earth, the geographical $\bar{\varphi}, \vartheta$-coordinate system is generally applied, where $\overline{\varphi} = \frac{\pi}{2} - \varphi$ denotes the latitude and ϑ the longitude of a respective point. The circle of latitude $\bar{\varphi} = 0$ describes the equator.

b) Ruled surfaces

Definition 2: The term ruled surface (derived from "surface réglée" = linear surface) or radial surface denotes each surface generated by moving a straight line g along a guide-line d (directrix) (Fig. 11.4). The single positions of the straight line are called the generatrices g of the ruled surface.

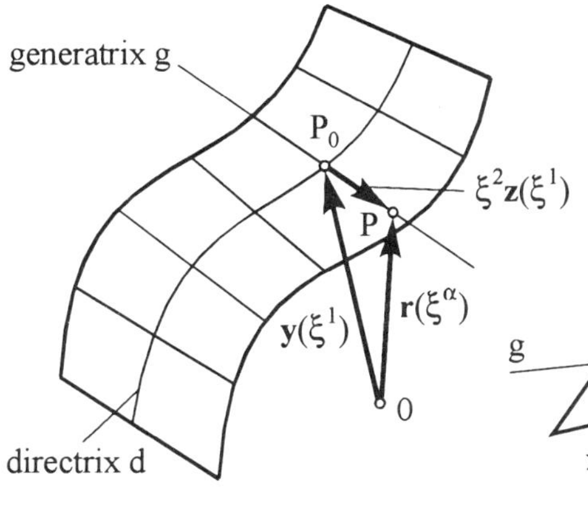

Fig. 11.4: Generation of a ruled surface

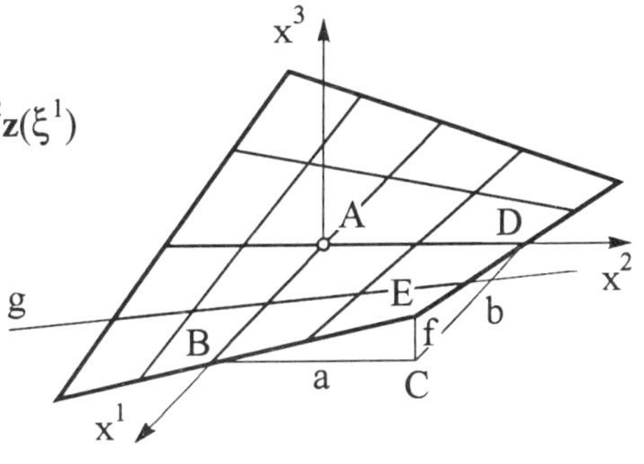

Fig. 11.5: Skew hyperbolical paraboloid surface

Such a movement can be described by defining the path $\mathbf{y} = \mathbf{y}(\xi^1)$ of a point P_0 of the straight line, and a unit vector $\mathbf{z}(\xi^1)$ which points in the direction of the straight line. The ruled surface is then formulated as

$$\mathbf{r} = \mathbf{r}(\xi^\alpha) = \mathbf{y}(\xi^1) + \xi^2 \mathbf{z}(\xi^1) \ , \qquad (11.5)$$

where ξ^2 is the distance of a point P on the surface (Fig. 11.4) to the point P_0 of intersection of the inherent generatrix g and the directrix d.

For this purpose, it is assumed that

$$\frac{d\mathbf{y}}{d\xi^1} \times \mathbf{z} = \mathbf{y}_{,1} \times \mathbf{z} \neq \mathbf{0} \ ,$$

i.e. the generatrix must not point into the direction of the directrix. The ξ^2 - lines ($\xi^1 = \text{const}$) are the linear generatrices of the surface. According to (*11.5*) the directrix $\mathbf{r}(\xi^1, 0) = \mathbf{y}(\xi^1)$ occurs for $\xi^2 = 0$ in the family of the ξ^1 - lines ($\xi^2 = \text{const}$).

Special cases:

– Cylindrical surfaces

We obtain a special type of a ruled surface if all generatrices are parallel to each other, and if thus the vector $\mathbf{z}$ is constant. In this case equation (*11.5*) reads:

$$\mathbf{r} = \mathbf{r}(\xi^\alpha) = \mathbf{y}(\xi^1) + \xi^2 \mathbf{z} \ . \qquad (11.6)$$

Surfaces of this type are called cylindrical surfaces. Depending on the form of the directrices we obtain different types of cylinder surfaces, e.g. elliptical cylinder surface:

$$\mathbf{y}(\vartheta) = a \cos\vartheta\, \mathbf{e}_1 + b \sin\vartheta\, \mathbf{e}_2 \quad \text{and} \quad \mathbf{z} = \mathbf{e}_3 \ . \qquad (11.7)$$

– Skew hyperbolical paraboloid surface

A so-called skew hyperbolic paraboloid is generated according to Fig. 11.5 by moving the straight line g along the x^1– or x^2–axis, respectively, as generatrices. Here, one also uses the term conoid surface, the explicit representation of which reads:

$$x^3 = \frac{f}{a\,b} x^1 x^2 \ . \qquad (11.8\,a)$$

The parametrical representation of this surface is

$$\mathbf{r} = \mathbf{r}(\xi^\alpha) = \xi^1 \mathbf{e}_1 + \xi^2 \mathbf{e}_2 + \frac{f}{a\,b} \xi^1 \xi^2 \mathbf{e}_3 \qquad (11.8\,b)$$

with the dimensions a, b and the height f (Fig. 11.5).

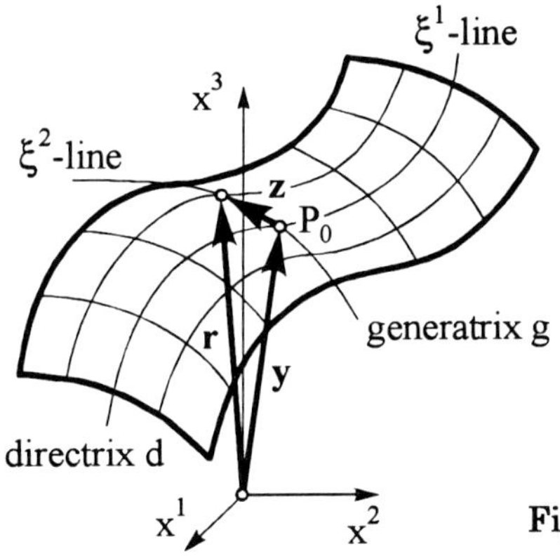

Fig. 11.6: Generation of a translation surface

c) *Translation or sliding surfaces*

Definition 3: A translation or sliding surface is generated by a parallel displacement of one curve g along a second, a so-called *guide curve* or directrix d. This surface can be described as follows:

$$\mathbf{r} = \mathbf{r}\,(\xi^\alpha) = \mathbf{y}(\xi^1) + \mathbf{z}(\xi^2) \;, \tag{11.9a}$$

where the curves shifted in parallel are called the generatrices g or sliding curves of the translation surface (Fig. 11.6). In a more extended definition ($\xi^2 \longrightarrow \xi^\alpha$), translation surfaces are represented by

$$\mathbf{r} = \mathbf{r}\,(\xi^\alpha) = \mathbf{y}(\xi^1) + \mathbf{z}(\xi^\alpha) \;. \tag{11.9b}$$

11.1.2 Fundamental quantities of first and second order

Base vectors on the surface

$$\mathbf{a}_\alpha = \frac{\partial\,\mathbf{r}}{\partial\,\xi^\alpha} = \mathbf{r}_{,\alpha} \;. \tag{11.10}$$

The base vectors $\mathbf{a}_\alpha$ are lying in the tangential plane which touches the surface at a point with the position vector $\mathbf{r}(\xi^\alpha)$ (Fig. 11.7a). From the base vectors (covariant surface tensors of order one) one obtains, in analogy to *(2.1a)*, the components of the covariant metric tensor or surface tensor.

Components of the covariant metric or surface tensor

Fundamental quantities of first order – *first fundamental form* of surface theory –

$$a_{\alpha\beta} = \mathbf{r}_{,\alpha} \cdot \mathbf{r}_{,\beta} = \mathbf{a}_\alpha \cdot \mathbf{a}_\beta \;. \tag{11.11}$$

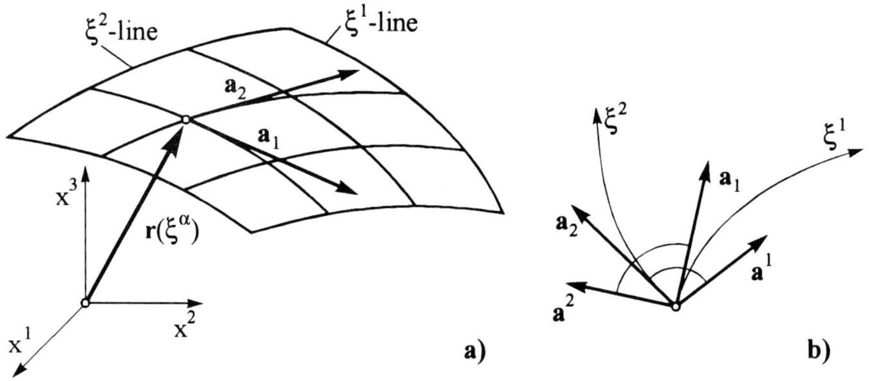

Fig. 11.7: Covariant and contravariant base vectors

Determinant of the surface tensor

$$a = \left| a_{\alpha\beta} \right| = \begin{vmatrix} a_{11} & a_{12} \\ a_{21} & a_{22} \end{vmatrix} \quad . \tag{11.12}$$

Length of an arc element

$$ds = \sqrt{a_{\alpha\beta} \dot{\xi}^{\alpha} \dot{\xi}^{\beta}} \; dt \tag{11.13}$$

(t = curve parameter, $(\dot{\ }) \mathrel{\hat{=}} d/dt$).

Area of a surface element

$$dA = \sqrt{a} \; d\xi^1 \, d\xi^2 \quad . \tag{11.14}$$

Contravariant components of the surface tensor from *(2.5c)*

$$a_{\alpha\gamma} \, a^{\gamma\beta} = \delta_{\alpha}^{\beta} \quad . \tag{11.15a}$$

Contravariant base vectors (Fig. 11.7b)

$$\mathbf{a}^{\beta} = a^{\beta\gamma} \, \mathbf{a}_{\gamma} \quad . \tag{11.15b}$$

Unit vector perpendicular to the surface (Fig. 11.8)

$$\mathbf{a}_3 = \frac{\mathbf{a}_1 \times \mathbf{a}_2}{\left| \mathbf{a}_1 \times \mathbf{a}_2 \right|} \quad . \tag{11.16}$$

Fundamental quantities of second order (Fig. 11.8) – tensor of curvature

$$d\mathbf{r} \cdot d\mathbf{a}_3 = - b_{\alpha\beta} \, d\xi^{\alpha} \, d\xi^{\beta} \tag{11.17}$$

with the curvature components $\quad b_{\alpha\beta} = \dfrac{\left[\mathbf{a}_{\alpha,\beta} , \mathbf{a}_1 , \mathbf{a}_2 \right]}{\sqrt{a}} \quad .$ (11.18)

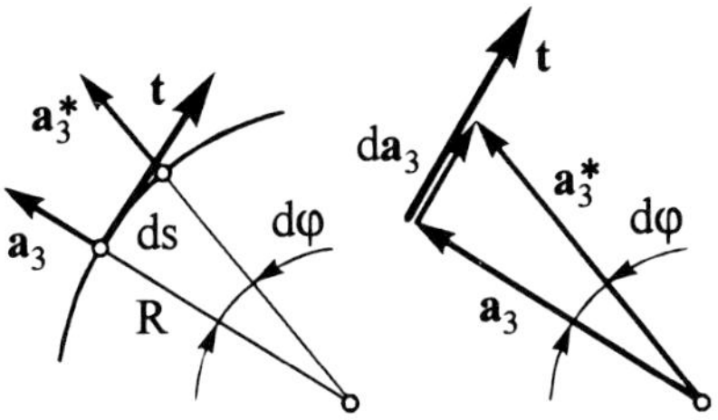

Fig. 11.8: Curvature of a surface

Determinant of the tensor of curvature

$$b = \left| b_{\alpha\beta} \right| = \begin{vmatrix} b_{11} & b_{12} \\ b_{21} & b_{22} \end{vmatrix} \quad . \tag{11.19}$$

Curvature in a point of a curve of the surface

$$\frac{1}{R} = \kappa \quad = -\frac{b_{\alpha\beta}\, d\xi^\alpha\, d\xi^\beta}{a_{\alpha\beta}\, d\xi^\alpha\, d\xi^\beta} \quad . \tag{11.20}$$

Characteristic equation for principal curvatures κ_i ($i = 1, 2$)

$$\kappa^2 - b^{\alpha\beta} a_{\alpha\beta}\, \kappa + \frac{b}{a} = \kappa^2 - 2H\kappa + K = 0 \quad . \tag{11.21}$$

Invariants

$$H = \frac{1}{2} a^{\alpha\beta} b_{\alpha\beta} = \frac{1}{2} b^{\alpha\beta} a_{\alpha\beta} = b^\alpha_\alpha \qquad \text{mean curvature} \; , \tag{11.22a}$$

$$K = \left| b^\alpha_\beta \right| = \frac{b}{a} \qquad \text{GAUSSIAN curvature} \; . \tag{11.22b}$$

CHRISTOFFEL symbols in surface theory

$$\Gamma^\alpha_{\beta\gamma} = \frac{1}{2} a^{\alpha\varrho} \left(a_{\varrho\beta,\gamma} + a_{\gamma\varrho,\beta} - a_{\beta\gamma,\varrho} \right) \quad . \tag{11.23a}$$

Special cases:

$$\left.\begin{aligned} \Gamma^\alpha_{\beta 3} &= -b^\alpha_\beta \; , \\ \Gamma^3_{\alpha\beta} &= b_{\alpha\beta} \; , \\ \Gamma^3_{3\alpha} &= \Gamma^\alpha_{33} = \Gamma^3_{33} = 0 \quad . \end{aligned}\right\} \tag{11.23b}$$

GAUSS-WEINGARTEN derivative equations

$$\left.\begin{aligned} \mathbf{a}_{\alpha,\beta} &= \Gamma^\gamma_{\alpha\beta} \mathbf{a}_\gamma + b_{\alpha\beta} \mathbf{a}_3 \\ \text{or} \qquad \mathbf{a}_\alpha\big|_\beta &= b_{\alpha\beta} \mathbf{a}_3 \; , \\ \mathbf{a}_{3,\alpha} &= -b^\beta_\alpha \mathbf{a}_\beta \quad . \end{aligned}\right\} \tag{11.24}$$

Example: *Application to surfaces of revolution*

Referring to Fig. 11.9, the arc length s and the angle ϑ are chosen as GAUSSIAN surface parameters $\longrightarrow \xi^1 \triangleq s \;,\; \xi^2 \triangleq \vartheta$.

According to (*11.3a*) the parametric representation of the surfaces of revolution then reads:

$$\mathbf{r}(s,\vartheta) = r(s)\cos\vartheta\,\mathbf{e}_1 + r(s)\sin\vartheta\,\mathbf{e}_2 + z(s)\,\mathbf{e}_3 \quad . \qquad (11.25)$$

With (*11.10*) the covariant base vectors result as

$$\left.\begin{aligned} \mathbf{a}_1 &= \mathbf{r}_{,s} = r_{,s}\cos\vartheta\,\mathbf{e}_1 + r_{,s}\sin\vartheta\,\mathbf{e}_2 + z_{,s}\,\mathbf{e}_3 \;, \\ \mathbf{a}_2 &= \mathbf{r}_{,\vartheta} = -r\sin\vartheta\,\mathbf{e}_1 + r\cos\vartheta\,\mathbf{e}_2 \;, \end{aligned}\right\} \qquad (11.26)$$

where the derivatives with respect to s are denoted by $\partial/\partial s \triangleq (\;)_{,s}$ and $\partial/\partial\vartheta \triangleq (\;)_{,\vartheta}$. Accordingly, $\mathbf{a}_2$ is parallel to the x^1, x^2 - plane.

With

$$(dr)^2 + (dz)^2 = (ds)^2 \quad \text{or} \quad r_{,s}^2 + z_{,s}^2 = 1$$

the derivative of z is

$$z_{,s} = -\sqrt{1 - r_{,s}^2} \quad . \qquad (11.27)$$

With the chosen measuring direction of s , z decreases with increasing s, i.e., $z_{,s} < 0$. Thus, the negative sign is valid for z.

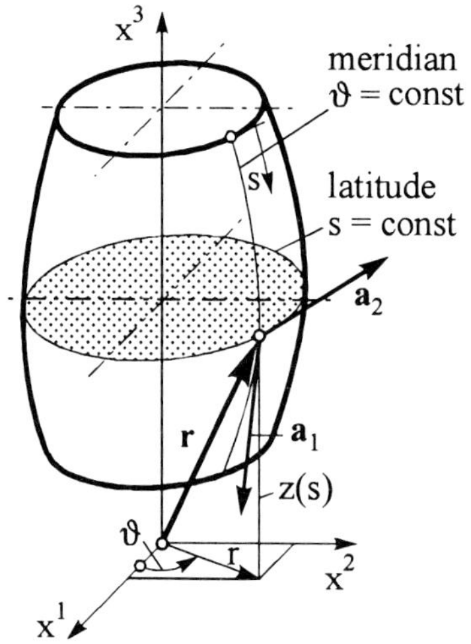

Fig. 11.9: Coordinates at a surface of revolution

The components of the covariant metric tensor then become

$$a_{11} = \mathbf{a}_1 \cdot \mathbf{a}_1 = r_{,s}^2 \cos^2\vartheta + r_{,s}^2 \sin^2\vartheta + z_{,s}^2 = r_{,s}^2 + z_{,s}^2 = 1 \ ,$$

$$a_{12} = \mathbf{a}_1 \cdot \mathbf{a}_2 = r_{,s} \cos\vartheta(-r\sin\vartheta) + r_{,s} \sin\vartheta(r\cos\vartheta) = a_{21} = 0 \ ,$$

$$a_{22} = \mathbf{a}_2 \cdot \mathbf{a}_2 = r^2 \sin^2\vartheta + r^2 \cos^2\vartheta = r^2 \ .$$

Thus, one obtains

$$(a_{\alpha\beta}) = \begin{bmatrix} 1 & 0 \\ 0 & r^2 \end{bmatrix} \ . \qquad (11.28)$$

Due to $a_{12} = a_{21} = 0$, the coordinate lines intersect each other perpendicularly; i.e. s and ϑ form an orthogonal system.

By means of the determinant

$$a = |a_{\alpha\beta}| = \begin{vmatrix} 1 & 0 \\ 0 & r^2 \end{vmatrix} = r^2 \ , \qquad (11.29)$$

the contravariant metric tensor can be calculated by inversion

$$(a^{\alpha\beta}) = \begin{bmatrix} 1 & 0 \\ 0 & 1/r^2 \end{bmatrix} \ . \qquad (11.30)$$

In order to determine the curvature tensor, the derivatives of the base vectors are required:

$$\mathbf{a}_{1,1} = \mathbf{r}_{,ss} = r_{,ss} \cos\vartheta\, \mathbf{e}_1 + r_{,ss} \sin\vartheta\, \mathbf{e}_2 + z_{,ss}\, \mathbf{e}_3 \ ,$$

$$\mathbf{a}_{1,2} = \mathbf{r}_{,\vartheta s} = -r_{,s} \sin\vartheta\, \mathbf{e}_1 + r_{,s} \cos\vartheta\, \mathbf{e}_2 = \mathbf{a}_{2,1} \ ,$$

$$\mathbf{a}_{2,2} = \mathbf{r}_{,\vartheta\vartheta} = -r\cos\vartheta\, \mathbf{e}_1 - r\sin\vartheta\, \mathbf{e}_2 \ .$$

We obtain from the scalar triple product *(11.18)*

$$b_{11} = \frac{1}{r} \begin{vmatrix} r_{,ss} \cos\vartheta & r_{,ss} \sin\vartheta & z_{,ss} \\ r_{,s} \cos\vartheta & r_{,s} \sin\vartheta & z_{,s} \\ -r\sin\vartheta & r\cos\vartheta & 0 \end{vmatrix}$$

or after differentiation of *(11.27)* with

$$z_{,ss} = \frac{r_{,s}\, r_{,ss}}{\sqrt{1 - r_{,s}^2}}$$

$$\longrightarrow \quad b_{11} = r_{,ss} \sqrt{1 - r_{,s}^2} + \frac{r_{,s}^2\, r_{,ss}}{\sqrt{1 - r_{,s}^2}} = \frac{r_{,ss}}{\sqrt{1 - r_{,s}^2}} \quad .$$

The other components are analogously determined as:

$$(b_{\alpha\beta}) = \begin{bmatrix} \dfrac{r_{,ss}}{\sqrt{1 - r_{,s}^2}} & 0 \\ 0 & -r\sqrt{1 - r_{,s}^2} \end{bmatrix} \tag{11.31a}$$

or the mixed tensor of curvature

$$(b_{\beta}^{\alpha}) = a^{\alpha\gamma}\, b_{\gamma\beta} = \begin{bmatrix} \dfrac{r_{,ss}}{\sqrt{1 - r_{,s}^2}} & 0 \\ 0 & -\dfrac{\sqrt{1 - r_{,s}^2}}{r} \end{bmatrix} \tag{11.31b}$$

and

$$b = \left| b_{\alpha\beta} \right| = -r_{,ss}\, r \quad . \tag{11.32}$$

Since $a_{12} = a_{21}$ as well as $b_{12} = b_{21}$ vanish, the coordinate lines are also curvature lines, i.e. lines with extremal curvature.

The mean curvature results from (*11.22a*) as

$$H = \frac{1}{2} b_{\alpha}^{\alpha} = \frac{1}{2}\left(\frac{r_{,ss}}{\sqrt{1 - r_{,s}^2}} - \frac{\sqrt{1 - r_{,s}^2}}{r} \right) = \frac{r\, r_{,ss} + r_{,s}^2 - 1}{2r\sqrt{1 - r_{,s}^2}} \; , \tag{11.33a}$$

and the GAUSSIAN curvature follows from *(11.22b)*

$$K = \frac{a}{b} = -\frac{r_{,ss}}{r} \quad . \tag{11.33b}$$

The CHRISTOFFEL symbols are determined by means of *(11.23a)*. We obtain for instance:

$$\Gamma_{22}^{1} = \frac{1}{2} a^{1\lambda} (a_{2\lambda,2} + a_{\lambda 2,2} - a_{22,\lambda}) = -\frac{1}{2}\, a^{11}\, a_{22,1} =$$

$$= -\frac{1}{2} \frac{d}{ds} (r^2) = -r\, r_{,s} \quad .$$

The further values are written without detailed calculation using *(11.23a)*:

$$(\Gamma^1_{\alpha\beta}) = \begin{bmatrix} 0 & 0 \\ 0 & -r\,r_{,s} \end{bmatrix} , \quad (\Gamma^2_{\alpha\beta}) = \begin{bmatrix} 0 & \dfrac{r_{,s}}{r} \\ \dfrac{r_{,s}}{r} & 0 \end{bmatrix} , \tag{11.34a}$$

$$\left.\begin{aligned} (\Gamma^3_{\alpha\beta}) &= \begin{bmatrix} \dfrac{r_{,ss}}{\sqrt{1-r_{,s}^2}} & 0 \\ 0 & -r\sqrt{1-r_{,s}^2} \end{bmatrix} , \\ (\Gamma^\alpha_{\beta 3}) &= \begin{bmatrix} -\dfrac{r_{,ss}}{\sqrt{1-r_{,s}^2}} & 0 \\ 0 & \dfrac{\sqrt{1-r_{,s}^2}}{r} \end{bmatrix} . \end{aligned}\right\} \tag{11.34b}$$

11.2 Basic theory of shells [C.9, C.10, C.11, C.12, C.14, C.17, C.18, C.19]

Geometry of shells

The shell continuum according to Fig. 11.10 is described by means of the mid-surface of the shell which halves the shell thickness t at each point. The shell space is presented by the GAUSSIAN surface parameters ξ^α of the mid-surface, and by a coordinate ζ perpendicular to the mid-surface.

Position vector $\mathbf{r}_P$ of an arbitrary point P of the shell space:

$$\mathbf{r}_P(\xi^\alpha, \zeta) = \mathbf{r}(\xi^\alpha) + \zeta\,\mathbf{a}_3(\xi^\alpha) \quad . \tag{11.35}$$

Covariant base in the three-dimensional space:

$$\mathbf{g}_j = \mathbf{r}_{P,j} \quad . \tag{11.36}$$

Eq. *(11.36)* with *(11.35)* and *(11.24)* yields:

$$\left.\begin{aligned} \mathbf{g}_\alpha &= (\delta^\beta_\alpha - \zeta b^\beta_\alpha)\,\mathbf{a}_\beta = \mu^\beta_\alpha\,\mathbf{a}_\beta , \\ \mathbf{g}_3 &= \mathbf{a}_3 \quad . \end{aligned}\right\} \tag{11.37}$$

The three-dimensional base $\mathbf{g}_i$ is transformed into the base $\mathbf{a}_\alpha$ of the mid-surface by means of the shell tensor or shell shifter μ^β_α introduced by NAGHDI [C.17].

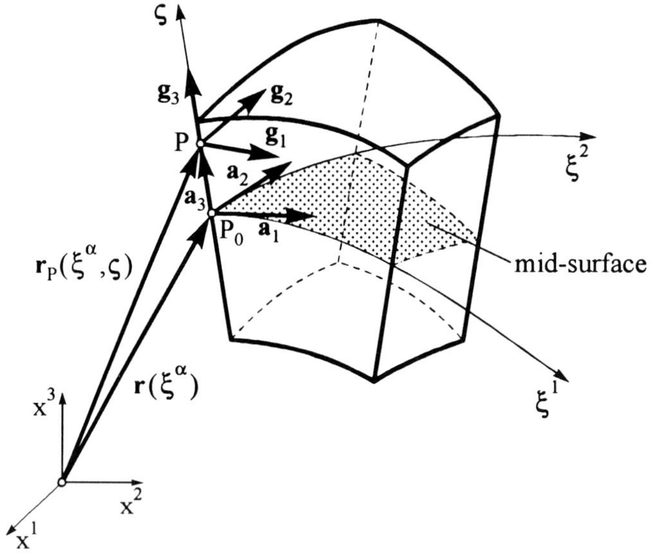

Fig. 11.10: Coordinates and base vectors

Contravariant base in the three-dimensional space:

$$\left.\begin{aligned} \mathbf{g}^{\alpha} &= (\mu^{-1})^{\alpha}_{\beta}\, \mathbf{a}^{\beta} \ , \\ \mathbf{g}^{3} &= \mathbf{a}^{3} \ . \end{aligned}\right\} \tag{11.38}$$

Determinant μ of the shell tensor:

$$\mu = \left|\mu^{\alpha}_{\beta}\right| = \begin{vmatrix} 1 - \zeta b^1_1 & -\zeta b^1_2 \\ -\zeta b^2_1 & 1 - \zeta b^2_2 \end{vmatrix} =$$

$$= 1 - \zeta (b^1_1 + b^2_2) + \zeta^2 (b^1_1 b^2_2 - b^1_2\, b^2_1) =$$

$$= 1 - 2H\zeta + K\zeta^2 \ , \tag{11.39a}$$

where H is the mean curvature, and K is the GAUSSIAN measure of curvature *(11.22a,b)*. The latter expression also denotes the ratio between the space metric g and the surface metric a [ET 1,2], i.e.,

$$\mu = \sqrt{\frac{g}{a}} \ . \tag{11.39b}$$

Kinematics of a shell continuum

According to Fig. 11.11, the base point P_0 on the mid-surface of the shell allocated to P is transformed into $\hat{P}_0$ (the state of deformation is denoted by $\hat{}$).

The position vector $\hat{\mathbf{r}}_p$ consists of the following parts:

$$\hat{\mathbf{r}}_p(\xi^\alpha, \zeta) = \mathbf{r}_p(\xi^\alpha, \zeta) + \mathbf{v}_p(\xi^\alpha, \zeta) \tag{11.40a}$$

with the vector of displacement $\mathbf{v}_p$

$$\mathbf{v}_p(\xi^\alpha, \zeta) = \mathbf{v}(\xi^\alpha) + \zeta \mathbf{w}(\xi^\alpha) \quad . \tag{11.40b}$$

The total distortion (normal rotation and shear deformation) is described by the vector $\mathbf{w}(\xi^\alpha)$.

The following assumptions are made:

a) Plane cross sections remain plane after loading (see *(11.40b)*).

b) Normal stresses τ^{33} in the ζ-direction are neglected (thin-walled shell), i.e. $\gamma_{33} = 0$.

Displacements of a shear-elastic shell are described by five independent components v_α, w, and w_α:

$$\mathbf{v} = (v_\alpha + \zeta w_\alpha)\mathbf{a}^\alpha + w\,\mathbf{a}^3 \quad . \tag{11.41}$$

This relation denotes a space tensor. Its components v_i are referred to the spatial base $\mathbf{g}^i$ at the point P of the undeformed shell.

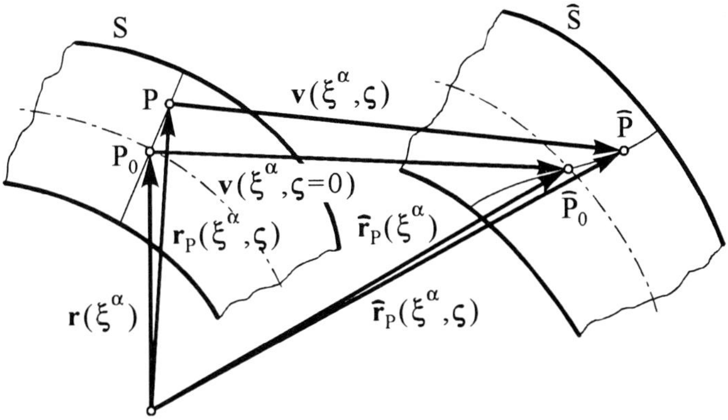

Fig. 11.11: Kinematics of the shell continuum

Strain tensor

CAUCHY-GREEN's strain tensor according to *(4.12b)* limited to small strains (linear theory):

$$\overset{S}{\gamma}_{ij} = \frac{1}{2}\left(\overset{S}{v}_i\big|_j + \overset{S}{v}_j\big|_i\right) \quad . \tag{11.42}$$

After transformation to the mid-surface of the shell we obtain

$$\alpha_{\alpha\beta} = \frac{1}{2}\left(v_\alpha\big|_\beta + v_\beta\big|_\alpha - 2\,b_{\alpha\beta}\,w\right) \tag{11.43}$$

≙ Normal and shear strains of the mid-surface of the shell;

$$\beta_{\alpha\beta} = \frac{1}{2}\left(w_\alpha\big|_\beta + w_\beta\big|_\alpha - b_\alpha^\varrho\, v_\varrho\big|_\beta - b_\beta^\varrho\, v_\varrho\big|_\alpha + 2\,b_\alpha^\varrho\, b_{\varrho\beta}\, w\right) \tag{11.44}$$

≙ Alterations of curvature and torsion;

$$\gamma_{3\alpha} = \frac{1}{2}\left(w_\alpha + w_{,\alpha} + b_\alpha^\varrho\, v_\varrho\right) \tag{11.45}$$

≙ Shear strains .

Stress Resultants

In analogy to the plate problem, the three-dimensional shell problem is reduced to a two-dimensional problem of the mid-surface of the shell. Resultant forces and moments are introduced instead of the stresses which are obtained by integrating the stresses over the shell thickness.

Membrane forces
$$N^{\alpha\beta} = \int_{-t/2}^{+t/2} \mu\, \mu_\varrho^\beta\, \tau^{\alpha\varrho}\, d\zeta \quad , \tag{11.46a}$$

Transverse shear forces
$$Q^{\alpha} = \int_{-t/2}^{+t/2} \mu\, \tau^{\alpha 3}\, d\zeta \quad , \tag{11.46b}$$

Moments
$$M^{\alpha\beta} = \int_{-t/2}^{+t/2} \mu\, \mu_\varrho^\beta\, \tau^{\alpha\varrho}\, \zeta\, d\zeta \quad . \tag{11.46c}$$

$N^{\alpha\beta}$ and $M^{\alpha\beta}$ are nonsymmetrical because of μ_ϱ^β.

In the theory of shallow shells one can approximate $\mu_\varrho^\beta \approx \delta_\varrho^\beta$, i.e.,

$$M^{\alpha\beta} = \int_{-t/2}^{+t/2} \mu\, \tau^{\alpha\beta}\, \zeta\, d\zeta \quad , \tag{11.47a}$$

$$\widetilde{N}^{\alpha\beta} = \int_{-t/2}^{+t/2} \mu\, \tau^{\alpha\beta}\, d\zeta \quad . \tag{11.47b}$$

$\tilde{N}^{\alpha\beta}$ is called the symmetrical "pseudo" tensor of resultant membrane forces, and it is valid that

$$\tilde{N}^{\alpha\beta} = N^{\alpha\beta} + b^{\beta}_{\varrho} M^{\alpha\varrho} \quad . \tag{11.48}$$

Equilibrium conditions [ET 2]

$$N^{\alpha\beta}|_{\alpha} - Q^{\alpha} b^{\beta}_{\alpha} + p^{\beta} = 0 \tag{11.48a}$$

≙ two equilibrium conditions of the resultant forces in the mid-surface;

$$Q^{\alpha}|_{\alpha} + N^{\alpha\beta} b_{\alpha\beta} + p = 0 \tag{11.48b}$$

≙ equilibrium conditions of forces perpendicular to the mid-surface;

$$M^{\alpha\beta}|_{\alpha} - Q^{\beta} = 0 \tag{11.48c}$$

≙ two equilibrium conditions of the resultant moments.

Constitutive equations for isotropic shells [C.5]

$$\tilde{N}^{\alpha\beta} = D\,H^{\alpha\beta\gamma\delta}\,\alpha_{\gamma\delta} \ , \tag{11.49a}$$

$$Q^{\alpha} = G\,t\,a^{\alpha\beta}\gamma_{\beta} \ , \tag{11.49b}$$

$$M^{\alpha\beta} = K\,H^{\alpha\beta\gamma\delta}\,\beta_{\gamma\delta} \tag{11.49c}$$

with strain stiffness $D = \dfrac{E\,t}{1 - \nu^2}$,

bending stiffness $K = \dfrac{E\,t^3}{12(1 - \nu^2)}$, (11.49d)

shear stiffness $G\,t = \dfrac{E\,t}{2(1 + \nu)}$,

and the elasticity tensor

$$H^{\alpha\beta\gamma\delta} = \frac{1-\nu}{2}\left(a^{\alpha\gamma} a^{\beta\delta} + a^{\alpha\delta} a^{\beta\gamma} + \frac{2\nu}{1-\nu} a^{\alpha\beta} a^{\gamma\delta} \right) \quad . \tag{11.49e}$$

11.3 Shear-rigid shells with weak curvature

Here, the following additional assumption is made:

- The shear deformation due to transverse forces is neglected. This means that points lying on a normal to the undeformed mid-surface after deformation lie on a normal to the deformed mid-surface (*normal hypothesis*).

This is one of the principle assumptions behind BERNOULLI's beam theory and KIRCHHOFF's plate theory (see Chapters 9, 10). It follows that

$$\gamma_{\alpha 3} = \frac{1}{2}\left(w_\alpha + w_{,\alpha} + b^\varrho_\alpha v_\varrho \right) = 0 \longrightarrow w_\alpha = -w_{,\alpha} - b^\varrho_\alpha v_\varrho \ . \qquad (11.50a)$$

Furthermore, for thin shells $\widetilde{N}^{\alpha\beta}$ is substituted by $N^{\alpha\beta}$, that means

$$\widetilde{N}^{\alpha\beta} \approx N^{\alpha\beta} = N^{\beta\alpha} \ . \qquad (11.50b)$$

A similar simplification is valid for the curvatures according to *(11.44)*:

$$\beta_{\alpha\beta} \approx \frac{1}{2}\left(w_\alpha\big|_\beta + w_\beta\big|_\alpha \right) = -\left[w\big|_{\alpha\beta} + \frac{1}{2}\left(b^\varrho_\alpha v_\varrho \right)\big|_\beta + \frac{1}{2}\left(b^\varrho_\alpha v_\varrho \right)\big|_\alpha \right] \mathrel{\hat{=}} \kappa_{\alpha\beta} \ . \qquad (11.51)$$

Neglecting the terms with b^ϱ_α in *(11.48a)* and *(11.51)* leads to the *basic equations*:

- Equilibrium conditions *(11.48a,b,c)*

$$\left.\begin{aligned} &N^{\alpha\beta}\big|_\alpha + p^\beta = 0 \ , \\ &Q^\alpha\big|_\alpha + N^{\alpha\beta} b_{\alpha\beta} + p = 0 \ , \\ &M^{\alpha\beta}\big|_\alpha - Q^\beta = 0 \ . \end{aligned}\right\} \qquad (11.52)$$

- Strain-displacement relations *(11.43, 11.44)*

$$\left.\begin{aligned} \alpha_{\alpha\beta} &= \frac{1}{2}\left(v_\alpha\big|_\beta + v_\beta\big|_\alpha - 2\, b_{\alpha\beta}\, w \right) \ , \\ \kappa_{\alpha\beta} &= \varrho_{\alpha\beta} \approx -w\big|_{\alpha\beta} \ . \end{aligned}\right\} \qquad (11.53)$$

- Constitutive equations *(11.49 a,c)*

$$\left.\begin{aligned} N^{\alpha\beta} &= D\, H^{\alpha\beta\gamma\delta}\, \alpha_{\gamma\delta} \ , \\ M^{\alpha\beta} &= K\, H^{\alpha\beta\gamma\delta}\, \varrho_{\gamma\delta} \ . \end{aligned}\right\} \qquad (11.54)$$

Equations *(11.52)*, *(11.53)* and *(11.54)* for the shear-rigid shell provide $3 + 3 + 3 + 6 = 15$ relations for the determination of the $3\,N^{\alpha\beta} + 3\,M^{\alpha\beta} + 3\,\alpha_{\alpha\beta} + 3\,\varrho_{\alpha\beta} + 2\,v_\alpha + w = 15$ unknowns.

12 Membrane theory of shells

12.1 General basic equations

Assumption: The stresses $\tau^{\alpha\beta}$ are uniformly distributed over the thickness, i.e. only so-called membrane forces occur, but *no* bending moments and *no* shear forces are found.

– Equilibrium conditions

$$\left.\begin{aligned} N^{\alpha\beta}\big|_{\alpha} + p^{\beta} &= 0 \ , \\ N^{\alpha\beta}\, b_{\alpha\beta} + p &= 0 \ . \end{aligned}\right\} \qquad (12.1)$$

These are three equations for three unknown resultant forces $N^{\alpha\beta}$, i.e. the membrane theory is *statically determinate.* The resultant forces $N^{\alpha\beta}$ can be calculated from the equilibrium conditions *(12.1)* alone and the deformations from *(12.2)* and *(12.3)*.

– Strain-displacement relations

$$\alpha_{\alpha\beta} = \frac{1}{2}\left(v_{\alpha}\big|_{\beta} + v_{\beta}\big|_{\alpha} - 2\, b_{\alpha\beta}\, w \right) \ . \qquad (12.2)$$

– Constitutive equations

$$N^{\alpha\beta} = D\, H^{\alpha\beta\gamma\delta}\, \alpha_{\gamma\delta} \qquad (12.3a)$$

with $H^{\alpha\beta\gamma\delta}$ defined by *(11.49e)* or, alternatively,

$$\left.\begin{aligned} \alpha_{\alpha\beta} &= \frac{1}{E\,t}\, D_{\alpha\beta\gamma\delta}\, N^{\gamma\delta} \\ \text{with } D_{\alpha\beta\gamma\delta} &= \frac{1+\nu}{2}\left(a_{\alpha\delta}\, a_{\beta\gamma} + a_{\alpha\gamma}\, a_{\beta\delta} \right) - \nu\, a_{\alpha\beta}\, a_{\gamma\delta} \ . \end{aligned}\right\} \qquad (12.3b)$$

– Specific deformation energy according to *(6.14)* or *(6.15b)*

$$\overline{U} = \frac{1}{2}\, N^{\alpha\beta}\, \alpha_{\alpha\beta} = \frac{1}{2\,E\,t}\, D_{\alpha\beta\gamma\delta}\, N^{\alpha\beta}\, N^{\gamma\delta} \ . \qquad (12.4)$$

12.2 Equilibrium conditions of shells of revolution

Coordinates: s or φ in meridional direction,
ϑ in circumferential (latitudinal) direction.

Derivatives: $\partial/\partial s \mathrel{\widehat{=}} (\ )_{,s}$, $\partial/\partial\vartheta \mathrel{\widehat{=}} (\ )_{,\vartheta}$.

– Equilibrium conditions

$$(r\,N_{ss})_{,s} + N_{s\vartheta\,,\vartheta} - \cos\varphi\, N_{\vartheta\vartheta} + r\,p_s = 0 \ , \qquad (12.5a)$$

$$(r\,N_{s\vartheta})_{,s} + N_{\vartheta\vartheta\,,\vartheta} + \cos\varphi\, N_{s\vartheta} + r\,p_{\vartheta} = 0 \ , \qquad (12.5b)$$

$$\frac{N_{ss}}{r_1} + \frac{N_{\vartheta\vartheta}}{r_2} = p \ . \qquad (12.5c)$$

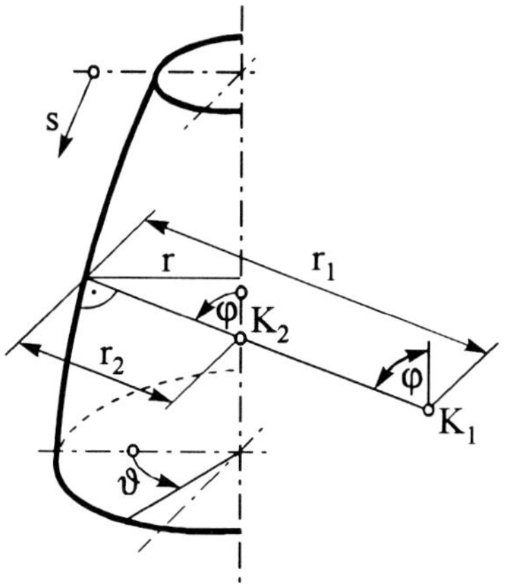

Fig. 12.1: Auxiliary quantities for a shell of revolution
$K_1 \triangleq$ centre point of curvature

Case 1: Axisymmetrical loading

$$p_\vartheta = 0 \quad , \quad \frac{\partial}{\partial \vartheta} \triangleq ,_\vartheta \equiv 0 \quad , \quad N_{s\vartheta} \equiv 0 \quad .$$

Introducing $ds = r_1 \, d\varphi$, one obtains

$$\left.\begin{aligned} & (r N_{\varphi\varphi})_{,\varphi} - r_1 \cos\varphi \, N_{\vartheta\vartheta} + r r_1 p_\varphi = 0 \quad , \\ & \frac{N_{\varphi\varphi}}{r_1} + \frac{N_{\vartheta\vartheta}}{r_2} - p \quad . \end{aligned}\right\} \qquad (12.6)$$

After elimination and integration, *(12.6)* leads to

$$N_{\varphi\varphi} = \frac{1}{r_2 \sin^2\varphi} \int\limits_{\bar\varphi = 0}^{\varphi} r_1 r_2 \left(p \cos\bar\varphi - p_\varphi \sin\bar\varphi \right) \sin\bar\varphi \, d\bar\varphi \quad , \qquad (12.7a)$$

$$N_{\vartheta\vartheta} = r_2 \, p - \frac{r_2}{r_1} N_{\varphi\varphi} \quad . \qquad (12.7b)$$

Case 2: Non-symmetrical loading

Expansion of loads in circumferential direction by means of FOURIER series :

$$\left.\begin{aligned} p_s(s,\vartheta) &= \sum_{m=1}^{\infty} p_{s_m}(s) \cos m\vartheta \quad , \\ p_\vartheta(s,\vartheta) &= \sum_{m=1}^{\infty} p_{\vartheta_m}(s) \sin m\vartheta \quad , \\ p(s,\vartheta) &= \sum_{m=1}^{\infty} p_m(s) \cos m\vartheta \quad . \end{aligned}\right\} \qquad (12.8)$$

A similar product expansion is chosen for the resultant forces:

$$\left.\begin{aligned} N_{ss} &= \sum_{m=1}^{\infty} N_{ss_m}(s)\cos m\vartheta \ , \\ N_{\vartheta\vartheta} &= \sum_{m=1}^{\infty} N_{\vartheta\vartheta_m}(s)\cos m\vartheta \ , \\ N_{s\vartheta} &= \sum_{m=1}^{\infty} N_{s\vartheta_m}(s)\sin m\vartheta \ . \end{aligned}\right\} \tag{12.9}$$

Substitution into *(12.5)* then yield the following set of ordinary differential equations:

$$\left.\begin{aligned} (r N_{ss_m})_{,s} + m N_{s\vartheta_m} - \cos\varphi N_{\vartheta\vartheta_m} + r p_{s_m} &= 0 \ , \\ (r N_{s\vartheta_m})_{,s} - m N_{\vartheta\vartheta_m} + \cos\varphi N_{s\vartheta_m} + r p_{\vartheta_m} &= 0 \ , \\ \frac{N_{ss_m}}{r_1} + \frac{N_{\vartheta\vartheta_m}}{r_2} &= p_m \ . \end{aligned}\right\} \tag{12.10}$$

Introduction of angle φ and elimination of $N_{\vartheta\vartheta_m}$ in *(12.10)* yield two ordinary differential equations:

$$\begin{aligned} (N_{\varphi\varphi_m})_{,\varphi} + \left(1 + \frac{r_1}{r_2}\right)\cot\varphi N_{\varphi\varphi_m} + m\frac{r_1}{r_2}\frac{N_{\varphi\vartheta_m}}{\sin\varphi} &= r_1 p_m \cot\varphi - r_1 p_{\vartheta_m} \ , \\ (N_{\varphi\vartheta_m})_{,\varphi} + 2\frac{r_1}{r_2}\cot\varphi N_{\varphi\vartheta_m} + \frac{N_{\varphi\varphi_m}}{\sin\varphi} &= m r_1 \frac{p_m}{\sin\varphi} - r_1 p_{\vartheta_m} \ . \end{aligned} \tag{12.11}$$

Special shells:

1) Spherical shell ($r_1 = r_2 = a$, $r = a\sin\varphi$)

$$(\sin\varphi N_{\varphi\varphi})_{,\varphi} + (N_{\varphi\vartheta})_{,\vartheta} - \cos\varphi N_{\vartheta\vartheta} + a\sin\varphi\, p_\varphi = 0 \ , \tag{12.12a}$$

$$(\sin\varphi N_{\varphi\vartheta})_{,\varphi} + (N_{\vartheta\vartheta})_{,\vartheta} + \cos\varphi N_{\varphi\vartheta} + a\sin\varphi\, p_\vartheta = 0 \ , \tag{12.12b}$$

$$N_{\varphi\varphi} + N_{\vartheta\vartheta} = p\,a \ . \tag{12.12c}$$

Boiler Formula: Spherical shell subjected to internal pressure $p_0 = \text{const}$

Stresses $$\sigma_{\varphi\varphi} = \frac{N_{\varphi\varphi}}{t} = \sigma_{\vartheta\vartheta} = \frac{N_{\vartheta\vartheta}}{t} = \frac{p_0 a}{2t} \ . \tag{12.13}$$

2) Circular cylindrical shell ($r = r_2 = a$, $r_1 d\varphi = dx$)

$$N_{xx,x} + \frac{1}{a} N_{x\vartheta,\vartheta} + p_x = 0 \;, \tag{12.14a}$$

$$N_{x\vartheta,x} + \frac{1}{a} N_{\vartheta\vartheta,\vartheta} + p_\vartheta = 0 \;, \tag{12.14b}$$

$$N_{\vartheta\vartheta} = p\,a \;. \tag{12.14c}$$

Boiler formula: Cylindrical shell with closed ends subjected to internal pressure $p_0 = \text{const}$

Stresses $$\sigma_{xx} = \frac{N_{xx}}{t} = \frac{p_0\,a}{2\,t} \;, \quad \sigma_{\vartheta\vartheta} = \frac{N_{\vartheta\vartheta}}{t} = \frac{p_0\,a}{t} \;. \tag{12.15}$$

3) Circular conical shell with semi-angle α ($r = s\sin\alpha$, $\cos\varphi = \sin\alpha$, $ds = r_1 d\varphi$; see Fig. 13.3)

$$(s N_{ss})_{,s} + \frac{1}{\sin\alpha} N_{s\vartheta,\vartheta} - N_{\vartheta\vartheta} + s\,p_s = 0 \;, \tag{12.16a}$$

$$(s N_{s\vartheta})_{,s} + \frac{1}{\sin\alpha} N_{\vartheta\vartheta,\vartheta} + N_{s\vartheta} + s\,p_\vartheta = 0 \;, \tag{12.16b}$$

$$N_{\vartheta\vartheta} = s\,p\tan\alpha \;. \tag{12.16c}$$

12.3 Equilibrium conditions of translation shells

Hyperbolical shell

Considering the class of translation shells we restrict our treatment to the special case of a hyperbolic shell. This type of shell has a wide-spread application especially in the design of cooling towers. Due to their negative (hyperbolic) curvature, these shells display a bearing behaviour that differs decisively from that of shells with a positive (elliptical) curvature (e.g. spherical shells, elliptical shells of revolution).

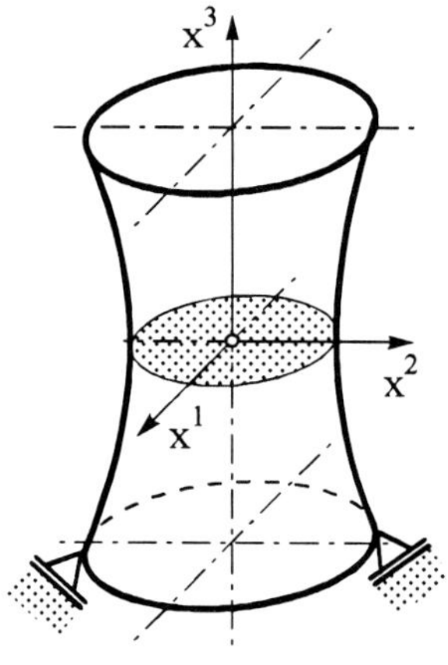

Fig. 12.2: Coordinates of an axisymmetrical hyperbolical shell

In order to illustrate this difference, we will proceed from the equilibrium conditions of the shell of revolution *(12.5)*. If we solve *(12.5c)* in terms of the resultant force $N_{\vartheta\vartheta}$ and substitute the solution into *(12.5a,b)*, we obtain a system of two first order partial differential equations in terms of the force resultants $N_{\varphi\varphi}$ and $N_{\varphi\vartheta}$. With $ds = r_1 d\varphi$ this system reads:

$$(r_2 \sin\varphi\, N_{\varphi\varphi})_{,\varphi} + r_2 \cos\varphi\, N_{\varphi\varphi} + r_1 N_{\varphi\vartheta,\vartheta} = r_1 r_2 (p\cos\varphi - p_\varphi \sin\varphi) \; ,$$

$$(r_2 \sin\varphi\, N_{\varphi\vartheta})_{,\varphi} + r_1 \cos\varphi\, N_{\varphi\vartheta} - r_2 N_{\varphi\varphi,\vartheta} = -r_1 r_2 (p_\vartheta \sin\varphi + p_{,\vartheta}) \; . \qquad (12.17)$$

By introducing the following substitutions

$$\left.\begin{aligned} U &= r_2 \sin^2\varphi\, N_{\varphi\varphi} \; , \\ V &= r_2^2 \sin^2\varphi\, N_{\varphi\vartheta} \; , \end{aligned}\right\} \qquad (12.18)$$

the equations are transformed into the simple form

$$V_{,\vartheta} + \frac{r_2^2}{r_1} \sin\varphi\, U_{,\varphi} = r_2^3 \sin^2\varphi\, (p\cos\varphi - p_\varphi \sin\varphi) \; , \qquad (12.19a)$$

$$V_{,\varphi} + \frac{r_2}{\sin\varphi} U_{,\vartheta} = -r_1 r_2^2 \sin\varphi\, (p_\vartheta \sin\varphi + p_{,\vartheta}) \; . \qquad (12.19b)$$

We now consider only the homogeneous part of the two differential equations *(12.19)*. V can be eliminated by differentiating *(12.19a)* with respect to φ and *(12.19b)* with respect to ϑ, and subtracting the equations. We then have

$$r_2^2 \sin^2\varphi\, U_{,\varphi\varphi} + r_1 r_2 U_{,\vartheta\vartheta} + r_1 \sin\varphi \left(\frac{r_2^2}{r_1} \sin\varphi\right)_{,\varphi} U_{,\varphi} = 0 \; . \qquad (12.20a)$$

Within the classification of linear partial differential equations of second order we write

$$A\, U_{,\varphi\varphi} + C\, U_{,\vartheta\vartheta} + a\, U_{,\varphi} = 0 \; , \qquad (12.20b)$$

where A, C, a are functions of φ. Depending on the sign of the discriminant

$$D = A\,C = r_1 r_2^3 \sin^2\varphi \; ,$$

eq. *(12.20b)* exhibits a different solution behaviour, where the decisive factor is whether the product of the radii of curvature $r_1 r_2$ is positive or negative. The following cases shall be considered:

a) $r_1 r_2 > 0$ $\longrightarrow$ Differential equation of an elliptical type (spherical shell, elliptical shell of revolution, etc.)

b) $r_1 r_2 < 0$ $\longrightarrow$ Differential equation of the hyperbolical type (hyperbolical shell)

Mechanical interpretation :

A principle way of solving this problem is to re-transform the partial differential equation into two ordinary differential equations by using a separation approach (see [C.18 , C.19]):

- *Solutions of differential equations of the elliptical type with* $r_1 r_2 > 0$ (spherical shell, elliptical shells of revolution) are such that discontinuities of the boundary values occuring in the case of point supports do not propagate into the inner regions but are confined to a narrow boundary zone.
- *Solutions of differential equations of the hyperbolical type with* $r_1 r_2 < 0$ display a completely different behaviour. These solutions are associated with curves on the shell surface, so-called characteristics, along which discontinuities of boundary conditions propagate over the entire shell [C.2]. This problem occurs in particular with hyperbolical shells with single supports. In this case, the membrane theory is not sufficient for determining the state of stress; bending deformations and resultant moments must then be considered by an extended theory.

12.4 Deformations of shells of revolution

Strain-displacement relations due to *(12.2)*

$$\varepsilon_{\varphi\varphi} = \frac{1}{r_1}(u_{,\varphi} + w) \ , \tag{12.21a}$$

$$\varepsilon_{\vartheta\vartheta} = \frac{1}{r}(v_{,\vartheta} + u\cos\varphi + w\sin\varphi) \ , \tag{12.21b}$$

$$\gamma_{\varphi\vartheta} = \frac{1}{r}u_{,\vartheta} + \frac{1}{r_1}v_{,\varphi} - \frac{v}{r}\cos\varphi \ . \tag{12.21c}$$

Special shells :

1) Spherical shell

$$\varepsilon_{\varphi\varphi} = \frac{1}{a}\left(u_{,\varphi} + w\right) \ , \tag{12.22a}$$

$$\varepsilon_{\vartheta\vartheta} = \frac{1}{a}\left(\frac{1}{\sin\varphi}v_{,\vartheta} + u\cot\varphi + w\right) \ , \tag{12.22b}$$

$$\gamma_{\varphi\vartheta} = \frac{1}{a}\left(\frac{1}{\sin\varphi}u_{,\vartheta} + v_{,\varphi} - v\cot\varphi\right) \ . \tag{12.22c}$$

2) Circular cylindrical shell

$$\varepsilon_{xx} = u_{,x} \ , \tag{12.23a}$$

$$\varepsilon_{\vartheta\vartheta} = \frac{1}{a}\left(v_{,\vartheta} + w\right) \ , \tag{12.23b}$$

$$\gamma_{x\vartheta} = \frac{1}{a}u_{,\vartheta} + v_{,x} \ . \tag{12.23c}$$

3) Circular conical shell

$$\varepsilon_{ss} = u_{,s} \ , \tag{12.24a}$$

$$\varepsilon_{\vartheta\vartheta} = \frac{1}{s}\left(\frac{1}{\sin\alpha} v_{,\vartheta} + u + w\cot\alpha\right) \ , \tag{12.24b}$$

$$\gamma_{s\vartheta} = \frac{1}{s\sin\alpha} u_{,\vartheta} + v_{,s} - \frac{v}{s} \ . \tag{12.24c}$$

12.5 Constitutive equations – material law

Regarding temperature fields ${}^{0}\Theta(\varphi,\vartheta)$ constant over the thickness

$$\left.\begin{aligned} N_{\varphi\varphi} &= D\left[\varepsilon_{\varphi\varphi} + \nu\,\varepsilon_{\vartheta\vartheta} - (1+\nu)\alpha_T\,{}^{0}\Theta\right] \ , \\ N_{\vartheta\vartheta} &= D\left[\varepsilon_{\vartheta\vartheta} + \nu\,\varepsilon_{\varphi\varphi} - (1+\nu)\alpha_T\,{}^{0}\Theta\right] \ , \\ N_{\varphi\vartheta} &= D\,\frac{1-\nu}{2}\,\gamma_{\varphi\vartheta} \qquad \text{with} \qquad D = \frac{E\,t}{1-\nu^2} \end{aligned}\right\} \tag{12.25}$$

or

$$\left.\begin{aligned} \varepsilon_{\varphi\varphi} &= \frac{1}{E\,t}(N_{\varphi\varphi} - \nu\,N_{\vartheta\vartheta}) + \alpha_T\,{}^{0}\Theta \ , \\ \varepsilon_{\vartheta\vartheta} &= \frac{1}{E\,t}(N_{\vartheta\vartheta} - \nu\,N_{\varphi\varphi}) + \alpha_T\,{}^{0}\Theta \ , \\ \gamma_{\varphi\vartheta} &= \frac{2(1+\nu)}{E\,t}N_{\varphi\vartheta} \ . \end{aligned}\right\} \tag{12.26}$$

Substitution of *(12.21)* into *(12.26)* generally leads to ordinary inhomogeneous differential equations of first order with variable coefficients. These equations have the general form:

$$\frac{dy}{dx} + P(x)y + Q(x) = 0$$

with the general solution:

$$y = e^{-\int P(x)dx}\left[-\int Q(x)e^{\int P(x)dx}dx + C\right] \ . \tag{12.27}$$

12.6 Specific deformation energy

General expression in membrane theory according to *(12.4)*

$$\overline{U} = \frac{1}{2}(N_{\vartheta\vartheta}\,\varepsilon_{\vartheta\vartheta} + N_{\varphi\varphi}\,\varepsilon_{\varphi\varphi} + N_{\varphi\vartheta}\,\varepsilon_{\varphi\vartheta}) \tag{12.28a}$$

or

$$\overline{U} = \frac{1}{2\,E\,t}\left[N_{\vartheta\vartheta}^2 + N_{\varphi\varphi}^2 - 2\nu\,N_{\varphi\varphi}N_{\vartheta\vartheta} + 2(1+\nu)N_{\varphi\vartheta}^2\right] \ . \tag{12.28b}$$

13 Bending theory of shells of revolution

13.1 Basic equations for arbitrary loads

Derivatives: $\partial/\partial\varphi \mathrel{\hat{=}} {}_{,\varphi}$ or $\partial/\partial s \mathrel{\hat{=}} {}_{,s}$, $\partial/\partial\vartheta \mathrel{\hat{=}} {}_{,\vartheta}$.

– Equilibrium conditions according to *(11.52)* and Fig. 13.1:

$$(r N_{\varphi\varphi})_{,\varphi} + r_1 N_{\varphi\vartheta,\vartheta} - r_1 \cos\varphi\, N_{\vartheta\vartheta} + r\, Q_\varphi = -r r_1 p_\varphi \ , \tag{13.1a}$$

$$(r N_{\varphi\vartheta})_{,\varphi} + r_1 N_{\vartheta\vartheta,\vartheta} + r_1 \cos\varphi\, N_{\varphi\vartheta} + r_1 \sin\varphi\, Q_\varphi = -r r_1 p_\vartheta \ , \tag{13.1b}$$

$$(r Q_\varphi)_{,\varphi} + r_1 Q_{\vartheta,\vartheta} - r N_{\varphi\varphi} - r_1 \sin\varphi\, N_{\vartheta\vartheta} = -r r_1 p \ , \tag{13.1c}$$

$$(r M_{\varphi\varphi})_{,\varphi} + r_1 M_{\varphi\vartheta,\vartheta} - r_1 \cos\varphi\, M_{\vartheta\vartheta} = r r_1 Q_\varphi \ , \tag{13.1d}$$

$$(r M_{\varphi\vartheta})_{,\varphi} + r_1 M_{\vartheta\vartheta,\vartheta} + r_1 \cos\varphi\, M_{\vartheta\varphi} = r r_1 Q_\vartheta \ . \tag{13.1e}$$

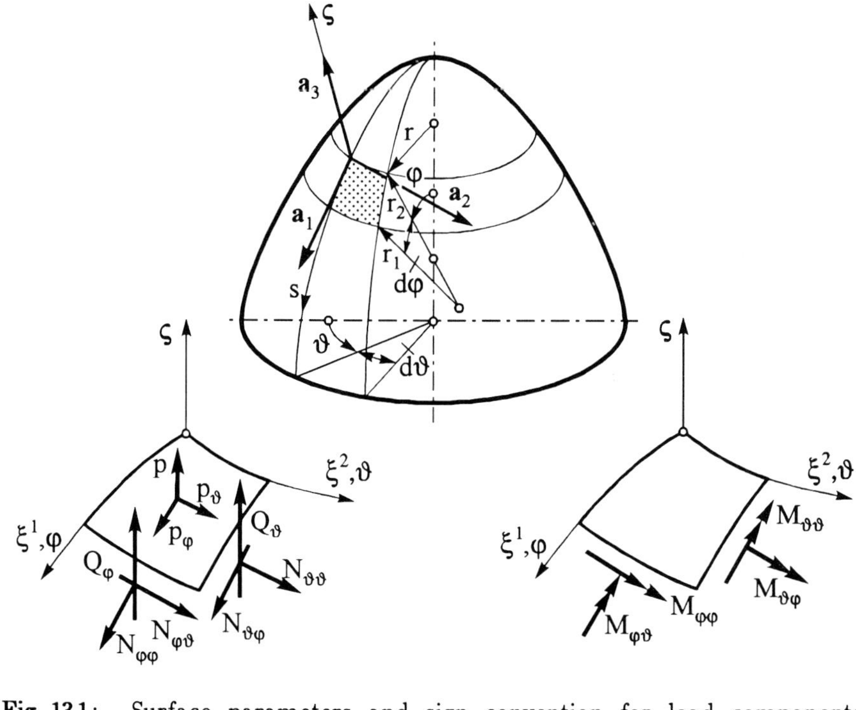

Fig. 13.1: Surface parameters and sign convention for load components and stress resultants of a shell element

– Strain–displacement relations due to *(11.51)*

$$\kappa_{ss} = \chi_{,s} \ , \tag{13.2a}$$

$$\kappa_{\vartheta\vartheta} = \frac{1}{r}\,\psi_{,\vartheta} + \frac{\cos\varphi}{r}\,\chi \ , \tag{13.2b}$$

$$\kappa_{s\vartheta} = \frac{1}{r}\Big[\chi_{,\vartheta} + (r\,\psi)_{,s} - 2\,\psi\cos\varphi\Big] \tag{13.2c}$$

with the two angular distortions ($\chi \mathrel{\widehat{=}} w_s \ , \ \psi \mathrel{\widehat{=}} w_\vartheta$) according to *(11.50a)*

$$\chi = \frac{u - w_{,\varphi}}{r_1} \qquad \text{in meridional direction ,} \tag{13.3a}$$

$$\psi = \frac{v\sin\varphi - w_{,\vartheta}}{r} \qquad \text{in circumferential direction .} \tag{13.3b}$$

Special relations for weakly curved shells of revolution with *(11.53)* [C.4]:

$$\varrho_{\varphi\varphi} = -\frac{1}{r_1}\Big(\frac{1}{r_1}\,w_{,\varphi}\Big)_{,\varphi} \ , \tag{13.4a}$$

$$\varrho_{\vartheta\vartheta} = -\frac{1}{r}\Big(\frac{1}{r}\,w_{,\vartheta\vartheta} + \frac{1}{r_1}\,w_{,\varphi}\cos\varphi\Big) \ , \tag{13.4b}$$

$$\varrho_{\varphi\vartheta} = -2\Big(\frac{1}{r_1 r}\,w_{,\varphi\vartheta} - \frac{\cos\varphi}{r^2}\,w_{,\vartheta}\Big) \ . \tag{13.4c}$$

– Material law

$$N_{ss} = D\big[\,\varepsilon_{ss} + \nu\,\varepsilon_{\vartheta\vartheta} - (1+\nu)\,\alpha_T\,{}^0\Theta\big] \ , \tag{13.5a}$$

$$N_{\vartheta\vartheta} = D\big[\,\varepsilon_{\vartheta\vartheta} + \nu\,\varepsilon_{ss} - (1+\nu)\,\alpha_T\,{}^0\Theta\big] \ , \tag{13.5b}$$

$$N_{s\vartheta} = D\,\frac{1-\nu}{2}\,\gamma_{s\vartheta} \ , \tag{13.5c}$$

$$M_{ss} = K\big[\,\kappa_{ss} + \nu\,\kappa_{\vartheta\vartheta} - (1+\nu)\,\alpha_T\,{}^1\Theta\big] \ , \tag{13.5d}$$

$$M_{\vartheta\vartheta} = K\big[\,\kappa_{\vartheta\vartheta} + \nu\,\kappa_{ss} - (1+\nu)\,\alpha_T\,{}^1\Theta\big] \ , \tag{13.5e}$$

$$M_{s\vartheta} = K\,\frac{1-\nu}{2}\,\kappa_{s\vartheta} \tag{13.5f}$$

with $D = \dfrac{E\,t}{1-\nu^2}$, $K = \dfrac{E\,t^3}{12\,(1-\nu^2)}$ according to *(11.50a)* .

Equations *(12.21)*, *(13.1)*, *(13.2)*, *(13.3)* and *(13.5)* altogether define a system of 19 equations for 19 unknowns (8 resultant forces, 6 strains and curvatures, 2 angular distortions, 3 displacements). They allow to calculate the stress and deformation states of shells of revolution.

Case 1: Axisymmetric loads

Assumptions: $N_{s\vartheta} = M_{s\vartheta} = Q_{\vartheta} = 0 \quad ; \quad v = \gamma_{s\vartheta} = \kappa_{s\vartheta} = 0 \quad .$ *(13.6)*

– Equilibrium conditions from *(13.1)*:

$$(r N_{\varphi\varphi})_{,\varphi} - r_1 \cos\varphi \, N_{\vartheta\vartheta} + r Q_{\varphi} = - r r_1 p_{\varphi} \ , \tag{13.7a}$$

$$(r Q_{\varphi})_{,\varphi} - r N_{\varphi\varphi} - r_1 \sin\varphi \, N_{\vartheta\vartheta} = - r r_1 p \ , \tag{13.7b}$$

$$(r M_{\varphi\varphi})_{,\varphi} - r_1 \cos\varphi \, M_{\vartheta\vartheta} = r r_1 Q_{\varphi} \ . \tag{13.7c}$$

– Strain–displacement relations from *(12.21)*, *(13.2)* and *(13.3)*

$$\varepsilon_{\varphi\varphi} = \frac{1}{r_1}(u_{,\varphi} + w) \ , \tag{13.8a}$$

$$\varepsilon_{\vartheta\vartheta} = \frac{1}{r}(u \cos\varphi + w \sin\varphi) \ , \tag{13.8b}$$

$$\kappa_{ss} = \chi_{,s} \quad \text{or} \quad \kappa_{\varphi\varphi} = \frac{1}{r_1}\chi_{,\varphi} \ , \tag{13.9a}$$

$$\kappa_{\vartheta\vartheta} = \frac{\cos\varphi}{r}\chi \tag{13.9b}$$

with

$$\chi = \frac{1}{r_1}(u - w_{,\varphi}) \ . \tag{13.10}$$

– Material law from *(13.5 a,b,d,e)*

$$N_{\varphi\varphi} = D[\varepsilon_{\varphi\varphi} + \nu \varepsilon_{\vartheta\vartheta} - (1 + \nu) \alpha_T {}^0\Theta] \ , \tag{13.11a}$$

$$N_{\vartheta\vartheta} = D[\varepsilon_{\vartheta\vartheta} + \nu \varepsilon_{\varphi\varphi} - (1 + \nu) \alpha_T {}^0\Theta] \ , \tag{13.11b}$$

$$M_{\varphi\varphi} = K[\kappa_{\varphi\varphi} + \nu \kappa_{\vartheta\vartheta} - (1 + \nu) \alpha_T {}^1\Theta] \ , \tag{13.11c}$$

$$M_{\vartheta\vartheta} = K[\kappa_{\vartheta\vartheta} + \nu \kappa_{\varphi\varphi} - (1 + \nu) \alpha_T {}^1\Theta] \ . \tag{13.11d}$$

Special shells:

1) Circular cylindrical shell subjected to axisymmetrical loads

Derivative: $d/d\xi \mathrel{\widehat{=}} {}_{,\xi}$ with $\xi = \dfrac{x}{a}$

- Equilibrium conditions

$$N_{xx,\xi} + a\,p_x = 0 \;, \tag{13.12a}$$

$$Q_{x,\xi} - N_{\vartheta\vartheta} + a\,p = 0 \;, \tag{13.12b}$$

$$M_{xx,\xi} - a\,Q_x = 0 \;. \tag{13.12c}$$

Elimination of the transverse force from *(13.12c)* and substitution into *(13.12b)*:

$$-M_{xx,\xi\xi} + a\,N_{\vartheta\vartheta} = a^2\,p \;. \tag{13.13}$$

- Strain-displacement relations

$$\varepsilon_{xx} = \frac{1}{a}\,u_{,\xi} \;, \quad \varepsilon_{\vartheta\vartheta} = \frac{w}{a} \;, \tag{13.14a}$$

$$\kappa_{xx} = \frac{1}{a}\,\chi_{,\xi} \;, \quad \kappa_{\vartheta\vartheta} = 0 \;. \tag{13.14b}$$

Bending angle

$$\chi = -\frac{1}{a}\,w_{,\xi} \;. \tag{13.14c}$$

- Material law

$$\left.\begin{aligned}
N_{xx} &= D[\,\varepsilon_{xx} + \nu\,\varepsilon_{\vartheta\vartheta} - (1+\nu)\,\alpha_T\,{}^0\Theta\,] \;,\\
N_{\vartheta\vartheta} &= D[\,\varepsilon_{\vartheta\vartheta} + \nu\,\varepsilon_{xx} - (1+\nu)\,\alpha_T\,{}^0\Theta\,] \;,\\
M_{xx} &= K[\,\kappa_{xx} + \nu\,\kappa_{\vartheta\vartheta} - (1+\nu)\,\alpha_T\,{}^1\Theta\,] \;,\\
M_{\vartheta\vartheta} &= K[\,\kappa_{\vartheta\vartheta} + \nu\,\kappa_{xx} - (1+\nu)\,\alpha_T\,{}^1\Theta\,] \;.
\end{aligned}\right\} \tag{13.15}$$

Eqs. *(13.14)* together with *(13.15)* in *(13.13)* yield the *boiler equation:*

$$\boxed{w_{,\xi\xi\xi\xi} + 4\,\kappa^4\,w = \frac{a^4\,p}{K} - (1+\nu)\,a^2\,\alpha_T\,{}^1\Theta_{,\xi\xi} + 4\,\kappa^4\,a\,\alpha_T\,{}^0\Theta} \tag{13.16a}$$

with $$\kappa^4 = 3\,(1-\nu^2)\left(\frac{a}{t}\right)^2 \quad (\kappa = \text{decay factor}) \;. \tag{13.16b}$$

Solution for a cylindrical shell with semi-infinite length subjected to boundary loads M, R (see Fig. 13.2):

$$w = \frac{a^2}{2\kappa^2 K}\left[\frac{a}{\kappa} R \cos\kappa\xi + M(\cos\kappa\xi - \sin\kappa\xi)\right] e^{-\kappa\xi} , \qquad (13.17a)$$

$$N_{\vartheta\vartheta} = \frac{2\kappa^2}{a}\left[\frac{a}{\kappa} R \cos\kappa\xi + M(\cos\kappa\xi - \sin\kappa\xi)\right] e^{-\kappa\xi} , \qquad (13.17b)$$

$$M_{xx} = -\left[\frac{a}{\kappa} R \sin\kappa\xi + M(\cos\kappa\xi + \sin\kappa\xi)\right] e^{-\kappa\xi} , \qquad (13.17c)$$

$$Q_x = -\left[R(\cos\kappa\xi - \sin\kappa\xi) - \frac{2\kappa}{a} M \sin\kappa\xi\right] e^{-\kappa\xi} . \qquad (13.17d)$$

2) Spherical shell – Method by GECKELER

This method utilizes the fast decay of boundary disturbances in a circular cylindrical shell ($\chi \approx e^{-\mu\varphi}$). The essential MEISSNER equations are the starting relations for the approximation method [C.7].

The following two uncoupled differential equations for the bending angle χ and the transverse force Q_φ are obtained ($d/d\varphi \mathrel{\widehat=} (\,)_{,\varphi}$):

$$\chi_{,\varphi\varphi\varphi\varphi} + 4\mu^4 \chi = 0 \qquad (13.18a)$$

$$Q_{\varphi,\varphi\varphi\varphi\varphi} + 4\mu^4 Q_\varphi = 0 \qquad (13.18b)$$

with $$4\mu^4 = \frac{E t a^2}{K} - \nu^2 .$$

ν^2 can be neglected in the case of thin-walled shells $\frac{a}{t} \gg 1$. Then, the decay factor given by *(13.16b)* can be used.

3) Circular conical shell – Method by GECKELER

In this case, it is also assumed that a fast decay of the disturbances from the boundaries occurs. The corresponding differential equation reads $\left(d/ds \mathrel{\widehat=} {}_{,s}\right)$:

$$Q_{s,ssss} + 4\bar{\kappa}_1^4 Q_s = 0 \qquad (13.19)$$

with $$\bar{\kappa}_1^4 = 3(1-\nu^2)\frac{\cot^2\alpha}{s^2 t^2} .$$

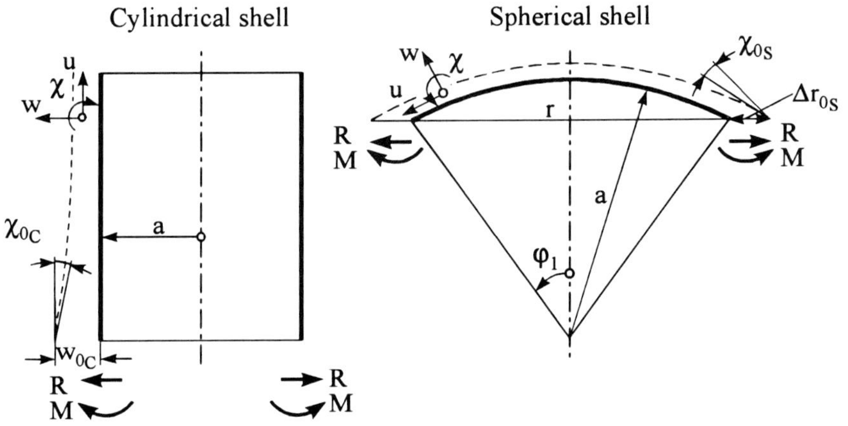

(13.20a,b)

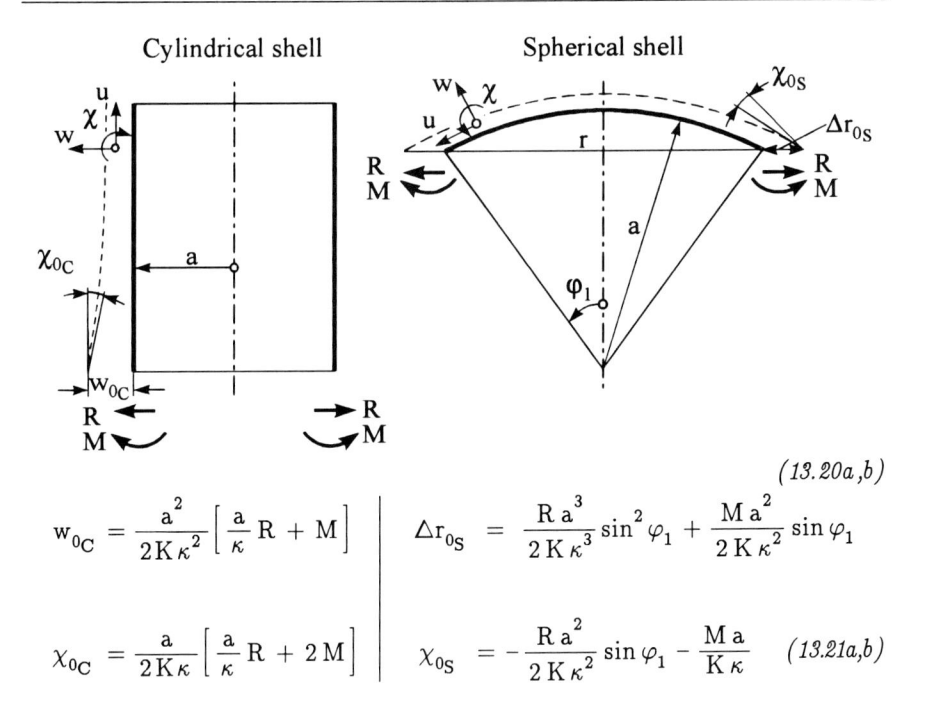

$$w_{0C} = \frac{a^2}{2K\kappa^2}\left[\frac{a}{\kappa}R + M\right] \qquad \Delta r_{0S} = \frac{R\,a^3}{2K\kappa^3}\sin^2\varphi_1 + \frac{M\,a^2}{2K\kappa^2}\sin\varphi_1$$

$$\chi_{0C} = \frac{a}{2K\kappa}\left[\frac{a}{\kappa}R + 2M\right] \qquad \chi_{0S} = -\frac{R\,a^2}{2K\kappa^2}\sin\varphi_1 - \frac{M\,a}{K\kappa} \qquad (13.21a,b)$$

Fig. 13.2: Displacements of the boundaries for cylindrical and spherical shells in dependence on the boundary loads R and M

Here, the decay factor depends on the variable s. Owing to the limitation to narrow boundary zones ($r_2 \approx$ const), we can assume the decay factor to be approximately constant in these areas. Thus, we obtain the same solution as when considering spherical shells.

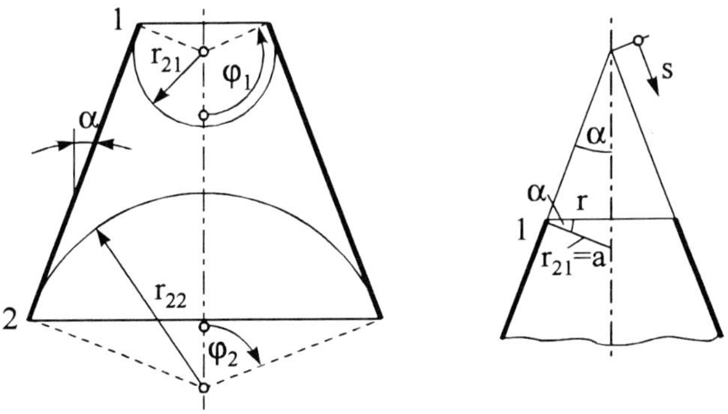

Fig. **13.3**: Substitute of a conical shell by spherical shells at the boundaries

4) Combined shell structures – Solution by *Method of Theory of Structures* [B.4, C.4, C.24]

The force-quantity procedure, the so-called *Method of Theory of Structures* for the analytical layout of combined structures can also be used successfully in shell statics. After a partitioning into single substructures ("0"-, "1"-, "2"-System etc..), compatibility conditions have to be formulated at the locations of transition between the subsystems.

Approximate determination of boundary disturbances for conical shells:

Cone-shaped joint units can be replaced by spherical shells with tangential joining. The wall thickness t of the substituting spherical shell is equal to that of the conical shell, and the radius a of the substituting shell is equal to the distance $r_{2i} = s_i \tan\alpha$ at the boundary of the conical shell (frustum) ($i = 1$ or 2) (Fig. 13.3).

13.2 Shells of revolution with arbitrary meridional shape - Transfer Matrix Method

Shells of revolution with arbitrarily variable meridional contours constitute an important group of components (casings, compensators, turbine disks, car wheels, etc.).

The structural behaviour of this type of shells can only be calculated iteratively. One way of solving the problem is to assume the shell to be assembled of single elements of shells of revolution, in the following abbreviated as *shell elements*. Here, a transfer procedure shall be introduced proceeding from the basic equations *(13.7)* to *(13.11)* of a shell subjected to axisymmetrical loading. These equations can be written as a system of differential equations of first order, i.e. the state equations.

If, on the right-hand side, we substitute into this system of differential equations $N_{\vartheta\vartheta}$, $M_{\vartheta\vartheta}$, $\varepsilon_{\varphi\varphi}$ and $\kappa_{\varphi\varphi}$ via the law of elasticity *(13.11 b,d)* and the strain-displacement relations *(13.8)*, then we obtain six differential equations of first order with six unknown state quantities:

$$\begin{aligned}
(r N_{\varphi\varphi})_{,\varphi} &= D\left[\frac{r_1}{r}(u\cos\varphi + w\sin\varphi)\cos\varphi + \right.\\
&\qquad \left. + \nu(u_{,\varphi} + w)\cos\varphi\right] - r Q_\varphi - r r_1 p_\varphi \;, \\
(r Q_\varphi)_{,\varphi} &= D\left[\frac{r_1}{r}(u\cos\varphi + w\sin\varphi)\sin\varphi + \right. \qquad (13.22)\\
&\qquad \left. + \nu(u_{,\varphi} + w)\sin\varphi\right] + r N_{\varphi\varphi} - r r_1 p \;,
\end{aligned}$$

$$
\begin{aligned}
(r M_{\varphi\varphi})_{,\varphi} &= K\left(\frac{r_1}{r}\chi\cos^2\varphi + \nu\cos\varphi\,\chi_{,\varphi}\right) + r r_1 Q_\varphi \;, \\
u_{,\varphi} &= \frac{r_1}{D} N_{\varphi\varphi} - \frac{r_1}{r}\nu\,(u\cos\varphi + w\sin\varphi) - w \;, \\
w_{,\varphi} &= u - r_1\chi \;, \\
\chi_{,\varphi} &= \frac{r_1}{K} M_{\varphi\varphi} - \frac{r_1}{r}\nu\,\chi\cos\varphi \;.
\end{aligned}
\tag{13.22}
$$

If we assemble the state quantities in a state vector

$$
\mathbf{y}^T = \left(r N_{\varphi\varphi},\, r Q_\varphi,\, r M_{\varphi\varphi},\, u,\, w,\, \chi\right) \;, \tag{13.23}
$$

we obtain the following symbolic notation for *(13.22)*:

$$
\mathbf{y}_{,\varphi} = \mathbf{A}\mathbf{y} + \mathbf{B}\mathbf{y}_{,\varphi} + \mathbf{p} \;, \tag{13.24}
$$

where the matrices **A**, **B** and **p** are given below:

$$
\mathbf{A} = \left[\begin{array}{ccc|ccc}
0 & -1 & 0 & D\frac{r_1}{r}\cos^2\varphi & D\left(\frac{r_1}{r}\sin\varphi + \nu\right)\cos\varphi & 0 \\
1 & 0 & 0 & D\frac{r_1}{r}\cos\varphi\sin\varphi & D\left(\frac{r_1}{r}\sin\varphi + \nu\right)\sin\varphi & 0 \\
1 & r_1 & 0 & 0 & 0 & K\frac{r_1}{r}\cos^2\varphi \\
\hline
\frac{1}{D}\frac{r_1}{r} & 0 & 0 & -\frac{r_1}{r}\nu\cos\varphi & -1-\frac{r_1}{r}\nu\sin\varphi & 0 \\
0 & 0 & 0 & 1 & 0 & -r_1 \\
0 & 0 & \frac{1}{K}\frac{r_1}{r} & 0 & 0 & -\frac{r_1}{r}\nu\cos\varphi
\end{array}\right]
$$

In the 6×6-matrix **B** only the following terms do not vanish:

$$
b_{14} = D\nu\cos\varphi \;, \quad b_{24} = D\nu\sin\varphi \;, \quad b_{36} = K\nu\cos\varphi \;.
$$

The load vector $\mathbf{p}$ reads:

$$\mathbf{p}^T = \left[-r\, r_1\, p_\varphi \,,\; -r\, r_1\, p,\; 0,\; 0,\; 0,\; 0 \right] \quad .$$

The shell is now subdivided into the above-mentioned *shell elements* with small angles $\Delta\varphi_i$ in such a way that the elements of the matrices within each single element are assumed to be constant (Fig. 13.4).

This task can be solved by substituting the first derivative for the i-th twill by the difference quotient:

$$(\mathbf{y}_{i-1})_{,\varphi} \approx \frac{1}{\Delta\varphi_i}(\mathbf{y}_i - \mathbf{y}_{i-1}) \quad .$$

All quantities at point i are expressed by values at point (i − 1). Equation (*13.24*) then reads

$$\mathbf{y}_i = \mathbf{y}_{i-1} + \Delta\varphi_i(\mathbf{I} - \mathbf{B}_i)^{-1}(\mathbf{A}_i\,\mathbf{y}_{i-1} + \mathbf{p}_i) \quad . \qquad (13.25\,a)$$

Owing to the suitable structure of matrix $\mathbf{B}_i$ (b_{14}, b_{24}, $b_{36} \neq 0$, all remaining $b_{ij} = 0$), a potential series expansion of the inverse of $(\mathbf{I} - \mathbf{B}_i)$ yields the following identity:

$$(\mathbf{I} - \mathbf{B}_i)^{-1} = \mathbf{I} + \mathbf{B}_i \quad .$$

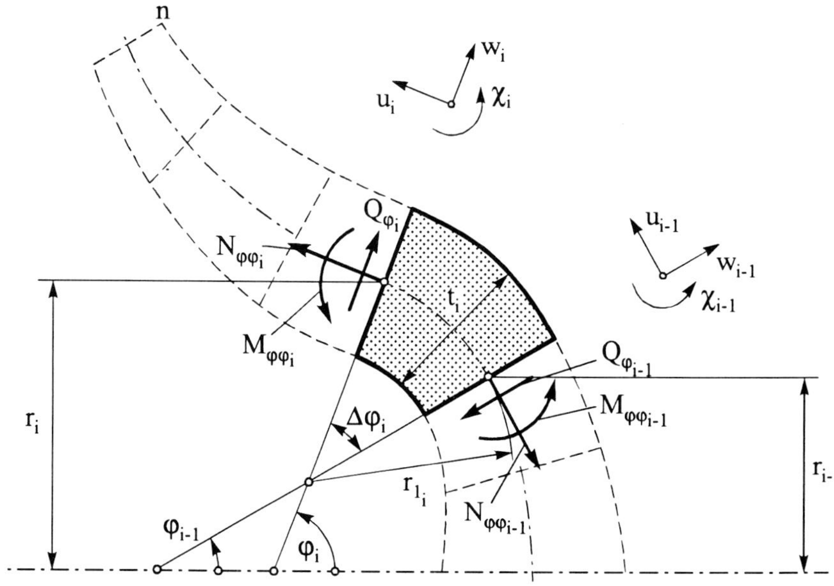

Fig. 13.4: *Shell element* for the transfer procedure

Equation *(13.25a)* then takes the form

$$\mathbf{y}_i = \mathbf{y}_{i-1} + \Delta\varphi_i(\mathbf{I} - \mathbf{B}_i)(\mathbf{A}_i\mathbf{y}_{i-1} + \mathbf{p}_i) \quad . \tag{13.25 b}$$

The *state vector* $\mathbf{y}_i$ still contains the radius r_i of the i-th subelement. As the radius may differ from one subelement to another, it must be eliminated from the vector when using the transfer matrix procedure. In addition, the load quantities must be included into the vector, and we therefore replace $\mathbf{y}_i$ by a new state vector $\mathbf{z}_i$ defined as

$$\mathbf{z}_i^T = (N_{\varphi\varphi}, Q_\varphi, M_{\varphi\varphi}, u, w, \chi, 1)_i \quad . \tag{13.26}$$

From *(13.25b)* we then obtain the transformation for the i-th subelement:

$$\mathbf{z}_i = \mathbf{C}_i\mathbf{z}_{i-1} \quad . \tag{13.27 a}$$

The transfer matrix $\mathbf{C}_i$ shall not be written explicitly here as it can be derived from *(13.25b)*.

The conditions of continuity and compatibility expressing that at the point of transition between two elements equal forces and moments are transferred and that equal deformations must occur, finally yields the transfer procedure between the boundaries i = 0 and i = n:

$$\mathbf{z}_n = \prod_i \mathbf{C}_i\mathbf{z}_{i-1} = \mathbf{C}\mathbf{z}_0 \quad . \tag{13.28}$$

The above matrix equation represents a set of linear equations containing six equations with 2 × 6 unknown state quantities at both boundaries. By giving 2 × 3 = 6 boundary conditions at the beginning and at the end of the shell, one obtains a solvable set of equations for the boundary quantities.

Extension to shells with large deflections

If large deformations are to be treated by a purely linear method, the single step procedure proves to be very suitable. Here, the load is applied incrementally, and the total transfer matrix is recalculated after each increment. When using the *transfer matrix procedure*, one proceeds from the equilibrium conditions of the undeformed structure, where the position vector $\mathbf{r}_i$ for the i-th shell element is assumed to be constant, but shall be treated as a function of the displacements u and w. Since the position vectors $\mathbf{r}_i$ of the deformed structure cannot be determined analytically, matrix $\mathbf{C}$ can only be calculated for an *undeformed structure*. Thus, equation *(13.27a)* becomes:

$$\tilde{\mathbf{z}}_i^0 = \mathbf{C}_i^0(\mathbf{r}_i^0, \Delta p^0)\tilde{\mathbf{z}}_{i-1}^0 \quad .$$

Thus, we obtain as a transfer rule (Fig. 13.5):

$$\tilde{\mathbf{z}}_i^k = \mathbf{C}_i^k(\underbrace{\mathbf{r}_i^{k-1} + \Delta\mathbf{r}_i^{k-1}}_{\mathbf{r}_i^k}, \Delta p^k)\tilde{\mathbf{z}}_{i-1}^k \tag{13.27b}$$

with

$$\Delta \mathbf{r}_i^k = \Delta \mathbf{r}_i^k \left(u_i^k , w_i^k \right) \quad .$$

The incremental procedure comprises the following steps:

Step 1: The structure is considered unloaded and is subjected to the load increment Δp (index 0).

Step 2: The resultant forces and moments as well as the deformations are calculated according to the linear theory.

Step 3: Forces and moments are summed up, and the contour subjected to the load is determined on the basis of the deformations. The deformed contour is then taken as the starting point for the next load step
$\longrightarrow$ Step 1 (index $0 \longrightarrow 1, 2, \ldots, k$ in (2)) .

This procedure is repeated until the sum of the load steps Δp^k equals the total load to be applied. Thus, the nonlinear load-deformation-curve is approximated by piecewise linear sections as shown in Fig. 13.6.

Since the equilibrium is established by the deformed structure, a correction is not carried out, and thus this procedure has the disadvantage that the approximated solution deviates from the exact solution with increasing loading. On the other hand, this procedure is characterized by numerical stability and by a simple realization since the structural analysis program does not require any manipulation.

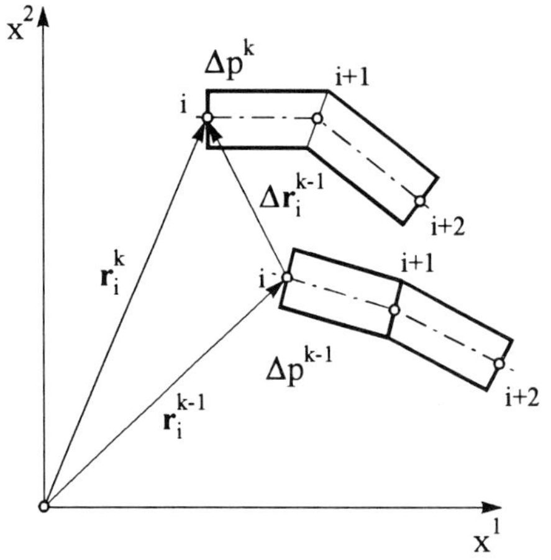

Fig. 13.5: Shell element of two successive load steps

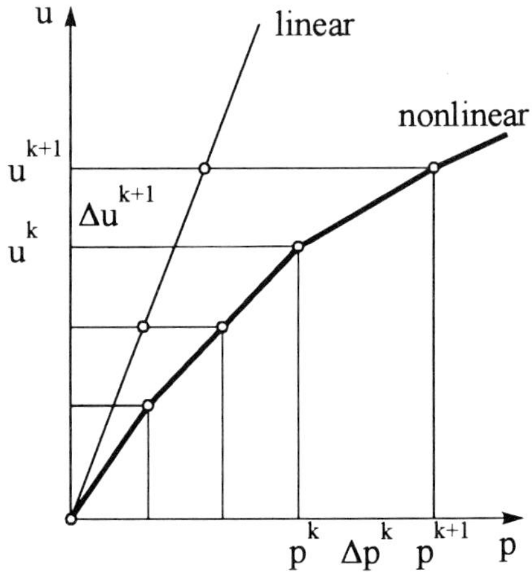

Fig. 13.6: Linear and nonlinear incremental procedure

13.3 Bending theory of a circular cylindrical shell

Derivatives: $\partial/\partial\xi \mathrel{\hat=} {}_{,\xi}$ with $\xi = \frac{x}{a}$, $\partial/\partial\vartheta \mathrel{\hat=} {}_{,\vartheta}$.

a) *General shear-rigid theory for an isotropic shell by FLUEGGE* [C.4]

- Equilibrium conditions from *(11.52)* after elimination of Q^α

$$\left.\begin{aligned} &N_{xx,\xi} + N_{\vartheta x,\vartheta} &&= -a\,p_x \;, \\ &N_{x\vartheta,\xi} + N_{\vartheta\vartheta,\vartheta} + \frac{1}{a}(M_{x\vartheta,\xi} + M_{\vartheta\vartheta,\vartheta}) &&= -a\,p_\vartheta \;, \\ &\frac{1}{a}(M_{xx,\xi\xi} + M_{x\vartheta,\xi\vartheta} + M_{\vartheta x,\xi\vartheta} + M_{\vartheta\vartheta,\vartheta\vartheta}) - N_{\vartheta\vartheta} &&= -a\,p \;. \end{aligned}\right\} \quad (13.29)$$

- Resultant forces – displacement relations from *(11.53)* and *(11.54)*

$$\left.\begin{aligned} N_{xx} &= \frac{D}{a}[u_{,\xi} + \nu(v_{,\vartheta} + w)] \;, \\ N_{x\vartheta} &= \frac{D}{a}\frac{1-\nu}{2}(u_{,\vartheta} + v_{,\xi}) - \frac{K(1-\nu)}{a^3}(w_{,\xi\vartheta} - v_{,\xi}) \;, \\ N_{\vartheta x} &= \frac{D}{a}\frac{1-\nu}{2}(u_{,\vartheta} + v_{,\xi}) \;, \\ N_{\vartheta\vartheta} &= \frac{D}{a}(v_{,\vartheta} + w + \nu u_{,\xi}) - \frac{K}{a^3}(w_{,\vartheta\vartheta} - 2v_{,\vartheta} - w + \nu w_{,\xi\xi}) \;, \\ M_{xx} &= -\frac{K}{a^2}[w_{,\xi\xi} + \nu(w_{,\vartheta\vartheta} - 2v_{,\vartheta} - w)] \;, \\ M_{x\vartheta} &= M_{\vartheta x} = -\frac{K}{a^2}(1-\nu)(w_{,\xi\vartheta} - v_{,\xi}) \;, \\ M_{\vartheta\vartheta} &= -\frac{K}{a^2}(w_{,\vartheta\vartheta} - 2v_{,\vartheta} - w + w_{,\xi\xi}) \;. \end{aligned}\right\} \quad (13.30)$$

On the basis of *(13.29)* and *(13.30)* we obtain within the general bending theory of shear-rigid circular cylindrical shells the following coupled system of three partial differential equations for the displacements u, v, w:

$$u_{,\xi\xi} + \frac{1-\nu}{2}u_{,\vartheta\vartheta} + \frac{1+\nu}{2}v_{,\xi\vartheta} + \nu w_{,\xi} = -\frac{a^2 p_x}{D} \;, \quad (13.31a)$$

$$\frac{1+\nu}{2}u_{,\xi\vartheta} + v_{,\vartheta\vartheta} + \frac{1-\nu}{2}v_{,\xi\xi} + 2k[(1-\nu)v_{,\xi\xi} + 2v_{,\vartheta\vartheta}] + {}$$
$$+ w_{,\vartheta} - 2k(w_{,\vartheta\vartheta\vartheta} - w_{,\vartheta} + w_{,\xi\xi\vartheta}) = -\frac{a^2 p_\vartheta}{D} \;, \quad (13.31b)$$

$$\nu u_{,\xi} + v_{,\vartheta} - 2k(v_{,\vartheta\vartheta\vartheta} - v_{,\vartheta} + v_{,\xi\xi\vartheta}) + w + \\ + k(\Delta\Delta w - 2w_{,\vartheta\vartheta} + w - 2\nu w_{,\xi\xi}) = \frac{a^2 p}{D} \tag{13.31c}$$

with
$$k = \frac{K}{a^2 D} = \frac{t^2}{12 a^2} \ . \tag{13.31d}$$

Boundary conditions using principle of total potential [ET 2]

- Clamped boundary at $x = \text{const}$:

$$u = v = w = w_{,\xi} = 0 \ . \tag{13.32 a}$$

- Free boundary at $x = \text{const}$:

$$N_{xx} = \overline{N}_{x\vartheta} = M_{xx} = \overline{Q}_x = 0 \ . \tag{13.32 b}$$

- Clamped boundary at $\vartheta = \text{const}$:

$$u = v = w = w_{,\vartheta} = 0 \ . \tag{13.33 a}$$

- Free boundary at $\vartheta = \text{const}$:

$$N_{\vartheta\vartheta} = N_{\vartheta x} = M_{\vartheta\vartheta} = \overline{Q}_\vartheta = 0 \ . \tag{13.33 b}$$

In cases of shear-rigid shells, only four boundary conditions (the differential equation is of the eighth order) can be fulfilled. Three of the existing five boundary stress resultants are re-defined as effective ones (similar to KIRCHHOFF's plate theory), namely the *effective transverse shear forces*

$$\overline{Q}_x = Q_x + \frac{M_{x\vartheta,\vartheta}}{a} \quad \text{or} \quad \overline{Q}_\vartheta = Q_\vartheta + \frac{M_{\vartheta x,\xi}}{a} \tag{13.34}$$

and the *effective in-plane shear force*

$$\overline{N}_{x\vartheta} = N_{x\vartheta} + \frac{M_{x\vartheta}}{a} \ . \tag{13.35}$$

b) *Simplified DONNELL's theory* [C.3, C.15]

- Equilibrium conditions without external loads:

$$\left.\begin{aligned} &N_{xx,\xi} + N_{\vartheta x,\vartheta} = 0 \ , \\ &N_{x\vartheta,\xi} + N_{\vartheta\vartheta,\vartheta} = 0 \ , \\ &\frac{1}{a}(M_{xx,\xi\xi} + 2M_{x\vartheta,\xi\vartheta} + M_{\vartheta\vartheta,\vartheta\vartheta}) - N_{\vartheta\vartheta} = 0 \ . \end{aligned}\right\} \tag{13.36}$$

– Resultant force – displacement relations

$$\left.\begin{aligned}
N_{xx} &= \frac{D}{a}\left[u_{,\xi} + \nu\,(v_{,\vartheta} + w)\right] ,\\
N_{\vartheta\vartheta} &= \frac{D}{a}\,(v_{,\vartheta} + w + \nu\,u_{,\xi}) ,\\
N_{x\vartheta} &= N_{\vartheta x} = \frac{D}{a}\,\frac{1-\nu}{2}\,(u_{,\vartheta} + v_{,\xi}) ,\\
M_{xx} &= -\frac{K}{a^2}\,(w_{,\xi\xi} + \nu\,w_{,\vartheta\vartheta}) ,\\
M_{\vartheta\vartheta} &= -\frac{K}{a^2}\,(w_{,\vartheta\vartheta} + \nu\,w_{,\xi\xi}) ,\\
M_{x\vartheta} &= M_{\vartheta x} = -\frac{K}{a^2}\,(1-\nu)\,w_{,\xi\vartheta} .
\end{aligned}\right\} \qquad (13.37)$$

From *(13.36)* we obtain, by substituting *(13.37)*, a simplified, coupled set of three differential equations for the displacements:

$$\boxed{\begin{aligned}
&u_{,\xi\xi} + \frac{1-\nu}{2}\,u_{,\vartheta\vartheta} + \frac{1+\nu}{2}\,v_{,\xi\vartheta} + \nu\,w_{,\xi} &&= 0\\
&\frac{1+\nu}{2}\,u_{,\xi\vartheta} + v_{,\vartheta\vartheta} + \frac{1-\nu}{2}\,v_{,\xi\xi} + w_{,\vartheta} &&= 0\\
&\nu\,u_{,\xi} + v_{,\vartheta} + w + k\,\Delta\Delta\,w &&= 0
\end{aligned}} \qquad . \qquad (13.38)$$

Solution with respect to w yields one differential equation of eighth order :

$$\boxed{k\,\Delta\Delta\Delta\Delta\,w + (1-\nu^2)\,w_{,\xi\xi\xi\xi} = 0} \qquad (13.39)$$

or a coupled system of two differential equations of fourth order for the displacement w and AIRY's stress function Φ (similarly to the coupled disk-plate problem):

$$\frac{K}{a^3}\,\Delta\Delta\,w + \Phi_{,\xi\xi} = 0 \qquad (13.40\,a)$$

$$\Delta\Delta\,\Phi - E\,\frac{t}{a}\,w_{,\xi\xi} = 0 \qquad . \qquad (13.40\,b)$$

The corresponding boundary conditions are analogous to *(13.32)* – *(13.35)*.

c) *Solution of closed shells under boundary loads*

- *Complete theory*

With $p_x = p_\vartheta = p = 0$, *(13.31a,b,c)* are transformed into a system of homogeneous differential equations. In the case of a closed shell, all displacements must be functions of the circumferential angle ϑ, since after one rotation at $\vartheta = 2\pi$ the same values as in the initial point ϑ must occur. The separation approach using FOURIER expansion series

$$u = \sum_{m=1}^{\infty} u_m \cos m\vartheta \ , \quad v = \sum_{m=1}^{\infty} v_m \sin m\vartheta \ , \quad w = \sum_{m=1}^{\infty} w_m \cos m\vartheta \qquad (13.41)$$

yields from *(13.31)* a coupled system of ordinary differential equations with constant coefficients for the unknown functions $u_m(\xi)$, $v_m(\xi)$, $w_m(\xi)$ $(\xi = x/a)$. The given problem is then treated further by applying exponential approximations:

$$u_m = U e^{\lambda\xi} \ , \quad v_m = V e^{\lambda\xi} \ , \quad w_m = W e^{\lambda\xi} \ . \qquad (13.42)$$

This leads to a homogeneous system of equations which only possesses non-trivial solutions provided that the determinant of the coefficients vanishes. If higher order terms $(k \ll 1)$ are neglected, one obtains the characteristic equation for the unknown eigenvalues λ:

$$\begin{aligned} &\lambda^8 - 2(2m^2 - \nu)\lambda^6 + \left[\frac{1-\nu^2}{k} + 6m^2(m^2-1)\right]\lambda^4 - \\ &\quad - 2m^2\left[2m^4 - (4+\nu)m^2 + (2+\nu)\right]\lambda^2 + m^4(m^2-1)^2 = 0 \ . \end{aligned} \qquad (13.43)$$

This fourth order equation in λ^2 has four complex roots. We thus obtain solutions for

$$m \geq 2 \ , \qquad (13.44a)$$

$$m = 0 \quad \text{and} \quad m = 1 \ . \qquad (13.44b)$$

The total solution consists of the single solutions of *(13.44a)* and *(13.44b)* (see [ET 2] for more details).

- *Simplified theory*

Here, we use the eighth order differential equation *(13.39)*.

With

$$w = W e^{\lambda\xi} \cos m\vartheta \ , \qquad (13.45)$$

we obtain the characteristic equation

$$\lambda^8 - 4m^2\lambda^6 + \left(\frac{1-\nu^2}{k} + 6m^4\right)\lambda^4 - 4m^6\lambda^2 + m^8 = 0 \ . \qquad (13.46)$$

Comparison of the eigenvalue equation *(13.46)* with *(13.43)* shows that in the simplified theory only the highest terms in the coefficients are retained. The two theories yield the same results for $m = 0$ and $m \geq 2$. For the case of $m = 1$, however, there is *no* agreement. MORLEY has solved this problem according to *(13.30)* by introducing higher order constitutive laws [C.16].

d) *Theory for fast decaying boundary disturbances*

If an arbitrary loading is given at the boundary, one has to calculate all partial amplitudes by means of the total solution, simultaneously considering the boundary conditions. The fast decaying partial solution (large λ) is predominantly removed via the circumferential force $N_{\vartheta\vartheta}$ and the bending moment M_{xx}, and one thus obtains an approximative theory with respect to the large roots. Here, the following simplifying assumptions are valid:

- $M_{x\vartheta} = M_{\vartheta\vartheta} = 0$ is set in the equilibrium conditions. Thus, *(13.29)* reduces to

$$\begin{aligned} N_{xx,\xi} + N_{\vartheta x,\vartheta} &= 0 \ , \\ N_{x\vartheta,\xi} + N_{\vartheta\vartheta,\vartheta} &= 0 \ , \\ \frac{1}{a} M_{xx,\xi\xi} - N_{\vartheta\vartheta} &= 0 \qquad (\partial/\partial\xi \mathrel{\widehat{=}} (\ )_{,\xi} \ , \ \partial/\partial\vartheta \mathrel{\widehat{=}} (\ )_{,\vartheta}) \ . \end{aligned} \tag{13.47}$$

- The strain ε_{xx} *(12.23a)* and the shear strain $\gamma_{x\vartheta}$ *(12.23c)* are set zero. It then holds for the derivatives of the displacements that

$$\begin{aligned} u_{,x} &= 0 \ , \\ u_{,\vartheta} + v_{,\xi} &= 0 \ . \end{aligned} \tag{13.48}$$

- If the influence of POISSON's ratio is neglected ($\nu = 0$) for the membrane force, the simplified material law *(13.37)* reads

$$\begin{aligned} N_{\vartheta\vartheta} &= \frac{E\,t}{a}(v_{,\vartheta} + w) \ , \\ M_{xx} &= \frac{K}{a^2} w_{,\xi\xi} \ . \end{aligned} \tag{13.49}$$

The seven equations *(13.47)* to *(13.49)* allow calculation of the seven unknowns

$$u \ , \ v \ , \ w \ , \ N_{xx} \ , \ N_{x\vartheta} \ , \ N_{\vartheta\vartheta} \ , \ M_{xx} \ .$$

Owing to *(13.48)*, no material law can be given for N_{xx} and $N_{x\vartheta}$. These two membrane forces are obtained from *(13.47)*.

Substitution of *(13.49)* into *(13.47)* yields after some re-calculation the sixth order differential equation

$$w_{,\xi\xi\xi\xi\xi\xi} + \frac{1-\nu^2}{k} w_{,\xi\xi} = 0 \ . \tag{13.50}$$

Introducing *(13.45)* leads to the eigenvalue equation

$$\lambda^4 + \frac{1-\nu^2}{k} = 0 \tag{13.51}$$

with the roots

$$\lambda_{1,2,3,4} = \pm\ \kappa_1 \pm \mathrm{i}\,\mu_1 \quad \text{with} \quad \kappa_1 = \mu_1 = \sqrt{\frac{1}{2}\sqrt{\frac{1-\nu^2}{k}}} \quad . \tag{13.52}$$

In the scope of this approximation the boundary disturbances thus decay independently of the number of circumferential waves, i.e. in the same manner as in the axisymmetric case (see Case 1 in Section 13.1).

e) *Theory for slowly decaying boundary disturbances*

This theory plays an important role in the case of small roots in the eigenvalue equation, since these roots extend over a large area of the shell. In this context, a special approximation theory has been developed which is called *Theory of Flexible Shells* or *Semi-Membrane Theory* [C.1]. As the theory omits the bending forces, it should more suitably be termed *Semi-Bending Theory*. In the total solution we have shown that, in the case of small roots, the moment $M_{\vartheta\vartheta}$ gains a decisive influence. A corresponding approximation theory can thus be determined on the basis of the following assumptions:

- For the conditions of equilibrium, $M_{xx} = M_{x\vartheta} = 0$ is set. Hence, *(13.29)* becomes

$$\begin{aligned} & N_{xx,\xi} + N_{\vartheta x,\vartheta} = 0 \quad , \\ & N_{x\vartheta,\xi} + N_{\vartheta\vartheta,\vartheta} + \frac{1}{a} M_{\vartheta\vartheta,\vartheta} = 0 \quad , \\ & \frac{1}{a} M_{\vartheta\vartheta,\vartheta\vartheta} - N_{\vartheta\vartheta} = 0 \quad . \end{aligned} \tag{13.53}$$

- The strain $\varepsilon_{\vartheta\vartheta}$ and the shear strain $\gamma_{x\vartheta}$ vanish. This requires statement of the following couplings between the displacements:

$$\left.\begin{aligned} & v_{,\vartheta} + w = 0 \quad , \\ & u_{,\vartheta} + v_{,\xi} = 0 \quad . \end{aligned}\right\} \tag{13.54}$$

- Considering *(13.54)* and neglecting ν in the membrane force, the material law *(13.30)* reduces to

$$\left.\begin{aligned} & N_{xx} = \frac{E\,t}{a} u_{,x} \quad , \\ & M_{\vartheta\vartheta} = \frac{K}{a^2}(w_{,\vartheta\vartheta} + w) \quad . \end{aligned}\right\} \tag{13..55}$$

The seven equations *(13.53)* to *(13.55)* allow calculation of the seven unknowns

$$u \quad , \quad v \quad , \quad w \quad , \quad N_{xx} \quad , \quad N_{x\vartheta} \quad , \quad N_{\vartheta\vartheta} \quad , \quad M_{\vartheta\vartheta} \quad .$$

Solution then leads to

$$\frac{1-\nu^2}{k} w_{,\xi\xi\xi\xi} + w_{,\vartheta\vartheta\vartheta\vartheta\vartheta\vartheta\vartheta\vartheta} + 2\,w_{,\vartheta\vartheta\vartheta\vartheta\vartheta\vartheta} + w_{,\vartheta\vartheta\vartheta\vartheta} = 0 \quad . \tag{13.56}$$

With *(13.45)*, the eigenvalue equation follows from *(13.56)*

$$\frac{1-\nu^2}{k}\lambda^4 + m^4(m^2-1)^2 = 0 \quad . \tag{13.57}$$

If, in accordance with DONNELL's approximation, w is neglected against $w_{,\vartheta\vartheta}$, we obtain from

$$\frac{1-\nu^2}{k}\lambda^4 + m^8 = 0 \tag{13.58a}$$

the small roots as:

$$\lambda_{5,6,7,8} = \pm\,\kappa_2 \pm i\,\mu_2 \quad \text{with} \quad \kappa_2 = \mu_2 = m^2\sqrt{\frac{1}{2}\sqrt{\frac{1-\nu^2}{k}}} \quad . \tag{13.58b}$$

The semi-bending theory can be further simplified if the bending-stiff shell (bending moments are transferred in circumferential direction only) is replaced by a membrane shell stiffened by discretely positioned ring stiffeners (Fig. 13.7).

Eqs. *(13.53)* to *(13.55)* then yield for each shell field (the bending stiffness K of the membrane shell is assumed to be zero):

$$\left.\begin{aligned} v_{,\xi} &= u_{,\vartheta} \ , \\ u_{,x} &= \frac{E\,t}{a} N_{xx} \ , \\ N_{xx,\xi} &= N_{x\vartheta,\vartheta} \ , \\ N_{x\vartheta,\xi} &= 0 \quad . \end{aligned}\right\} \tag{13.59}$$

According to that, the in-plane shear $N_{x\vartheta}$ has to be constant in each field (*shear field theory*), while N_{xx} is linear with respect to x. The shear in the longitudinal direction is then changed at the stiffener ring. If the ring is also considered as a shell with the length l_r, we obtain from the second equation of *(13.53)* with *(13.54)* and *(13.55)*

$$N_{x\vartheta,x} \approx \frac{\left((N_{x\vartheta})_{i+1} - (N_{x\vartheta})_i\right)a}{l_r} = \frac{K}{a^3}\left(v_{,\vartheta\vartheta\vartheta\vartheta\vartheta\vartheta} + 2\,v_{,\vartheta\vartheta\vartheta\vartheta} + v_{,\vartheta\vartheta}\right) \tag{13.60}$$

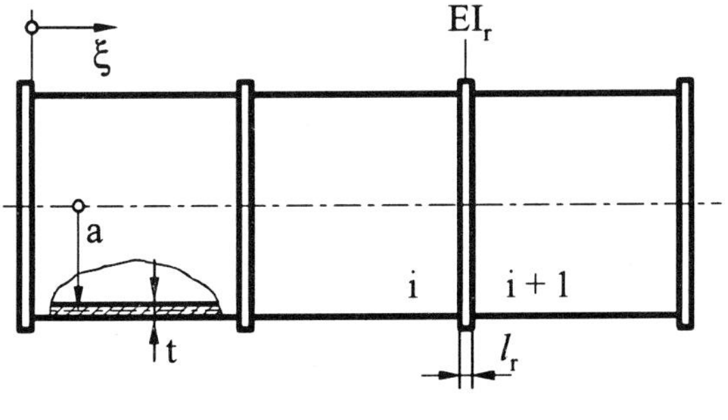

Fig. 13.7: Stiffened Shell

or, with the bending stiffness of a ring $E I_r = l_r K_r$, the step condition at the transition from the i-th to the (i + 1)-th field

$$N_{x\vartheta_{i+1}} = N_{x\vartheta_i} - \frac{E I_R}{a^4} \left(v_{,\vartheta\vartheta\vartheta\vartheta\vartheta\vartheta} + 2 v_{,\vartheta\vartheta\vartheta\vartheta} + v_{,\vartheta\vartheta} \right) . \qquad (13.61)$$

Equations *(13.59)* and *(13.61)* suggest assemblage of the essential field quantities $u, v, N_{xx}, N_{x\vartheta}$ in a state vector, and to solve the problem by means of the transfer matrix procedure as described in Section 13.2 [ET2].

f) *Orthotropic cylindrical shells*

In analogy with the orthotropic plates considered in Section 9.1, the corresponding material laws can also be stated for orthotropic shells. Here, the principal stiffness directions are perpendicular to each other (e.g. sandwich shell, shells made of fibre composite materials, stiffened shells). Assuming DONNELL's simplifications, the material law reads as follows:

$$\left.\begin{aligned}
N_{xx} &= \frac{D_x}{a} u_{,\xi} + \frac{D_\nu}{a} (v_{,\vartheta} + w) , \\
N_{\vartheta\vartheta} &= \frac{D_\vartheta}{a} (v_{,\vartheta} + w) + \frac{D_\nu}{a} u_{,\xi} , \\
N_{x\vartheta} &= \frac{D_{x\vartheta}}{a} (u_{,\vartheta} + v_{,\xi}) , \\
M_{xx} &= \frac{K_x}{a^2} w_{,\xi\xi} - \frac{K_\nu}{a^2} w_{,\vartheta\vartheta} , \\
M_{\vartheta\vartheta} &= \frac{K_\vartheta}{a^2} w_{,\vartheta\vartheta} - \frac{K_\nu}{a^2} w_{,\xi\xi} , \\
M_{x\vartheta} &= \frac{K_{x\vartheta}}{a^2} w_{,\xi\vartheta} .
\end{aligned}\right\} \qquad (13.62)$$

Depending on the given material or on the considered construction, the strain stiffnesses D_x, D_ϑ, D_ν, the shear stiffness $D_{x\vartheta}$, the bending stiffnesses K_x, K_ϑ, K_ν, as well as the torsional stiffness $K_{x\vartheta}$ have to be calculated or to be determined by experiments.

Substitution of *(13.62)* into *(13.29)* yields a system of equations that is analogous with *(13.31)* and which contains eight independent characteristic values for the stiffness as parameters: D_x, D_ϑ, D_ν, $D_{x\vartheta}$, K_x, K_ϑ, K_ν, $K_{x\vartheta}$. Depending on the problem formulation, the system can be solved by means of approximations of the type *(13.45)*. Further details concerning stiffened shells can be found in [B.7, B.9, C.6, ET2].

14 Theory of shallow shells

14.1 Characteristics of shallow shells

Shallow shells possess a very large *characteristic* shell radius or, in other words, a very small, non-vanishing shell curvature. Therefore, a typical behaviour of such shells also occurs, namely the support of transverse loads on the mid-surface by means of membrane forces. This effect has already been described within the scope of membrane theory in Chapter 12.

In addition, the theory of shallow shells does not neglect completely the transverse forces and bending moments, but considers them in the equations of equilibrium of forces perpendicular to the mid-surface, as well as in the equilibrium of moments. Thus, we are no longer dealing with a statically determinate system, as was the case in the membrane theory, and the computational effort for solving the shell problem therefore increases. In the following, however, it will be shown that the effort does not exceed an acceptable limit in comparison with a treatment by the complete shell theory [C.7, C.8, C.20].

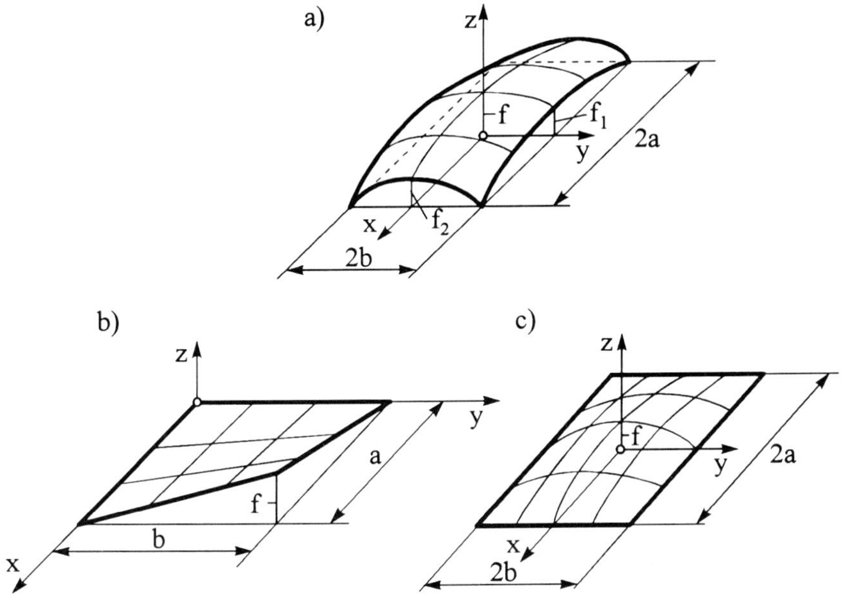

Fig. 14.1: Typical forms of shallow shells

a) Elliptical paraboloid over a rectangular base

b) Hyperbolical paraboloid

c) Shells with horizontal boundaries over a rectangular base (*soap-film shells*)

Typical examples of important shapes of shallow shells (Fig. 14.1)

Let an *elliptical paraboloid surface* be extended over a rectangular base (Fig. 14.1a). The explicit form can be derived from [ET 2] and by a coordinate transformation as

$$z = f_1\left(1 - \frac{x^2}{a^2}\right) + f_2\left(1 - \frac{y^2}{b^2}\right) \quad .$$

Fig. 14.1b presents the form of a *hyperbolical paraboloid*. In Section 11.1. this type of surface has already been treated under the heading of surfaces. In explicit notation these surfaces can be described as follows:

$$z = f\,\frac{x}{a}\,\frac{y}{b} \quad .$$

Finally, Fig. 14.1c illustrates a so-called *soap-film shell*, i.e. a shell with horizontal boundaries extended over a rectangular base.

14.2 Basic equations and boundary conditions

The following notations of approximation are valid (projections onto the plane are denoted by $\overline{\ }$) [C.11]:

$$\left.\begin{aligned} N^{\alpha\beta} &\approx \overline{N}^{\alpha\beta} \;, \quad M^{\alpha\beta} \approx \overline{M}^{\alpha\beta} \;, \quad Q^{\alpha} \approx \overline{Q}^{\alpha} \;, \\ p &\approx \overline{p} \;, \quad p^{\alpha} \approx \overline{p}^{\alpha} \;, \end{aligned}\right\} \qquad (14.1)$$

$$v_\alpha = \overline{v}_\alpha + \overline{w}\, z|_\alpha \;, \quad w \approx \overline{w} \quad . \qquad (14.2)$$

– Equilibrium conditions according to *(11.52)*

$$\left.\begin{aligned} N^{\alpha\beta}|_\alpha + p^\beta &= 0 \;, \\ Q^\alpha|_\alpha + N^{\alpha\beta}\, z|_{\alpha\beta} + p &= 0 \;, \\ M^{\alpha\beta}|_\alpha - Q^\beta &= 0 \;. \end{aligned}\right\} \qquad (14.3)$$

– Strain-displacement relations due to *(11.53)* and *(14.1)*, *(14.2)*

$$\left.\begin{aligned} \alpha_{\alpha\beta} &= \frac{1}{2}\left(\overline{v}_\alpha|_\beta + \overline{v}_\beta|_\alpha + z|_\alpha w|_\beta + z|_\beta w|_\alpha\right) \;, \\ \varrho_{\alpha\beta} &= -w|_{\alpha\beta} \;. \end{aligned}\right\} \qquad (14.4)$$

– Constitutive equations due to *(11.54)*

$$N^{\alpha\beta} = D\, H^{\alpha\beta\gamma\delta}\, \alpha_{\gamma\delta} \;, \qquad (14.5a)$$

$$M^{\alpha\beta} = K\, H^{\alpha\beta\gamma\delta}\, \varrho_{\gamma\delta} \;. \qquad (14.5b)$$

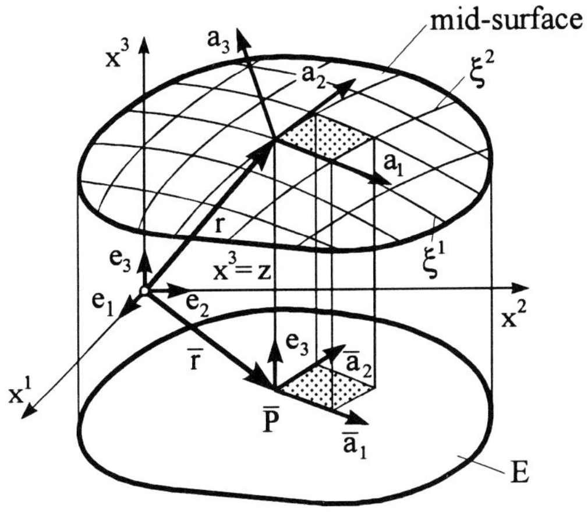

Fig. 14.2: Projection of a mid-surface onto the x^1, x^2-plane E of the three-dimensional space

Reduction of the number of basic equations follows in analogy with *(7.13)*

$$N^{\alpha\beta} = \varepsilon^{\alpha\gamma}\,\varepsilon^{\beta\delta}\,\Phi|_{\gamma\delta} - P^{\alpha\beta} \tag{14.6}$$

and $$P^{\alpha\beta}|_{\alpha} = p^{\beta} \;. \tag{14.7}$$

We obtain two coupled differential equations with w and Φ for a curvilinear system of coordinates:

$$\boxed{\begin{aligned} &K\,a^{\alpha\beta}\,a^{\gamma\delta}\,w|_{\alpha\beta\gamma\delta} - \varepsilon^{\alpha\gamma}\,\varepsilon^{\beta\delta}\,z|_{\alpha\beta}\,\Phi|_{\gamma\delta} - p + P^{\alpha\beta}\,z|_{\alpha\beta} &= 0\\ &a^{\alpha\beta}\,a^{\gamma\delta}\,\Phi|_{\alpha\beta\gamma\delta} + E\,t\,\varepsilon^{\alpha\delta}\,\varepsilon^{\beta\gamma}\,z|_{\alpha\beta}\,w|_{\delta\gamma} - \varepsilon^{\alpha\beta}\,\varepsilon^{\gamma\delta}\,D_{\alpha\gamma\mu\nu}\,P^{\mu\nu}|_{\beta\delta} &= 0\end{aligned}} \tag{14.8}$$

Eqs. *(14.8)* expressed in Cartesian coordinates ($\partial/\partial x \,\hat{=}\, (\;)_{,x}$, $\partial/\partial y \,\hat{=}\, (\;)_{,y}$) read as follows:

$$\boxed{\begin{aligned} &K\,\Delta\Delta\,w - \Diamond^4(z,\Phi) = F_1(x,y)\\ &\Delta\Delta\,\Phi + E\,t\,\Diamond^4(z,w) = F_2(x,y)\end{aligned}} \tag{14.9}$$

with $\Delta\Delta = (\;)_{,xxxx} + 2(\;)_{,xxyy} + (\;)_{,yyyy}$ bipotential operator,

$$\left.\begin{aligned} \Diamond^4(f,g) &= f_{,xx}\,g_{,yy} + f_{,yy}\,g_{,xx} - 2f_{,xy}\,g_{,xy} \;, \\ \text{and}\quad F_1(x,y) &= p - z_{,xx}\int p_x\,dx - z_{,yy}\int p_y\,dy \;, \\ F_2(x,y) &= \left(\int p_x\,dx\right)_{,yy} + \left(\int p_y\,dy\right)_{,xx} - \nu\,(p_{x,x} + p_{y,y}) \;. \end{aligned}\right\} \qquad (14.10)$$

Special cases:

1) Curvature and distortion vanish in one direction

$$\longrightarrow \quad z_{,xx} = z_{,xy} = 0 \;, \quad z_{,yy} = \kappa_y = \text{finite} \;.$$

The differential equation *(14.9)* can be simplified with $p_x = p_y = 0$:

$$K\,\Delta\Delta\,w - \kappa_y\,\Phi_{,xx} = p \;, \qquad (14.11a)$$

$$\Delta\Delta\,\Phi + E\,t\,\kappa_y\,w_{,xx} = 0 \;. \qquad (14.11b)$$

The above equations correspond to the differential equations of a cylindrical shell (see *(13.40a,b)*).

2) No curvature or distortion occur in both directions

$$\longrightarrow \quad z_{,xx} = z_{,yy} = z_{,xy} = 0 \;.$$

The system of differential equations then splits into the two uncoupled differential equations

$$K\,\Delta\,\Delta\,w = p \;, \qquad (14.12a)$$

$$\Delta\,\Delta\,\Phi = 0 \;. \qquad (14.12b)$$

The first relation *(14.12a)* is the differential equation of KIRCHHOFF's plate theory *(9.13)* while *(14.12b)* is a special case of the differential equations of the theory of disks *(8.1)* following from the compatibility condition.

Boundary conditions

At each of the four boundaries of the reference plane, boundary stress resultants (N_{xx}, N_{xy}, M_{xx}, M_{xy}, Q_x, or N_{yy}, N_{xy}, M_{yy}, M_{xy}, Q_y) or boundary displacements or -slopes (u, v, w, $w_{,x}$, $w_{,y}$) can be described. However, since the order of the system of differential equations only possesses four boundary conditions, so-called effective transverse shear forces *(13.34)* and one effective in-plane shear force have to be introduced in analogy with KIRCHHOFF's plate theory (see 9.1). The effective forces read as follows:

$$\left.\begin{aligned} Q_{x_e} &= Q_x + M_{xy,y} \ , \\ Q_{y_e} &= Q_y + M_{xy,x} \ . \end{aligned}\right\} \tag{14.13}$$

In case of a shallow shell, the effective in-plane shear force N_{xy_e} can be replaced by the shear force N_{xy}. In order to avoid confusion with projected forces according to *(14.1)*, the effective forces are here indicated by $(\ )_e$.

The following boundary conditions may be formulated for a boundary x = const:

- Clamped edge

$$u = v = w = w_{,x} = 0 \ . \tag{14.14a}$$

- Simply supported

$$N_{xx} = M_{xx} = v = w = 0 \ . \tag{14.14b}$$

- Free edge

$$N_{xx} = N_{xy} = M_{xx} = Q_{x_e} = 0 \ . \tag{14.14c}$$

14.3 Shallow shell over a rectangular base with constant principal curvatures

This tye of shell can often be found in civil engineering applications, e.g. as a typical roof construction extended over a rectangular base (length 2a, width 2b). The mid-surface of the shell is defined by $z = z(x, y)$, where the following characteristical values are assumed:

$$z_{,xx} = \kappa_x = \text{const} \quad , \quad z_{,yy} = \kappa_y = \text{const} \quad , \quad z_{,xy} = 0 \ . \tag{14.15}$$

The shell is simply supported at all boundaries, and is subjected to a vertical surface load $p(x, y)$ $(p_x = p_y = 0)$.

From the system *(14.9)* we obtain with *(14.15)*

$$K \Delta \Delta w - \kappa_x \Phi_{,yy} - \kappa_y \Phi_{,xx} = p \ , \tag{14.16a}$$

$$\Delta \Delta \Phi + E t (\kappa_x w_{,yy} + \kappa_y w_{,xx}) = 0 \ . \tag{14.16b}$$

Using an auxiliary function $\psi(x, y)$ and the approaches

$$w = \Delta \Delta \psi \ , \tag{14.17a}$$

$$\Phi = -E t \Diamond^4 (z, \psi) = -E t (\kappa_x \psi_{,yy} + \kappa_y \psi_{,xx}) = 0 \ , \tag{14.17b}$$

(14.16) is now transformed into a partial differential equation of eighth order. The differential operator $\lozenge^4$ transforms into a modified LAPLACE-operator with constant coefficients $\Delta^* \psi = \kappa_x \psi_{,yy} + \kappa_y \psi_{,xx}$. The approaches for w and Φ *(14.17)* identically fulfill *(14.16b)*. Substitution into *(14.16a)* then yields

$$\boxed{\Delta\Delta\Delta\Delta\psi + \frac{Et}{K}\Delta^*\Delta^*\psi = \frac{p}{K}} \quad . \tag{14.18}$$

The approaches for the auxiliary functions ψ *(14.18)* are also substituted into the relations for the stress resultants *(14.6)* and *(14.5b)*, and we thus obtain in Cartesian coordinates

$$\left.\begin{aligned} N_{xx} &= \Phi_{,yy} = -Et(\kappa_y \psi_{,yyxx} + \kappa_x \psi_{,yyyy}) = -Et\Delta^*\psi_{,yy} \ , \\ N_{yy} &= \Phi_{,xx} = -Et(\kappa_y \psi^{IV} + \kappa_x \psi_{,yyxx}) \quad = -Et\Delta^*\psi_{,xx} \ , \\ N_{xy} &= -\Phi_{,xy} = Et(\kappa_y \psi_{,yxxx} + \kappa_x \psi_{,yyyx}) \quad = Et\Delta^*\psi_{,xy} \ , \end{aligned}\right\} \tag{14.19}$$

$$\left.\begin{aligned} M_{xx} &= -K\Delta\Delta(\psi_{,xx} + \psi_{,yy}) \ , \\ M_{yy} &= -K\Delta\Delta(\psi_{,yy} + \psi_{,xx}) \ , \\ M_{xy} &= -K(1-\nu)\Delta\Delta\psi_{,xy} \ . \end{aligned}\right\} \tag{14.20}$$

In the follwoing, a solution shall be given for a shell that is simply supported at all edges. For this purpose we draw on the treatment of the simply supported, shear-rigid plate (see Section 9.2). This problem was solved using a FOURIER double series expansion that strictly fulfilled both the KIRCHHOFF plate equation and the boundary conditions. The shallow shell is treated analogously by choosing a FOURIER double series expansion for the auxiliary function ψ:

$$\psi(x,y) = \sum_{m=1}^{\infty}\sum_{n=1}^{\infty} \psi_{mn} \sin\frac{m\pi x}{a} \sin\frac{n\pi x}{b} \ , \tag{14.21}$$

where ψ_{mn} are free FOURIER-coefficients $(m,n = 1,2,3,\ldots)$.

It can be shown that the above approach fulfills the boundary conditions of the simply supported shell according to *(14.14b)*.

C.2 Exercises

Exercise C-11-1:

A circular conical surface constitutes a special case of an elliptic conical surface, and belongs to those conical surfaces that can be described by moving a generatrix (parameter) along a directrix $y(\vartheta)$ (circle with radius a) parallel to the x^1, x^2-plane (see Fig. C-1). The position vector $\mathbf{r}$ of a point P on the surface reads in parametric presentation:

$$\mathbf{r} = \mathbf{r}(s,\vartheta) = s\sin\alpha\cos\vartheta\,\mathbf{e}_1 + s\sin\alpha\sin\vartheta\,\mathbf{e}_2 + s\cos\alpha\,\mathbf{e}_3$$

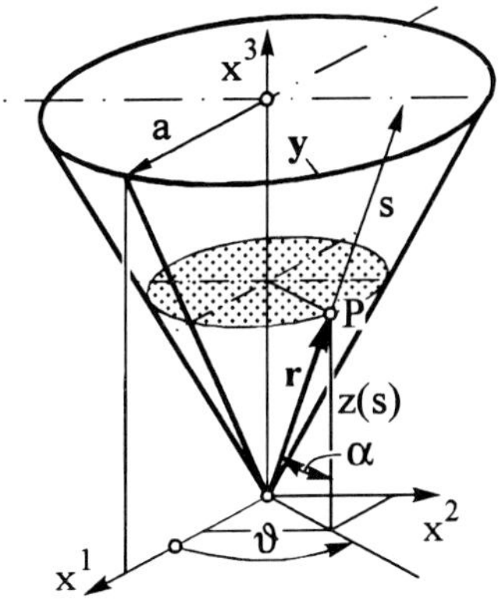

Fig. C-1: Circular conical surface

with s,ϑ GAUSSIAN parameters ,
$\alpha = \text{const}$ semi-angle of a cone .

Determine

a) the fundamental quantities of first and second order ,

b) the equilibrium conditions for the membrane theory of a circular conical shell.

Solution :

a) *Fundamental quantity of first order – surface tensors*

By means of the given parametric representation of a circular conical surface

$$\mathbf{r}(s,\vartheta) = \begin{bmatrix} s\sin\alpha\cos\vartheta \\ s\sin\alpha\sin\vartheta \\ s\cos\alpha \end{bmatrix} \tag{1}$$

we determine from *(11.10)* the covariant base vectors :

$$\mathbf{a}_\alpha = \frac{\partial\mathbf{r}}{\partial\xi^\alpha} = \mathbf{r}_{,\alpha} \quad , \qquad \text{where} \quad \xi^1 \longrightarrow s \;,\; \xi^2 \longrightarrow \vartheta \;.$$

It then follows

$$\mathbf{a}_1 = \mathbf{r}_{,s} = \begin{bmatrix} \sin\alpha\cos\vartheta \\ \sin\alpha\sin\vartheta \\ \cos\alpha \end{bmatrix} \;,\quad \mathbf{a}_2 = \mathbf{r}_{,\vartheta} = \begin{bmatrix} -s\sin\alpha\sin\vartheta \\ s\sin\alpha\cos\vartheta \\ 0 \end{bmatrix} \;. \tag{2a,b}$$

By means of (2a,b) and according to *(11.11)*, the covariant components of the surface tensor (*first fundamental form for the surface*) are calculated as:

$$\mathrm{a}_{\alpha\beta} = \mathbf{a}_\alpha \cdot \mathbf{a}_\beta \quad \longrightarrow \quad \begin{aligned} \mathrm{a}_{11} &= \mathbf{a}_1 \cdot \mathbf{a}_1 = 1 \ , \\ \mathrm{a}_{22} &= \mathbf{a}_2 \cdot \mathbf{a}_2 = \mathrm{s}^2 \sin^2\alpha \ , \\ \mathrm{a}_{12} &= \mathbf{a}_1 \cdot \mathbf{a}_2 = 0 \ . \end{aligned}$$

The covariant surface tensor thus reads :

$$(\mathrm{a}_{\alpha\beta}) = \begin{bmatrix} 1 & 0 \\ 0 & \mathrm{s}^2 \sin^2\alpha \end{bmatrix} \tag{3a}$$

and the determinant due to *(11.12)*

$$\mathrm{a} = \left| \mathrm{a}_{\alpha\beta} \right| = \mathrm{s}^2 \sin^2\alpha \ . \tag{3b}$$

The diagonal form of (3a) ($\mathrm{a}_{12} = 0$) implies that the parametric lines are mutually perpendicular (orthogonal mesh). The contravariant surface tensor can be obtained by forming the reciprocal values of the elements of the principal diagonal, i.e.

$$(\mathrm{a}^{\alpha\beta}) = (\mathrm{a}_{\alpha\beta})^{-1} = \begin{bmatrix} 1 & 0 \\ 0 & \dfrac{1}{\mathrm{s}^2 \sin^2\alpha} \end{bmatrix} \ . \tag{4}$$

b) Fundamental quantity of second order – curvature tensor

The curvature tensor constitutes the *second fundamental form for the surface.* The single components are calculated by means of *(11.18)* :

$$\mathrm{b}_{\alpha\beta} = \frac{[\mathbf{a}_{\alpha,\beta}, \mathbf{a}_1, \mathbf{a}_2]}{\sqrt{\mathrm{a}}}$$

with the derivatives

$$\mathbf{a}_{1,1} = \frac{\partial \mathbf{a}_1}{\partial \mathrm{s}} = \begin{pmatrix} 0 \\ 0 \\ 0 \end{pmatrix} \ , \quad \mathbf{a}_{2,2} = \frac{\partial \mathbf{a}_2}{\partial \vartheta} = \begin{pmatrix} -\mathrm{s} \sin\alpha \cos\vartheta \\ -\mathrm{s} \sin\alpha \sin\vartheta \\ 0 \end{pmatrix} \ ,$$

$$\mathbf{a}_{1,2} = \mathbf{a}_{2,1} = \frac{\partial \mathbf{a}_1}{\partial \vartheta} = \frac{\partial \mathbf{a}_2}{\partial \mathrm{s}} = \begin{pmatrix} -\sin\alpha \sin\vartheta \\ \sin\alpha \cos\vartheta \\ 0 \end{pmatrix} \ .$$

One obtains the components of the curvature tensor by formulating the scalar triple products :

$$\mathrm{b}_{11} = \frac{1}{\mathrm{s} \sin\alpha} \begin{vmatrix} 0 & 0 & 0 \\ \sin\alpha \cos\vartheta & \sin\alpha \sin\vartheta & \cos\alpha \\ -\mathrm{s} \sin\alpha \sin\vartheta & \mathrm{s} \sin\alpha \cos\vartheta & 0 \end{vmatrix} = 0 \ ,$$

$$b_{22} = \frac{1}{s \sin\alpha} \begin{vmatrix} -s\sin\alpha\cos\vartheta & -\sin\alpha\sin\vartheta & 0 \\ \sin\alpha\cos\vartheta & \sin\alpha\sin\vartheta & \cos\alpha \\ -s\sin\alpha\sin\vartheta & s\sin\alpha\cos\vartheta & 0 \end{vmatrix} = s\sin\alpha\cos\alpha \ ,$$

$$b_{12} = \frac{1}{s \sin\alpha} \begin{vmatrix} -s\sin\alpha\sin\vartheta & \sin\alpha\cos\vartheta & 0 \\ \sin\alpha\cos\vartheta & \sin\alpha\sin\vartheta & \cos\alpha \\ -s\sin\alpha\sin\vartheta & s\sin\alpha\cos\vartheta & 0 \end{vmatrix} = 0 \ .$$

The curvature tensor thus reads

$$(b_{\alpha\beta}) = \begin{bmatrix} 0 & 0 \\ 0 & s\sin\alpha\cos\alpha \end{bmatrix} \tag{5a}$$

with the determinant $b = \left| b_{\alpha\beta} \right| = 0$. (5b)

The form of the fundamental quantities allows us to draw the following conclusions :

- $a_{12} = 0$ and $b_{12} = 0$ mean that the parametric lines are simultaneously lines of principal curvature.
- $b_{11} = 0$ implies that the curvature is zero along the parametric line s .

The curvature at a point P of the surface can be calculated according to *(11.20)*

$$\frac{1}{R} = -\frac{b_{\alpha\beta}\,d\xi^\alpha\,d\xi^\beta}{a_{\alpha\beta}\,d\xi^\alpha\,d\xi^\beta} \quad \Longrightarrow \quad \left. \begin{aligned} \frac{1}{R_1} &= \frac{1}{R_s} = -\frac{b_{11}}{a_{11}} = 0 \ , \\ \frac{1}{R_2} &= \frac{1}{R_\vartheta} = -\frac{b_{22}}{a_{22}} = -\frac{1}{s}\cot\alpha \ . \end{aligned} \right\} \tag{6}$$

The two invariants describe the curvature properties of a surface (see *(11.22a,b)*):

$$H = \frac{1}{2} a^{\alpha\beta} b_{\alpha\beta} \qquad \text{mean curvature} \ ,$$

$$K = \frac{b}{a} \qquad \text{GAUSSIAN curvature} \ .$$

This yields

$$H = -\frac{1}{2} s \cot\alpha \ , \tag{7a}$$

$$K = 0 \ . \tag{7b}$$

Surfaces with an equal measure of GAUSSIAN curvature $K = \text{const}$ can be mapped isometrically onto each other, i.e. they are developable on each other. Owing to the fact that $K = 0$ due to (7b), the circular conical surface can be developed on the plane, just as is the case with any cylindrical surface.

b) *Equilibrium conditions for the membrane theory of a circular conical shell*

We proceed from the equations *(12.1)*

$$N^{\alpha\beta}\big|_\alpha + p^\beta = 0 \; ,$$

$$N^{\alpha\beta}\, b_{\alpha\beta} + p = 0 \; .$$

As an example, the first equilibrium condition ($\beta = 1$), i.e.,

$$N^{11}\big|_1 + N^{21}\big|_2 + p^1 = 0 \; , \tag{8a}$$

shall be written in expanded form. The resultant normal forces N^{11}, N^{21} are tensors of the second order, and their covariant derivatives are to be formed according to *(2.35b)*:

$$N^{11}{}_{,1} + \Gamma^1_{1\varrho} N^{\varrho 1} + \Gamma^1_{1\varrho} N^{1\varrho} + N^{21}{}_{,2} + \Gamma^2_{2\varrho} N^{\varrho 1} + \Gamma^1_{2\varrho} N^{2\varrho} + p^1 = 0 \; . \tag{8b}$$

In a first step, the CHRISTOFFEL symbols of the surface have to be determined, using *(11.23a)*:

$$\Gamma^\alpha_{\beta\gamma} = \frac{1}{2} a^{\alpha\varrho} \left(a_{\varrho\beta,\gamma} + a_{\gamma\varrho,\beta} - a_{\beta\gamma,\varrho} \right) \; .$$

One thus obtains the following CHRISTOFFEL symbols:

$$\left(\Gamma^1_{\alpha\beta}\right) = \begin{bmatrix} 0 & 0 \\ 0 & -s\sin^2\alpha \end{bmatrix} \; , \quad \left(\Gamma^2_{\alpha\beta}\right) = \begin{bmatrix} 0 & 1/s \\ 1/s & 0 \end{bmatrix} . \tag{9a,b}$$

By substituting (9a,b) into (8b) one obtains:

$$N^{11}_{,1} + N^{21}_{,2} + \frac{1}{s} N^{11} - s\sin^2\alpha\, N^{22} + p^1 = 0 \; . \tag{10}$$

Finally, the physical components are introduced into (10) by *(2.17)*:

$$N^{*11} \mathrel{\widehat{=}} N_{ss} = N^{11} \; , \tag{11a}$$

$$N^{*12} \mathrel{\widehat{=}} N_{s\vartheta} = N^{12}\, s \sin\alpha \; , \tag{11b}$$

$$N^{*22} \mathrel{\widehat{=}} N_{\vartheta\vartheta} = N^{22} s^2 \sin^2\alpha \; . \tag{11c}$$

From (10) and (11) now follows

$$\frac{\partial N_{ss}}{\partial s} + \frac{\partial}{\partial \vartheta}\left(\frac{N_{s\vartheta}}{s\sin\alpha}\right) + \frac{1}{s} N_{ss} - \frac{1}{s} N_{\vartheta\vartheta} + p_s = 0$$

$$\longrightarrow \quad \left.\begin{aligned} & s\frac{\partial N_{ss}}{\partial s} + N_{ss} + \frac{1}{\sin\alpha}\frac{\partial N_{s\vartheta}}{\partial\vartheta} - N_{\vartheta\vartheta} + s\,p_s = 0 \\ \text{or} \quad & \left(s N_{ss}\right)_{,s} + \frac{1}{\sin\alpha} N_{s\vartheta,\vartheta} - N_{\vartheta\vartheta} + s\,p_s = 0 \; . \end{aligned}\right\} \tag{12}$$

The above equation is identical with equilibrium condition *(12.16a)* where $(\;)_{,s} \mathrel{\widehat{=}} \partial/\partial s$ and $(\;)_{,\vartheta} \mathrel{\widehat{=}} \partial/\partial\vartheta$.

Finally, the equilibrium condition *(12.16c)* is checked, i.e.,

$$N^{11} b_{11} + N^{22} b_{22} + p = 0 \quad \longrightarrow \quad N^{22} s \sin\alpha \cos\alpha + p = 0 \, .$$

Using (11c) it follows that

$$N_{\vartheta\vartheta} \frac{1}{s^2 \sin^2 \alpha} s \sin\alpha \cos\alpha + p = 0 \quad \longrightarrow \quad N_{\vartheta\vartheta} = -p\, s \tan\alpha \, .$$

Exercise C-12-1:

A shell of revolution with an elliptic meridional shape (Fig. C-2) is subjected to a constant internal overpressure p_0 .

Determine the membrane forces in the shell.

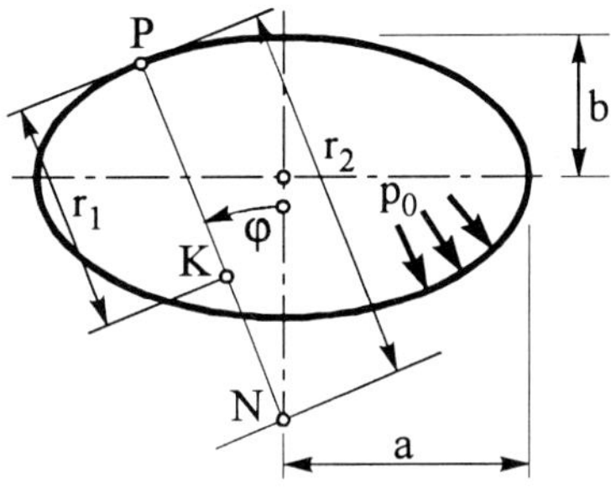

Fig. C-2: Shell of revolution with elliptical meridional shape

Solution :

We take from analytical geometry the radius of curvature r_1 for a point P of the ellipse

$$r_1 = \frac{a^2\, b^2}{(a^2 \sin^2\varphi + b^2 \cos^2\varphi)^{3/2}}$$

and the distance $r_2 \triangleq PN$ to the axis of revolution

$$r_2 = \frac{a^2}{(a^2 \sin^2\varphi + b^2 \cos^2\varphi)^{1/2}} \, .$$

Assuming that $p_\varphi = 0$, we obtain according to *(12.7a)*

$$N_{\varphi\varphi} = \frac{(a^2 \sin^2\varphi + b^2 \cos^2\varphi)^{1/2}}{a^2 \sin^2\varphi} \int\limits_{\overline{\varphi}=0}^{\varphi} \frac{a^4\, b^2}{(a^2 \sin^2\overline{\varphi} + b^2 \cos^2\overline{\varphi})^2}\, p_0 \cos\overline{\varphi}\, d\overline{\varphi} \, .$$

By means of the substitution

$$\sin^2\overline{\varphi} = z - \frac{b^2}{a^2 - b^2} \quad , \quad 2 \sin\overline{\varphi} \cos\overline{\varphi}\, d\overline{\varphi} = d\, z \quad ,$$

the integral can be transformed into a basic integral.

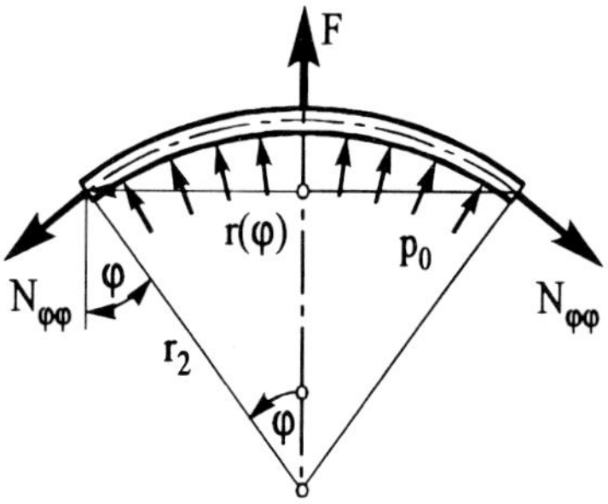

Fig. C-3: Equilibrium at large

However, the above results can be obtained more easily if we consider the *equilibrium at large* for a thin top section, cut symmetrically from the shell of revolution at arbitrary angles φ (see Fig. C-3). The vertical load F results from the pressure acting on the horizontal projection of the shell (circular surface of radius $r(\varphi)$, since the horizontal components of p_0 counterbalance each other):

$$F = \pi r^2(\varphi)\, p_0 \ .$$

From the *equilibrium at large* follows that

$$2\pi r(\varphi) N_{\varphi\varphi} \sin\varphi = F = \pi r^2(\varphi)\, p_0 \ ,$$

and by assuming that $r(\varphi) = r_2 \sin\varphi$, one obtains the membrane force in the meridional direction

$$N_{\varphi\varphi} = \frac{p_0 r_2}{2} \ ,$$

and the membrane force in the latitudinal direction by *(12.7b)*

$$N_{\vartheta\vartheta} = r_2 p_0 - \frac{r_2}{r_1}\frac{p_0 r_2}{2} = p_0 r_2 \left(1 - \frac{r_2}{2 r_1}\right) \ .$$

At the top $(\varphi = 0)$ holds with $r_1 = r_2 = \frac{a^2}{b}$ that

$$N_{\varphi\varphi} = N_{\vartheta\vartheta} = \frac{p_0 a^2}{2b} \ ,$$

and at the equator $(\varphi = \frac{\pi}{2})$ follows with $r_1 = \frac{b^2}{a}$, $r_2 = a$ that

$$N_{\varphi\varphi} = \frac{p_0 a}{2} \ , \quad N_{\vartheta\vartheta} = p_0 a \left(1 - \frac{a^2}{2b^2}\right) \ .$$

For $a > \sqrt{2}\, b$, i.e. in cases of more shallow shells, a compressive stress occurs in the circumferential direction at the equator. An elliptic shell bottom reduces its diameter when subjected to overpressure. In the special case of a spherical shell with $r_1 = r_2 = a = b$, the boiler formula *(12.13)* is verified in the form:

$$N_{\varphi\varphi} = N_{\vartheta\vartheta} = \frac{p_0 a}{2} \ .$$

A *spherical* shell subjected to internal overpressure only exhibits tensile stresses. The same applies for a cylinder.

Exercise C-12-2:

A spherical boiler (radius a, wall thickness t) subjected to internal overpressure p_0 is supported in bearings at its top and bottom points (Fig. C-4). The boiler rotates around the vertical axis A-A with a constant angular velocity ω.

Determine the rotational speed for unset of yielding, assuming that only a membrane state of stress exists and that the deadweight can be neglected.

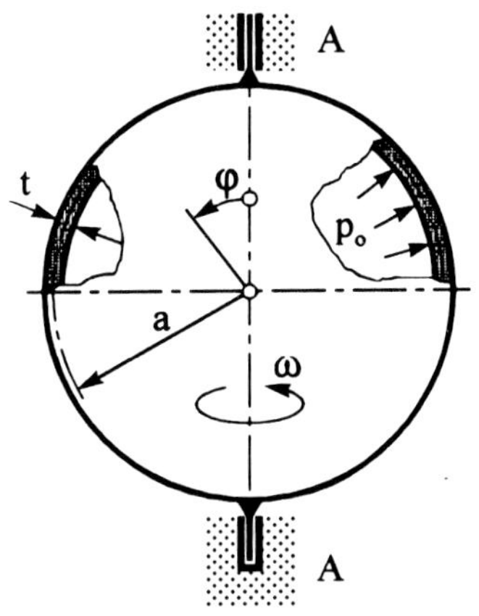

Fig. C-4: Spherical boiler

Numerical values:

$a = 1\,m \quad , \quad t = 2 \cdot 10^{-3} m \quad ,$

$\sigma_y = 360\,MPa$ (yield stress) , $p_0 = 0.8\,MPa$, $\varrho = 7.86\,kg/m^3$.

Solution :

Besides the internal overpressure, a centrifugal load occurs in this problem. With $r = a \sin\varphi$, the resulting load components in the meridional and the normal direction become :

$$p = p_0 + \rho\, t\, \omega^2 a \sin^2\varphi \ , \tag{1a}$$

$$p_\varphi = \rho\, t\, \omega^2 a \sin\varphi \cos\varphi \ . \tag{1b}$$

Substitution of (1a,b) into *(12.7a)* yields

$$N_{\varphi\varphi} = \frac{a}{\sin^2\varphi} \int_{\bar\varphi=0}^{\varphi} (p \cos\bar\varphi - p_\varphi \sin\bar\varphi) \sin\bar\varphi \, d\bar\varphi = \frac{a}{\sin^2\varphi} \int_{\bar\varphi=0}^{\varphi} p_0 \cos\bar\varphi \sin\bar\varphi \, d\bar\varphi \ .$$

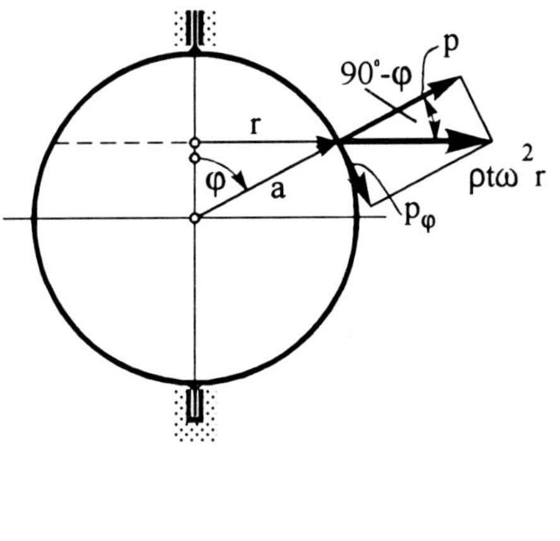

Fig. C-5: Components of the centrifugal load

All terms with ω vanish so that the meridional resultant force $N_{\varphi\varphi}$ only depends on the internal pressure p_0. After integration we obtain

$$N_{\varphi\varphi} = \frac{a}{\sin^2\varphi}\left(-\frac{p_0\cos 2\varphi}{4} + C\right) . \tag{2}$$

Since the meridional resultant force $N_{\varphi\varphi}$ has to be finite for $\varphi = 0$, we get

$$N_{\varphi\varphi} = \frac{a}{\sin^2\varphi}\left(-\frac{p_0\cos 2\varphi}{4} + C\right)\Bigg|_{\varphi=0} \longrightarrow \text{finite} \Longrightarrow \quad C = \frac{p_0}{4} .$$

Substitution of C into (2) yields :

$$N_{\varphi\varphi} = \frac{p_0 a}{2} . \tag{3a}$$

The resultant forces in the latitudinal direction are calculated by means of equation *(12.12c)* and by superposing the two load cases :

$$N_{\vartheta\vartheta} = \frac{p_0 a}{2} + \rho t \omega^2 a^2 \sin^2\varphi . \tag{3b}$$

The stresses in the latitudinal and meridional direction then become:

$$\sigma_{\vartheta\vartheta} = \frac{p_0 a}{2t} + \rho\omega^2 a^2 \sin^2\varphi \quad , \quad \sigma_{\varphi\varphi} = \frac{p_0 a}{2t} .$$

The maximum stress occurs at $\pi/2$. Following the von MISES hypothesis, the maximum stress can be expressed as follows :

$$\sigma_r = \sqrt{\sigma_1^2 + \sigma_2^2 - \sigma_1\sigma_2} \qquad \longrightarrow$$

$$\sigma_{rmax} = \sqrt{\left(\frac{p_0 a}{2t}\right)^2 + (\rho\omega^2 a^2)^2 + \frac{p_0 a}{2t}\rho\omega^2 a^2} \leq \sigma_y . \tag{4}$$

With $\omega = \frac{\pi n}{30}$, relation (4) allows us to calculate the rotational speed n for unset of yielding:

$$n = \frac{30}{\pi a}\sqrt{\frac{1}{\rho}\left(\sqrt{\sigma_y^2 - 3\left(\frac{p_0 a}{4t}\right)^2} - \frac{p_0 a}{4t}\right)} =$$

$$= \frac{30}{\pi\cdot 1000}\sqrt{\frac{1}{7.86\cdot 10^{-9}}\left(\sqrt{360^2 - 3\cdot 100^2} - 100\right)}$$

$$n \approx 26.4 \text{ rev/sec} .$$

Exercise C-12-3:

Calculate the membrane forces in a spherical shell (radius a) subjected to a wind pressure described by the approximate distribution

$$p = -p_0 \sin\varphi \cos\vartheta \quad .$$

Tangential frictional forces occur in practice but will be neglected here.

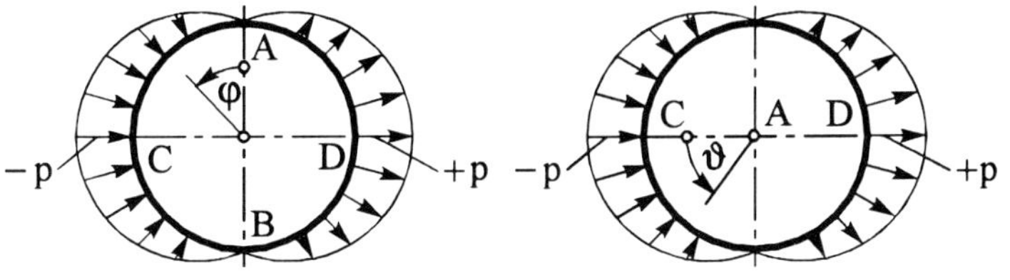

Fig. C-6: Spherical shell subjected to wind pressure load

Solution:

Assuming that $p_\varphi = p_\vartheta = 0$, the equilibrium conditions *(12.12)* read:

$$\begin{aligned}
&\sin\varphi\,(N_{\varphi\varphi})_{,\varphi} + \cos\varphi\, N_{\varphi\varphi} + (N_{\varphi\vartheta})_{,\vartheta} - \cos\varphi\, N_{\vartheta\vartheta} = 0\,, \\
&\sin\varphi\,(N_{\varphi\vartheta})_{,\varphi} + 2\cos\varphi\, N_{\varphi\vartheta} + (N_{\vartheta\vartheta})_{,\vartheta} = 0\,, \qquad (1)\\
&N_{\varphi\varphi} + N_{\vartheta\vartheta} = -p_0\, a \sin\varphi \cos\vartheta\,.
\end{aligned}$$

By a product approach according to *(12.9)*

$$N_{\varphi\varphi} = \Phi(\varphi)\cos\vartheta\,, \quad N_{\varphi\vartheta} = \Psi(\varphi)\sin\vartheta\,, \quad N_{\vartheta\vartheta} = \Theta(\varphi)\cos\vartheta\,, \qquad (2)$$

we transform the system of partial differential equations (1) into a system of ordinary differential equations ($_{,\varphi} \mathrel{\widehat{=}} (\;)'$):

$$\sin\varphi\,\Phi' + \cos\varphi\,\Phi + \Psi - \cos\varphi\,\Theta = 0\,, \qquad (3a)$$

$$\sin\varphi\,\Psi' + 2\cos\varphi\,\Psi - \Theta = 0\,, \qquad (3b)$$

$$\Phi + \Theta = -p_0\, a \sin\varphi\,. \qquad (3c)$$

By eliminating from (3c)

$$\Theta = -\Phi - p_0\, a \sin\varphi\,,$$

we obtain

$$\left.\begin{aligned}
&\sin\varphi\,\Phi' + 2\cos\varphi\,\Phi + \Psi + p_0\, a \sin\varphi\cos\varphi = 0\,,\\
&\sin\varphi\,\Psi' + 2\cos\varphi\,\Psi + \Phi + p_0\, a \sin\varphi = 0\,.
\end{aligned}\right\} \qquad (4)$$

The form of (4) suggests introduction of the sum and the difference of the unknown functions as new functions :

$$F_1 = \Phi + \Psi \quad , \quad F_2 = \Phi - \Psi \ . \tag{5}$$

If we now divide (4) by $\sin\varphi$, (5) yields by addition and subtraction, respectively, of the two equations (4)

$$F'_{1,2} + \lambda_{1,2} F_{1,2} + P_{1,2} = 0 \tag{6}$$

with $$\lambda_{1,2} = 2\cot\varphi \pm \frac{1}{\sin\varphi} \quad , \quad P_{1,2} = p_0 a(\cos\varphi \pm 1) , \tag{7}$$

where the index 1 implies " + " and the index 2 implies " – ".

The ordinary inhomogeneous differential equations of the first order with variable coefficients (6) have the following solutions according to *(12.27)* :

$$F_{1,2} = \left(C_{1,2} - \int P_{1,2}\, e^{\int \lambda_{1,2} d\varphi}\, d\varphi \right) e^{-\int \lambda_{1,2} d\varphi} \ . \tag{8}$$

The integrals are evaluated by means of (7):

$$\int \lambda_1 d\varphi = \int \left(2\cot\varphi + \frac{1}{\sin\varphi} \right) d\varphi = 2\ln\sin\varphi + \ln\tan\frac{\varphi}{2} ,$$

$$e^{\int \lambda_1 d\varphi} = e^{2\ln\sin\varphi + \ln\tan\varphi/2} = \sin^2\varphi \tan\frac{\varphi}{2} \ .$$

In a similar way we determine

$$e^{-\int \lambda_1 d\varphi} = \frac{\cot\frac{\varphi}{2}}{\sin^2\varphi} \quad , \quad e^{\int \lambda_2 d\varphi} = \sin^2\varphi \cot\frac{\varphi}{2} \quad , \quad e^{-\int \lambda_2 d\varphi} = \frac{\tan\frac{\varphi}{2}}{\sin^2\varphi} \ .$$

For F_1 we then obtain

$$F_1 = \left[C_1 - \int p_0 a (\cos\varphi + 1) \sin^2\varphi \tan\frac{\varphi}{2} d\varphi \right] \frac{\cot\frac{\varphi}{2}}{\sin^2\varphi} \ . \tag{9}$$

By means of

$$1 + \cos\varphi = 2\cos^2\frac{\varphi}{2} \quad , \quad \sin^2\varphi = 4\sin^2\frac{\varphi}{2}\cos^2\frac{\varphi}{2} ,$$

the integral can be determined as follows :

$$\int (\cos\varphi + 1)\sin^2\varphi \tan\frac{\varphi}{2} d\varphi = \int 8\cos^3\frac{\varphi}{2}\sin^3\frac{\varphi}{2} d\varphi = \int \sin^3\varphi \, d\varphi =$$

$$= -\cos\varphi + \frac{1}{3}\cos^3\varphi \ .$$

If we substitute

$$\cot\frac{\varphi}{2} = \frac{2\cos^2\frac{\varphi}{2}}{2\sin\frac{\varphi}{2}\cos\frac{\varphi}{2}} = \frac{1 + \cos\varphi}{\sin\varphi} ,$$

we obtain from (9)

$$F_1 = \left[C_1 + p_0 a \left(\cos\varphi - \frac{1}{3}\cos^3\varphi \right) \right] \frac{1 + \cos\varphi}{\sin^3\varphi} \quad ,$$

and analogously

$$F_2 = \left[C_2 - p_0 a \left(\cos\varphi - \frac{1}{3}\cos^3\varphi \right) \right] \frac{1 - \cos\varphi}{\sin^3\varphi} \quad .$$

Substitution into (5) and solving leads, after introduction of two new integration constants $D_1 = C_1 + C_2$ and $D_2 = C_1 - C_2$, to

$$\Phi = \frac{1}{2}(F_1 + F_2) =$$

$$= \frac{1}{2}\left[D_1 + D_2 \cos\varphi + 2 p_0 a \cos\varphi \left(\cos\varphi - \frac{1}{3}\cos^3\varphi \right) \right] \frac{1}{\sin^3\varphi} \, , \qquad (10)$$

$$\Psi = \frac{1}{2}(F_1 - F_2) = \frac{1}{2}\left[D_2 + D_1 \cos\varphi + 2 p_0 a \left(\cos\varphi - \frac{1}{3}\cos^3\varphi \right) \right] \frac{1}{\sin^3\varphi} \, .$$

In order to ensure finiteness of the resultant forces at the top ($\varphi = 0$), we demand that

$$D_1 + D_2 + 2 p_0 a \frac{2}{3} = 0 \quad . \qquad (11)$$

Since $\sin^3\varphi$ occurs in the denominator, not only the numerator but also its first and second derivative have to vanish at the point $\varphi = 0$. We obtain, from the second equation (10), for the first derivative of the term in square brackets

$$\left[- D_1 \sin\varphi + 2 p_0 a \left(- \sin\varphi + \cos^2\varphi \sin\varphi \right) \right] \Big|_{\varphi = 0} = 0$$

and for the second derivative

$$\left[- D_1 \cos\varphi + 2 p_0 a \left(- \cos\varphi - 2\cos\varphi \sin^2\varphi + \cos^3\varphi \right) \right] \Big|_{\varphi = 0} = 0 \quad .$$

Whereas the first condition is fulfilled directly for $\varphi = 0$, the second derivative for $\varphi = 0$ yields :

$$- D_1 + 2 p_0 a (-1 + 1) = 0 \quad \longrightarrow \quad D_1 = 0$$

and thus, according to (11),

$$D_2 = -\frac{4}{3} p_0 a \quad .$$

With (10) and (2) the following expressions for the membrane forces are obtained :

$$N_{\varphi\varphi} = p_0 a \left(-\frac{2}{3} + \cos\varphi - \frac{1}{3}\cos^3\varphi \right) \frac{\cos\varphi}{\sin^3\varphi} \cos\vartheta \quad ,$$

$$N_{\varphi\vartheta} = p_0 a \left(-\frac{2}{3} + \cos\varphi - \frac{1}{3}\cos^3\varphi \right) \frac{1}{\sin^3\varphi} \sin\vartheta \quad , \qquad (12)$$

$$N_{\vartheta\vartheta} = p_0 a \left(\frac{2}{3}\cos\varphi - \sin^2\varphi - \frac{2}{3}\cos^4\varphi \right) \frac{1}{\sin^3\varphi} \cos\vartheta \quad .$$

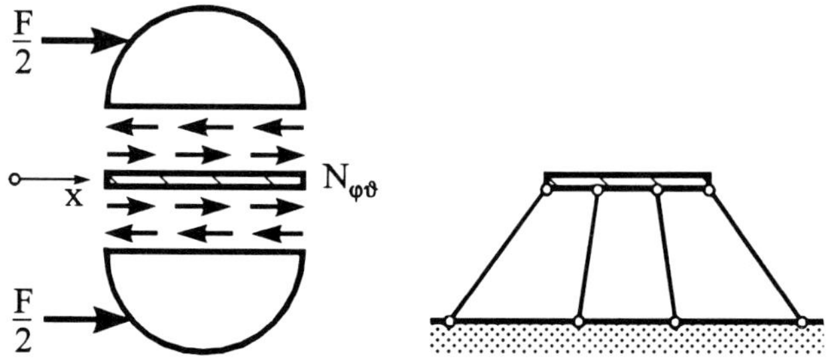

Fig. C-7: Support of the spherical shell at the ground

The wind load $p(\varphi, \vartheta)$ possesses a resultant F in the x-direction which can be equilibrated by the resultant of the shear forces $N_{\varphi\vartheta}$ at the cut $\varphi = \pi/2$. At other cuts defined by φ, components of $N_{\varphi\varphi}$ contribute to the *equilibrium at large*. However, since the shear forces at the two semi-spheres act in the same direction and therefore add up, their resulting force has to be provided by the ground through a stiffening ring (Fig. C-7). Without this or a similar type of support, the spherical shell would be *blown away*. Thus, the support disturbes the membrane state of the shell which can therefore only be considered as an approximation.

Exercise C-12-4:

A hanging conical shell (height h, conical semi-angle α) supported as depicted in Fig. C-8 is filled with liquid of mass density ϱ.

Determine expressions for the membrane forces in the ranges I and II shown in Fig. C-8. The deadweight of the shell can be disregarded.

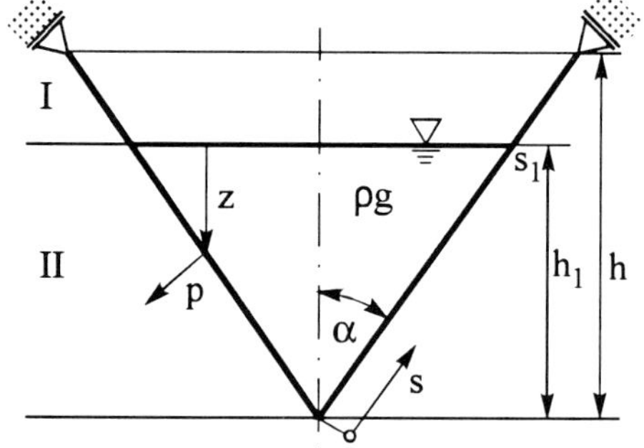

Fig. C-8: Hanging conical shell filled with liquid

Solution :

The loads are axisymmetrical, and can be written as follows for the two ranges :

Range I: $\quad p = 0 \, , \qquad p_s = 0 \quad ,$ (1a)

Range II: $\quad p = \rho g z = \rho g (h_1 - s \cos\alpha) \, , \qquad p_s = 0 \quad .$ (1b)

The expressions for the membrane forces can be determined by means of the equilibrium conditions *(12.16)* for the axisymmetrical load case

$$\frac{d}{ds}(s\,N_{ss}) = N_{\vartheta\vartheta} \;, \tag{2a}$$

$$N_{\vartheta\vartheta} = p\,s\tan\alpha \;. \tag{2b}$$

Range I: $$N_{\vartheta\vartheta} = 0 \longrightarrow N_{ss} = \frac{C}{s}\;. \tag{2c}$$

In order to determine the constant C, we proceed from the "equilibrium at large" at the transition between range I and II. We demand according to Fig. C-9 that

$$(N_{ss}\cos\alpha)\,2\pi h_1\tan\alpha = \frac{1}{3}\rho g\pi(h_1^2\tan^2\alpha)\,h_1 \longrightarrow N_{ss} = \frac{1}{6}\rho g h_1^2\frac{\sin\alpha}{\cos^2\alpha}\;. \tag{3}$$

We determine the constant C from the boundary conditions for $s = s_1 = h_1/\cos\alpha$ with (3) as follows:

$$N_{ss}(s_1) = \frac{\cos\alpha}{h_1}\,C = \frac{1}{6}\rho g h_1^2\frac{\sin\alpha}{\cos^2\alpha} \longrightarrow C = \frac{1}{6}\rho g h_1^3\frac{\sin\alpha}{\cos^3\alpha}\;.$$

Substitution into (2c) then yields the following expression for the membrane force N_{ss} in range I:

$$N_{ss} = \frac{1}{6}\rho g\frac{h_1^3}{\cos^3\alpha}\,\frac{\sin\alpha}{s}\;. \tag{4}$$

Range II: By including (1b), we obtain from (2b)

$$N_{\vartheta\vartheta} = s\rho g(h_1 - s\cos\alpha)\tan\alpha$$

and from (2a) after integration

$$N_{ss} = \frac{\rho g}{s}\left(h_1\frac{s^2}{2} - \frac{s^3}{3}\cos\alpha + C\right)\tan\alpha\;. \tag{5a}$$

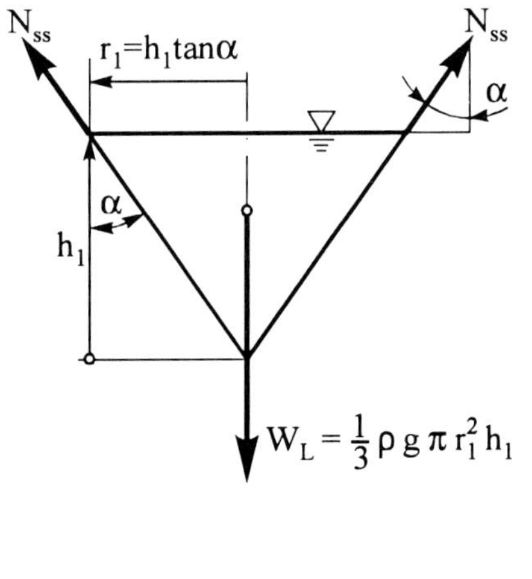

Fig. C-9: *Equilibrium at large* for range II of the conical shell

At the boundary $s = s_1$ we have

$$N_{ss}(s_1) = \frac{\rho g \sin\alpha}{h_1}\left[\frac{h_1^3}{2\sin^2\alpha} - \frac{h_1^3}{3\sin^2\alpha}\right] + C\tan\alpha = \frac{1}{6}\rho g h_1^2 \frac{\sin\alpha}{\cos^2\alpha}$$

$$\rightarrow \quad C = 0 \quad .$$

Thus, we determine the following expression for the membrane force N_{ss} in range II:

$$N_{ss} = \frac{\rho g s}{6}(3h_1\tan\alpha - 2s\sin\alpha) \quad . \tag{5b}$$

Exercise C-12-5:

A section of a casing has the shape of a circular toroidal shell as shown in Fig. C-10 (radius of the circular section a, radius from centre point r_0, wall thickness t).

At the boundary $\varphi = \varphi_0$ the shell is subjected to a uniformly distributed boundary load N_0 acting in the tangential direction.

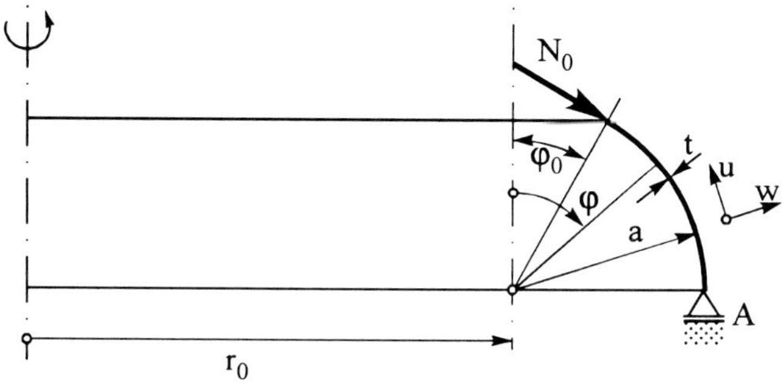

Fig. C-10: Section of a casing with toroidal shell shape

a) Determine the membrane forces and the stresses in the shell.

b) State the basic equations for determining the displacements u and w for the section of the casing.

Solution :

a) We proceed from the equilibrium conditions for shells of revolution with arbitrary contours *(12.6)* subject to an axisymmetrical loading ($p_\vartheta = 0$; $\frac{\partial}{\partial\vartheta} = 0$):

$$(r N_{\varphi\varphi})_{,\varphi} - r_1\cos\varphi\, N_{\vartheta\vartheta} + r r_1 p_\varphi = 0 \quad , \tag{1a}$$

$$\frac{N_{\varphi\varphi}}{r_1} + \frac{N_{\vartheta\vartheta}}{r_2} = p \quad . \tag{1b}$$

With the angle φ relative to the axis of rotational symmetry, the radius of curvature $r_1 = a$, the distance $r = a \sin\varphi + r_0$ from the centre line, and the auxiliary radius $r_2 = a + r_0/\sin\varphi$ resulting from the projection onto the centre line, the following system of equations is obtained:

$$\left[(a \sin\varphi + r_0) N_{\varphi\varphi}\right]_{,\varphi} - a \cos\varphi \, N_{\vartheta\vartheta} = 0 \,, \tag{2a}$$

$$\frac{N_{\varphi\varphi}}{a} + \frac{\sin\varphi}{r_0 + a \sin\varphi} N_{\vartheta\vartheta} = 0 \ . \tag{2b}$$

Differentiation of (2a) and transformation of (2b) yield

$$N_{\varphi\varphi} a \cos\varphi + N_{\varphi\varphi,\varphi}(a \sin\varphi + r_0) - N_{\vartheta\vartheta} a \cos\varphi = 0$$

$$\longrightarrow \quad N_{\varphi\varphi,\varphi} + \frac{a \cos\varphi}{r_0 + a \sin\varphi}(N_{\varphi\varphi} - N_{\vartheta\vartheta}) = 0 \,, \tag{3a}$$

$$N_{\vartheta\vartheta} = -\frac{r_0 + a \sin\varphi}{a \sin\varphi} N_{\varphi\varphi} \ . \tag{3b}$$

If we substitute (3b) into (3a), we get

$$N_{\varphi\varphi,\varphi} + \frac{a \cos\varphi}{r_0 + a \sin\varphi}\left(1 + \frac{r_0 + a \sin\varphi}{a \sin\varphi}\right) N_{\varphi\varphi} = 0$$

$$\longrightarrow \quad N_{\varphi\varphi,\varphi} + \underbrace{\left[\frac{a \cos\varphi}{r_0 + a \sin\varphi} + \cot\varphi\right]}_{P(\varphi)} N_{\varphi\varphi} = 0 \,.$$

The general solution of the differential equation of type $N_{\varphi\varphi,\varphi} + P(\varphi) N_{\varphi\varphi} = 0$ reads

$$N_{\varphi\varphi} = C \, e^{-\int P(\varphi)\, d\varphi} \ .$$

Evaluation of the integral leads to:

$$\int P(\varphi)\, d\varphi = \int \frac{a \cos\varphi}{r_0 + a \sin\varphi} d\varphi + \int \cot\varphi \, d\varphi =$$

$$= \int \frac{\cos\varphi}{\frac{r_0}{a} + \sin\varphi} d\varphi + \int \cot\varphi \, d\varphi = \ln\left(\frac{r_0}{a} + \sin\varphi\right) + \ln(\sin\varphi) \tag{4}$$

$$\longrightarrow \quad N_{\varphi\varphi} = C e^{-\left[\ln\left(\frac{r_0}{a} + \sin\varphi\right) + \ln(\sin\varphi)\right]}$$

or $$N_{\varphi\varphi} = C\left[\frac{1}{\frac{r_0}{a} + \sin\varphi} \, \frac{1}{\sin\varphi}\right] = C^* \frac{1}{\sin\varphi \,(r_0 + a \sin\varphi)} \ . \tag{5}$$

Boundary condition: $$N_{\varphi\varphi}(\varphi = \varphi_0) = -N_0 \ . \tag{6}$$

From (5) follows that $$- N_0 = C^* \frac{1}{\sin \varphi_0 (r_0 + a \sin \varphi_0)}$$

$$\longrightarrow \quad C^* = - \sin \varphi_0 (r_0 + a \sin \varphi_0) N_0 \ .$$

Thus we obtain

$$N_{\varphi\varphi} = - \frac{\sin \varphi_0 (r_0 + a \sin \varphi_0)}{\sin \varphi (r_0 + a \sin \varphi)} N_0 \tag{7}$$

and by including (3b):

$$N_{\vartheta\vartheta} = \frac{\sin \varphi_0 (r_0 + a \sin \varphi_0)}{a \sin^2 \varphi} N_0 \ . \tag{8}$$

The stresses are given by

$$\sigma_{\varphi\varphi} = \frac{N_{\varphi\varphi}}{t} \quad \text{and} \quad \sigma_{\vartheta\vartheta} = \frac{N_{\vartheta\vartheta}}{t} \ . \tag{9}$$

b) With axisymmetrical loading and support conditions, we apply the following strain-displacement relations *(12.21)* with $\partial / \partial \vartheta \mathrel{\hat{=}} (\ )_{,\vartheta} = 0$ and $v = 0$:

$$\varepsilon_{\varphi\varphi} = \frac{u_{,\varphi} + w}{r_1} = \frac{1}{a} (u_{,\varphi} + w) \ , \tag{10a}$$

$$\varepsilon_{\vartheta\vartheta} = \frac{u \cos \varphi + w \sin \varphi}{r} = \frac{u \cos \varphi + w \sin \varphi}{a \sin \varphi + r_0} \ , \tag{10b}$$

$$\gamma_{\varphi\vartheta} = 0 \ . \tag{10c}$$

According to *(12.26)* the constitutive equations read :

$$\varepsilon_{\varphi\varphi} = \frac{1}{E t} (N_{\varphi\varphi} - \nu N_{\vartheta\vartheta}) \ , \tag{11a}$$

$$\varepsilon_{\vartheta\vartheta} = \frac{1}{E t} (N_{\vartheta\vartheta} - \nu N_{\varphi\varphi}) \ . \tag{11b}$$

Solution of (10) with respect to w yields :

(10a) $\longrightarrow$ $$w = \varepsilon_{\varphi\varphi} a - u_{,\varphi} \ ,$$

(10b) $\longrightarrow$ $$w = \frac{\varepsilon_{\vartheta\vartheta} (a \sin \varphi + r_0) - u \cos \varphi}{\sin \varphi} \ .$$

By comparing we obtain

$$\varepsilon_{\varphi\varphi} a \sin \varphi - u_{,\varphi} \sin \varphi = \varepsilon_{\vartheta\vartheta} (a \sin \varphi + r_0) - u \cos \varphi$$

$$\longrightarrow \quad u_{,\varphi} - u \cot \varphi = \varepsilon_{\varphi\varphi} a - \varepsilon_{\vartheta\vartheta} \left(a + \frac{r_0}{\sin \varphi} \right) \ . \tag{12}$$

We now substitute (11) into (12) and get

$$u_{,\varphi} - u \cot\varphi = \frac{1}{Et}\left[(N_{\varphi\varphi} - \nu N_{\vartheta\vartheta}) a - (N_{\vartheta\vartheta} - \nu N_{\varphi\varphi})\left(a + \frac{r_0}{\sin\varphi}\right)\right] =$$

$$= \frac{1}{Et}\left[N_{\varphi\varphi}\left(a + \nu a + \nu\frac{r_0}{\sin\varphi}\right) - N_{\vartheta\vartheta}\left(a + \nu a + \frac{r_0}{\sin\varphi}\right)\right] \quad . \qquad (13)$$

Finally, substitution of the membrane forces (7) and (8) into (13) yields :

$$u_{,\varphi} - u \cot\varphi = \frac{N_0}{Et}\left[- \frac{\sin\varphi_0 (r_0 + a \sin\varphi_0)\left[a(1+\nu) + \nu\frac{r_0}{\sin\varphi}\right]}{\sin\varphi\,(r_0 + a\sin\varphi)} - \right.$$

$$\left. - \frac{\sin\varphi_0 (r_0 + a\sin\varphi_0)\left[a(1+\nu) + \frac{r_0}{\sin\varphi}\right]}{a\sin^2\varphi}\right] \quad . \qquad (14)$$

The linear, first order differential equation (14) reads in abbreviated form

$$u_{,\varphi} + P(\varphi)u = Q(\varphi)$$

with $P(\varphi) = -\cot\varphi$ and $Q(\varphi)$ = right-hand side of (14) .

With *(12.27)*, the general solution is

$$u(\varphi) = e^{-\int P(\varphi)\,d\varphi}\left[-\int Q(\varphi)\, e^{\int P(\varphi)\,d\varphi}\, d\varphi + C\right] \quad . \qquad (15)$$

Calculation of the integrals :

$$e^{\int P(\varphi)\,d\varphi} = e^{-\int \cot\varphi\,d\varphi} = e^{-\ln\sin\varphi} = \frac{1}{\sin\varphi} \; ,$$

$$e^{-\int P(\varphi)\,d\varphi} = e^{\int \cot\varphi\,d\varphi} = e^{\ln\sin\varphi} = \sin\varphi \; ,$$

$$\int Q(\varphi)\, e^{\int P(\varphi)\,d\varphi}\, d\varphi =$$

$$= -\frac{N_0 l_0}{Et}\int \frac{a(1+\nu) + \nu\frac{r_0}{\sin\varphi}}{(r_0 + a\sin\varphi)\sin^2\varphi}\, d\varphi - \frac{N_0 l_0}{Et}\int \frac{a(1+\nu) + \frac{r_0}{\sin\varphi}}{a\sin^2\varphi}\, d\varphi$$

with $l_0 = \sin\varphi_0 (r_0 + a\sin\varphi_0)$.

In order to determine the constants of integration , we write the boundary conditions at point A

$$u\left(\varphi = \frac{\pi}{2}\right) = 0 \quad . \qquad (16)$$

Thus we obtain the meridional displacement $u(\varphi)$ by means of which we can determine the normal displacement $w(\varphi)$ from (10a). For reasons of brevity, the integrals will not be determined here.

Exercise C-12-6:

A thin-walled circular cylindrical shell with one end clamped as shown in Fig. C-11 is subjected to a sinusoidal distribution of tangential membrane forces at its free end with the shown vertical force F_R as resultant.

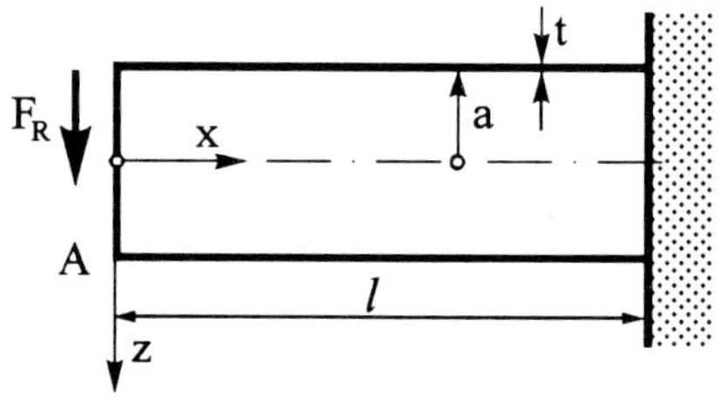

Fig. C-11: Circular cylindrical shell subjected to an end load

a) How large are the membrane forces?

b) Determine the vertical displacement w of the bottom point A of the free end of the shell.

c) Check this displacement by means of the first theorem of CASTIGLIANO.

Solution:

a) We assume that the vertical force F_R at the free end of the shell stems from the following sinusoidal distribution (see Fig. C-12):

$$-N_{x\vartheta} = k \sin\vartheta \ .$$

Then

$$F_R = 4\int_0^{\pi/2} -N_{x\vartheta} \sin\vartheta \, a \, d\vartheta = 4ka\int_0^{\pi/2} \sin^2\vartheta \, d\vartheta = 4ka\frac{\pi}{4} \tag{1}$$

must hold. From (1) follows that $k = F_R/\pi a$, and according to *(12.14)* with $p_\vartheta = p_x = p = 0$ we obtain the resultant forces as follows

$$N_{\vartheta\vartheta} = 0 \quad , \quad N_{x\vartheta} = -\frac{F_R}{\pi a}\sin\vartheta \quad , \quad N_{xx} = \frac{1}{a}\frac{F_R}{\pi a}\, x\cos\vartheta + C_1(\vartheta) \ .$$

Owing to the boundary condition $N_{xx}(x=0) = 0$, the constant $C_1(\vartheta)$ vanishes. Thus, the final result reads

$$N_{\vartheta\vartheta} = 0 \quad , \quad N_{x\vartheta} = -\frac{F_R}{\pi a}\sin\vartheta \quad , \quad N_{xx} = \frac{F_R}{\pi a^2}\, x\cos\vartheta \ . \tag{2}$$

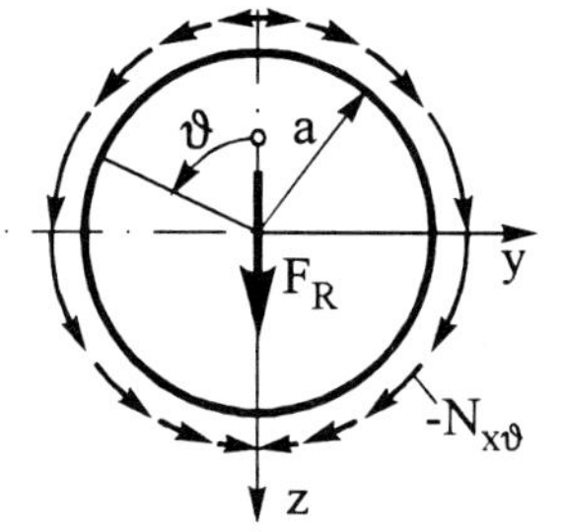

Fig. C-12: Relationship between vertical load F_R and tangential membrane forces $N_{x\vartheta}$

This corresponds to the solution that would be obtained by the elementary beam theory. By defining the moment of inertia for a thin-walled circular section as $I_y = \pi a^3 t$ and the bending moment of the cantilever beam $M_y = -F_R x$, the normal stress at sections x = const is

$$\sigma_{xx} = \frac{M_y}{I_y} z = \frac{F_R x}{\pi a^3 t} a \cos\vartheta = \frac{F_R}{\pi a^2} x \frac{1}{t} \cos\vartheta = \frac{N_{xx}}{t} \ .$$

b) The deformations are calculated by means of the equations of the constitutive equations *(12.26)* after substituting the strain-displacement relations of the circular cylindrical shell *(12.23)*:

$$u_{,x} = \frac{1}{Et} (N_{xx} - \nu N_{\vartheta\vartheta}) \ , \qquad (3a)$$

$$v_{,\vartheta} + w = \frac{a}{Et} (N_{\vartheta\vartheta} - \nu N_{xx}) \ , \qquad (3b)$$

$$\frac{1}{a} u_{,\vartheta} + v_{,x} = \frac{2(1+\nu)}{Et} N_{x\vartheta} \ . \qquad (3c)$$

After substituting the resultant forces (2) into equations (3), we calculate the axial displacement u by integrating (3a), the tangential displacement from (3c) and, finally, by a simple transformation the radial displacement w from (3b). We then obtain

$$u = \frac{1}{Et} \left[\frac{F_R}{\pi a^2} \frac{x^2}{2} \cos\vartheta + C_2(\vartheta) \right] \ , \qquad (4a)$$

$$v = \frac{1}{Et} \left[-2(1+\nu) \frac{F_R}{\pi a} x \sin\vartheta + \frac{F_R}{\pi a^3} \frac{x^3}{6} \sin\vartheta - \frac{x}{a} C_{2,\vartheta} + C_3(\vartheta) \right] \ , \qquad (4b)$$

$$w = \frac{1}{Et} \left[\frac{F_R}{\pi a} (2+\nu) x \cos\vartheta - \frac{F_R}{\pi a^3} \frac{x^3}{6} \cos\vartheta + \frac{x}{a} C_{2,\vartheta\vartheta} - C_{3,\vartheta} \right] \ . \qquad (4c)$$

The *two* arbitrary functions $C_2(\vartheta)$ and $C_3(\vartheta)$ only allow the fulfillment of *two* boundary conditions, e.g. $u(l) = v(l) = 0$, instead of the *four* boundary conditions for the clamped boundary $u(l) = v(l) = w(l) = w_{,x}(l) = 0$. Thus,

$$(4a) \text{ yields} \quad u(l) = 0 \quad \longrightarrow \quad C_2(\vartheta) = -\frac{F_R}{\pi a^2} \frac{l^2}{2} \cos\vartheta \ , \qquad (5a)$$

$$(4b) \text{ yields} \quad v(l) = 0 \quad \longrightarrow$$

$$C_3(\vartheta) = 2(1+\nu) \frac{F_R}{\pi a} l \sin\vartheta - \frac{F_R}{\pi a^3} \frac{l^3}{6} \sin\vartheta + \frac{l}{a} \frac{F_R}{\pi a^2} \frac{l^2}{2} \sin\vartheta$$

$$\longrightarrow \quad C_3(\vartheta) = \frac{F_R}{\pi} \sin\vartheta \left[\frac{l^3}{3a^3} + 2(1+\nu) \frac{l}{a} \right] \ . \qquad (5b)$$

We then obtain the radial displacement w from (4c) with (5a,b)

$$w = \frac{F_R}{Et\pi} \cos\vartheta \left[(2+\nu) \frac{x}{a} - \frac{x^3}{6a^3} + \frac{x}{a} \frac{l^2}{2a^2} - \frac{l^3}{3a^3} - 2(1+\nu) \frac{l}{a} \right] . \qquad (5c)$$

Since no function is available to fulfill $w(l) = 0$, this condition cannot be complied with. It holds that

$$w(l) = -\frac{\nu F_R}{E t \pi} \frac{l}{a} \cos\vartheta \neq 0 \quad . \quad \text{Similarly,} \quad w_{,x}(l) \neq 0 \quad .$$

The membrane theory cannot meet these essential boundary conditions and therefore only yields an approximate solution as we have already seen in various examples. In order to fulfill the essential boundary conditions $w(l) = w_{,x}(l) = 0$, a bending solution has to be superposed onto the approximate solution.

From (5c) we obtain for the displacement of point A:

$$w(x = 0, \vartheta = \pi) = w_{max} = \frac{F_R}{E t \pi}\left[\frac{l^3}{3 a^3} + 2(1 + \nu)\frac{l}{a}\right] . \tag{6}$$

When compared to TIMOSHENKO beam theory, the first term represents the contribution from bending, and the second term the contribution from shear deformation.

c) *Comparison by means of the Theorem of CASTIGLIANO*

The displacement of the point of load application can be calculated by means of the first theorem of CASTIGLIANO *(6.27a)* as follows:

$$v_i = \frac{\partial U^*(F^j)}{\partial F^i} = \frac{\partial U(F^j)}{\partial F^i} \quad . \tag{7}$$

Equation (7) applies to a linearly elastic structure. In the present case, the deformation energy according to *(12.28b)* can be employed. For the circular cylindrical shell $x \mathrel{\hat{=}} \varphi$, so:

$$U = \frac{1}{2 E t} \iint \left[N_{xx}^2 + N_{\vartheta\vartheta}^2 - 2\nu N_{xx} N_{\vartheta\vartheta} + 2(1 + \nu) N_{x\vartheta}^2 \right] dA \quad . \tag{8a}$$

According to (2), $N_{\vartheta\vartheta} = 0$ holds in the present case, i.e. (8a) reduces to:

$$U = \frac{1}{2 E t} \iint_A \left[N_{xx}^2 + 2(1 + \nu) N_{x\vartheta}^2 \right] dA \quad . \tag{8b}$$

We then obtain the displacement w by (7) with (8b) as

$$w = \frac{1}{E t} \int_{x=0}^{l} \int_{\vartheta=0}^{2\pi} \left[N_{xx} \frac{\partial N_{xx}}{\partial F_R} + 2(1 + \nu) N_{x\vartheta} \frac{\partial N_{x\vartheta}}{\partial F_R} \right] a\, d\vartheta\, dx \quad . \tag{9}$$

We now substitute into (9) the resultant forces N_{xx} and $N_{x\vartheta}$ from (2) and their derivatives:

$$w = \frac{a}{E t} \int_{x=0}^{l} \int_{\vartheta=0}^{2\pi} \left[\frac{F_R}{\pi^2 a^4} x^2 \cos^2\vartheta + 2(1 + \nu) \frac{F_R}{\pi^2 a^2} \sin^2\vartheta \right] dx\, d\vartheta \quad .$$

After integration we obtain the same result as given in (6)

$$w_{max} = \frac{F_R}{E t \pi}\left[\frac{l^3}{3 a^3} + 2(1 + \nu)\frac{l}{a}\right] .$$

Exercise C-12-7:

A type of shell often found in civil and mechanical engineering is a ruled shell as shown in Fig. C-13. Its mid-surface has the form of a special hyperbolical paraboloid which is generated by moving a straight line g along a rectangle ABCD. The rectangle lies in the x^1, x^2-plane and has the side lengths l_1, l_2. The straight line moves along the line AD and along the hypotenuse of the triangle BEC. This so-called skew hyperbolical paraboloid shell is also termed a *hypar shell* and its parametric description is given by

$$\mathbf{r}(x,y) = x\,\mathbf{e}_1 + y\,\mathbf{e}_2 + \frac{xy}{c}\,\mathbf{e}_3 \qquad \left(x \mathrel{\hat=} \xi^1 \;,\; y \mathrel{\hat=} \xi^2 \;,\; c = \frac{l_1 l_2}{l_3}\right) .$$

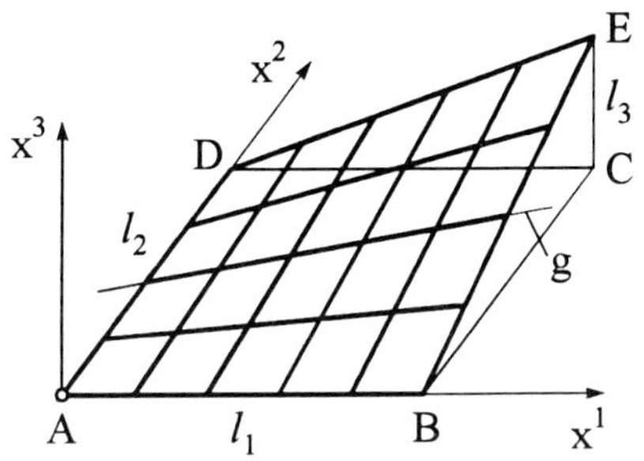

Fig. C-13: Coordinates of a hyperbolical paraboloid shell

a) Set up the equilibrium conditions of this shell according to membrane theory.

b) Determine the resultant forces and moments for a shell subjected to the deadweight g per unit surface area, i.e. its physical load components in the global Cartesian coordinate system x^i $(i = 1,2,3)$ are given as:

$$p_1 = p_2 = 0 \quad , \quad p_3 = -g \quad .$$

Solution

a) Equilibrium conditions

First, the fundamental quantities of first and second order as well as the CHRISTOFFEL-symbols have to be determined. Proceeding from the given parameter description

$$\mathbf{r}(x,y) = x\,\mathbf{e}_1 + y\,\mathbf{e}_2 + \frac{xy}{c}\,\mathbf{e}_3 \ ,$$

the base vectors are determined as:

$$\mathbf{a}_1 = \mathbf{r}_{,x} = \mathbf{e}_1 + \frac{y}{c}\,\mathbf{e}_3 \ , \tag{1a}$$

$$\mathbf{a}_2 = \mathbf{r}_{,y} = \mathbf{e}_2 + \frac{x}{c}\,\mathbf{e}_3 \quad . \tag{1b}$$

Similarly, the metric tensors are calculated according to *(11.11)*, the determinant according to *(11.12)*, and the covariant tensor of curvature according to *(11.18)*:

$$(a_{\alpha\beta}) = \begin{bmatrix} 1+\left(\frac{y}{c}\right)^2 & \frac{xy}{c^2} \\ \frac{xy}{c^2} & 1+\left(\frac{x}{c}\right)^2 \end{bmatrix} , \tag{2a}$$

$$a = |a_{\alpha\beta}| = 1 + \left(\frac{x}{c}\right)^2 + \left(\frac{y}{c}\right)^2 , \tag{2b}$$

$$(a^{\alpha\beta}) = (a_{\alpha\beta})^{-1} = \frac{1}{a}\begin{bmatrix} 1+\left(\frac{x}{c}\right)^2 & -\frac{xy}{c^2} \\ -\frac{xy}{c^2} & 1+\left(\frac{y}{c}\right)^2 \end{bmatrix} , \tag{2c}$$

$$b_{\alpha\beta} = \begin{bmatrix} 0 & \frac{1}{c\sqrt{\alpha}} \\ \frac{1}{c\sqrt{\alpha}} & 0 \end{bmatrix} . \tag{3}$$

From *(11.23a)*, the CHRISTOFFEL-symbols of the second kind result as

$$(\Gamma^1_{\alpha\beta}) = \begin{bmatrix} 0 & \frac{y}{c^2 a} \\ \frac{y}{c^2 a} & 0 \end{bmatrix} , \quad (\Gamma^2_{\alpha\beta}) = \begin{bmatrix} 0 & \frac{x}{c^2 a} \\ \frac{x}{c^2 a} & 0 \end{bmatrix} . \tag{4}$$

In order to formulate the equilibrium conditions *(12.1)*

$$\left.\begin{aligned} & N^{11}|_1 + N^{21}|_2 + p^1 = 0 , \\ & N^{12}|_1 + N^{22}|_2 + p^2 = 0 , \\ & N^{11} b_{11} + 2N^{12} b_{12} + N^{22} b_{22} + p = 0 , \end{aligned}\right\} \tag{5}$$

the covariant derivatives of the stress resultants are required. With the relations *(2.35b)* and (4) they become

$$\left.\begin{aligned} & N^{11}|_1 = N^{11}{}_{,1} + \frac{2y}{c^2 a} N^{12} , \\ & N^{12}|_1 = N^{12}{}_{,1} + \frac{y}{c^2 a} N^{22} + \frac{x}{c^2 a} N^{12} , \\ & N^{21}|_2 = N^{21}{}_{,2} + \frac{x}{c^2 a} N^{11} + \frac{y}{c^2 a} N^{21} , \\ & N^{22}|_2 = N^{22}{}_{,2} + \frac{2x}{c^2 a} N^{12} . \end{aligned}\right\} \tag{6}$$

Substitution of the derivatives (6) into (5) yields :

$$\left.\begin{aligned} N^{11}{}_{,1} + N^{21}{}_{,2} + \frac{3\,y}{c^2 a} N^{12} + \frac{x}{c^2 a} N^{11} + p^1 &= 0 \ , \\ N^{12}{}_{,1} + N^{22}{}_{,2} + \frac{3\,x}{c^2 a} N^{12} + \frac{y}{c^2 a} N^{22} + p^2 &= 0 \ , \\ \frac{2}{c\sqrt{a}} N^{12} + p &= 0 \ . \end{aligned}\right\} \qquad (7)$$

The solution of this system of equations requires a transformation into physical components. Owing to the occuring non-orthogonal surface coordinate system (metric (2a) is fully occupied), the relations *(2.17)* cannot be used for determining the physical components. On the basis of [C.6 , C.11] we therefore define, as physical components of a stress vector τ^{ij}, the components of the stress vector in the direction of the unit vectors that are parallel to the base vectors and that are thus not perpendicular to a tetrahedron cut plane. We obtain from the equilibrium of the tetrahedron

$$\tau^{*ij} = \sqrt{\frac{g_{(jj)}}{g^{(ii)}}}\ \tau^{ij} \quad . \qquad (8a)$$

In transition to the shell, (8a) yields the physical components of the membrane forces

$$N^{*\alpha\beta} = \sqrt{\frac{a_{(\beta\beta)}}{a^{(\alpha\alpha)}}}\ N^{\alpha\beta} \quad . \qquad (8b)$$

Substitution of (8b) into (7) requires formation of the following derivatives :

$$\frac{\partial}{\partial x}\sqrt{\frac{a^{11}}{a_{11}}} = \frac{x\,y^2}{c^4 a\sqrt{a}\ \sqrt{a_{11}\,a_{22}}} \ , \qquad \frac{\partial}{\partial x}\sqrt{\frac{a^{11}}{a_{22}}} = -\frac{x}{c^2 a\sqrt{a}} \ ,$$

$$\frac{\partial}{\partial y}\sqrt{\frac{a^{22}}{a_{22}}} = \frac{x^2\,y}{c^4 a\sqrt{a}\ \sqrt{a_{11}\,a_{22}}} \ , \qquad \frac{\partial}{\partial y}\sqrt{\frac{a^{22}}{a_{11}}} = \frac{y}{c^2 a\sqrt{a}} \ .$$

By introducing the physical components of the surface loads

$$p^{*\alpha} = \sqrt{a_{(\alpha\alpha)}}\ p^\alpha \ , \quad p^* = p \ ,$$

and by denoting the physical components of the membrane forces by subscripts

$$N^{*11} \mathrel{\widehat{=}} N_{xx} \ , \quad N^{*22} \mathrel{\widehat{=}} N_{yy} \ , \quad N^{*12} \mathrel{\widehat{=}} N_{xy} \ .$$

Eqs. (7) finally yield the equilibrium conditions of the skew hyperbolic paraboloid shell ($\partial/\partial x \mathrel{\widehat{=}} {}_{,x}$, $\partial/\partial y \mathrel{\widehat{=}} {}_{,y}$):

$$N_{xx,x}\sqrt{a_{22}} + N_{xy,y}\sqrt{a_{11}} + \frac{x}{c^2\sqrt{a_{22}}} N_{xx} + 2\,\frac{y\sqrt{a_{11}}}{c^2 a} N_{xy} + p^{*1}\sqrt{a} = 0 \ , \qquad (9a)$$

$$N_{xy,x}\sqrt{a_{22}} + N_{yy,y}\sqrt{a_{11}} + \frac{y}{c^2\sqrt{a_{11}}}N_{yy} + 2\,\frac{y\sqrt{a_{22}}}{c^2\,a}N_{xy} + p^{*2}\sqrt{a} = 0 \quad , \qquad (9b)$$

$$\frac{2}{ca}N_{xy} + p^* = 0 \quad . \qquad (9c)$$

Eqs. (9a,b) are a system of first order partial differential equations with variable coefficients. Eq. (9c) yields the membrane shear force

$$N_{xy} = -\frac{c}{2}\,a\,p^* \quad .$$

By formulating the derivatives (note (2b))

$$N_{xy,x} = -\frac{x}{c}\,p^* - \frac{c}{2}\,a\,p^* \quad , \quad N_{xy,y} = -\frac{y}{c}\,p^* - \frac{c}{2}\,a\,p^* \quad ,$$

and by substituting them together with the metric (2a) into (9a,b), we obtain after re-formulation two uncoupled differential equations for the two unknown membrane forces:

$$\left(N_{xx}\sqrt{a_{22}}\right)_{,x} = \frac{2y}{c}\sqrt{a_{11}}\;p^* + \frac{c}{2}\,a\sqrt{a_{11}}\;p^*_{,y} - \sqrt{a}\;p^{*1} \quad , \qquad (10a)$$

$$\left(N_{yy}\sqrt{a_{11}}\right)_{,y} = \frac{2x}{c}\sqrt{a_{22}}\;p^* + \frac{c}{2}\,a\sqrt{a_{11}}\;p^*_{,x} - \sqrt{a}\;p^{*2} \quad . \qquad (10b)$$

b) Resultant membrane forces

First, the physical load components in the global Cartesian coordinate system x^i ($i = 1, 2, 3$) have to be decomposed into components both in the direction of the local surface parameters and perpendicular to them.

$\mathbf{a}_3$ is calculated from $\mathbf{a}_1$ and $\mathbf{a}_2$ (1a,b) by forming the vector product according to *(11.16)*:

$$\mathbf{a}_1 = \mathbf{e}_1 + \frac{y}{c}\,\mathbf{e}_3 \quad , \qquad (11a)$$

$$\mathbf{a}_2 = \mathbf{e}_2 + \frac{x}{c}\,\mathbf{e}_3 \quad , \qquad (11b)$$

$$\mathbf{a}_3 = -\frac{y}{c}\,\mathbf{e}_1 - \frac{x}{c}\,\mathbf{e}_2 + \mathbf{e}_3 \quad . \qquad (11c)$$

The above vector equations constitute the transformation between the local base vectors and the base vectors in the Cartesian coordinate system. The latter vectors can be written in abbreviated form as

$$\mathbf{a}_{i'} = \beta^{j}_{i'}\,\mathbf{e}_j \quad .$$

Correspondingly, the vector can be written in different bases. The covariant components of the load vector with *(2.9a)* read, for instance,

$$p^{i'} = \beta^{i'}_{k}\,p^{k} \quad .$$

With *(2.8)*

$$\beta^{i'}_{k}\,\beta^{j}_{i'} = \delta^{j}_{k} \quad ,$$

the transformation coefficients $\beta^{i'}_{k}$ are determined by inverting ($\beta^{j}_{i'}$):

$$(\beta_k^{i'}) = \frac{1}{a}\begin{bmatrix} 1+\frac{x^2}{c^2} & -\frac{xy}{c^2} & \frac{y}{c} \\ -\frac{xy}{c^2} & 1+\frac{y^2}{c^2} & \frac{x}{c} \\ -\frac{y}{c} & -\frac{x}{c} & 1 \end{bmatrix} . \tag{12}$$

By substituting the load components we obtain with *(2.10)* the physical components of the load vectors:

$$p^{*1} = -\frac{y}{c}\, g \frac{\sqrt{a_{11}}}{a} \quad , \quad p^{*2} = -\frac{x}{c}\, g \frac{\sqrt{a_{22}}}{a} \quad , \quad p^{*} = -g \frac{1}{\sqrt{a}} . \tag{13}$$

Substitution of these transformed loads into (10a) yields after re-formulation

$$\left(N_{xx}\sqrt{1+\left(\frac{x}{c}\right)^2}\,\right)_{,x} = -\frac{g}{2}\frac{y}{c}\frac{\sqrt{1+\left(\frac{x}{c}\right)^2}}{\sqrt{a}} . \tag{14}$$

By integration we obtain

$$N_{xx} = -\frac{g}{2}\, y \frac{\sqrt{1+\left(\frac{y}{c}\right)^2}}{\sqrt{1+\left(\frac{x}{c}\right)^2}}\left[\ln\left(x+\sqrt{c^2+x^2+y^2}\right)+C(y)\right] . \tag{15}$$

From the boundary condition $N_{xx}(x,0) = 0$, the integration function $C(y)$ follows as

$$C(y) = -\ln\sqrt{c^2+y^2}$$

and thus
$$N_{xx} = \frac{1}{2}\, g\, y \frac{\sqrt{1+\left(\frac{y}{c}\right)^2}}{\sqrt{1+\left(\frac{x}{c}\right)^2}} \ln \frac{\sqrt{1+\left(\frac{y}{c}\right)^2}}{\frac{x}{c}+\sqrt{a}} . \tag{16a}$$

From (10b) with the boundary condition $N_{yy}(x,0) = 0$, one analogously obtains the membrane force in the y-direction:

$$N_{yy} = \frac{1}{2}\, g\, x \frac{\sqrt{1+\left(\frac{x}{c}\right)^2}}{\sqrt{1+\left(\frac{y}{c}\right)^2}} \ln \frac{\sqrt{1+\left(\frac{x}{c}\right)^2}}{\frac{y}{c}+\sqrt{a}} . \tag{16b}$$

Eq. (9c) finally yields the membrane shear force

$$N_{xy} = \frac{cg}{2}\sqrt{a} . \tag{16c}$$

TIMOSHENKO [C.24] and other authors have treated the same problem by projecting the forces onto the x , y-plane, and then formulating the equilibrium. Their results can be transformed, by respective measures (e.g. $N_{xx} = \sqrt{a_{11}/a_{22}}\, N_{11\,TIM}$) into eq. (16). Given the prescribed boundary conditions, the load at the boundaries $x = 0$ and $y = 0$ only acts via shear. Thus, boundary stiffeners are required, a fact that leads to incompatibilities between the deformations of the stiffeners and of the shell boundaries. For this reason, the membrane solution has to be augmented by a solution from bending theory. Further examples are treated in [C.2 , C.8].

Exercise C-13-1:

A circular water tank (radius a, height h) has a linearly varying wall thickness (t_0 = maximum wall thickness)

$$t(x) = t_0\left(1 - \frac{x}{h}\right)$$

as shown in Fig. C-14.

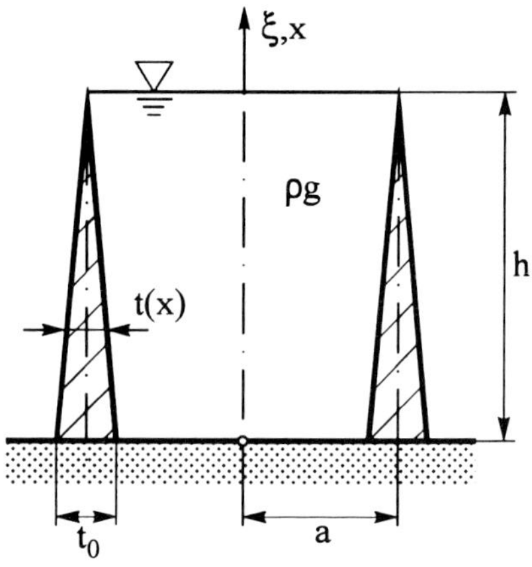

Fig. C-14: Water tank clamped at the bottom

Given values : $a = 4.0\ m$, $h = 5.0\ m$, $t_0 = 0.35\ m$, $\nu = 0.3$,

$E = 2.1 \cdot 10^5\ MPa$, $\varrho g = 1 \cdot 10^4\ N/m^3$.

a) Derive the differential equation and the boundary conditions for the circular water tank by means of a variational principle.

b) Determine the radial displacement w by a RITZ approach. For this purpose,

$$f_k(x) = \left(\frac{x}{h}\right)^2 \left(1 - \frac{x}{h}\right)^k \qquad (k = 1, 2)$$

shall be chosen as coordinate functions for the approximation of w, and the calculation shall be performed using a two-term approach .

Note: The deadweight of the tank can be disregarded. The assumptions of the technical shell theory are valid.

Solution :

a) The total potential energy is composed of the deformation energy of the shell and the potential energy of the external loads (see [C.11]). With the approximation $\widetilde{N}^{\alpha\beta} \approx N^{\alpha\beta}$, we obtain the total potential energy expression

$$\Pi = \frac{1}{2} \int_A \left(N^{\alpha\beta} \alpha_{\alpha\beta} + M^{\alpha\beta} \omega_{\alpha\beta} \right) dA - \int_A \left(p^\alpha v_\alpha + p\, w \right) dA \ . \qquad (1)$$

For a cylindrical shell we write in physical components :

$$\Pi = \frac{1}{2}\int_A \Big(N_{xx}\,\varepsilon_{xx} + N_{\vartheta\vartheta}\,\varepsilon_{\vartheta\vartheta} + 2\,N_{x\vartheta}\,\varepsilon_{x\vartheta} + M_{xx}\,\omega_{xx} +$$

$$+ M_{\vartheta\vartheta}\,\omega_{\vartheta\vartheta} + 2\,M_{x\vartheta}\,\omega_{x\vartheta}\Big)\,dA - \int_A \Big(p_x\,u + p_\vartheta\,v + p\,w\Big)\,dA \ . \qquad (2)$$

An axisymmetrical load case is given in the present problem, and the longitudinal force N_{xx} vanishes. Thus, (2) reduces to

$$\Pi = \frac{1}{2}\int_A \Big(N_{\vartheta\vartheta}\,\varepsilon_{\vartheta\vartheta} + M_{xx}\,\omega_{xx}\Big)\,dA - \int_A p\,w\,dA \ . \qquad (3)$$

With *(13.14)*

$$\varepsilon_{\vartheta\vartheta} = \frac{w}{a} \ , \quad \kappa_{xx} \mathrel{\widehat{=}} \omega_{xx} = -\frac{w_{,\xi\xi}}{a^2} \qquad \left(\xi \mathrel{\widehat{=}} \frac{x}{a}\right)$$

and

$$N_{\vartheta\vartheta} = D(1-\nu^2)\frac{w}{a} = E\,t(\xi)\frac{w}{a} \ , \qquad (4a)$$

$$M_{xx} = -\frac{K(\xi)}{a^2}\,w_{,\xi\xi} \ . \qquad (4b)$$

From (3) follows that

$$\Pi = \Pi(\xi, w, w_{,\xi\xi}) = \int_{\vartheta=0}^{2\pi}\int_{\xi=0}^{h/a}\left\{\frac{1}{2}\left[E\,t\left(\frac{w}{a}\right)^2 + K\left(\frac{w_{,\xi\xi}}{a^2}\right)^2\right] - p\,w\right\}dA \ . \ (5)$$

By (5) we have determined a variational functional for which we now have to find an extremum according to *(6.34)*. Therefore, we formulate

$$\delta\,\Pi = \delta\int L(\xi, w, w_{,\xi\xi})\,dA = 0 \ .$$

We then obtain an EULER differential equation in accordance with *(6.35)* as a necessary condition:

$$\left(\frac{\partial L}{\partial w_{,\xi\xi}}\right)_{,\xi\xi} + \frac{\partial L}{\partial w} = 0$$

$$\longrightarrow \quad \frac{1}{a^2}\left(\frac{K}{a^2}\,w_{,\xi\xi}\right)_{,\xi\xi} + E\,t\,\frac{w}{a^2} - p = 0 \ . \qquad (6)$$

For a constant wall thickness t follows

$$w_{,\xi\xi\xi\xi} + \underbrace{\frac{E\,t}{K}\,a^2}_{4\kappa^4}\,w = \frac{p\,a^4}{K}$$

as the differential equation of a circular cylindrical boiler *(13.16 a)*.

We obtain as *boundary conditions*

$$\frac{\partial L}{\partial w_{,\xi\xi}}\,\delta w_{,\xi}\Big|_{\xi=\text{const}} = 0 \ \longrightarrow \ \frac{K}{a^2}\,w_{,\xi\xi} = 0 \quad \text{or} \quad \delta w_{,\xi} = 0 \ , \qquad (7a)$$

$$-\left(\frac{\partial L}{\partial w_{,\xi\xi}}\right)_{,\xi}\delta w\Big|_{\xi=\text{const}} = 0 \quad\longrightarrow\quad \frac{K}{a} w_{,\xi\xi\xi} = 0 \quad \text{or} \quad \delta w = 0\,. \qquad (7b)$$

b) In order to calculate the radial displacement w by means of the RITZ method, we employ the energy expression (5). For this purpose, we introduce the linearly increasing pressure

$$p = g\rho h\left(1 - \frac{a}{h}\xi\right)$$

and the varying bending stiffness

$$K = \underbrace{\frac{E\,t_0^3}{12\,(1-\nu^2)}}_{K_0}\left(1 - \frac{a}{h}\xi\right)^3 .$$

With $dA = 2\pi a dx = 2\pi a^2 d\xi$ we obtain

$$\Pi = 2\pi \int_{\xi=0}^{h/a}\left[\frac{1}{2}E\,t_0\left(1-\frac{a}{h}\xi\right)w^3 + \frac{K_0}{a^2}\left(1-\frac{a}{h}\xi\right)^3\left(w_{,\xi\xi}\right)^2 - g\rho h\left(1-\frac{a}{h}\xi\right)a^2 w\right]d\xi \,. \qquad (8)$$

The application of the RITZ method (cf. Section 6.7) requires that we choose an approximation to w with linearly independent coordinate functions in such a way that the essential, i.e. geometrical, boundary conditions are fulfilled. According to *(6.36)* we choose an approximation

$$w^* = \sum_{n=1}^{N} c_n f_n(\xi) \quad (n = 1,2,\dots,N)\,, \qquad (9a)$$

where the coordinate functions in the problem formulation are given as

$$f_n(\xi) = \left(\frac{a}{h}\xi\right)^2\left(1-\frac{a}{h}\xi\right)^n . \qquad (9b)$$

The coefficients c_n are the free, yet unknown coefficients.

The approximation (9) obviously fulfills the geometrical boundary conditions $(w(0) = w_{,\xi}(0) = 0)$. In addition, the dynamic boundary conditions are also satisfied since $K(x = h) = 0$.

Based upon *(6.37)*

$$\frac{\partial \Pi}{\partial c_n} = 0 \quad , \quad n = 1,2,\dots,N \,,$$

we derive a linear system of equations for determination of the coefficients c_n with

$$\sum_{k=1}^{N} c_k \int_{\xi=0}^{h/a}\left[E\,t_0\left(1-\frac{a}{h}\xi\right)f_k f_n + \frac{K_0}{a^2}\left(1-\frac{a}{h}\xi\right)^3 f_{k,\xi\xi}\,f_{n,\xi\xi}\right]d\xi - g\rho h a^2\int_{\xi=0}^{h/a}\left(1-\frac{a}{h}\xi\right)f_n\,d\xi = 0 \,. \qquad (10)$$

For a two-termed approximation n = 1,2, this system of equations reads:

$$n = 1: \quad c_1 \int_{\xi=0}^{h/a} \left[E t_0 (1 - \tfrac{a}{h}\xi) f_1^2 + \frac{K_0}{a^2}(1 - \tfrac{a}{h}\xi)^3 f_{1,\xi\xi}^2 \right] d\xi +$$

$$+ c_2 \int_{\xi=0}^{h/a} \left[E t_0 (1 - \tfrac{a}{h}\xi) f_2 f_1 + \frac{K_0}{a^2}(1 - \tfrac{a}{h}\xi)^3 f_{2,\xi\xi} f_{1,\xi\xi} \right] d\xi =$$

$$= g \varrho h a^2 \int_{\xi=0}^{h/a} (1 - \tfrac{a}{h}\xi) f_1 d\xi ,$$

$$n = 2: \quad c_1 \int_{\xi=0}^{h/a} \left[E t_0 (1 - \tfrac{a}{h}\xi) f_1 f_2 + \frac{K_0}{a^2}(1 - \tfrac{a}{h}\xi)^3 f_{1,\xi\xi} f_{2,\xi\xi} \right] d\xi +$$

$$+ c_2 \int_{\xi=0}^{h/a} \left[E t_0 (1 - \tfrac{a}{h}\xi) f_2^2 + \frac{K_0}{a^2}(1 - \tfrac{a}{h}\xi)^3 f_{2,\xi\xi}^2 \right] d\xi =$$

$$= g \rho h a^2 \int_{\xi=0}^{h/a} (1 - \tfrac{a}{h}\xi) f_2 d\xi .$$

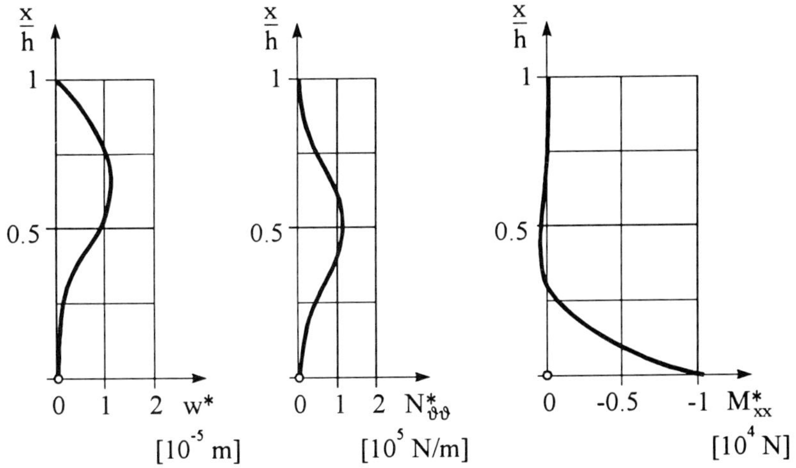

Fig. C-15: Approximate displacement w^*, membrane force $N^*_{\vartheta\vartheta}$, and bending moment M^*_{xx} of a cylindrical tank with variable thickness

After integration and solution of the linear system of equations we obtain the following coefficients:

$$c_1 = \frac{g\rho(1-\nu^2)h^5}{Et_0^3}\,\frac{1008}{468 + 87\lambda + \lambda^2}\,,$$

$$c_2 = -\frac{g\rho(1-\nu^2)h^5}{Et_0^3}\,\frac{21(36-\lambda)}{468 + 87\lambda + \lambda^2}$$

with
$$\lambda = (1-\nu^2)\left(\frac{h^2}{a t_0}\right)^2 .$$

By (9a) we thus approximate the radial displacement w* as

$$w^* = \frac{g\rho(1-\nu^2)h^5}{Et_0^3}\left(\frac{a}{h}\xi\right)^2 \frac{1008\left(1-\frac{a}{h}\xi\right) - 21(36-\lambda)\left(1-\frac{a}{h}\xi\right)^2}{468 + 87\lambda + \lambda^2}. \qquad (11)$$

Finally, the curves for $N^*_{\vartheta\vartheta}$ and M^*_{xx} are calculated by means of (4a,b). Fig. C-15 presents the w*-curve and the approximations for the resultant forces of the numerical example.

Exercise C-13-2:

A reinforcing ring 2 (cross-section $b \cdot 3t$, $b \ll 1$) is to be positioned in the middle of a thin-walled, long pressure tube 1 made of sheet steel (radius a, wall thickness t). For this purpose, the ring is warmed up in such a way that it can be slided into its position on the unloaded tube (see Fig. C-16).

At a temperature $T_2 = 50$°C the ring just fits the tube in stress-free contact. Cooling of the ring to the tube temperature of $T_1 = 20$°C leads to shrinking of the ring.

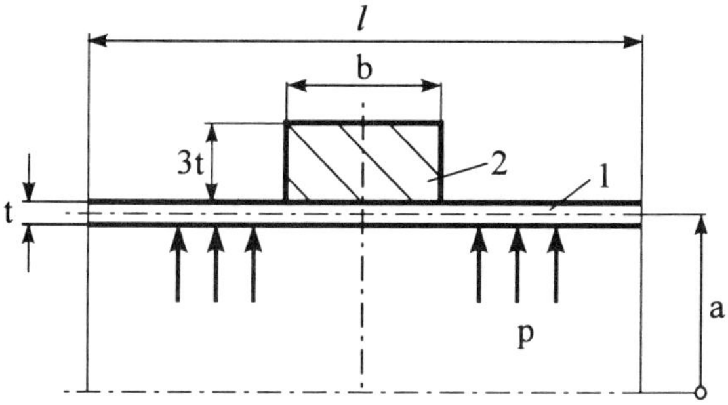

Fig. C-16: Pressurized tube with shrinked reinforcing ring

Determine the resultant quantities $N_{\vartheta\vartheta}$ and M_{xx} in the tube as well as the stresses in the tube and the ring, when the tube is subjected to a constant internal overpressure p.

Numerical values: $a = 1\ m$, $b = 1\ m$, $t = 1.5 \cdot 10^{-2}\ m$,

$p = 1.5\ MPa$, $\alpha_{T2} = 1.1 \cdot 10^{-5} / {}^\circ C$, $\nu = 0.3$,

$E_1 = E_2 = E = 2.1 \cdot 10^5\ MPa$.

Solution :

The problem will be solved by means of the well-known, so-called *Method of Theory of Structures* (Section 13.1.4). For this purpose, we partition the pressure tube and the reinforcing ring into three subsystems ("0"–, "1"– and "2"–system) according to Fig. C-17. We can now formulate the compatibility conditions :

$$w_1^{(0)} + w_1^{(1)} + w_1^{(2)} = w_2^{(0)} + w_2^{(1)} + w_2^{(2)} \ , \qquad (1a)$$

$$\chi_1^{(0)} + \chi_1^{(1)} + \chi_1^{(2)} = \chi_2^{(0)} + \chi_2^{(1)} + \chi_2^{(2)} \ . \qquad (1b)$$

Here, the subscript denotes tube 1 or ring 2, respectively, (including tube element). The parenthesized superscript refers to the "0"–, "1"– and "2"–system.

We can now compile the values of deformation for the tube and the ring, where the membrane solution follows from *(12.14)*, *(12.23)* and *(12.26)*. The values for the partitioned tube subjected to the boundary force and boundary moment M are derived from Fig. 13.2 :

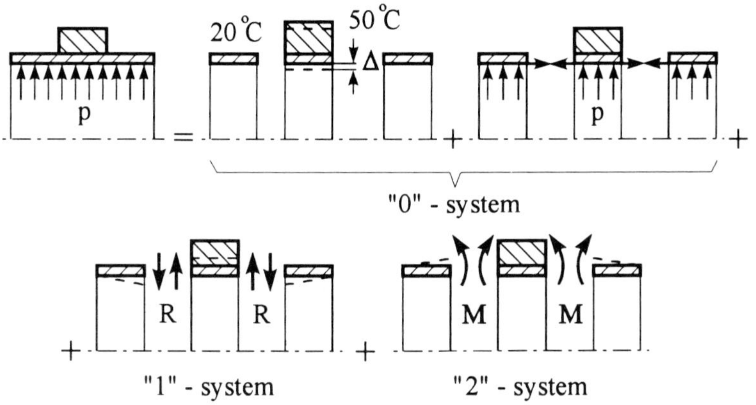

Fig. C-17: Partitioning of the pressure tube in single subsystems

Tube 1:

$$w_1^{(0)} = \frac{p\,a^2}{E\,t} \quad , \quad w_1^{(1)} = -\frac{R\,a^3}{2\,K\,\kappa^3} \quad , \quad w_1^{(2)} = \frac{M\,a^2}{2\,K\,\kappa^2} \quad , \qquad (2a)$$

$$\chi_1^{(0)} = 0 \quad , \quad \chi_1^{(1)} = -\frac{R\,a^2}{2\,K\,\kappa^2} \quad , \quad \chi_1^{(2)} = \frac{M\,a}{K\,\kappa} \quad . \qquad (2b)$$

Ring 2:

$$w_2^{(0)} = w_2^{(01)} + w_2^{(02)} = \Delta + \frac{p\,b\,a^2}{E\,A} = \Delta + \frac{p\,a^2}{4\,E\,t} \quad ,$$

$$w_2^{(1)} = \frac{2\,R\,a^2}{E\,A} = \frac{R\,a^2}{2\,E\,t\,b} \quad , \quad w_2^{(2)} = 0 \quad , \qquad (3a)$$

$$\chi_2^{(0)} = \chi_2^{(1)} = \chi_2^{(2)} = 0 \quad . \qquad (3b)$$

Δ in (3a) denotes the shrinking measure.

We now substitute (2) and (3) into (1), and obtain a system of linear equations by means of which we can determine the unknown boundary loads :

$$\frac{p\,a^2}{E\,t} - \frac{R\,a^3}{2\,K\,\kappa^3} + \frac{M\,a^2}{2\,K\,\kappa^2} = \Delta + \frac{p\,a^2}{4\,E\,t} + \frac{R\,a^2}{2\,E\,t\,b} + 0 \quad , \qquad (4a)$$

$$0 - \frac{R\,a^2}{2\,K\,\kappa^2} + \frac{M\,a}{K\,\kappa} = 0 \quad . \qquad (4b)$$

Eq. (4b) leads to

$$R = \frac{2\,\kappa}{a}\,M \qquad (5a)$$

and (4a) correspondingly to

$$M = -\frac{\Delta - \dfrac{3}{4}\dfrac{p\,a^2}{E\,t}}{\dfrac{a^2}{2\,K\,\kappa^2} + \dfrac{a\,\kappa}{E\,t\,b}} \quad . \qquad (5b)$$

The shrinking measure Δ has to be determined by an additional calculation. For this purpose, we separate ring 2 from tube element 1 according to Fig. C-18 and insert the forces acting on the single parts. Then, the following circumferential strains are determined :

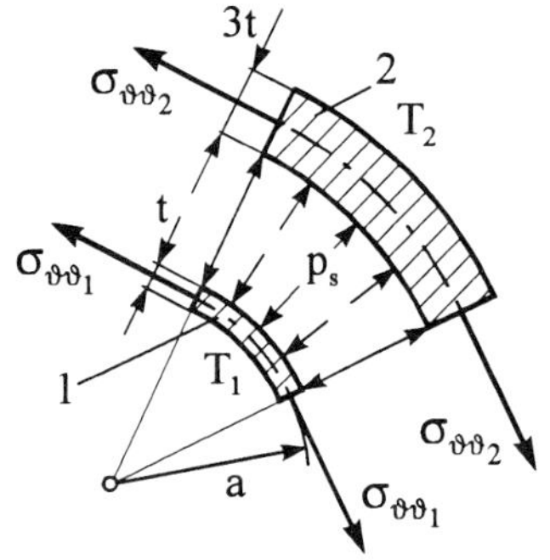

Fig. C-18: Free-body-diagram of ring and tube element

$$\varepsilon_{\vartheta\vartheta_1} = \frac{\sigma_{\vartheta\vartheta 1}}{E} = -\frac{p_s\,a}{E\,t} \quad , \qquad (6a)$$

$$\varepsilon_{\vartheta\vartheta_2} = \frac{\sigma_{\vartheta\vartheta 2}}{E} - \alpha_{T2}(T_2 - T_1) = $$

$$= \frac{p_s(a + 2\,t)}{3\,E\,t} - \alpha_{T2}\,\Theta \quad , \qquad (6b)$$

where p_s denotes the shrinking pressure and $\Theta = T_2 - T_1$ the temperature difference. After the ring has been mounted and cooled to T_1, the circumferential extensions and hence strains in (6a,b) must be equal, i.e.,

$$\varepsilon_{\vartheta\vartheta_1} = \varepsilon_{\vartheta\vartheta_2} \quad . \tag{7}$$

Substitution of (6a,b) into (7) yields with $\frac{a}{t} \gg 1$:

$$-\frac{p_s\, a}{E\, t} \approx \frac{p_s\, a}{3\, E\, t} - \alpha_{T2}\, \Theta \quad \longrightarrow \quad p_s = \frac{3}{4}\frac{E\, t}{a}\alpha_{T2}\, \Theta \quad . \tag{8}$$

We then calculate the circumferential strain from (6a) with (8) as

$$\varepsilon_{\vartheta\vartheta\, 1} = -\frac{3}{4}\alpha_{T2}\, \Theta \quad ,$$

and with $\varepsilon_{\vartheta\vartheta_1} = \frac{\Delta}{a}$, the shrinking measure Δ is determined as

$$\Delta = -\frac{3}{4}\, a\, \alpha_{T2}\, \Theta \quad . \tag{9}$$

By substituting (9) into (5b) we obtain the boundary moment :

$$M = \frac{3}{4}\,\frac{\dfrac{p\, a^2}{E\, t} + a\, \alpha_{T2}\, \Theta}{\dfrac{a^2}{2\, \kappa^2\, K} + \dfrac{a\, \kappa}{E\, t\, b}} \quad .$$

We are now able to calculate the circumferential membrane force $N_{\vartheta\vartheta}$ and the bending moment M_{xx} from *(13.17c)*. For this purpose, the membrane solution w_p of a circular cylindrical shell subjected to internal pressure has to be superposed. The total deformation then reads as follows:

$$w = \frac{p\, a^2}{E\, t} + \frac{a^2}{2\, \kappa^2\, K}\left[-\frac{a}{\kappa}\, R \cos\kappa\,\xi + M\left(\cos\kappa\,\xi - \sin\kappa\,\xi\right)\right] e^{-\kappa\,\xi} \quad . \tag{10}$$

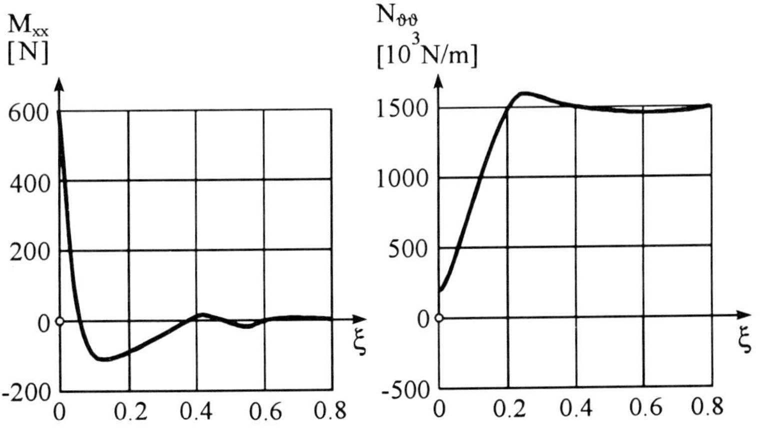

Fig. C-19: Bending moment M_{xx} and circumferential force $N_{\vartheta\vartheta}$ in the pressure tube

Now we replace R by (5a) and substitute it into the relation for the circumferential force $N_{\vartheta\vartheta}$:

$$N_{\vartheta\vartheta} = \frac{E\,t}{a}\,w = p\,a - \frac{2\,\kappa^2}{a}\,M\,(\cos\kappa\,\xi + \sin\kappa\,\xi)\,e^{-\kappa\,\xi} \quad . \tag{11a}$$

From *(13.17c)* we obtain for the bending moment M_{xx} with $R = -\frac{2\,\kappa}{a}\,M$ (here opposite to the assumed direction of R)

$$M_{xx} = -M\,(\cos\kappa\,\xi - \sin\kappa\,\xi)\,e^{-\kappa\,\xi} \quad . \tag{11b}$$

Fig. C-19 depicts the curves of the resultant moments and forces for the given numerical values.

Finally, we calculate the stresses in the tube and the ring :

- *Tube 1*

Longitudinal stress :

$$\sigma_{xx_1} = \pm\frac{6\,M_{xx}}{t^2} + \frac{p\,a}{2\,t} \quad . \tag{12a}$$

The second term in (12a) only applies for a tube closed at both ends, since in this case an additional longitudinal load occurs.

Equation (11b) substituted into (12a) yields the maximum stress

$$\sigma_{xx_{max}} = \frac{-6\,M_{xx}(\,\xi = 0\,)}{t^2} + \frac{p\,a}{2\,t} = \frac{6\,M}{t^2} + \frac{p\,a}{2\,t} \quad . \tag{12b}$$

Circumferential stress :

$$\sigma_{\vartheta\vartheta_1} = \frac{N_{\vartheta\vartheta 1}}{t} \pm \nu\,\frac{6\,M_{xx1}}{t^2} = \frac{p\,a}{t} - \frac{2\,\kappa^2}{a\,t}\,M\,(\cos\kappa\,\xi + \sin\kappa\,\xi)\,e^{-\kappa\,\xi} \pm$$

$$\pm\,\nu\,\frac{6\,M}{t^2}\,(\cos\kappa\,\xi - \sin\kappa\,\xi)\,e^{-\kappa\,\xi} \quad .$$

From

$$\frac{d\sigma_{\vartheta\vartheta 1}}{d\xi} = 0 \longrightarrow \xi = 0.35$$

we obtain

$$\sigma_{\vartheta\vartheta_1}(\,\xi = 0.35\,) = \sigma_{\vartheta\vartheta_{max}} \quad .$$

Numerical values:

$$\sigma_{xx_{max}} \approx 156\ MPa\ ,$$

$$\sigma_{\vartheta\vartheta_{max}} \approx 102\ MPa\ .$$

- *Ring 2*

Circumferential stress :

$$\sigma_{\vartheta\vartheta_2} \approx \frac{p_s\,a}{3\,t} + \frac{p\,a}{4\,t} + \frac{2\,R\,a}{4\,b\,t} = \frac{E\,\alpha_{T2}\,\Theta}{4} + \frac{p\,a}{4\,t} + \frac{\kappa\,M}{b\,t} \quad .$$

Numerical value :

$$\sigma_{\vartheta\vartheta_2} \approx 82\ MPa\ .$$

Exercise C-13-3:

A pressure boiler made of steel consists of a circular cylindrical shell (radius a, wall thickness t) closed at each end by two semi-spherical shells (Fig. C-20). The boiler is subjected to a constant internal overpressure p (the deadweight of the boiler can be neglected).

Determine the curves for the stress resultants both in the cylindrical shell and the semi-spherical shells.

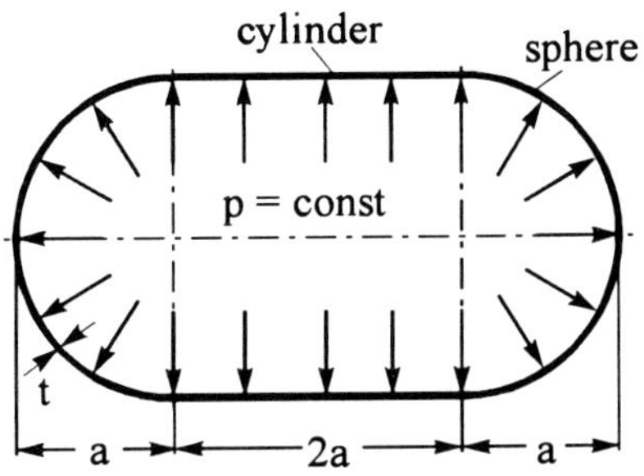

Fig. C-20: Pressure boiler

Numerical values: $\quad p = 10\ MPa \quad , \quad a = 2\ m \quad , \quad t = 0.1\ m \quad ,$

$$E = 2.1 \cdot 10^5\ MPa \quad , \quad \nu = \frac{1}{3} \quad .$$

Solution:

Owing to the symmetry we only consider one half of the pressure boiler. As in the previous exercise **C-13-2**, we partition the spherical shell from the cylindrical shell and mark the single loads according to the "0"-, "1"- and "2"-systems in Fig. C-21.

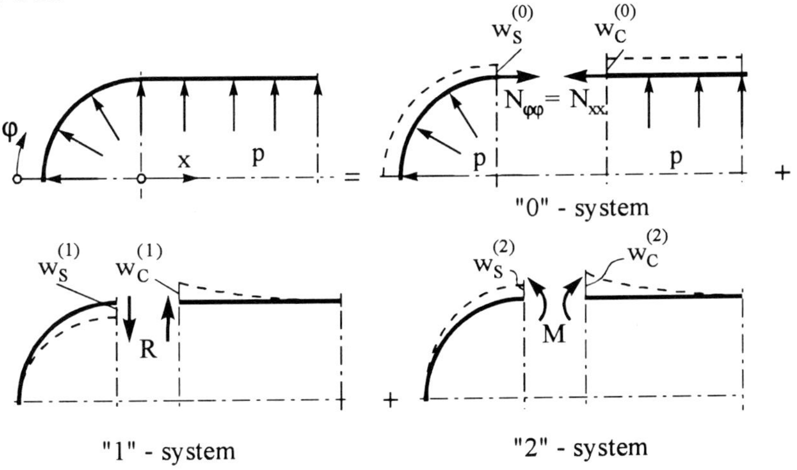

Fig. C-21: Partitioning of the pressure boiler in subsystems

Here, the compatibility conditions for displacements and rotations at the interface become:

$$w_S^{(0)} + w_S^{(1)} + w_S^{(2)} = w_C^{(0)} + w_C^{(1)} + w_C^{(2)} \quad , \tag{1a}$$

$$\chi_S^{(0)} + \chi_S^{(1)} + \chi_S^{(2)} = \chi_C^{(0)} + \chi_C^{(1)} + \chi_C^{(2)} \quad . \tag{1b}$$

We substitute the single deformation values for the boundary loads (see Fig. C-21); the deformations of the "0"-system are membrane solutions of the cylindrical and the spherical shell:

$$\frac{p\,a^2}{2\,E\,t}(1-\nu) - \frac{R\,a^3}{2\,K\,\kappa^3} + \frac{M\,a^2}{2\,K\,\kappa^2} = \frac{p\,a^2}{2\,E\,t}(2-\nu) + \frac{R\,a^3}{2\,K\,\kappa^3} + \frac{M\,a^2}{2\,K\,\kappa^2} \quad , \tag{2a}$$

$$0 + \frac{R\,a^2}{2\,K\,\kappa^2} - \frac{M\,a}{K\,\kappa} = 0 + \frac{R\,a^2}{2\,K\,\kappa^2} + \frac{M\,a}{K\,\kappa} \quad . \tag{2b}$$

Eq. (2b) immediately yields

$$M = 0 \quad . \tag{3a}$$

Owing to the fact that the semi-sphere and the cylindrical shell exhibit the same deformation behaviour at their boundaries when subjected to boundary forces, and because no twisting angle χ of the boundaries occurs subject to internal compression, the compatibility of the deformations can be introduced by the transverse boundary forces alone. Eq. (2a) then leads to:

$$R = -\frac{p\,a}{8\,\kappa} \quad . \tag{3b}$$

The curves for the resultant forces as a function of $\xi = x/a$ can be determined by means of the relations *(13.17)*:

$$N_{\vartheta\vartheta} = p\,a\left(1 - \frac{1}{4}e^{-\kappa\xi}\cos\kappa\xi\right) \quad , \quad Q_x = \frac{p\,a}{8\,\varkappa}e^{-\kappa\xi}(\cos\kappa\xi - \sin\kappa\xi) \quad ,$$

$$M_{xx} = \frac{p\,a^2}{8\,\kappa^2}e^{-\kappa\xi}\sin\kappa\xi \quad , \qquad N_{xx} = \frac{p\,a}{2} \quad .$$

We then calculate the resultant forces in the semi-spheres by means of *(13.18)*:

$$N_{\vartheta\vartheta} = \frac{p\,a}{2}\left(1 + \frac{1}{2}e^{-\kappa\omega_1}\cos\kappa\,\omega_1\right) \quad ,$$

$$Q_\varphi = -\frac{p\,a}{8\,\varkappa}e^{-\kappa\omega_1}(\cos\kappa\,\omega_1 - \sin\kappa\,\omega_1) \quad ,$$

$$M_{\varphi\varphi} = \frac{p\,a^2}{8\,\kappa^2}e^{-\kappa\omega_1}\sin\kappa\,\omega_1 \quad , \qquad N_{\varphi\varphi} \approx \frac{p\,a}{2} \quad .$$

Fig. C-22 shows the behaviour of the stress resultants around the transition between the cylindrical and the semi-spherical shell.

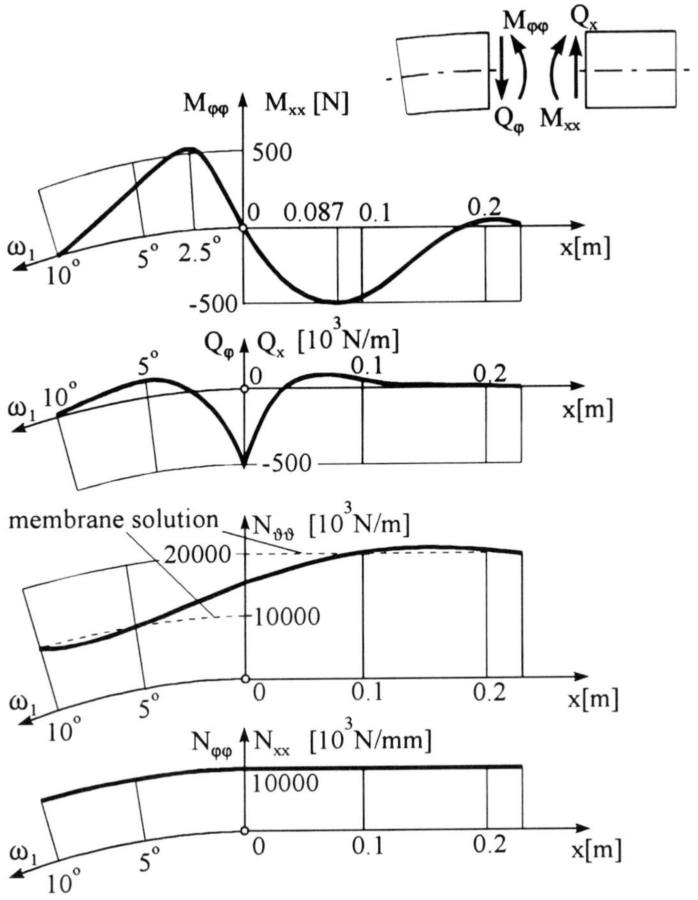

Fig. C-22: Resultant forces and moments in cylindrical and semi-spherical shell

Exercise C-13-4:

A thin-walled circular cylindrical tube made of steel (radius a, wall thickness t) as shown in Fig. C-23 is horizontally supported between two rigid walls in such a way that the cross-sections at both ends of the tube are completely clamped.

Determine the stresses in the tube due to its specific deadweight ϱg, after removal of the mounting equipment which ensures an initial stress-free state of the tube. Use the following numerical values :

$$l = 10\ m\ , \qquad a = 1\ m\ , \qquad t = 1 \cdot 10^{-2}\ \mathrm{m}\ ,$$

$$E = 2.1 \cdot 10^{5}\ MPa\ , \qquad \nu = 0.3\ , \qquad \varrho g = 8 \cdot 10^{4}\ N/m^{3}\ .$$

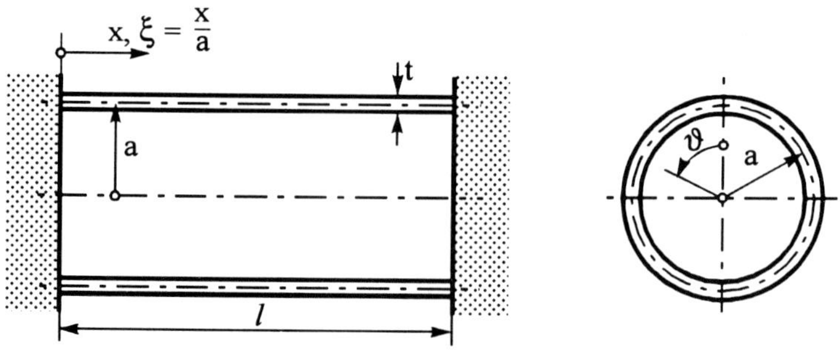

Fig. C-23: Circular cylindrical tube clamped horizontally at both ends

Solution :

The complete solution is determined by superposition of a membrane solution (Ch. 12) and the solution of the boundary disturbance problem (Ch. 13).

Membrane solution (denoted by superscript 0)

Using the abbreviated notation γ for the deadweight $\rho g t$ per unit area of the midsurface, the following surface loads are acting on the shell (see Fig. C-24).

$$p_x = 0 \quad ,$$

$$p_\vartheta = \gamma \sin \vartheta \quad ,$$

$$p = -\gamma \cos \vartheta \quad .$$

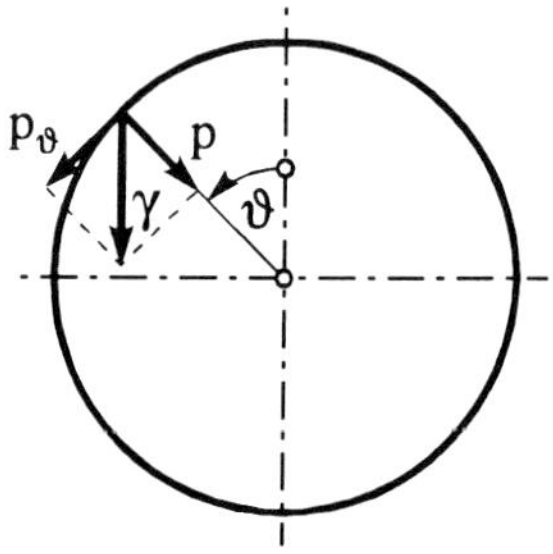

Fig. C-24: Components of the deadweight within the shell

We obtain the following resultant forces by substituting the loads into the equilibrium conditions *(12.14)* and by defining $\xi = \frac{x}{a}$

$$N^0_{\vartheta\vartheta} = -\gamma a \cos \vartheta \quad , \tag{1a}$$

$$N^0_{x\vartheta} = -(2\gamma a \xi + D_1) \sin \vartheta \quad , \tag{1b}$$

$$N^0_{xx} = (\gamma a \xi^2 + D_1 \xi + D_2) \cos \vartheta \quad . \tag{1c}$$

Based on *(12.23)* and *(12.26)*, we write

$$u_{,\xi} = \frac{a}{Et}(N_{xx} - \nu N_{\vartheta\vartheta}) \quad , \tag{2a}$$

$$v_{,\vartheta} + w = \frac{a}{Et}(N_{\vartheta\vartheta} - \nu N_{xx}) \quad , \tag{2b}$$

$$u_{,\vartheta} + v_{,\xi} = \frac{2(1+\nu)a}{Et} N_{x\vartheta} \quad . \tag{2c}$$

Substituting (1) into (2) and integrating, we obtain the membrane displacements

$$u^0 = \frac{a}{Et}\left[\gamma a\left(\frac{\xi^3}{3} + \nu\xi\right) + D_1\frac{\xi^2}{2} + D_2\xi + D_3\right]\cos\vartheta \quad , \tag{3a}$$

$$v^0 = \frac{a}{Et}\left[\gamma a\left(\frac{\xi^4}{12} - \frac{\xi^2}{2}(4 + 3\nu)\right) + D_1\left(\frac{\xi^3}{6} - 2(1+\nu)\xi\right) + D_2\frac{\xi^2}{2} + \right.$$
$$\left. + D_3\xi + D_4\right]\sin\vartheta \quad , \tag{3b}$$

$$w^0 = -\frac{a}{Et}\left[\gamma a\left(\frac{\xi^4}{12} - \frac{\xi^2}{2}(4+\nu) + 1\right) + D_1\left(\frac{\xi^3}{6} - (2+\nu)\xi\right) + \right.$$
$$\left. + D_2\left(\frac{\xi^2}{2} + \nu\right) + D_3\xi + D_4\right]\cos\vartheta \quad . \tag{3c}$$

The integration constants D_i $(i = 1, \ldots, 4)$ can only be determined from the complete solution of the problem.

Bending solution (denoted by superscript 1)

Since the membrane solution depends on the circumferential coordinate ϑ via $\cos\vartheta$ or $\sin\vartheta$, respectively, the bending solution of a shell clamped at its boundaries possesses terms with $m = 1$ only. The eigenvalue equation thus reduces to the following characteristic equation dealt with in detail in [ET2 | 11.3.2]:

$$\lambda^8 - 2(2-\nu)\lambda^6 + \frac{1-\nu^2}{k}\lambda^4 = 0 \tag{4}$$

with the shell parameter k defined by *(13.31d)*.

The characteristic equation has the roots

$$\lambda^2_{1,2} = 2 - \nu \pm \sqrt{(2-\nu)^2 - \frac{1-\nu^2}{k}} = 2 - \nu \pm i\sqrt{\frac{1-\nu^2}{k} - (2-\nu)^2} \quad .$$

Since $\frac{1-\nu^2}{k} \gg (2-\nu)^2$, these roots may be approximated by

$$\lambda_{1,2,3,4} = \pm\mu_1 \pm i\mu_1 \quad \text{with} \quad \mu_1 = \sqrt{\frac{1}{2}\sqrt{\frac{1-\nu^2}{k}}} \quad . \tag{5}$$

The characteristic equation (4) has four additional eigenvalues $\lambda_{5,6,7,8} = 0$. The corresponding solutions are already included in the membrane solution (3), and therefore they do not need to be considered in the homogeneous solution.

The shell shall have a sufficient length so that no mutual influence of the boundary disturbances occurs. We therefore exclusively consider the boundary $\xi = 0$, and by including (5) we obtain the following homogeneous solution:

$$u^1 = \left(A_1 e^{i\mu_1\xi} + A_2 e^{-i\mu_1\xi}\right)e^{-\mu_1\xi}\cos\vartheta \quad ,$$

$$v^1 = \left(B_1 e^{i\mu_1\xi} + B_2 e^{-i\mu_1\xi}\right)e^{-\mu_1\xi}\sin\vartheta \quad , \tag{6}$$

$$w^1 = \left(C_1 e^{i\mu_1\xi} + C_2 e^{-i\mu_1\xi}\right)e^{-\mu_1\xi}\cos\vartheta \quad .$$

The complex constants A_j, B_j, C_j $(j = 1, 2)$ are coupled to each other via a homogeneous system of equations. The first equation (for $m = 1$, $k \ll 1$) yields

$$\left(\lambda_j^2 - \frac{1-\nu}{2}\right) A_j + \frac{1+\nu}{2}\lambda_j B_j + \nu\lambda_j C_j = 0 \ ,$$

$$\frac{1+\nu}{2}\lambda_j A_j + \left(1 - \frac{1-\nu}{2}\lambda_j^2\right) B_j + C_j = 0 \ ,$$

and for $j = 1$ with $\lambda_1 = -\mu_1 + i\mu_1$, we obtain the following dependencies of the constants:

$$\begin{aligned} A_1 &= \frac{1}{4\mu_1^3}\left[-1 + 2\nu\mu_1^2 + i(1 + 2\nu\mu_1^2)\right] C_1 = (\alpha_1 + i\alpha_2) C_1 \ , \\ B_1 &= \frac{1}{4\mu_1^2}\left[\frac{1}{\mu_1^2} + i(2+\nu)\right] C_1 \qquad\qquad = (\beta_1 + i\beta_2) C_1 \ . \end{aligned} \tag{7}$$

Since $\lambda_2 = -\mu_1 - i\mu_1$, the conjugate complex relations for $j = 2$ follow as

$$A_2 = (\alpha_1 - i\alpha_2) C_2 \quad , \quad B_2 = (\beta_1 - i\beta_2) C_2 \ . \tag{8}$$

If we substitute (7) and (8) into (6), all displacements depend on C_1 and C_2 only.

Boundary conditions

If we consider the boundary $\xi = 0$ only in the case of the membrane solution, the two boundary conditions for $\xi = 0$ and $\xi = l/a$ have to be replaced by two symmetry conditions for $\xi = l/2a$. We thus obtain from

$$N_{x\vartheta}^0\left(\frac{l}{2a}\right) = 0 \quad \text{und} \quad u^0\left(\frac{l}{2a}\right) = 0$$

with (1b) and (3a)

$$D_1 = -\gamma l \quad , \quad D_3 = -D_2\frac{l}{2a} + \gamma a\left[\frac{1}{12}\left(\frac{l}{a}\right)^3 - \nu\frac{l}{2a}\right] \ .$$

The remaining four constants C_1, C_2, D_2 and D_4 result from the four boundary conditions

$$\begin{aligned} u(0) &= u^0(0) + u^1(0) &&= 0 \ , \\ v(0) &= v^0(0) + v^1(0) &&= 0 \ , \\ w(0) &= w^0(0) + w^1(0) &&= 0 \ , \\ w_{,\xi}(0) &= w^0_{,\xi}(0) + w^1_{,\xi}(0) &&= 0 \ . \end{aligned}$$

After carrying out the numerical calculation with the given values we obtain the circumferential force as

$$N_{\vartheta\vartheta} = N^0_{\vartheta\vartheta} + N^1_{\vartheta\vartheta} = \left[-8 + (47.3\cos 12.9\,\xi + 5.02\sin 12.9\,\xi)\, e^{-12.9\xi}\right]\cos\vartheta \ .$$

Fig. C-25 shows the membrane forces according to (1) and the bending moments M^1_{xx} und $M^1_{\vartheta\vartheta}$ acting along the top longitudinal line $\vartheta = 0$ of the shell. One can see how fast the bending disturbance has decayed already at a distance of ~ 0.4 m from the boundary. The stresses are calculated from

$$\sigma_{xx} = \frac{N_{xx}}{t} \pm \frac{6\,M_{xx}}{t^2} \quad , \quad \sigma_{\vartheta\vartheta} = \frac{N_{\vartheta\vartheta}}{t} \pm \frac{6\,M_{\vartheta\vartheta}}{t^2} \quad . \qquad (9)$$

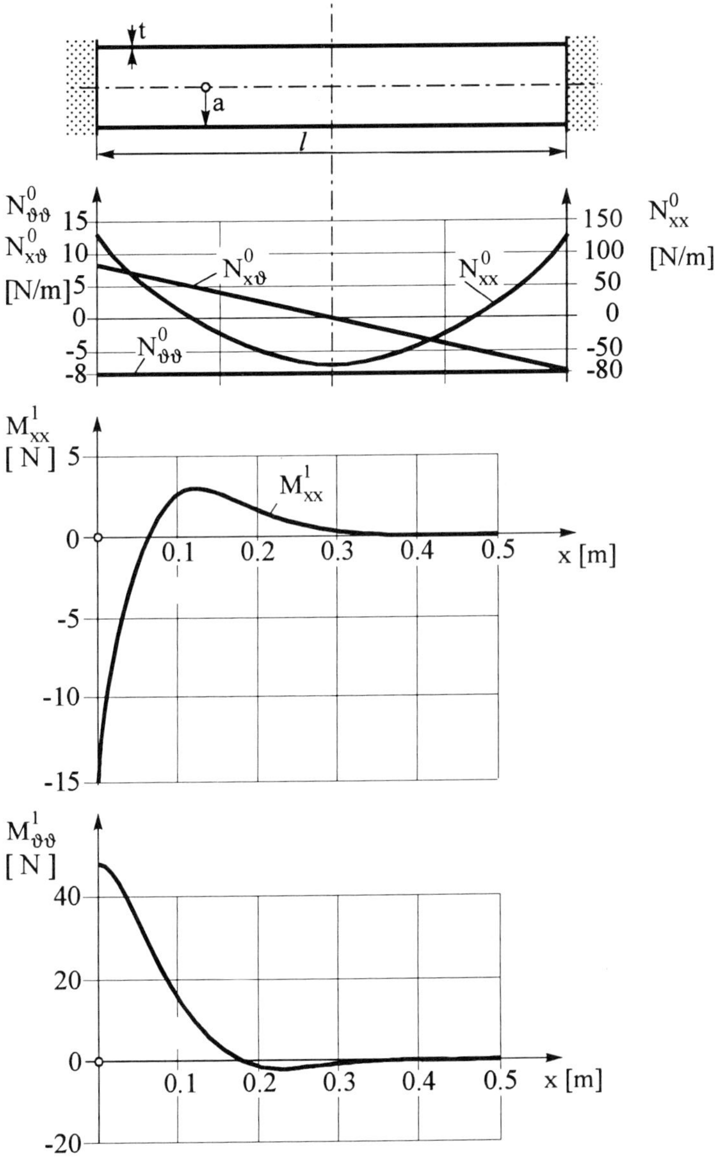

Fig. C-25: Membrane forces and bending moments along the top longitudinal line of the cylindrical tube under deadweight

The maximum stresses at the boundary are due to (9)

$$\sigma_{xx_{max}} = 1.31 \ (\pm) \ 0.92 = 2.23 \ MPa \quad ,$$

$$\sigma_{\vartheta\vartheta_{max}} = 0.39 \ (\pm) \ 0.28 = 0.67 \ MPa \quad .$$

The numerical values show that both the longitudinal and the circumferential stresses due to the boundary disturbances are of similar magnitude as the membrane stresses.

Exercise C-13-5:

A circular cylindrical shell (a , l = 4 a , t = a / 400) is subjected to a constant external pressure p (Fig. C-26).

Formulate the basic equation for shell buckling in analogy with the basic equation of plate buckling (see *(10.17)*).

Determine then the critical load for the special case of a shell which is simply supported at both ends.

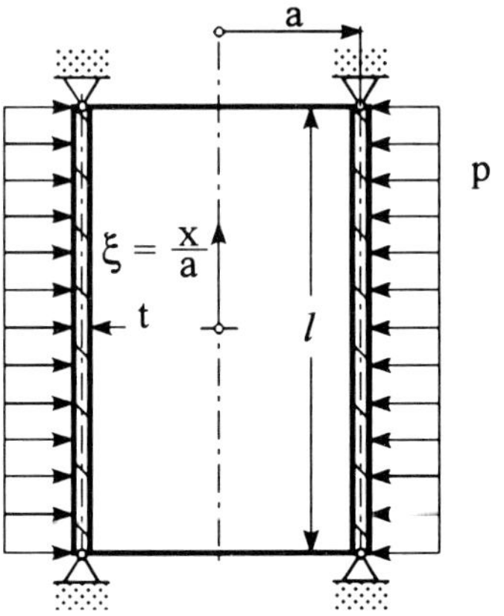

Fig. C-26: Circular cylindrical shell under external pressure

Solution :

We proceed from the simplified basic equations for a shear-rigid shell (DONNELL´s theory). Before buckling the initial stress state prevails within the shell

$$N_{xx} = 0 \quad , \quad N_{yy} = -\,p\,a \quad , \quad N_{xy} = 0 \quad . \tag{1}$$

At buckling the component $N_{yy}\, w_{,\vartheta\vartheta}$ must be included in the equilibrium condition in the radial direction of the deformed shell . Then, u and v can be eliminated, and we obtain in analogy with *(13.39)*

$$k\,\Delta\Delta\Delta\Delta\, w + (1-\nu^2)\, w_{,\xi\xi\xi\xi} + \frac{pa}{D}\Delta\Delta\, w_{,\vartheta\vartheta} = 0 \tag{2}$$

as the basic *equation of shell buckling under external pressure.*

In order to determine the critical load, we put the coordinate x in the centre of the cylinder. The approximation

$$w = W \cos\frac{m\pi a \xi}{l} \cos\frac{ny}{a} \qquad (m\,,n = 1\,,2\,,3\ldots) \quad , \tag{3}$$

fulfills the boundary conditions of the simply supported shell in the longitudinal direction

$$w\left(\pm\frac{l}{2a}\right)=0\ ,\quad M_{xx}\left(\pm\frac{l}{2a}\right)=0\ ,\quad N_{xx}\left(\pm\frac{l}{2a}\right)=0\ ,\quad v\left(\pm\frac{l}{2a}\right)=0\ ,$$

and the condition of periodicity in the circumferential direction

$$w(2\pi a)=w(0)\ . \qquad (4)$$

By substituting (3) into (2) and by defining $\lambda=\frac{m\pi a}{l}$, we obtain the relation for the critical load as

$$\frac{pa}{D}(\lambda^2+n^2)n^2=k(\lambda^2+n^2)^4+(1-\nu^2)\lambda^4 \qquad (5a)$$

or with $\quad \bar{p}=\frac{pa}{D}\ ,\quad \bar{p}=k\frac{(\lambda^2+n^2)^2}{n^2}+(1-\nu^2)\frac{\lambda^4}{n^2(\lambda^2+n^2)^2}\ . \qquad (5b)$

We now have to determine that combination of m and n for which $\bar{p}$ has the smallest value. We can immediately see from (5b) that λ will attain its smallest value for $m=1$. The shell therefore buckles with one wave in the longitudinal direction.

Assuming that $\quad \alpha=\frac{\lambda^2}{n^2}=\left(\frac{\pi a}{nl}\right)^2\ ,$

we obtain

$$\bar{p}=k\,n^2(1+\alpha)^2+(1-\nu^2)\frac{\alpha^2}{n^2(1+\alpha)^2}=$$

$$=\frac{1-\nu^2}{\lambda^2}\left[\frac{k\lambda^4}{1-\nu^2}\frac{(1+\alpha)^2}{\alpha}+\frac{\alpha^3}{(1+\alpha)^2}\right]\ . \qquad (6a)$$

Assuming that many waves occur in the circumferential direction ($n\gg 1$), then it is valid for long shells $l\gg a$ that

$$\alpha\ll 1\ ,$$

and (6a) then reduces to

$$\bar{p}=\frac{1-\nu^2}{\lambda^2}\left(\frac{k\lambda^4}{1-\nu^2}\frac{1}{\alpha}+\alpha^3\right)\ . \qquad (6b)$$

The minimum value follows by differentiating $\bar{p}$

$$\frac{d\bar{p}}{d\alpha}=\frac{1-\nu^2}{\lambda^2}\left[\frac{k\lambda^4}{1-\nu^2}\left(-\frac{1}{\alpha^2}\right)+3\alpha^2\right]=0\quad\Longrightarrow\quad \alpha^*=\lambda\sqrt[4]{\frac{k}{3(1-\nu^2)}}\ . \qquad (7)$$

Substituting (7) into (6b) yields after elementary re-transformations:

$$p_{crit}=p(\alpha^*)=\frac{Et}{1-\nu^2}\,\frac{4}{3}\,\frac{\pi}{l}\,\sqrt[4]{3(1-\nu^2)}\left(\frac{t^2}{12a^2}\right)^{3/4} \qquad (8a)$$

or with $\nu=0.3$

$$p_{crit}\approx 0.92\,E\,\frac{a}{l}\left(\frac{t}{a}\right)^{5/2}\ . \qquad (8b)$$

Eq. (7) delivers the corresponding number of waves as

$$\alpha^* = \frac{\lambda^2}{n^2} = \lambda \sqrt[4]{\frac{k}{3(1-\nu^2)}} \quad \longrightarrow \quad n^2 \approx 7.3 \frac{a}{l} \sqrt{\frac{a}{t}} \quad .$$

Using the given numerical values $l = 4a$ and $t = a/400$, we obtain

$$n^2 \approx 7.3 \frac{1}{4} \sqrt{400} = 36.5 \quad , \quad n = 6.04 \text{ and } p_{crit} \approx 7.19\, E \cdot 10^{-8}\, MPa \quad .$$

Owing to the necessary periodicity in the circumferential direction, the following adjacent integer number at buckling is $n = 6$; in the numerical example follows $\alpha^* \approx 0.017$ and therefore $\ll 1$.

Exercise C-13-6:

Determine the eigenfrequencies of the free vibrations of a circular cylindrical shell with simply supported ends as shown in Fig. C-26 (without external pressure p). Assume small vibration amplitudes (linear theory) and solve the exercise using DONNELL's theory.

Note: The coordinate system has in this case been moved to the lower boundary.

Solution:

In order to treat the circular cylindrical shell, the basic equations *(13.31)* are simplified in accordance with DONNELL's theory (see *(13.38)*):

$$u_{,\xi\xi} + \frac{1-\nu}{2} u_{,\vartheta\vartheta} + \frac{1+\nu}{2} v_{,\xi\vartheta} + \nu w_{,\xi} = -\frac{a^2 p_x}{D} \quad , \tag{1a}$$

$$\frac{1+\nu}{2} u_{,\xi\vartheta} + v_{,\vartheta\vartheta} + \frac{1-\nu}{2} v_{,\xi\xi} + w_{,\vartheta} = -\frac{a^2 p_y}{D} \quad , \tag{1b}$$

$$\nu u_{,\xi} + v_{,\vartheta} + w + k \Delta\Delta w = \frac{a^2 p}{D} \quad . \tag{1c}$$

As "loadings" we write D'ALEMBERT's inertia forces:

$$p_x = -\rho t \frac{\partial^2 u}{\partial \tau^2} \quad , \quad p_y = -\rho t \frac{\partial^2 v}{\partial \tau^2} \quad , \quad p = -\rho t \frac{\partial^2 w}{\partial \tau^2} \quad , \tag{2}$$

where ρ denotes the mass density and τ the time (τ is introduced in order to avoid confusion with the wall-thickness t). This approximate theory neglects the rotational inertia of the shell; its consideration would result in additional, very high frequencies, whereas its influence on the lower frequencies considered here is negligible (see [C.21, C.22, C.23]).

The eigenfrequencies are determined via separation of variables:

$$\left.\begin{aligned} u &= \bar{u}(x,\varphi) \sin \omega\tau \quad , \\ v &= \bar{v}(x,\varphi) \sin \omega\tau \quad , \\ w &= \bar{w}(x,\varphi) \sin \omega\tau \quad . \end{aligned}\right\} \tag{3}$$

Substitution of (3) into (1a,b,c) and (2) yields elimination of time, and we obtain the equations

$$\left.\begin{aligned} &\lambda\bar{u} + \bar{u}_{,\xi\xi} + \frac{1-\nu}{2}\bar{u} + \frac{1+\nu}{2}\bar{v}_{,\xi} + \nu\bar{w}_{,\xi} = 0 \ , \\ &\frac{1+\nu}{2}\bar{u}_{,\xi} + \lambda\bar{v} + \frac{1-\nu}{2}\bar{v}_{,\xi\xi} + \bar{v} + \bar{w} = 0 \ , \\ &\nu\bar{u} + \bar{v}_{,\xi} - \lambda\bar{w} + k\Delta\Delta\bar{w} = 0 \end{aligned}\right\} \tag{4}$$

with the frequency parameter

$$\lambda = \frac{\rho a^2 (1-\nu^2)}{E}\omega^2 \ . \tag{5}$$

In (4), $\bar{u}$, $\bar{v}$, $\bar{w}$ are position functions $\bar{u}(x,\varphi)$, etc. The following assumption regarding the form of these functions

$$\left.\begin{aligned} &\bar{u} = U\cos n\varphi \cos \bar{m}\frac{x}{a} \ , \\ &\bar{v} = V\sin n\varphi \sin \bar{m}\frac{x}{a} \ , \\ &\bar{w} = W\cos n\varphi \sin \bar{m}\frac{x}{a} \qquad \text{with} \quad \bar{m} = \frac{m\pi a}{l} \end{aligned}\right\} \tag{6}$$

fulfills all boundary conditions for a shell with simply supported ends ($\bar{w} = \bar{v} = \bar{M}_x = \bar{N}_x = 0$ for $x = 0$ and $x = l$). Substitution into (4) yields a homogeneous system of equations for the unknown amplitudes U, V, W. By setting the determinant of the coefficients equal to zero, one obtains the characteristic equation of an eigenvalue problem

$$\lambda^3 - A\lambda^2 + B\lambda - C = 0 \ , \tag{7}$$

where the coefficients A, B, and C depend on the dimensions of the shell as well as on m and n.

Numerical evaluation of (7) shows that each pair of values of m and n defines one lower and two higher eigenvalues which exceed the lower ones by powers of ten. The technically relevant lower eigenvalue can thus be approximated from (7):

$$\lambda_1 = \frac{C}{B} \ . \tag{8}$$

The numerical values additionally show that λ_1 is associated with a pronounced transverse vibration ($W \gg U, V$), while longitudinal vibrations are associated with λ_2 and λ_3 ($U, V \gg W$). Based upon this observation, the lowest frequency can be governed by a single formula.

If the amplitudes U and V are very small, the terms of inertia forces in the longitudinal and circumferential direction can be neglected in equation (4). Consequently, the displacements u and v can be eliminated from the first two equations, using the same procedure as in *(13.38)*. Thereby, *(13.39)* is augmented by the w-inertia term, and we obtain considering the vibration approach

$$k\Delta\Delta\Delta\Delta\bar{w} + (1-\nu^2)\bar{w}^{IV} - \lambda\Delta\Delta\bar{w} = 0 \ . \tag{9}$$

With (6) this yields an approximation for the lowest eigenfrequency:

$$\lambda_1 = (1-\nu^2)\frac{\bar{m}^4}{(\bar{m}^2 + n^2)^2} + k(\bar{m}^2 + n^2)^2 \ . \tag{10}$$

The results according to (8) and (10) are numerically almost identical.

The eigenfrequency equation (10) consists of two termes, where the first stems from extensional vibrations and the second from bending vibrations. Fig. C-27 presents a numerical example, where both terms are drawn separately in dependence on n. The curves clearly illustrate that for a small number n of circumferential waves extensional vibrations predominantly occur, and for large n bending vibrations, respectively. Close to the minimum, the two terms are approximately equal. Therefore, *no* further simplifications must be performed for the simply supported shell. Elimination of the first term in (9), for instance, cannot be admitted since this would correspond to an inextensional vibration. If deformations which are incompatible with the assumption of inextensional bending are prescribed at the boundary, bending and extension will act jointly, and thus have to be considered by a complete shell theory.

The minimum of λ_1 can be calculated by a formula. Eq. (10) implies that λ_1 increases with $\overline{m}$. It attains its smallest value for m = 1, i.e. the shell vibrates with one wave in the longitudinal direction. λ_1 then only depends on n. If the actually discrete number of waves n is assumed to be continuous, (10) can be differentiated with respect to n:

$$\frac{d\lambda_1}{d(\overline{m}^2+n^2)^2} = -(1-\nu^2)\frac{\overline{m}^4}{(\overline{m}^2+n^2)^4} + k = 0 \quad .$$

It follows that

$$(\overline{m}^2+n^2)^2 = \overline{m}^2\sqrt{\frac{(1-\nu^2)}{k}} \quad . \tag{11}$$

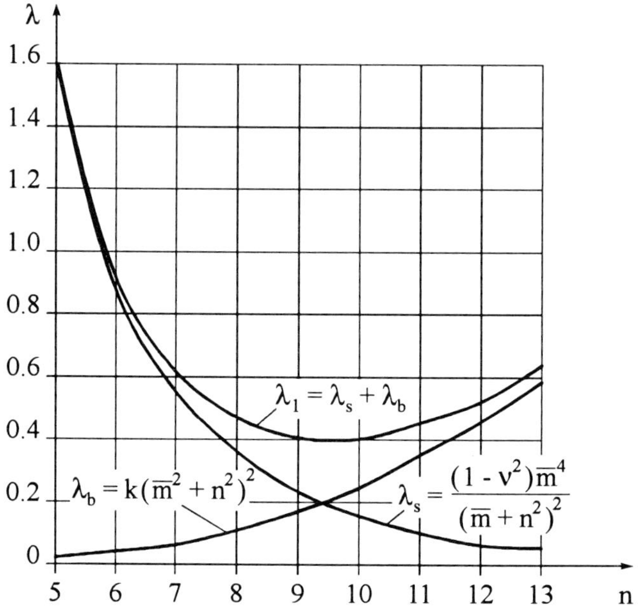

Fig. C-27: Lowest eigenfrequency of a simply supported cylindrical shell

Substitution yields (the two terms in (10) result in the same value)

$$\lambda_{1_{Min}} = 2\sqrt{(1-\nu^2)k}\;\bar{m}^2$$

or, with the frequency parameter

$$\omega^2_{Min} = 2\,\frac{E}{\rho a^2}\sqrt{\frac{k}{(1-\nu^2)}}\left(\frac{\pi a}{l}\right)^2 \quad . \qquad (12)$$

This proves that the lowest frequency depends on all dimensions and on the material data. From (11) we determine the number of waves n assigned to the minimum. In practice, the adjacent integer value of n would occur. In our example, the shell vibrates with nine circumferential waves; using the given numerical values, (11) would give the value n = 9.22.

For a shell with free boundaries, inextensional bending may be assumed as a possible vibration mode. For this case, Lord RAYLEIGH determined, by equalling kinetic and elastic energy, that

$$\lambda = k\,\frac{n^2(n^2-1)^2}{n^2+1} \quad . \qquad (13)$$

We obtain the same result by defining $m \longrightarrow 0$ in (10). Here, the differences in dependence on n are a result of the DONNELL simplifications. For larger n, these differences are unimportant ($\lambda = k \cdot n^4$).

Exercise C-14-1:

A spherical cap (Fig. C-28) is extended over a circular base (radius r_0, $r_0 << a$), and is assumed to be subjected to a constant surface pressure load p. The height of the cap is given as f.

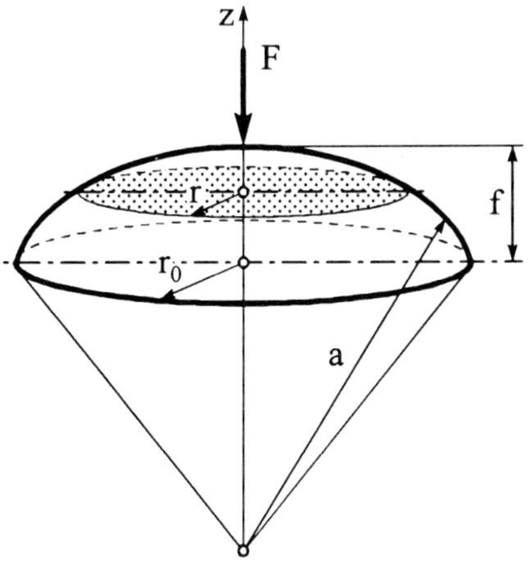

Fig. C-28: Spherical cap over a circular base

As $r_0 << a$, the mid-surface of the spherical cap can be approximated by

$$z(r) \approx f - \frac{x^2 + y^2}{2a} = f - \frac{r^2}{2a} , \tag{1}$$

implying constant curvatures everywhere, i.e. $\kappa_x = \kappa_y = -\frac{1}{a}$.

a) State the differential equation and obtain the solution of the homogeneous equation for an axisymmetrical problem.

b) Calculate the deflection of the shallow spherical cap when subjected to a concentrated force F at the top point and assuming that $r \longrightarrow 0$.

Solution:

a) With the given assumptions, the system of differential equations *(14.9)* reads

$$\Delta\Delta w + \frac{1}{Ka}\Delta\Phi = \frac{p}{K} , \tag{2a}$$

$$\Delta\Delta\Phi - \frac{Et}{a}\Delta w = 0 . \tag{2b}$$

We now multiply (2b) by a factor λ and add it to (2a). We thus obtain

$$\Delta\Delta(w + \underline{\lambda\Phi}) - \frac{Et}{a}\lambda\Delta\left(w - \underline{\frac{1}{\lambda EtK}}\Phi\right) = \frac{p}{K} . \tag{3}$$

If the underscored terms in (3) are equal, one can formulate a differential equation for

$$F = w + \lambda\Phi . \tag{4}$$

Introducing i as the imaginary unit, we write

$$\lambda = \frac{i}{Et^2}\sqrt{12(1-\nu^2)} . \tag{5a}$$

With the abbreviation k we obtain

$$\frac{Et}{a}\lambda = \frac{i}{ta}\sqrt{12(1-\nu^2)} = ik^2 . \tag{5b}$$

By (5b) and (4), (3) transforms into

$$\boxed{\Delta\Delta F - ik^2\Delta F = \frac{p}{K}} . \tag{6}$$

Here, our considerations will be restricted to the homogeneous solution of (6). Then, the differential equation can be split as follows:

$$(\Delta - ik^2)\Delta F = \Delta(\Delta - ik^2)F = 0 . \tag{7}$$

In the present case it is sensible to use polar coordinates owing to the axisymmetry of shell and load. The LAPLACE-operator is then independent of the angular coordinate ϑ and hence

$$\Delta = \frac{d^2}{dr^2} + \frac{1}{r}\frac{d}{dr} \quad .$$

We can thus determine partial solutions from the two differential equations:

$$\Delta F = 0 \quad , \tag{8a}$$

$$\Delta F - i k^2 F = 0 \quad . \tag{8b}$$

The solution to (8a) can be stated immediately as

$$F_1 = C_1 + C_2 \ln r \quad , \tag{9a}$$

while (8b) is a BESSEL differential equation [B.3] of the form

$$\frac{d^2F}{dr^2} + \frac{1}{r}\frac{dF}{dr} - i k^2 F = 0 \quad . \tag{10}$$

Solutions to (10) are modified cylinder functions of first and second type, $I_0(kr\sqrt{i}\,)$ and $K_0(kr\sqrt{i}\,)$, respectively, that are linearly independent [B.3]:

$$F_2 = C_3 I_0(kr\sqrt{i}\,) + C_4 K_0(kr\sqrt{i}\,) \quad , \tag{9b}$$

where C_3 and C_4 are complex constants.

According to KELVIN, two new functions $\mathrm{ber}(kr)$ and $\mathrm{bei}(kr)$ can be introduced that correspond to the real and the imaginary part of $I_0(kr\sqrt{i}\,)$, respectively, as well as the functions $\mathrm{ker}(kr)$ and $\mathrm{kei}(kr)$ which are equal to the real and imaginary part of $K_0(kr\sqrt{i}\,)$ [B.3]:

$$\left.\begin{aligned} I_0(kr\sqrt{i}\,) &= \mathrm{ber}(kr) + i\,\mathrm{bei}(kr) \quad , \\ K_0(kr\sqrt{i}\,) &= \mathrm{ker}(kr) + i\,\mathrm{kei}(kr) \quad . \end{aligned}\right\} \tag{11}$$

The reader is referred to standard tables, e.g. [B.3], for graphs of the KELVIN functions.

The general solution to (6) then consists of a linear combination of F_1 according to (9a) and of F_2 according to (9b). If we substitute the solution into (4) and compare the coefficients, considering the complex constants, we obtain from (11) the following terms for the bending w and for AIRY's stress function Φ:

$$\begin{aligned} w = {} & B_1\,\mathrm{ber}(kr) + B_2\,\mathrm{bei}(kr) + B_3\,\mathrm{ker}(kr) + \\ & + B_4\,\mathrm{kei}(kr) + B_5 + B_6 \ln r \quad , \end{aligned} \tag{12a}$$

$$\begin{aligned} \Phi = \frac{E t^2}{\sqrt{12(1-\nu^2)}} & \Big[- B_1\,\mathrm{bei}(kr) + B_2\,\mathrm{ber}(kr) - B_3\,\mathrm{kei}(kr) + \\ & + B_4\,\mathrm{ker}(kr) + B_7 + B_8 \ln r \Big] \quad , \end{aligned} \tag{12b}$$

where $B_1 \dots B_8$ are real constants.

b) In the following, the deformation in the middle of the shallow spherical cap shall be considered. For this purpose it is assumed that the boundary of the shell is very remote from the top point ($r \longrightarrow \infty$), and that the displacement w and its higher derivatives vanish at the boundary.
Under the given assumptions we write for $r \longrightarrow \infty$ ($d/dr \triangleq (\ )_{,r}$):

$$w = w_{,r} = w_{,rr} = 0 \ . \tag{13a}$$

In addition, for $r = 0$, i.e. at the top point

$$w\,,\ w_{,r}\,,\ N_{rr}\,,\ N_{\varphi\varphi} \quad \text{have to be finite}. \tag{13b}$$

The concentrated force F at the top point ($r = 0$) is equilibrated by a total vertical shear force V_z along any circle of radius r. Thus,

$$V_z = \frac{F}{2\pi r} \ , \tag{14}$$

where $V_z = Q_r + \frac{r}{a} N_{rr}$.

After evaluation of all conditions, we obtain the constants as follows:

$$B_1 = B_2 = B_3 = B_4 = B_5 = B_6 = B_7 = 0\,,\ \ B_4 = B_8 \ ,$$

$$B_8 = \frac{F\,a}{2\,\pi} \frac{\sqrt{12(1-\nu^2)}}{E\,t^2} \ .$$

The deflection function then reads

$$w = \frac{F\,a}{2\,\pi} \frac{\sqrt{12(1-\nu^2)}}{E\,t^2} \operatorname{kei}(k\,r) \ . \tag{15a}$$

The maximum deflection occurs at the top point where the load acts. For $r = 0$ we have $\operatorname{kei}(0) = -\pi/4$. This yields:

$$w_{max} = -\frac{1}{4}\sqrt{3(1-\nu^2)}\ \frac{F\,a}{E\,t^2} \ . \tag{15b}$$

In his fundamental papers, REISSNER has treated problems of shallow spherical shells with a number of load cases. For further details refer to [C.20].

Exercise C-14-2:

The eigenfrequencies of a hypar shell projected against a rectangular base (Fig. C-29) shall be determined. The distance f between base and shell is assumed to be small.

a) Set up the fundamental equations for the eigenfrequencies.

b) Which eigenfrequencies appear for the special case of a simply supported shell? Derive an approximate formula for the lowest frequency.

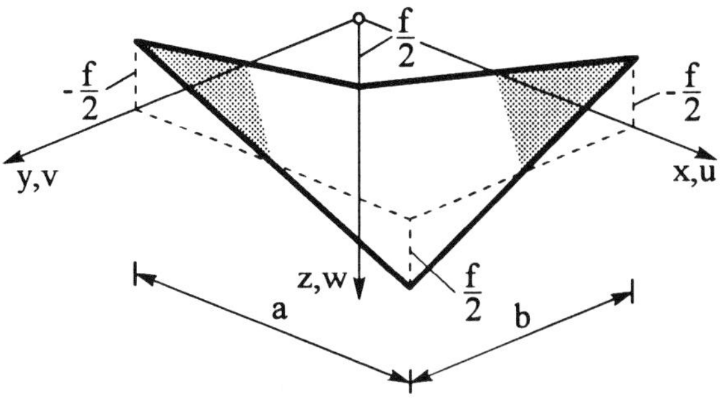

Fig. C-29: Hypar shell against a rectangular base

Solution :

a) For $f << a, b$ we can apply the fundamental equations from the theory of shallow shells. In this example, we replace the external loads in the equilibrium conditions *(14.3)* by D'ALEMBERT's forces of inertia (density ρ, time τ). Neglecting the rotational inertia, we obtain

$$\begin{aligned} N^{\alpha\beta}\big|_\alpha &= \rho\, t \frac{\partial^2 v_\beta}{\partial \tau^2} \ , \\ Q^\alpha\big|_\alpha + N^{\alpha\beta} z\big|_{\alpha\beta} &= \rho\, t \frac{\partial^2 w}{\partial \tau^2} \ , \\ M^{\alpha\beta}\big|_\alpha - Q^\beta &= 0 \ . \end{aligned} \tag{1}$$

Here, the only difference relative to the equilibrium conditions for the shallow cylindrical shell *(13.36)* is that instead of the circumferential force a component of the shear force occurs perpendicular to the shell.

The kinematic relations are obtained from *(14.4)*. By transforming the first equation of *(14.4)* by means of *(14.2)* we write

$$\alpha_{\alpha\beta} = \frac{1}{2}\left(v_\alpha\big|_\beta + v_\beta\big|_\alpha - 2\, z\big|_{\alpha\beta}\, w \right) \ , \quad \rho_{\alpha\beta} = -\, w\big|_{\alpha\beta} \ . \tag{2}$$

The relations between the resultant forces, moments, and the strains are described by the material law *(14.5)*

$$N^{\alpha\beta} = D\, H^{\alpha\beta\gamma\delta} \alpha_{\gamma\delta} \ , \qquad M^{\alpha\beta} = K\, H^{\alpha\beta\gamma\delta} \rho_{\gamma\delta} \ . \tag{3}$$

We assume that the boundaries of the hypar shell belong to the linear generatrices. The mid-surface can then be described in Cartesian coordinates as

$$z = \frac{2f}{ab}\left(x - \frac{a}{2}\right)\left(y - \frac{b}{2}\right) \ . \tag{4}$$

When expressed in Cartesian coordinates, all covariant derivatives simply become partial derivatives. By introducing the dimensionless coordinates $\xi = x/a$ and $\eta = y/b$, and denoting the derivatives by

$$\frac{\partial}{\partial \xi}(\,) \mathrel{\hat=} (\,)_{,\xi} \quad , \quad \frac{\partial}{\partial \eta}(\,) \mathrel{\hat=} (\,)_{,\eta}$$

we obtain with $\quad z|_{xx} = z|_{yy} = 0 \quad , \quad z|_{xy} = \frac{2f}{ab} \quad ,$

the fundamental equations (1) through (3) as

$$\left.\begin{aligned}
&\frac{1}{a} N_{xx,\xi} + \frac{1}{b} N_{xy,\eta} = \rho\, t \frac{\partial^2 u}{\partial \tau^2} \quad , \\
&\frac{1}{a} N_{xy,\xi} + \frac{1}{b} N_{yy,\eta} = \rho\, t \frac{\partial^2 v}{\partial \tau^2} \quad , \\
&\frac{1}{a} Q_{x,\xi} + \frac{1}{b} Q_{y,\eta} + \frac{4f}{ab} N_{xy} = \rho\, t \frac{\partial^2 w}{\partial \tau^2} \quad , \\
&\frac{1}{a} M_{xx,\xi} + \frac{1}{b} M_{xy,\eta} - Q_x = 0 \quad , \\
&\frac{1}{a} M_{xy,\xi} + \frac{1}{b} M_{yy,\eta} - Q_y = 0 \quad ;
\end{aligned}\right\} \qquad (5)$$

$$\left.\begin{aligned}
&\varepsilon_{xx} = \frac{1}{a} u_{,\xi} \quad , \quad \varepsilon_{yy} = \frac{1}{b} v_{,\eta} \quad , \\
&\varepsilon_{xy} = \frac{1}{2}\left(\frac{1}{b} u_{,\eta} + \frac{1}{a} v_{,\xi} - \frac{4f}{ab} w\right) \quad , \\
&\rho_{xx} = -\frac{1}{a^2} w_{,\xi\xi} \quad , \quad \rho_{yy} = -\frac{1}{b^2} w_{,\eta\eta} \quad , \\
&\rho_{xy} = -\frac{1}{ab} w_{,\xi\eta} \quad ;
\end{aligned}\right\} \qquad (6)$$

$$\left.\begin{aligned}
&N_{xx} = D(\varepsilon_{xx} + \nu\, \varepsilon_{yy}) \quad , \quad N_{yy} = D(\varepsilon_{yy} + \nu\, \varepsilon_{xx}) \quad , \\
&N_{xy} = D(1 - \nu)\, \varepsilon_{xy} \quad , \\
&M_{xx} = K(\rho_{xx} + \nu\, \rho_{yy}) \quad , \quad M_{yy} = K(\rho_{yy} + \nu\, \rho_{xx}) \quad , \\
&M_{yx} = K(1 - \nu)\, \rho_{xy} \quad .
\end{aligned}\right\} \qquad (7)$$

Here, the physical components of the originally tensorial quantities are denoted by the usual indices x, y. The displacements of the mid-surface of the shell are denoted by u and v.

The kinematic relations (6) differ from the kinematic relations of the shallow cylindrical shell (see *(13.37)*) by two terms only:

1) In the case of the cylindrical shell the radial displacement is a part of the circumferential strain ε_{yy} , and
2) in the case of the hypar shell the deflection w contributes to the shear strain ε_{xy} .

By eliminating the transverse forces and by substituting (6) and (7) into (5), we obtain the following three partial differential equations for the three displacements from the 17 equations (8 resultant forces and moments, 6 strain quantities, 3 displacements) :

$$\left.\begin{aligned} &\frac{b}{a}u_{,\xi\xi} + \frac{1+\nu}{2}v_{,\xi\eta} + \frac{1-\nu}{2}\frac{a}{b}\left(u_{,\eta\eta} - 4\frac{f}{a}w_{,\eta}\right) = \frac{\rho t a b}{D}\frac{\partial^2 u}{\partial\tau^2} \ , \\ &\frac{a}{b}v_{,\eta\eta} + \frac{1+\nu}{2}u_{,\xi\eta} + \frac{1-\nu}{2}\frac{b}{a}\left(v_{,\xi\xi} - 4\frac{f}{b}w_{,\xi}\right) = \frac{\rho t a b}{D}\frac{\partial^2 v}{\partial\tau^2} \ , \\ &-\frac{h^2}{12ab}\left(\frac{b^2}{a^2}w_{,\xi\xi\xi\xi} + 2w_{,\xi\xi\eta\eta} + \frac{a^2}{b^2}w_{,\eta\eta\eta\eta}\right) + \\ &\qquad + 2(1-\nu)\frac{f}{ab}(a u_{,\eta} + b v_{,\xi} - 4fw) = \frac{\rho t a b}{D}\frac{\partial^2 w}{\partial\tau^2} \ . \end{aligned}\right\} \quad (8)$$

b) The following boundary conditions have to be fulfilled for a shell with all sides simply supported :

$$\begin{aligned} &u = w = 0 \ , \quad N_{xy} = M_x = 0 \quad \text{for} \quad \xi = 0 \quad \text{and} \quad \xi = 1 \ , \\ &v = w = 0 \ , \quad N_{xy} = M_y = 0 \quad \text{for} \quad \eta = 0 \quad \text{and} \quad \eta = 1 \ . \end{aligned} \quad (9)$$

These conditions are satisfied if we assume displacement functions u , v , w of the form

$$\left.\begin{aligned} u &= U \sin m\pi\xi \cos n\pi\eta \sin\omega\tau \ , \\ v &= V \cos m\pi\xi \sin n\pi\eta \sin\omega\tau \ , \\ w &= W \sin m\pi\xi \sin n\pi\eta \sin\omega\tau \quad (m, n \ \text{integer}) \ . \end{aligned}\right\} \quad (10)$$

Substitution of (10) into (8) then yields a homogeneous, linear algebraic system of equations for the unknown amplitudes U, V, and W. Vanishing of the determinant of the coefficients leads to a cubic equation for the eigenvalue

$$\lambda^2 = \frac{\rho t a b}{D}\omega^2 \ ,$$

where the solutions depend on the dimensions and the integers m and n . Numerical evaluation shows that there exist one lower and two substantially higher eigenvalues for each pair of values of m and n . The numerical values clearly show that the lowest frequency corresponds to pronounced transverse vibration ($W \gg U, V$), and we therefore obtain a good approximation to the smallest eigenvalue provided that the terms of inertia forces tangential to the mid-surface of the shell are neglected in (8) :

$$\rho t\frac{\partial^2 u}{\partial\tau^2} \approx 0 \ , \quad \rho t\frac{\partial^2 v}{\partial\tau^2} \approx 0 \ .$$

After substitution of (10) into (8), we can determine the amplitude ratios U / W and V / W by means of the first two equations (8). Substitution of the ratios into the third equation of (8) finally allows us to approximate analytically the smallest eigenvalue as

$$\lambda^2_{1mn} = \frac{t^2\pi^4}{12ab}\left(\frac{1}{\alpha}m^2 + \alpha n^2\right)^2 + 16(1-\nu^2)\frac{f^2\alpha^2 m^2 n^2}{ab(m^2 + \alpha^2 n^2)^2} \ , \quad (11)$$

where $\alpha = a/b$ denotes the side aspect ratio. Multiplying (11) by α, and introducing the scaled frequency $\bar{\omega}^2 = \frac{\rho t a^2}{D}\omega^2$ then yields:

$$\bar{\omega}_{mn}^2 = \frac{\pi^4}{12}\left(\frac{t}{a}\right)^2 (m^2 + \alpha^2 n^2)^2 + 16(1-\nu^2)\left(\frac{f}{a}\right)^2 \frac{\alpha^4 m^2 n^2}{(m^2+\alpha^2 n^2)^2} \quad . \tag{12}$$

We now need to determine that combination of m and n for the given dimensions which provides the smallest values of $\bar{\omega}_{mn}$.

If we, at this point, limit our considerations to values $\alpha < 1$ (all solutions for $\alpha > 1$ can be obtained by suitably exchanging the sides), we can deduce from (12) that n always has to take the value 1. Calculations show that $m > 2$ is always valid for the lowest frequencies. We can therefore approximately assume $m^2 \gg \alpha^2 n^2$, and we thus obtain from (12)

$$\bar{\omega}_{m1}^2 \approx \frac{\pi^4}{12}\left(\frac{t}{a}\right)^2 m^4 + 16(1-\nu^2)\left(\frac{f}{a}\right)^2 \frac{\alpha^4}{m^2} \quad . \tag{13}$$

In order to determine the dependence on m of the smallest value of $\bar{\omega}_{m1}$, we assume the number of waves to vary continuously and differentiate:

$$\frac{d\bar{\omega}_{m1}^2}{dm^2} = 2\frac{\pi^4}{12}\left(\frac{t}{a}\right)^2 m^2 - 16(1-\nu^2)\left(\frac{f}{a}\right)^2 \frac{\alpha^4}{m^4} = 0$$

$$\longrightarrow \quad m_{min}^2 = \left[\frac{96(1-\nu^2)}{\pi^4}\right]^{1/3} \left(\frac{f}{t}\right)^{2/3} \alpha^{4/3} \quad .$$

Substitution into (13) and re-formulation yield:

$$\bar{\omega}_{min} = \left[12\pi^2(1-\nu^2)\frac{t}{a}\left(\frac{f}{a}\right)^2 \alpha^4\right]^{1/3} \quad . \tag{14}$$

Thus we have obtained the desired approximate formula for the lowest frequency in dependence on the dimensions.

D Structural optimization

– Chapter 15 to 18 –

D.1 Definitions – Formulas – Concepts

15 Fundamentals of structural optimization

15.1 Motivation – aim – development

The previous chapters have introduced fundamentals for determining the structural behaviour required for the dimensioning and design of a structure, i.e. calculation of deformations, stresses, natural vibration frequencies, buckling loads, etc. In view of the development and construction of machines and system components the question arises which measures must be taken in order to reduce costs *and* to improve quality and reliability; in other words this means that an *optimization* of the properties is being aimed at. In terms of this demand, the topic *Structural Optimization* has emerged, over the past years, an extensive field of research that can be described by the following formulation [D.29]:

> *Structural optimization* may be defined as the rational establishment of a structural design that is the best of all possible designs within a prescribed objective and a given set of geometrical and/or behavioral limitations.

Current research in optimal structural design may very roughly be said to follow two main paths. Along the first, the research is primarily devoted to studies of *fundamental aspects of structural optimization.* Broad conclusions may be drawn on the basis of mathematical properties of governing equations for optimal design. These properties are not only studied analytically in order to derive qualitative results of general validity, but are also often investigated numerically via example problems. Along the other main path of research, the emphasis is laid on the *development of effective numerical solution procedures* for optimization of complex practical structures [D.3, D.12, D.21, D.22, D.30].

The constant flow of general reviews, surveys of subfields, conference proceedings, and new textbooks on optimal structural design testify the strong activity, recent advances, and increasing importance of the field. A selection of such recent publications is listed as references at the end of this book.

15.2 Single problems in a design procedure

In this section, the distinctions between usual structural analysis, redesign or sensitivity analysis, and optimization of structures will be made clear, and the basic steps pertaining to optimal design will be outlined.

In a usual *structural analysis problem*, the structural design is given, together with relevant properties of the material(s) to be used and the support conditions for the structure. Also, one set or more of loading is specified, that is, completely specified in deterministic problems, or given in terms of probabilities in probabilistic problems. For each set of loading, the relevant set of equilibrium (or state) equations, constitutive equations, compatibility conditions, and boundary conditions, are then used for determining the structural *response*, e.g., the state of stress, strain and deflection, natural vibration frequencies, and load factors for elastic instability or plastic collapse (see Main Chapters A, B, C).

Redesign or *sensitivity analysis* (Ch. 18) refers to the type of problem where some of the design, material, or support parameters are changed (or varied), and where the corresponding changes (or variations) of the structural response are determined via a repeated (or special) analysis. It is worth noting that a conventional design procedure normally consists of a series of repeated changes of the structural parameters followed by analysis, i.e., a series of redesign analyses, which is carried out until a structure is found that fulfills the behavioral requirements and is reasonable in costs. If the changes of the structural parameters prior to a given redesign analysis are determined rationally from the earlier analyses as the best possible ones, the procedure would identify one of optimal design. Such procedures are of much higher significance than the traditional design procedure where usually the design changes are only decided by guesswork based on experience or information obtained from previous analysis, and a structure obtained by the traditional design procedure will therefore not necessarily be better than other possible alternatives.

The label *structural optimization* identifies the type of design problem where the set of structural parameters is subdivided into so-called *preassigned parameters* and *design variables*, and the problem consists in determining the optimal values of the design variables such that they *maximize* or *minimize* a specific function termed the *objective function* (*criterion* or *cost function*) while satisfiying a set of *geometrical* and/or *behavioural requirements*, which are specified prior to design, and are called *constraints*.

According to the manner in which the *design variables* are assumed to *depend on the spatial variables*, optimal design problems may be roughly categorized as

(1) *Continuous* (or *distributed parameter*) optimization problems, and

(2) *discrete* (or *parameter*) optimization problems.

Usually the design variables of structural elements like rods, beams, arches, disks, plates, and shells are considered to vary continuously over the length or domain of the element, and such problems then fall into category (1), while problems of optimizing inherently discrete structures like trusses, grillages, frames, or complex practical structures belong to category (2).

15.3 Design variables – constraints – objective function

Design variables may describe the configuration of a structure, element quantities like cross-sections, wall-thicknesses, shapes, etc., and physical properties of the material.

Optimization problems can best be classified in terms of their design variables. Based upon the example of a truss-like structure according to L.A. SCHMIT/R.H. MALLET [D.41], and N. OLHOFF/J.E. TAYLOR [D.29] possible design variables can be divided into five (or six, respectively) different classes (Fig. 15.1). In the following, these groups are briefly described in terms of the degree of complexity with which they enter into the design process:

a) Constructive layout

The determination of the most suitable layout is on principle only possible by investigation into all existing types and by comparing the calculated optima.

b) Topology

The topology or arrangement of the elements in a structure is often described by parameters that can be modified in discrete steps only (e.g. number of trusses at the supporting structure of a reflector, number of sections of a continuous girder). Different topologies can also be obtained by eliminating nodes and linking elements.

c) Material properties

In terms of their properties, conventional materials possess variables like specific weight, YOUNG's moduli, mechanical strength properties, etc. which can usually only attain certain discrete values. For brittle materials, often the stochastic character of the properties has to be regarded.

d) Geometry – shape

The geometry of trusses or frames is described either by the nodal coordinates or by the bar lengths, while in the case of load carrying structures with plane or curved surfaces (plates, shells) the geometry is given by spans, curvatures, and the thickness distributions. Usually, these variables are continuously variable quantities.

e) Supports - loadings

Design variables of this type describe the support (or boundary) conditions or the distribution of loading on a structure. Thus, either the location, number, and type of support or the external forces may be varied in order to yield a more effective design. The design variables of this category may be continuous or discrete.

f) Cross-sections

This class of design variables has been used most frequently in optimization tasks (cross-sectional areas, moments of inertia).

When considering structural parameters, we distinguish between *independent* and *dependent variables*, as well as pre-assigned (constant) parameters, where a structure is uniquely characterized by stating the values of its independent variables, the so-called *design variables*. Most often, certain cross-sectional characteristics (thicknesses, diameters) are employed as independent variables, from which all other values can be calculated. As previously mentioned, consideration of the constant parameters and the determination of the dependent variables are made in algorithms for analysis of the structural design.

The i-th design variable will be denoted by x_i, and all n design variables are composed in a vector $\mathbf{x}$ which lies in the *design space*, an n-dimensional EUCLIDEAN space:

$$\mathbf{x}^T = [x_1, x_2, \ldots, x_i, \ldots, x_n]\,. \tag{15.1}$$

Any set of design variables defines a design of the structure and may be represented as a point in the *design space*.

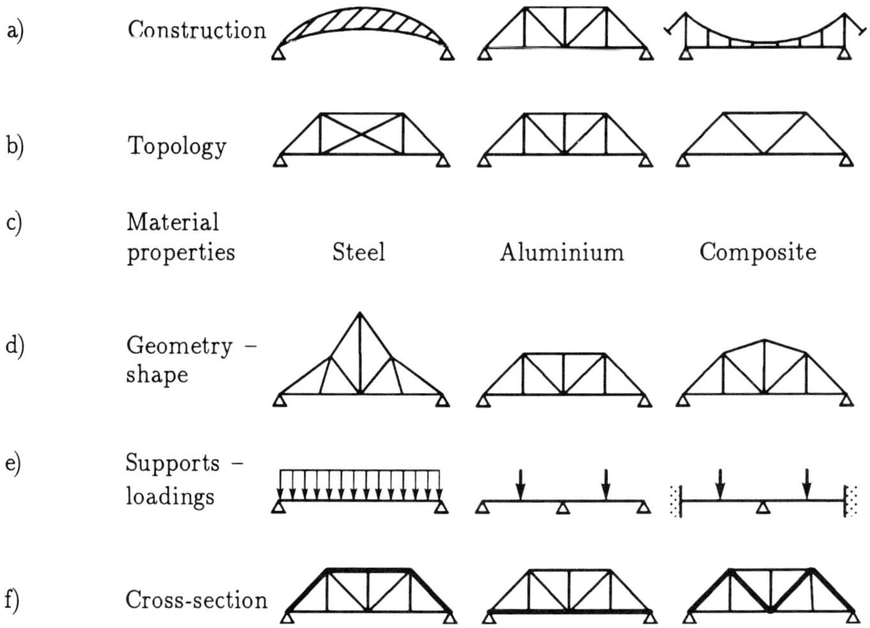

Fig. 15.1: Classification of design optimization problems for truss-like structures in terms of different types of design variables.

Many designs from the totality of possible designs will generally not be acceptable in terms of various design and performance requirements. To exclude such designs as candidates for an optimal solution, the design and performance requirements are expressed mathematically as constraints prior to optimization. The constraints may be of two following types:

– *Geometrical* (or *side*) *constraints* are restrictions imposed explicitly on the design variables due to considerations such as manufacturing limitations, physical practicability, aesthetics, etc. Constraints of this kind are typically *inequality constraints* that specify lower or upper bounds on the design variables, but they may also be *equality constraints* like, e.g., *linkage constraints* that prescribe given proportions between a group of design variables.

– *Behavioral constraints* are generally nonlinear and implicit in terms of the design variables, and they may be of two types. The first type consists of *equality constraints* such as state and compatibility equations governing the structural response associated with the loading condition(s) under consideration. The second type of behavioral constraints comprises *inequality constraints* that specify restrictions on those quantities that characterize the response of the structure. These constraints may impose bounds on *local* quantities like stresses and deflections, or on *global* quantities such as compliance, natural vibration frequencies, etc.

The constraints are formulated in the form of *equality* and/or *inequality constraints*:

$$h_i(\mathbf{x}) = 0 \qquad (i = 1,\ldots,q)\,, \tag{15.2a}$$

$$g_j(\mathbf{x}) = \leq 0 \qquad (j = 1,\ldots,p)\,. \tag{15.2b}$$

Each inequality constraint *(15.2b)* is represented by a surface in the design space which comprises all points $\mathbf{x}$ for which the condition is satisfied as an equality constraint $g_i(\mathbf{x}) = 0$. One distinguishes between feasible, or admissible, and infeasible, or inadmissible designs. All the feasible (or admissible) designs lie within a subdomain of the design space defined by the constraint surfaces as indicated in Fig. 15.2 for the special case of a plane design space (only two design variables). We generally assume that the constraints are such that the design space is *open*, i.e. that a set of feasible designs exists.

The *objective function*, which is also termed the *cost* or the *criterion function*, must be expressed in terms of the design variables in such a way that its value can be determined for any point in the design space. It is this function whose value is to be minimized or maximized by the optimal set of values of the design variables within the feasible design space, and it may represent the structural weight or cost, or it may be taken to represent some local or global measure of the structural performance like stress, displacement, stress intensity factor, stiffness, plastic collapse load, fatigue life, buckling load, natural vibration frequency, aeroelastic divergence, or flutter speed, etc.

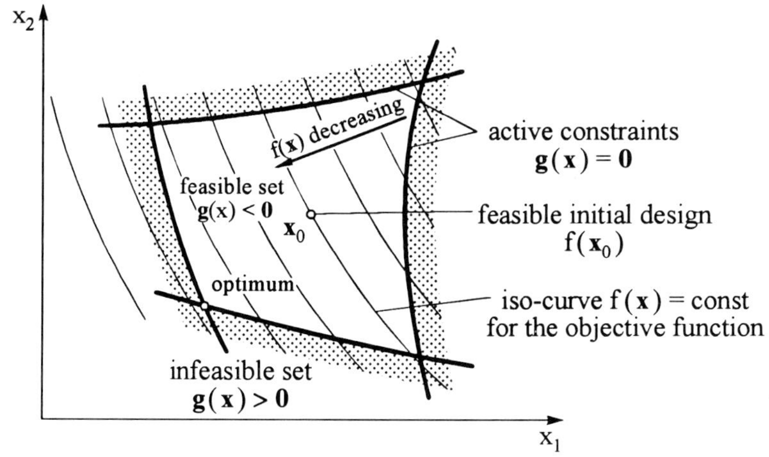

Fig. 15.2: Concepts of structural optimization

The objective function is in most cases a scalar function f of the design variables $\mathbf{x}$ defined as follows:

$$f := f(\mathbf{x}) \quad . \tag{15.3}$$

15.4 Problem formulation – Task of structural optimization

The usual problem of optimal structural design consists in determining the values of the design variables x_i $(i = 1, \ldots, n)$ such that the objective function attains an extreme value while simultaneously all constraints are satisfied. Minimization of the objective function, i.e. Min f, is considered in the conventional mathematical formulation. If an objective function f is to be *maximized*, one simply substitutes f by $-f$ in the formulation, since $\text{Max}\, f \Longleftrightarrow \text{Min}\,(-f)$.

Mathematical formulation:

$$\min_{\mathbf{x} \in \mathbf{R}^n} \left\{ f(\mathbf{x}) \,\middle|\, \mathbf{h}(\mathbf{x}) = \mathbf{0} \;,\; \mathbf{g}(\mathbf{x}) \leq \mathbf{0} \right\} \tag{15.4}$$

with

with	$\mathbf{R}^n$	n-dimensional set of real numbers,
	$\mathbf{x}$	vector of the n design variables,
	$f(\mathbf{x})$	objective function,
	$\mathbf{g}(\mathbf{x})$	vector of the p inequality constraints,
	$\mathbf{h}(\mathbf{x})$	vector of the q equality constraints.

Feasible domain:

$$X := \left\{ \mathbf{x} \in \mathbf{R}^n \,\middle|\, \mathbf{h}(\mathbf{x}) = \mathbf{0} \;,\; \mathbf{g}(\mathbf{x}) \leq \mathbf{0} \right\} \quad . \tag{15.5}$$

Structural optimization generally deals with the solution of *Non-Linear Optimization Problems* (NLOP).

15.5 Definitions in Mathematical Optimization

Definition 1: *Global and local minima*

1) A *global* minimal point $\mathbf{x}^* \in X$ is characterized by

$$f(\mathbf{x}^*) \leq f(\mathbf{x}) \quad \forall \quad \mathbf{x} \in X . \qquad (15.6a)$$

2) A *local* minimal point $\mathbf{x}^* \in X$ is characterized by

$$f(\mathbf{x}^*) \leq f(\mathbf{x}) \quad \forall \quad \mathbf{x} \in X \cap U_\varepsilon(\mathbf{x}^*) , \qquad (15.6b)$$

where $U_\varepsilon(\mathbf{x}^*)$ denotes the ε - neighbourhood of point $\mathbf{x}^*$.

Definition 2: *Conditions for a minimum of an unconstrained problem*

1) Necessary condition

$$\nabla f(\mathbf{x}^*) = \left(\frac{\partial f}{\partial x_1}, \frac{\partial f}{\partial x_2}, \ldots, \frac{\partial f}{\partial x_n}\right)_{\mathbf{x}^*} = \mathbf{0} . \qquad (15.7)$$

2) Sufficient condition

If *(15.7)* is fulfilled, and if the HESSIAN matrix

$$\mathbf{H}^* := \mathbf{H}(\mathbf{x}^*) = \left[\frac{\partial^2 f}{\partial x_i \, \partial x_j}\right]_{\mathbf{x}^*} \qquad (15.8)$$

is strictly positive definite, then $\mathbf{x}^*$ is a local minimum of $f(\mathbf{x})$.

Definition 3: *Conditions for a minimum of a constrained problem*

Determination of optimality conditions by means of the LAGRANGE function

$$L(\mathbf{x}, \boldsymbol{\alpha}, \boldsymbol{\beta}) = f(\mathbf{x}) + \sum_{i=1}^{q} \alpha_i h_i(\mathbf{x}) + \sum_{j=1}^{p} \beta_j g_j(\mathbf{x}) \qquad (15.9)$$

with the LAGRANGEAN multipliers α_i, β_j .

1) Necessary conditions for a local minimum $\mathbf{x}^*$ to the problem *(15.4)*

KUHN-TUCKER conditions [D.26]:

$$\nabla L(\mathbf{x}^*) = \nabla f(\mathbf{x}^*) + \sum_{i=1}^{q} \alpha_i^* \nabla h_i(\mathbf{x}^*) + \sum_{j=1}^{p} \beta_j^* \nabla g_j(\mathbf{x}^*) = \mathbf{0} \qquad (15.10a)$$

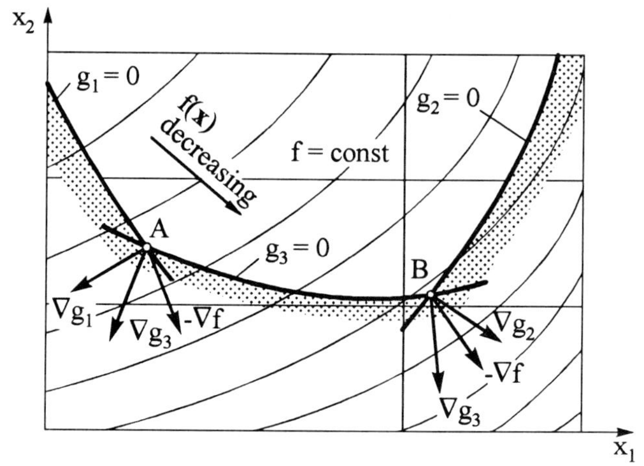

Fig. 15.3: Geometrical interpretation of the KUHN-TUCKER-conditions for a problem with three inequality constraints

and

$$h_i(\mathbf{x}^*) = 0 \quad (i = 1,\ldots,q) \ , \tag{15.10b}$$

$$g_j(\mathbf{x}^*) \leq 0 \quad (j = 1,\ldots,p) \ , \tag{15.10c}$$

$$\beta_j^* \geq 0 \qquad (j = 1,\ldots,p) \ . \tag{15.10d}$$

For each value of j $(j = 1,\ldots,p)$ it holds that $\beta_j^* = 0$ if $g_j(\mathbf{x}^*) < 0$ and that $\beta_j^* \geq 0$ if $g_j(\mathbf{x}^*) = 0$. The values of the LAGRANGIAN multipliers α_i^* associated with the equality constraints *(15.10b)* may both be positive and negative (or zero).

2) Sufficient conditions

For a convex problem, the KUHN-TUCKER conditions are also sufficient. A *geometrical interpretation* for a problem with three inequality constraints (and no equality constraints) is illustrated in Fig. 15.3. For this problem, the following equations must be valid for points A and B according to *(15.10)* in order to be minimum points:

a) Point A: $$-\nabla f(\mathbf{x}_A^*) = \beta_1^* \nabla g_1(\mathbf{x}_A^*) + \beta_3^* \nabla g_3(\mathbf{x}_A^*) \ . \tag{15.11a}$$

The negative gradient of the objective function does not lie within the subset defined by the gradients of the constraint functions, i.e. *(15.10d)* is violated by either β_1^* or β_3^*. $\mathbf{x}_A^*$ is not a minimum point as the function value $f(\mathbf{x}^*)$ can be reduced in the feasible set.

b) Point B: $$-\nabla f(\mathbf{x}_B^*) = \beta_2^* \nabla g_2(\mathbf{x}_B^*) + \beta_3^* \nabla g_3(\mathbf{x}_B^*) \ . \tag{15.11b}$$

The considered point $\mathbf{x}_B^*$ is a local minimum point with all KUHN-TUCKER conditions satisfied. Note that for point B there is no direction in the feasible set in which the function value $f(\mathbf{x}^*)$ can be reduced.

15.6 Treatment of a Structural Optimization Problem (SOP)

An optimization problem can generally be treated by proceeding in accordance with the *Three-Columns-Concept* [D.12] combined in a so-called optimization loop (Fig. 15.4). This concept is based on the fundamental concepts presented in 15.1 and 15.2, and will be described briefly in the following.

Column 1: *Structural Model – Structural Analysis*

One of the most important assumptions within an optimization process consists in transferring a real-life model into a structural model. Thus, any structural optimization problem is based on a mathematical description of the physical behaviour of a structure. In the case of mechanical systems, these are typical structural responses to static and dynamic loads like deformations, stresses, eigenfrequencies, buckling loads, etc. The state variables required for the formulation of objective functions and constraints are computed in the structural model by means of efficient analysis procedures like Finite-Element-Methods, Transfer Matrices Methods, etc. In order to achieve a wide range of application for an optimization procedure, various structural analysis methods, e.g. hybrid methods, should be available.

Column 2: *Optimization Algorithm*

Mathematical programming procedures are predominantly applied for the solution of nonlinear, constrained optimization problems (NLOP). These algorithms are based on iteration procedures which, proceeding from an initial design $\mathbf{x}_0$, generally yield an improved design variable vector $\mathbf{x}_k$ after each iteration cycle k. The optimization calculation is terminated when a predefined convergence criterion becomes satisfied during an iteration (for more details see Chapter 16). Numerous studies have shown that the choice of the most appropriate optimization algorithm is problem-dependent, a fact that is of particular importance in terms of a reliable optimization flow and high efficiency (computational time, rate of convergence).

Column 3: *Optimization Model*

From an engineering point of view, the optimization model constitutes a bridge between the structural model and the optimization algorithms, and is a very considerable column within the optimization process. First, the analysis variables are chosen from the structural parameters as those quantities that are varied during the optimization process. The *design model* to be determined describes the mathematical relation between the analysis variables and the design variables. By additionally transforming the design variables into transformation variables, the optimization problem is adapted to the special requirements of the optimization algorithm in order to increase efficiency and convergence of the optimization calculation. The *evaluation model* determines the values of the objective functions and constraints from the values of the state variables. The *sensitivity analysis* modules compute the sensitivities of the objective functions and the constraints with respect to small changes of the design variables; all this information is transferred to the optimization algorithm or to the decision maker, respectively, for judging the design.

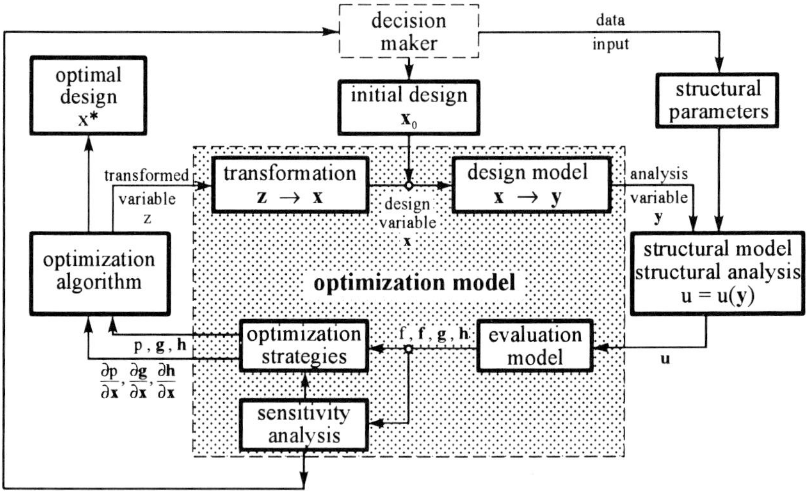

Fig. 15.4: Structure of an optimization loop

The accommodation of the modules for the *sensitivity analysis* (see Chapter 17) and further important modules for *optimization strategies* (see Chapter 18) like Multicriteria Optimization, Shape and Topology Optimization, Multilevel Optimization, etc. within the optimization model, will be mentioned in Section 18.3.

16 Algorithms of Mathematical Programming (MP)

In the following, solution algorithms for optimization problems cast in the standard MP form *(15.4)* will be considered. One distinguishes between *optimization algorithms of zeroth, first and second order* depending whether the solution algorithm only requires the function values, or also the first and second derivatives of the functions. It will be assumed in this Chapter that in general the functions f, h_i, g_j in *(15.4)* are continuous and at least twice continuously differentiable.

The majority of solution algorithms is of an iterative character, i.e. starting from an initial vector $\mathbf{x}_0$ one obtains improved vectors $\mathbf{x}_1, \mathbf{x}_2, \ldots,$ by successive application of the algorithm. Iterative solution procedures are necessary since practical problems of structural optimization are generally highly nonlinear.

16.1 Problems without constraints

(a) *Methods of one-dimensional minimization steps*

Iteration rule:

$$\mathbf{x}_{i+1} = \mathbf{x}_i + \alpha_i \mathbf{s}_i \quad (i = 1, 2, 3, \ldots) \qquad (16.1\,a)$$

with arbitrary search directions $\mathbf{s}_i$ in the design space and an optimal step length α_i for the i-th iteration step (Fig. 16.1).

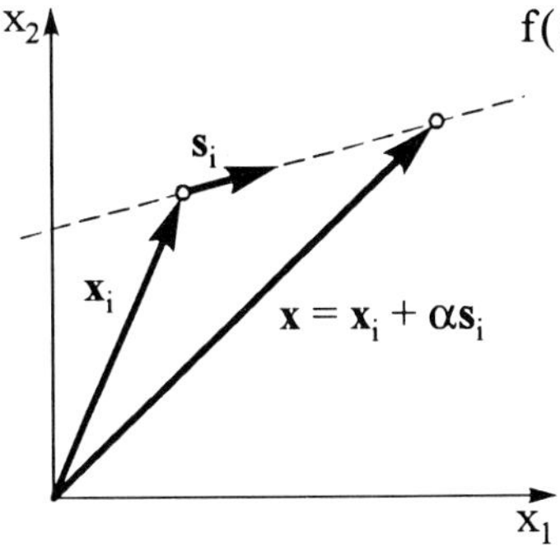

Fig. 16.1: One-dimensional minimization step along a straight line

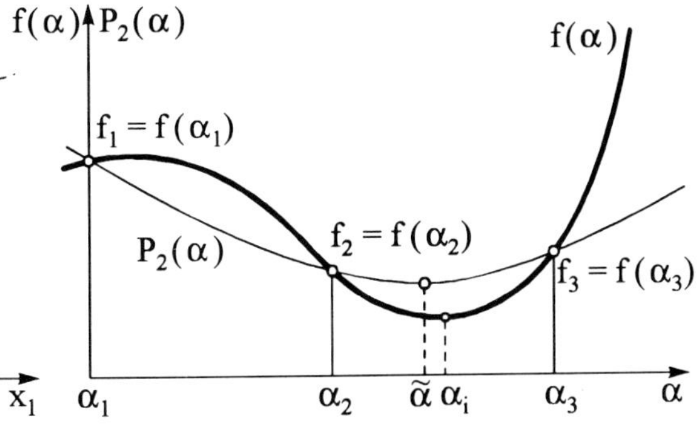

Fig. 16.2: Approximation of $f(\alpha)$ by means of a quadratic polynomium $P_2(\alpha)$

In case of a variable step length α

$$\mathbf{x}(\alpha) = \mathbf{x}_i + \alpha\,\mathbf{s}_i\ , \tag{16.1b}$$

a straight line occurs in the design space, and we have the following form of the objective function:

$$f\left[\mathbf{x}(\alpha)\right] = f(\mathbf{x}_i + \alpha\,\mathbf{s}_i) = f(\alpha)\ , \tag{16.2}$$

where that value α_i of α is to be determined which minimizes f in a given direction $\mathbf{s}_i$. This implies a reduction of the problem to a series of one-dimensional minimization problems where the result depends on the choice of the search direction $\mathbf{s}_i$ (*Line-Search-Method*).

The following procedures can be applied for determining minimal points [D.10, D.21, D.22, D.24, D.40].

- Quadratic polynomia (LAGRANGEAN interpolation) $P_2(\alpha)$ (Fig. 16.2),
- Cubic polynomia (HERMITE interpolation) $P_3(\alpha)$,
- Regula falsi ,
- Interval reduction by means of FIBONACCI-search [D.25].

By applying mathematical optimization methods, it is intended that $\tilde{\alpha}$ yields a good approximation of f in the neighbourhood of the point of minimum α_i (Fig. 16.2). Different convergence criteria can be used, for instance:

$$f(\tilde{\alpha}) \le \begin{cases} f(\tilde{\alpha} + \varepsilon) \\ f(\tilde{\alpha} - \varepsilon) \end{cases} \tag{16.3a}$$

or

$$\frac{|P_2(\tilde{\alpha}) - f(\tilde{\alpha})|}{|P_2(\tilde{\alpha})|} < \varepsilon \quad \text{with a limit of accuracy } \varepsilon \approx 0.01 \ . \qquad (16.3b)$$

(b) *First POWELL method of conjugate directions – method of 0th order*

In order to calculate an iteration point $\mathbf{x}_{i+1}$, information on the previous points $\mathbf{x}_1,\ldots,\mathbf{x}_i$ is used. For this purpose, we need the notion of conjugate directions. The vectors $\mathbf{s}_j$ and $\mathbf{s}_k$ are called conjugate if they satisfy the condition [D.37]

$$\mathbf{s}_j^T \mathbf{H} \mathbf{s}_k = 0 \quad (j \neq k) \ , \qquad (16.4)$$

where $\mathbf{H}$ is the HESSIAN matrix *(15.8)*. In POWELL's method, $\mathbf{H}$ in *(16.4)* is replaced by an approximation matrix $\mathbf{A}$, and vectors that fulfill this condition are called $\mathbf{A}$-conjugate.

In case of a quadratic function, the conjugate direction $\mathbf{s}_k$ belonging to an arbitrary tangent $\mathbf{s}_j$ always leads to the centre of a family of ellipses.

Solution strategy:

Proceeding from an initial point $\mathbf{x}_0$, n one-dimensional minimization iterations (steps) are carried out along n linearly independent search directions $\mathbf{s}_0,\ldots,\mathbf{s}_n$, e.g. along the n unit vectors $\mathbf{e}_0,\ldots,\mathbf{e}_n$ of the coordinate axes in the design space, i.e.

$$\mathbf{x}_{i+1} = \mathbf{x}_i + \alpha_i \mathbf{s}_i \quad (i = 0,1,2,\ldots,n) \ . \qquad (16.1c)$$

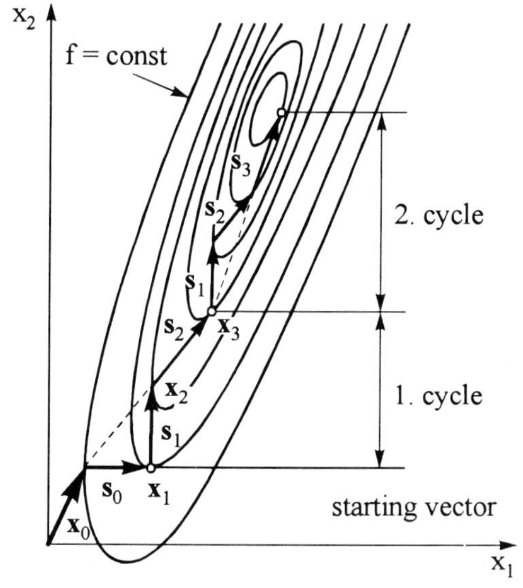

Fig. 16.3: Application of the First POWELL method of conjugate directions

The search cycle is considered by the (n + 1)-th one-dimensional minimization step in the direction of

$$\mathbf{s}_{n+1} := \mathbf{x}_{n+1} - \mathbf{x}_{n-1} \; , \tag{16.1d}$$

which leads to point $\mathbf{x}_{n+2}$. For the following search cycle, $\mathbf{s}_1$ is eliminated, the index of the remaining directions $\mathbf{s}_2, \ldots, \mathbf{s}_{n+1}$ is decreased by one, and the described procedure is then applied to all subsequent search directions (Fig. 16.3).

In many cases the convergence behaviour of this procedure is insufficient due to generation of almost linearly dependent search directions. Certain modifications, however, lead to improvements [D.18 , D.38].

(c) *Gradient method of steepest descent – method of 1st order*

By this method we choose the search vector $\mathbf{s}_i$ in the direction of the steepest descent of $f(\mathbf{x})$ at the point $\mathbf{x}_i$, i.e. in the negative gradient direction (Fig. 16.4):

$$\mathbf{s}_i = -\nabla f(\mathbf{x}_i) \; . \tag{16.5}$$

By means of *(16.5)* we determine the optimum point along this direction, using the *iteration rule*

$$\mathbf{x}_{i+1} = \mathbf{x}_i + \alpha_i \mathbf{s}_i = \mathbf{x}_i - \alpha_i \nabla f(\mathbf{x}_i) \quad (i = 1, 2, 3 \ldots) \; , \tag{16.6}$$

where the optimal step length α_i can be calculated by one of the methods described under (a).

(d) *FLETCHER-REEVES-Method of conjugate gradients* [D.19 , D.23] – *method of 1st order*

The first one-dimensional minimization step is carried out in the direction of the steepest descent according to *(16.5)*

$$\mathbf{s}_1 = -\nabla f(\mathbf{x}_1) \; .$$

and we thus reach point $\mathbf{x}_2$ (Fig. 16.5).

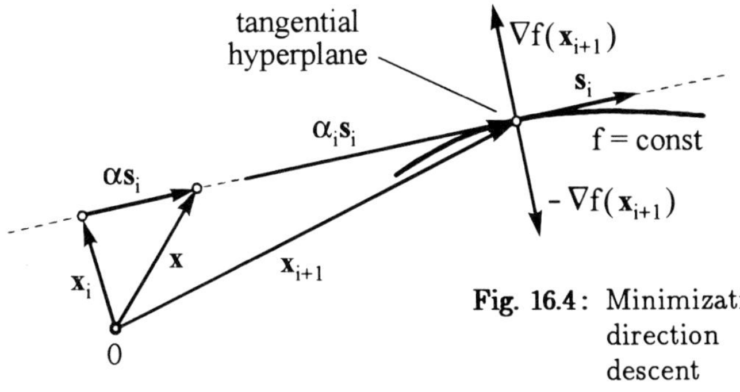

Fig. 16.4: Minimization step in the direction of the steepest descent

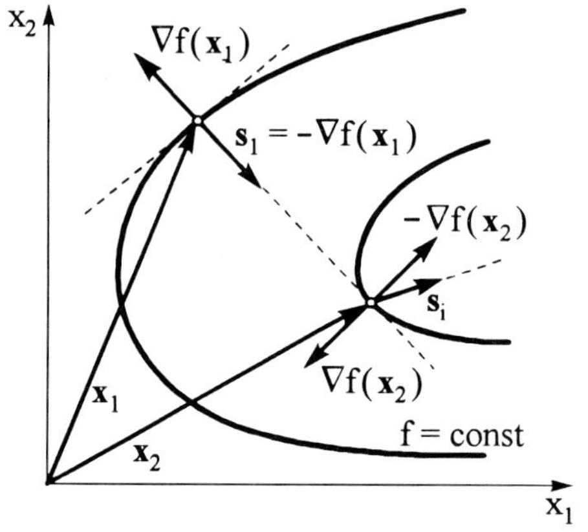

Fig.16.5: FLETCHER-REEVES-Method of conjugate gradients

Proceeding from this point, modified search directions are generated:

$$\mathbf{s}_{i+1} = -\nabla f(\mathbf{x}_{i+1}) + \frac{|\nabla f(\mathbf{x}_{i+1})|^2}{|\nabla f(\mathbf{x}_i)|^2}\mathbf{s}_i \quad (i = 1,2,3\ldots). \tag{16.7}$$

(e) *Special Quasi-NEWTON procedure SQNP – method of 2nd order*

In this method, the search direction $\mathbf{s}_k$ of the k-th iteration is defined as

$$\mathbf{s}_k = -\mathbf{H}_k \nabla f(\mathbf{x}_k) , \tag{16.8}$$

where the matrix $\mathbf{H}_k$ (which is not the HESSIAN) is calculated by means of variable metrics according to DAVIDON, FLETCHER, POWELL (DFP) [D.11]:

$$\mathbf{H}_k = \mathbf{H}_{k-1} + \lambda^*_{k-1}\frac{\mathbf{s}_{k-1}\mathbf{s}^T_{k-1}}{\mathbf{s}^T_{k-1}\mathbf{y}_{k-1}} - \frac{(\mathbf{H}_{k-1}\mathbf{y}_{k-1})(\mathbf{H}_{k-1}\mathbf{y}_{k-1})^T}{\mathbf{y}^T_{k-1}\mathbf{H}_{k-1}\mathbf{y}_{k-1}} \tag{16.9a}$$

with $\mathbf{H}_0 = \mathbf{I}$,

$$\mathbf{y}_{k-1} = \nabla f(\mathbf{x}_k) - \nabla f(\mathbf{x}_{k-1}), \tag{16.9b}$$

λ^*_{k-1} step length along the search direction $\mathbf{s}_{k-1}$.

16.2 Problems with constraints

16.2.1 Reduction to unconstrained problems

(a) General remarks on penalty functions

These methods have originally been developed by FIACCO and McCORMICK who chose the name SUMT (Sequential Unconstrained Minimization Techniques) [D.17].

The optimization problem *(15.4)* contains p inequality constraints $g_j(\mathbf{x}) \le 0$

$\longrightarrow$ Formulation of a modified objective function for each iteration:

$$\Phi_i(\mathbf{x}, R_i) := f(\mathbf{x}) + R_i \sum_{j=1}^{p} G[g_j(\mathbf{x})] \quad (i = 1,2,3,\ldots) \qquad (16.10)$$

with a penalty function $G[g_i(\mathbf{x})]$ and the penalty parameters R_i.

Irrespective of explicit constraints, minimization is then carried out by means of the previously described methods for unconstrained problems.

Function G is chosen in such a way that during minimization for a series of values for R_i the solution converges towards that of the original problem with constraints.

(b) Frequently applied methods

– *Method of exterior penalty functions*

Penalty objective function in *(16.10)*:

$$G[g_j(\mathbf{x})] := \left(\max[0, g_j(\mathbf{x})]\right)^2 . \qquad (16.11a)$$

Modified objective function

$$\Phi_i(\mathbf{x}, R_i) = f(\mathbf{x}) + R_i \sum_{j=1}^{p} \left(\max[0, g_j(\mathbf{x})]\right)^2 \quad (i = 1,2,3,\ldots) . \qquad (16.11b)$$

Φ_i is to be minimized for a set of increasing values of R_i (e.g. $R_1 = 10^{-3}$, $R_2 = 10^{-2}$, $R_3 = 10^{-1}, \ldots$).

Owing to the inequality constraints $g_j(\mathbf{x}) \le 0$ $(j = 1,2,3,\ldots)$ in *(16.11a)* we obtain:

$$G[g_j(\mathbf{x})] = 0 \qquad \text{in the feasible domain ,}$$

$$G[g_j(\mathbf{x})] = g_j(\mathbf{x})^2 > 0 \qquad \text{in the infeasible domain .}$$

Functions f and Φ are identical in the feasible domain, whereas f is *penalized* in the infeasible domain by a summation of the non-negative terms *(16.11b)*.

– *Method of interior penalty functions or barrier functions*

Penalty function in *(16.10)*:

$$G[g_j(\mathbf{x})] := -\frac{1}{g_j(\mathbf{x})} . \qquad (16.12a)$$

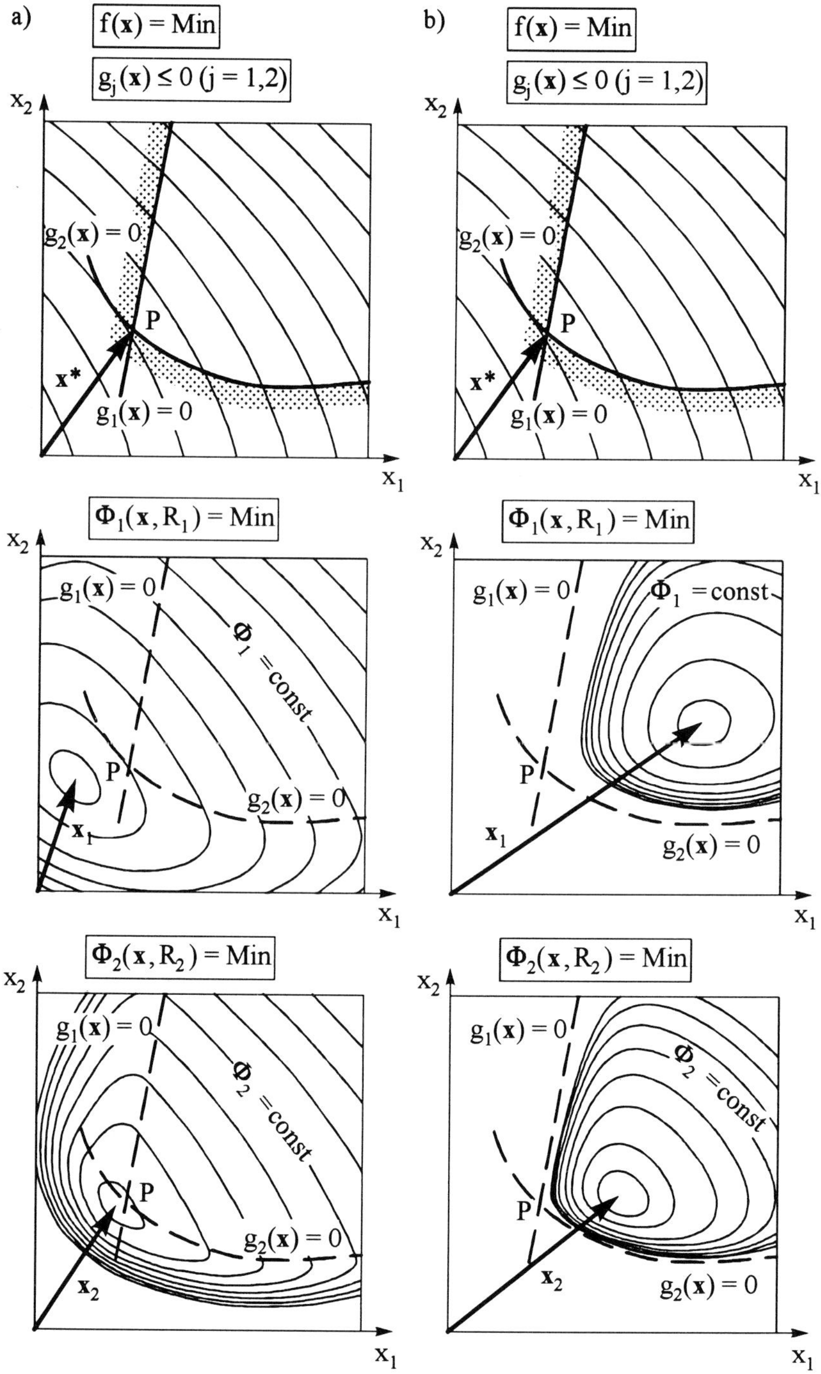

Fig. 16.6: Method of a) exterior penalty functions
b) interior penalty functions

Modified objective function

$$\Phi_i(\mathbf{x}, R_i) = f(\mathbf{x}) - R_i \sum_{j=1}^{p} \frac{1}{g_j(\mathbf{x})} \quad (i = 1, 2, 3, \ldots) . \qquad (16.12b)$$

Φ_i is to be minimized for a set of decreasing values of R_i (e.g. $R_1 = 10^5$, $R_2 = 10^4$, $R_3 = 10^3$ etc.).

In this case, the objective function f is penalized in the feasible domain by summation of the positive terms.

For details on further procedures refer to [D.24].

16.2.2 General nonlinear problems – direct methods

In the past, a substantial number of *direct algorithms* of Mathematical Programming (MP-algorithms) have been developed for the solution of general, nonlinear optimization problems. When applied to problems of component optimization, these algorithms must be able to reduce the number of optimization steps during optimization since often extensive structural analyses have to be carried out at each iteration. In addition, a sufficiently good convergence behaviour as well as reliability and robustness must be demanded of these procedures. These characteristics largely depend on the degree of nonlinearity of the posed problem.

In the following, two frequently used procedures shall be briefly described as typical algorithms:

(a) *Sequential Linearization Procedure SLP*

The efficiency of linear methods can also be utilized for nonlinear design problems by successively solving linear substitute problems in the form of a so-called *sequential linearization* [D.20, D.35, D.38].

By introducing upper and lower bounds (hypercube, move limits) for all design variables, GRIFFITH and STEWARD have augmented the range of application to problems where the solutions are not at the intersection of constraints but, more general, on a curved hypersurface [D.20]. The objective function and constraints of the nonlinear, scalar initial problem *(15.4)* are expanded in a TAYLOR-series in the vicinity of a point $\mathbf{x}^k$. By maintaining the linear terms only we obtain

$$f(\mathbf{x}^k + \Delta\mathbf{x}) \approx f(\mathbf{x}^k) + \nabla f(\mathbf{x}^k)\Delta\mathbf{x} , \qquad (16.13a)$$

$$h_i(\mathbf{x}^k + \Delta\mathbf{x}) \approx h_i(\mathbf{x}^k) + \nabla h_i(\mathbf{x}^k)\Delta\mathbf{x} \quad (i = 1, 2, \ldots, q) , \qquad (16.13b)$$

$$g_j(\mathbf{x}^k + \Delta\mathbf{x}) \approx g_j(\mathbf{x}^k) + \nabla g_j(\mathbf{x}^k)\Delta\mathbf{x} \quad (j = 1, 2, \ldots, p) . \qquad (16.13c)$$

In addition, the design space of the linearized problem is bounded by a hypercube according to Fig. 16.7

$$x_{i_l}^k \leq x_i^k \leq x_{i_u}^k \quad (i = 1, 2, \dots, n) , \tag{16.14}$$

since the TAYLOR-expansion is only valid for small $\Delta\mathbf{x}$. Here, it is convenient to use superscripts k for the approximation steps, whereas the subscripts i denote the number of design variables, and u or l the upper and lower bounds, respectively.

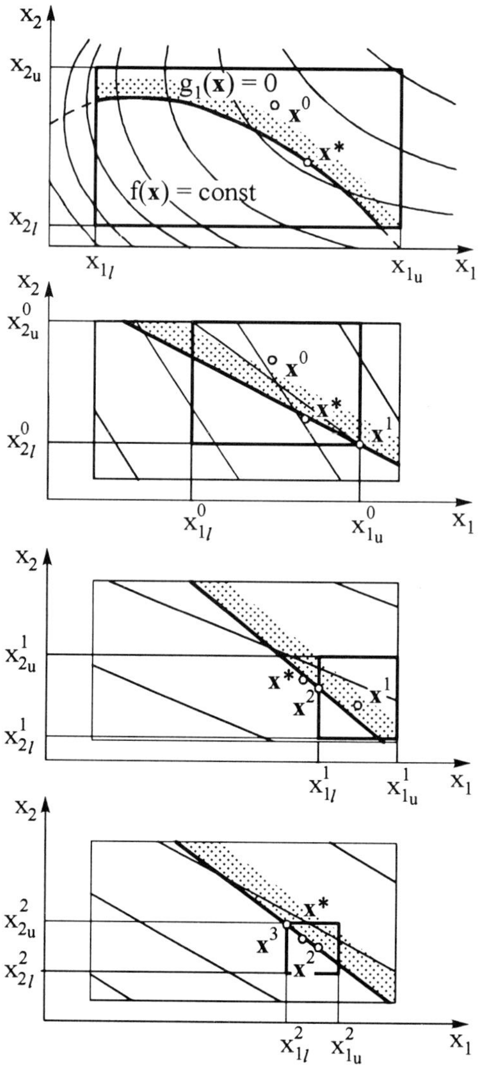

Fig. 16.7: Optimization flow of SLP in the two-dimensional case

The linearized problem according to *(16.13)* is solved by means of the SIMPLEX-procedure by DANTZIG [D.20]. Since for that purpose all variables have to be larger than zero, a linear transformation of variables has to be carried out:

$$y_i = \Delta x_i + \left(x_i^k - x_{il}^k \right) \quad i = 1,2,\ldots,n \; . \qquad (16.15)$$

The linearized problem then reads:

$$\underset{\mathbf{y}}{\mathrm{Min}} \left\{ \mathbf{c}^T \mathbf{y} \right\} = \mathbf{c}^T \mathbf{y}^* \quad \text{with} \quad \mathbf{c} = \nabla f(\mathbf{x}^k) \, , \qquad (16.16)$$

together with the linearized constraints $h_i(\mathbf{y})$ and $g_j(\mathbf{y})$, and the hypercube according to *(16.14)*.

The solution $\mathbf{y}^*$ of the linearized problem yields an improved x^{k+1} for the nonlinear problem. The hypercube is reduced by means of correction rules, thus the side length of the hypercube decreases during the optimization. Fig. 16.7 illustrates how SLP works in the two-dimensional case. It becomes obvious that the optimization flow strongly depends on the choice of the initial hypercube.

(b) *Augmented LAGRANGE–Function Procedure LPNLP*

Equality and inequality constraints are included in an augmented LAGRANGE–function. The unconstrained problem is then solved by means of search techniques [D.36].

PIERRE and LOWE have developed the optimization procedure LPNLP (LAGRANGE Penalty Method for Non-Linear Problems). Using the LAGRANGE–function (see *(15.9)*) directly as an objective function has certain immanent disadvantages as the KUHN-TUCKER-conditions are not necessarily sufficient. Even if $\nabla L(\mathbf{x}^*, \boldsymbol{\alpha}^*, \boldsymbol{\beta}^*) = \mathbf{0}$ is valid, $\nabla^2 L^*$ may not be positive definite. Therefore, the LAGRANGE-function defined in *(15.9)* is augmented by *penalty-terms* with the special weighting factors w_i ($i = 1,2,3$)

$$L_a(\mathbf{x}, \boldsymbol{\alpha}, \boldsymbol{\beta}, \mathbf{w}) = L(\mathbf{x}, \boldsymbol{\alpha}, \boldsymbol{\beta}) + w_1 P_1 + w_2 P_2 + w_3 P_3 \qquad (16.17)$$

with

$$P_1 = \sum_{i=1}^{q} [h_i(\mathbf{x})]^2 \; ,$$

$$P_2 = \sum_{j \in C_a} [g_j(\mathbf{x})]^2 \; , \quad C_a = \{ j \,|\, \beta_j > 0 \} \, ,$$

$$P_3 = \sum_{j \in C_b} [g_j(\mathbf{x})]^2 \; , \quad C_b = \{ j \,|\, \beta_j > 0 \;\text{ and }\; g_j \geq 0 \} \, .$$

The optimization problem is then solved sequentially by unconstrained optimization and correction steps. For further details refer to [D.36]. Fig. 16.8. illustrates the flow-chart of the LPNLP-procedure.

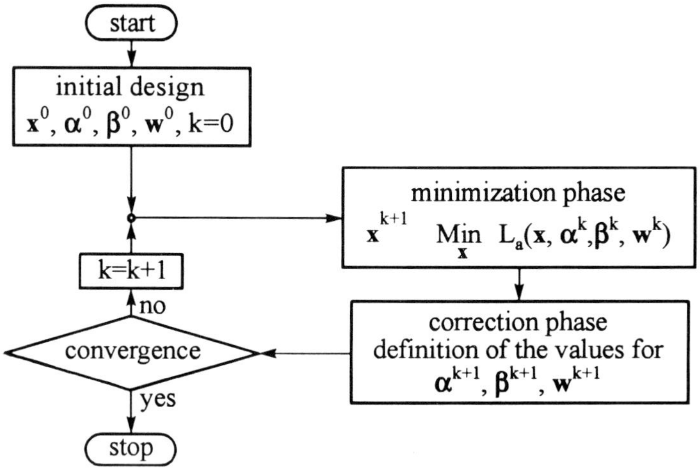

Fig. 16.8: Flow-chart of the LPNLP-procedure

(c) *Further algorithms*

In the following, some of the algorithms currently applied in structural optimization shall be briefly introduced.

• *Sequential Quadratic Programming (SQP)* [D.38]

Based upon the LAGRANGE–function, this method utilizes the sequential linearization and quadratic approximation of a nonlinear problem by means of the BFGS (BROYDEN, FLETCHER, GOLDFARB, SHANNO)-formula [D.21] of the HESSIAN-matrix. The quadratic subproblem is then solved in order to generate a search direction; for the one-dimensional search, the optimal step length is determined by a penalty function and a quadratic interpolation.

• *Method of Generalized Reduced Gradients (GRG)* [D.1]

A subset of the design variables is eliminated from the objective function by means of active constraints. The "reduced gradient" is then calculated in order to generate a search direction. The optimal step length is found by employing the Quasi-NEWTON - Algorithm.

• *Hybrid procedure consisting of SQP- and GRG-methods (QPRLT)* [D.1, D.35]

In a first step, a search direction is determined by means of the SQP-algorithm; then, the optimal step length is calculated by the GRG-algorithm. Thus, the advantages of SQP and GRG are combined in one single algorithm.

• *Method of Moving Asymptotes (MMA)* [D.45]

Here, a sequential, convex aproximation of the nonlinear problem is carried out. First, the problem is transformed into a dual problem with specific characteristics (separable, convex problem) and is then solved by means of a conventional gradient algorithm or by a NEWTON-algorithm. This solution point serves as the starting point for the subsequent approximation.

A compilation of direct MP-algorithms for the solution of general, nonlinear optimization problems can be found in numerous books, among others [D.24 , D.33 , D.39 , D.40 , D.46].

17 Sensitivity analysis of structures

17.1 Purpose of sensitivity analysis

The objective of *design sensitivity analysis* is to calculate gradients of the structural responses and cost functions with respect to small changes of the design variables. The determination of the gradients of the objective function and the constraints is a highly important step in the optimization process (see Fig. 15.4), since these gradients are not only a prerequisite for the majority of optimization algorithms (see Chapter 16), but they also provide important information on the structural sensitivity when changing arbitrary structural parameters. The choice of an appropriate method of sensitivity analysis strongly influences the numerical efficiency and thus has impact on the entire course of the optimization. For this reason, the treatment of the fundamentals of structural optimization shall be dealt with separately with some remarks on frequently applied techniques for determination of gradients.

A complete overview of sensitivity methods in structural optimization is given in [D.2 , D.21 , D.22]. In addition to simple numerical finite difference procedures, analytical or semi-analytical methods and formulations derived from the variational principle, are increasingly applied.

Proceeding from an m-dimensional vector function $\boldsymbol{\varphi}$ (e.g. vector of objective function and constraints)

$$\boldsymbol{\varphi} = \boldsymbol{\varphi}[\mathbf{x}, \mathbf{u}(\mathbf{x})] \tag{17.1}$$

with $\mathbf{x} \in \mathrm{R}^{\mathrm{n}}$ design variable vector ,

$\mathbf{u} \in \mathrm{R}^{\mathrm{n_u}}$ vector of state quantities ,

we obtain the total differential of *(17.1)* as:

$$\mathrm{d}\boldsymbol{\varphi}[\mathbf{x}, \mathbf{u}(\mathbf{x})] = \frac{\partial \boldsymbol{\varphi}}{\partial \mathbf{x}} \mathrm{d}\mathbf{x} + \frac{\partial \boldsymbol{\varphi}}{\partial \mathbf{u}} \mathrm{d}\mathbf{u} = \left(\frac{\partial \boldsymbol{\varphi}}{\partial \mathbf{x}} + \frac{\partial \boldsymbol{\varphi}}{\partial \mathbf{u}} \frac{\partial \mathbf{u}}{\partial \mathbf{x}} \right) \mathrm{d}\mathbf{x} \tag{17.2}$$

$$\longrightarrow \quad d\boldsymbol{\varphi}[\mathbf{x},\mathbf{u}(\mathbf{x})] = \mathbf{A}\,d\mathbf{x} \ , \tag{17.3a}$$

where $\mathbf{A} = [a_{ij}]_{m\times n}$ is called the *sensitivity matrix*. The following assumptions are made:

$$\frac{\partial \boldsymbol{\varphi}}{\partial \mathbf{x}} = \left[\frac{\partial \varphi_i}{\partial x_j}\right]_{m\times n} ; \ \frac{\partial \boldsymbol{\varphi}}{\partial \mathbf{u}} = \left[\frac{\partial \varphi_i}{\partial u_k}\right]_{m\times n_u} ; \ \frac{\partial \mathbf{u}}{\partial \mathbf{x}} = \left[\frac{\partial u_k}{\partial x_j}\right]_{n_u\times n} . \tag{17.3b}$$

In the sequel, different ways of determining $\mathbf{A}$ will be introduced.

17.2 Overall Finite Difference (OFD) sensitivity analysis

The Overall Finite Difference (OFD) - approach implies that the entries in the sensitivity matrices are approximated by simple finite difference quotients:

$$a_{ij} \approx \frac{\varphi_i[\widetilde{\mathbf{x}}_j,\mathbf{u}(\widetilde{\mathbf{x}}_j)] - \varphi_i[\mathbf{x},\mathbf{u}(\mathbf{x})]}{\Delta x_j} \qquad \begin{matrix}(i = 1,\ldots,m)\\(j = 1,\ldots,n\ )\end{matrix} \tag{17.4}$$

with

$\widetilde{\mathbf{x}}_j = (x_1, x_2, \ldots, x_j + \Delta x_j, \ldots, x_n)$,

Δx_j perturbation of the j-th design variable .

The OFD-method is very easily implemented, and is completely independent of the structural model. The method is also applicable for tasks beyond the field of optimization. However, the OFD-analysis has the immanent disadvantage that for n design variables, n + 1 complete structural analyses of the total structural response are required, a fact that leads to very extensive computation times in case of larger optimization problems. In addition, the occurrence of round-off errors does not allow the relative perturbation $\varepsilon = \Delta x_j / x_j$ to be chosen arbitrarily small. Based on experience, values of $\varepsilon \approx 10^{-5}$ to 10^{-3} are suitable.

17.3 Analytical and semi-analytical sensitivity analyses

In order to reduce the extensive computational effort, analytical or semi-analytical methods are increasingly employed for gradient calculation [D.2, D.15, D.21, D.22, D.30, D.31]. These procedures are closely linked with the applied structural analysis procedure, and their realization thus renders manipulations in the source code necessary. These methods have originally been developed for the FE-methods, but they can be generalized to all other structural analysis procedures which transform the differential equations of the considered mechanical system into a set of algebraic equations (transfer matrix methods, difference procedures, analytical solution methods according to RAYLEIGH/RITZ, etc.). Here, our considerations will be limited to those linear systems that have a system matrix equation of the form:

$$\mathbf{F}\,\mathbf{u} = \mathbf{r} \tag{17.5}$$

with $\mathbf{F} = \mathbf{F}(\mathbf{x})$ global system matrix ,

$\mathbf{u} = \mathbf{u}(\mathbf{x})$ vector of state quantities ,

$\mathbf{r} = \mathbf{r}(\mathbf{x})$ load vector .

The solution of *(17.5)* for the state vector **u** subjected to a given load vector **r**, is normally carried out by GAUSSIAN elimination performed as a two-phase process of factorization of the system matrix **F** which does not require simultaneous modification of **r**, and thus makes it possible to solve *(17.5)* for additional load cases, i.e. several right-hand sides, without much additional computational effort.

By implicit differentiation of *(17.5)* with respect to any of the design variables x_j $(j = 1,\ldots,n)$, rearrangement of terms, and multiplying the equation by $\partial\boldsymbol{\varphi}/\partial\mathbf{u}$, one can replace the derivatives of the state quantities in *(17.2)* by the following expression:

$$\frac{\partial \boldsymbol{\varphi}}{\partial \mathbf{u}}\frac{\partial \mathbf{u}}{\partial x_j} = \frac{\partial \boldsymbol{\varphi}}{\partial \mathbf{u}}\mathbf{F}^{-1}\underbrace{\left(\frac{\partial \mathbf{r}}{\partial x_j} - \frac{\partial \mathbf{F}}{\partial x_j}\mathbf{u}\right)}_{\mathbf{p}_j} , \tag{17.6}$$

where $\mathbf{p}_j$ is the so-called *pseudo load vector* associated with the design variable x_j.

With **r** known and **u** obtained by solution of *(17.5)*, computation of the pseudo load vectors $\mathbf{p}_j$ $(j = 1,\ldots,n)$ in *(17.6)* only requires that the design sensitivities $\partial\mathbf{r}/\partial x_j$ and $\partial\mathbf{F}/\partial x_j$ of the load and the system matrix are known. Note here that the former sensitivities vanish if the load is design independent.

If in *(17.6)* the design sensitivities of the global system matrix $\partial\mathbf{F}/\partial x_j$ are determined analytically before their numerical evaluation, the approach is called the method of *analytical* sensitivity analysis, and if they are determined by numerical differentiation, cf. *(17.4)*, the label *semi-analytical* sensitivity analysis is used. While the analytical method is expedient for problems with cross-sectional design variables (see Section 15.3), it is usually a formidable task to implement the method when shape design variables (see Sections 15.3 and 18.2) are encountered. Thus, a large amount of analytical work and programming may be required in order to develop analytical expressions for derivatives of, for instance, various finite element stiffness matrices with respect to a large number of possible shape parameters. For problems involving shape design variables, it is much more attractive to apply the semi-analytical method because it is easier to implement as it treats different types of finite elements and design variables in a unified way.

Eq. *(17.6)* can be treated in two different ways:

(a) *Direct method (Design Space Method)*

First, we computate the pseudo load vector $\mathbf{p}_j$ associated with each x_j as described above, and then determine from

$$\mathbf{F}\frac{\partial \mathbf{u}}{\partial x_j} = \left(\frac{\partial \mathbf{r}}{\partial x_j} - \frac{\partial \mathbf{F}}{\partial x_j}\mathbf{u}\right) = \mathbf{p}_j \tag{17.7}$$

the corresponding gradients of the state quantities $\partial \mathbf{u}/\partial x_j$. Since the form of *(17.7)* is analogous to that of *(17.5)*, each of these gradients ($j = 1,\dots,n$) can be solved from *(17.7)* using the same factorization of the global system matrix $\mathbf{F}$ as was used when solving *(17.5)* for the vector $\mathbf{u}$. Thus, *(17.7)* has to be solved for n right-hand sides $\mathbf{p}_j$ for each load case $\mathbf{r}$ in *(17.5)*.

Finally, the gradients $\partial \boldsymbol{\varphi}/\partial \mathbf{u}$ and $\partial \boldsymbol{\varphi}/\partial \mathbf{x}$ are calculated in order to establish the total differential according to *(17.2)*.

(b) *Auxiliary variable method (State Space Method)*

Here, we introduce an auxiliary variable vector $\boldsymbol{\lambda}_i$, the transpose of which we define as the following product according to *(17.6)*:

$$\boldsymbol{\lambda}_i^T = \frac{\partial \varphi_i}{\partial \mathbf{u}}\mathbf{F}^{-1} \quad (i = 1,\dots,m)\,. \tag{17.8a}$$

This yields the equation system

$$\mathbf{F}^T\boldsymbol{\lambda}_i = \left(\frac{\partial \varphi_i}{\partial \mathbf{u}}\right)^T. \tag{17.8b}$$

for $\boldsymbol{\lambda}_i$ that is to be solved for the m right-hand sides $\frac{\partial \varphi_i}{\partial \mathbf{u}}$ ($i = 1, \dots, m$).

Generally, analytical and semi-analytical methods of sensitivity analysis are able to reduce the computational time to a fraction of that necessary for the OFD-approach (Section 17.2). This is mainly due to the fact that the system matrix of the former methods has to be factorized or inverted only once for a sensitivity analysis.

Depending on the number of equation systems to be solved in *(17.7)* and *(17.8b)*, the direct method is preferred in those cases where fewer design variables than constraints are defined in the optimization model, whereas the auxiliary variable method should be applied to problems with a prevailing number of design variables.

Although the sensitivity methods described above primarily focus on gradient calculation for structures subjected to static loading, the methodology can be extended to other problems in a straight-forward manner.

18 Optimization strategies

In order to treat different types of optimization tasks like shape and topology optimization problems as well as multicriteria or multilevel optimization tasks, specific optimization strategies have to be integrated into the optimization loop according to Fig. 15.4. These strategies are sub-parts of optimization modeling, and they transfer arbitrary optimization problems into so-called *substitute problems* by way of transformation or decomposition so that the given tasks can be solved by usual scalarized parameter optimization procedures. In the following, two of these strategies will be briefly treated, namely

- vector, multicriteria or multiobjective optimization, and
- shape optimization,

where a transformation into parameter optimization problems for both strategies is carried out.

18.1 Vector, multiobjective or multicriteria optimization – PARETO-optimality [D.14, D.34, D.43, D.44]

In contrast to problems where a single criterion governs, for *multicriteria optimization* the optimal design reflects simultaneous minimization on two or more criteria. The labels *vector optimization* or *multiobjective optimization* are also used for such problems. Problems of this kind are of particular relevance to practice where, in general, several structural response modes and failure criteria must be taken into account in the design process.

Ordinarily in vector optimization there exists a *trade-off among criteria*, i.e. a change in design may result in *improvement* according to one or more criteria, but only at the expense of a *worsening* as measured by others. One alternative is to apply the concept of *PARETO-optimality* (see Def. 1 below) according to which a given multicriteria optimization problem may have anything from one to an infinity of PARETO - solutions. It is then up to the designer to identify the optimal design within this set. This step requires the application by the designer of judgment or some other basis of choice.

This state of affairs reflects the fact that it is only possible to obtain a unique optimal design if a single, scalar objective function $f(\mathbf{x})$ is encountered in the optimization problem, cf. *(15.4)*. This fact, however, suggests another option that is available for the treatment of a multicriteria optimization problem, namely to interpret it into a form with a single, scalar objective function. As it is shown in Section 18.2, several options exist for scalarization of multicriteria optimization problems.

The form of a *Vector Optimization Problem* (VOP) is in analogy with *(15.4)*

$$\text{"Min"}_{\mathbf{x} \in \mathbf{R}^n} \left\{ \mathbf{f}(\mathbf{x}) \,\middle|\, \mathbf{h}(\mathbf{x}) = \mathbf{0} \quad , \quad \mathbf{g}(\mathbf{x}) \le \mathbf{0} \right\} , \tag{18.1a}$$

where $\mathbf{f}(\mathbf{x})$ is a so-called *vector objective function* of the design variables

$$\mathbf{f}(\mathbf{x}) := \begin{pmatrix} f_1(\mathbf{x}) \\ \vdots \\ f_m(\mathbf{x}) \end{pmatrix} \quad . \tag{18.1b}$$

Problems with multiple objective functions are characterized by the occurence of an *objective conflict*, i.e. *none* of the possible solutions allows for simultaneous optimal fulfillment of *all* objectives (denoted by "Min" in *(18.1a)*).

Definition 1: *Functional-efficiency or PARETO-optimality* [D.34, D.43]

A vector $\mathbf{x}^* \in X$ is then – and only then – termed PARETO-optimal or functional-efficient or p-efficient for the VOP *(18.1)* if no vector $\mathbf{x} \in X$ exists for which

$$\left.\begin{array}{lll} & f_j(\mathbf{x}) \le f_j(\mathbf{x}^*) & \text{for all } \; j \in \{1,\dots,m\} \\ \text{and} & f_j(\mathbf{x}) < f_j(\mathbf{x}^*) & \text{for at least one } \; j \in \{1,\dots,m\} \;. \end{array}\right\} \tag{18.2}$$

Fig. 18.1 depicts, as an example, a projection from the two-dimensional design space X into the objective function or criteria space Y. The PARETO-optimal solutions then lie on the sections of the arc AB ($\partial X^* \rightleftarrows \partial Y^*$). The designer may now choose one of these solutions depending on how he or she, from practical considerations, assesses the relative merits of the two objective functions.

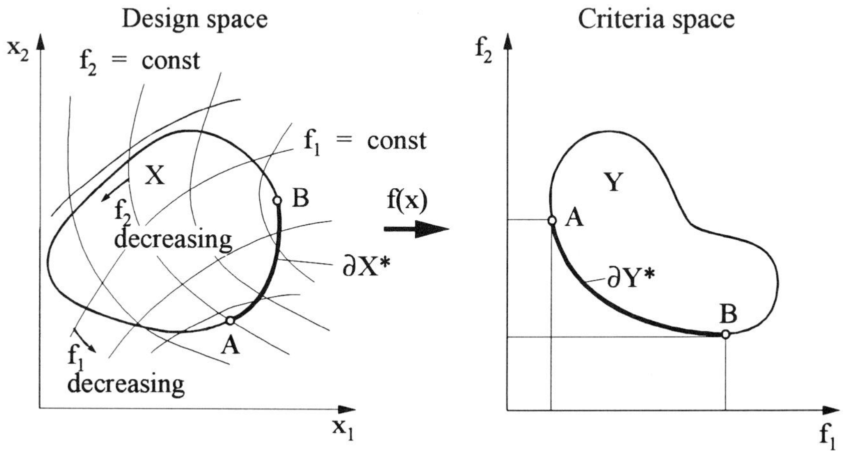

Fig. 18.1: Mapping of a feasible design space into the criteria space

Definition 2: *Substitute problem and preference function – Scalarization of multicriteria optimization problems*

Nonlinear multicriteria, vector, or multiobjective optimization problems can be scalarized, i.e. reduced to usual optimization problems with a single, scalar objective function by formulating a ***substitute problem*** [D.6, D.14, D.32, D.44].

The problem

$$\underset{\mathbf{x} \in \mathbf{R}^n}{\operatorname{Min}} \; p[\mathbf{f}(\mathbf{x})] \tag{18.3}$$

is called a scalarized substitute problem for a multicriteria optimization problem if there exists a $\tilde{\mathbf{x}} \in X$ such that

$$p[\mathbf{f}(\tilde{\mathbf{x}})] = \underset{\mathbf{x} \in \mathbf{R}^n}{\operatorname{Min}} \; p[\mathbf{f}(\mathbf{x})] \quad .$$

Here, p is called the *preference function* or scalarized objective function.

Preference functions:

– *Sum of weighted objectives*

$$p[\mathbf{f}(\mathbf{x})] := \sum_{j=1}^{m} [w_j f_j(\mathbf{x})] \quad , \quad \mathbf{x} \in \mathbf{R}^n \tag{18.5a}$$

with weighting factors chosen by the designer

$$0 \le w_j \le 1 \quad , \quad \sum_{j=1}^{m} w_j = 1 \; . \tag{18.5b}$$

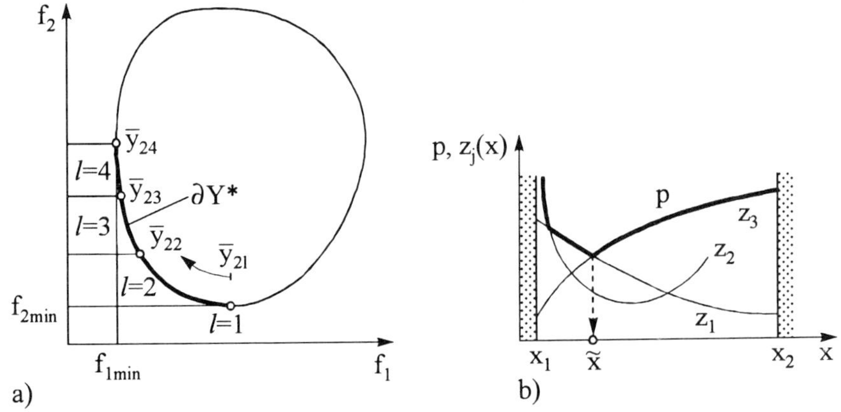

Fig. 18.2: Preference function a) Trade-off-formulation
b) Min-max-formulation

– *Sum of distance functions*

$$p[\mathbf{f}(\mathbf{x})] := \Big(\sum_{j=1}^{m} |f_j(\mathbf{x}) - \bar{y}_j|^r\Big)^{1/r} \quad , \quad \mathbf{x} \in \mathbf{R}^n \tag{18.6}$$

with the vector $\bar{\mathbf{y}}$ designating given goal values or demand levels for criteria f_j $(j = 1,\ldots,m)$. Here, the values of the components of $\bar{\mathbf{y}}$ and the exponent $r\,(1 \le r \le \infty)$ are at the choice of the designer.

– *Constraint-oriented transformation (Trade-off method)*

$$p[\mathbf{f}(\mathbf{x})] = f_1(\mathbf{x}) \quad , \quad f_j(\mathbf{x}) \le \bar{y}_j \quad , \quad j = 2,\ldots,m \quad , \quad \mathbf{x} \in \mathbf{R}^n \tag{18.7}$$

with $f_1(\mathbf{x})$ as the principal or main objective, and $f_2,\ldots,f_m$ as secondary or sides objectives (constraints). $\bar{y}_j$ denotes the corresponding respondence levels which are chosen by the designer.

– *Min-Max-formulation*

$$p[\mathbf{f}(\mathbf{x})] := \operatorname*{Max}_j[z_j(\mathbf{x})] \quad , \quad \mathbf{x} \in \mathbf{R}^n \tag{18.8}$$

$$\text{with} \quad z_j(\mathbf{x}) = \frac{f_j(\mathbf{x}) - \bar{f}_j}{\bar{f}_j} \quad , \quad \bar{f}_j > 0 \quad , \quad j = 1,\ldots,m \ ,$$

where $\bar{f}_j$ denote values specified separately by the designer for each objective function. For further details refer to [D.14, D.32].

Extended Min-Max by weighted objectives:

$$p[\mathbf{f}(\mathbf{x})] := \operatorname*{Max}_j[w_j \cdot f_j(\mathbf{x})] \quad , \quad \mathbf{x} \in \mathbf{R}^n \tag{18.9}$$

with weighting factors analogously to *(18.5b)*.

Note that the full formulation of the Min-Max problem with weighted objectives is (substitute *(18.9)* into *(18.3)*):

$$\operatorname*{Min}_{\mathbf{x} \in \mathbf{R}^n} \Big(\operatorname*{Max}_j[w_j \cdot f_j(\mathbf{x})]\Big) \ . \tag{18.10}$$

This Min-Max problem can be given as the equivalent *Bound Formulation* [D.6, D.30, D.31]

$$\operatorname*{Min}_{\beta,\,\mathbf{x} \in \mathbf{R}^n} \beta \qquad \text{subject to} \qquad w_j \cdot f_j(\mathbf{x}) \le \beta \ . \tag{18.11}$$

Here, β is an additional, scalar parameter termed the *bound variable* which executes the task *(18.10)*. Thus, β constitutes a variable upper bound on each of the weighted objectives (now transformed into constraints) while at the same time subject to minimization since adopted as the objective function of the scalarized optimization problem *(18.11)*.

In fact, the full set of PARETO-solutions, cf. Section 18.1, can be generated by application of the preference functions covering an appropriate range of values for the weighting factors w_j. Thus, the designer's choice of values for w_j is related to the application of judgment in a PARETO-approach.

18.2 Shape optimization

The term *shape optimization* denotes the optimal shaping of components by simultaneously considering given requirements. In order to achieve this goal, functions have to be determined which describe the shape to be optimized (*shape functions*). Hence, in general, shape optimization problems lead to the formulation of *objective functionals* F, and similarly, general *constraint operators* G have to be considered. As the shape is continuously varied during the optimization, the respective model (e.g. partitioning into structure or shell elements) must be adapted accordingly, and this often requires re-discretization.

(a) *Indirect methods* [D.4, D.5, D.7, D.9, D.21, D.22, D.27, D.28, A.3]

The above methods incorporate two steps:

1) Derivation of *optimality conditions* as necessary conditions for the optimal design,
2) Fulfilment of the *optimality conditions* by means of suitable solution procedures.

The curvilinear coordinates ξ^α define the area A of a load-bearing structure with the boundary Γ. The optimal shape function $\mathbf{R}$ with the components R^j $(j = 1,2,3)$ shall be determined. The derivatives $\partial R^j / \partial \xi^\alpha$ are abbreviated as $R^j_{,\alpha}$. In addition, EINSTEIN's summation convention (see Section 2.2) will be used.

The following considerations will be limited to such shape optimization problems for which both the optimization objective and the constraints can be expressed in the form of integrals [A.3, A.6, D.5]:

$$\operatorname{Min} F = \operatorname{Min} \int_A f(\xi^\alpha, R^j, R^j_{,\alpha})\, dA \qquad (j = 1,2,3)\,, \tag{18.12a}$$

where

$$\int_A f_k(\xi^\alpha, R^j, R^j_{,\alpha})\, dA = 0 \quad (k = 1,\ldots,m)\,, \tag{18.12b}$$

$$\int_A f_l(\xi^\alpha, R^j, R^j_{,\alpha})\, dA \le 0 \quad (l = 1,\ldots,r) \tag{18.12c}$$

are assumed to be given.

In order to derive the necessary conditions for the present problem, the inequality operators *(18.12c)* are first transformed into equality operators, using slack variables η_l:

$$\int_A f_l(\xi^\alpha, R^j, R^j_{,\alpha})\, dA + \eta_l^2 = 0 \qquad (l = 1,\ldots,r)\,. \tag{18.13}$$

Using the LAGRANGEAN multipliers λ_k, λ_l and the abbreviation

$$\begin{aligned}\Phi = {} & f(\xi^\alpha, R^j, R^j_{,\alpha}) + \sum_{k=1}^{m} \lambda_k f_k(\xi^\alpha, R^j, R^j_{,\alpha}) + \\ & + \sum_{l=1}^{r} \lambda_l f_l(\xi^\alpha, R^j, R^j_{,\alpha})\,,\end{aligned} \tag{18.14a}$$

we obtain the LAGRANGE-functional as

$$I = \int_A \Phi(\xi^\alpha, R^j, R^j_{,\alpha}, \lambda_k, \lambda_l)\, dA + \sum_{l=1}^{r} \lambda_l \eta_l^2\,. \tag{18.14b}$$

The EULER equations for the variational problem

$$\text{Min}\; I(\xi^\alpha, R^j, R^j_{,\alpha}, \lambda_k, \lambda_l, \eta_l) \tag{18.15}$$

will now be very briefly set up and solved.

Owing to the demand that the first variation δI has to vanish ($\delta I = 0$) for arbitrary variations of R^j, we obtain the following partial differential equation including boundary conditions when considering the GAUSSIAN rule of integration and component-wise application of the *fundamental lemma of variational calculus*:

$$\frac{\partial \Phi}{\partial R^j} - \left(\frac{\partial \Phi}{\partial R^j_{,\alpha}}\right)_{,\alpha} = 0\,, \qquad \int n^\alpha \frac{\partial \Phi}{\partial R^j_{,\alpha}}\, \delta R^j\, d\Gamma = 0\,, \tag{18.16}$$

where n^α are the components of the normal unit vector on the boundary Γ.

$\delta I = 0$ recovers the constraints *(18.12b)* and *(18.13)* for arbitrary admissible variation of λ_k and λ_l, and variation of η_l yields

$$2\lambda_l \eta_l = 0 \qquad (l = 1,\ldots,r)\,. \tag{18.17}$$

By separating *(18.13)* by means of *(18.17)* into the cases of active ($\eta_l = 0$, $\lambda_l \neq 0$) and non-active ($\eta_l \neq 0, \lambda_l = 0$) inequality operators, the η_l can be eliminated as follows:

$$\lambda_l \left[\int_A f_l(\xi^\alpha, R^j, R^j_{,\alpha})\, dA\right] \leq 0 \qquad (l = 1,\ldots,r)\,. \tag{18.18}$$

With EULER equations according to *(18.16)*, the relations *(18.18)*, and the equality constraints *(18.12b)*, all required equations are available for determining the unknown quantities R^j, λ_k, and λ_l for the present problem.

(b) *Direct methods* [D.16, D.47, D.48]

In the direct methods the shape optimization problems are transformed into parameter optimization problems which are then treated by means of MP-algorithms according to Fig. 8.6.

One determines an optimal shape function R^* for which the objective functional F attains a minimum

$$\operatorname*{Min}_{R \in \boldsymbol{\Gamma}_2} F(\mathbf{R}) \longrightarrow F(\mathbf{R}^*) \qquad (18.19)$$

with $\boldsymbol{\Gamma}_2$ denoting the set of all shape functions.

The feasible variational domain is defined by the constraint operators H_i, G_j:

$$\left.\begin{aligned} H_i \mathbf{R} &= \varphi_i \quad (i = 1,\ldots,q) , \\ G_j \mathbf{R} &\leq \chi_j \quad (j = 1,\ldots,p) . \end{aligned}\right\} \qquad (18.20)$$

The unknown functions $\mathbf{R}$ are approximated by suitable functions $\widetilde{\mathbf{R}}$, so-called shape approximation functions.

In recent years, the progress of CAD-techniques in the design and construction departments has substantially increased the importance of *geometrical modeling* also in application to structural optimization. Basically, geometrical modeling deals with computer-based design and manipulation of geometrical shapes [D.8, D.30].

The choice of suitable approximation functions for optimal geometries is problem-dependent. The chosen function is to approximate the course to be followed as precisely as possible, a demand that leads to a large amount of shape parameters and thus to increased computational effort. A reduction of this effort can be achieved by decreasing the number of parameters, which, however, requires some a-priori knowledge and experience concerning the choice of a given type of approximation. If this information does not exist, optimization should proceed with simple approximations to be refined with increasing level of knowledge.

In the following, we will introduce some of the most important approximation functions for geometric modeling of shapes of components:

1) Shape functions depending on a single variable

This type of approximation function is chosen if the shape optimization can be reduced to optimization of curves R that only depend on one coordinate ξ, i.e. curves that can be described by either a continuous function $\widetilde{R}$ or by the sum of single continuous functions within the defined domain.

A *general polynomial function* describes the dependence of a shape function $\widetilde{R}^j$ on the local coordinate ξ in the following form:

$$\widetilde{R}^j(\xi, \mathbf{x}) = x_1 + x_2\xi + x_3\xi^2 + \ldots + x_n\xi^{n+1} . \qquad (18.21)$$

Monotonically increasing or decreasing functions can be suitably approximated by polynomia. However, for $n > 3$ very undesirable, strong oscillatory behaviour generally occurs. In addition, a precise representation of particularly interesting boundaries, edges, or transitions is often not possible.

CHEBYCHEV-functions are polynomia with special properties [A.3]:

$$\widetilde{R}^{j}(\xi, \mathbf{x}) = x_1 T_0(\xi) + x_2 T_1(\xi) + x_3 T_2(\xi) \dots \quad , \qquad (18.22)$$

where T_i ($i = 0, \dots, n-1$) describe the CHEBYCHEV-polynomia from the 0-th to the ($n-1$)-th degree. The T_1-polynomia depending on ξ are calculated as follows for the range of $\xi_l \leq \xi \leq \xi_u$:

$$T_0 = 1 \ , \quad T_1 = 2\frac{\xi - \xi_l}{\xi_u - \xi_l} - 1 \ , \quad T_2 = 2T_1 \cdot T_1 - T_0 \ , \quad \dots,$$

$$T_k = 2T_1 \cdot T_{k-1} - T_{k-2} \quad .$$

Besides being orthogonal, the CHEBYCHEV system of polynomia also possesses the favourable properties *uniform convergence* and *optimality*. However, even the CHEBYCHEV-polynomia only allow for a limited precise representation of single domains.

A nonlinear, parametric *B-spline-function* is defined by $n+1$ control points which define a so-called control polygon (Fig. 18.3). With the exception of the starting point and the end point of the control polygon, the control points do not lie on the B-spline curve. The curve $\mathbf{r}(\xi)$ is defined by

$$\mathbf{r}(\xi) = \sum_{i=0}^{n} \mathbf{p}_i B_{ik}(\xi) \qquad (18.23a)$$

with

$\mathbf{p}_i$ vector of the i-th control point in the given x^1,x^2 coordinate system ,

$B_{ik}(\xi)$ mixed function .

The mixed functions $B_{ik}(\xi)$, or base functions of B-splines, are polynomial-parameter functions of degree $k-1$ which can be calculated by means of the following recursive formula:

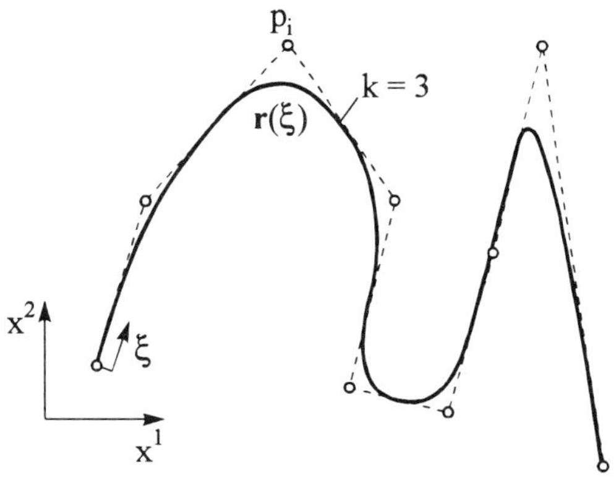

Fig. 18.3: B-spline-curve of degree k = 3 with 9 control points

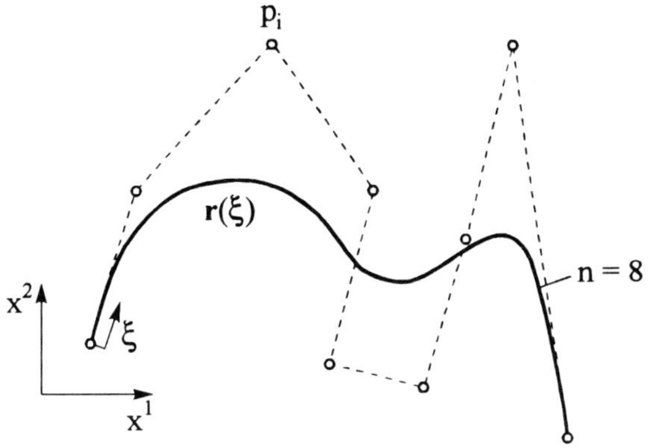

Fig. 18.4: BÉZIER-curve of degree n = 8 (9 control points)

$$B_{i1}(\xi) = \begin{cases} 1 & \text{for } t_1 \le \xi \le t_{i+1} \\ 0 & \text{for all other } \xi \end{cases} \quad , \tag{18.23b}$$

$$B_{ik}(\xi) = \frac{(\xi - t_i)}{t_{i+k-1} - t_i} B_{ik-1}(\xi) + \frac{t_{i+k} - \xi}{t_{i+k} - t_{i+1}} B_{i+1\,k-1}(\xi) \quad . \tag{18.23c}$$

The quantities t_i are called course node quantities, and they assign the value of the parameter ξ to the control points and thus influence the shape of the curve. For more details see [D.8].

BÉZIER-curves are defined in analogy with the description of the B-splines, using n + 1 control points:

$$\mathbf{r}(\xi) = \sum_{i=0}^{n} \mathbf{p}_i B_{ik}(\xi) \tag{18.24a}$$

with $\mathbf{p}_i$ vector of the i-th control point ,

$B_{ik}(\xi)$ scalar mixed function of degree k .

In contrast to the non-periodical B-splines, the parameters here range between $0 \le \xi \le 1$. As is the case with the B-spline-curves, the control points of BÉZIER-curves generally do not lie on the curve, with the exception of the first and last control point (Fig. 18.4). The mixed function $B_{ik}(\xi)$ is a scalar polynomial-parameter function, a so-called BERNSTEIN-polynomium of k-th degree, weighted by binomial coefficients [D.8]:

$$B_{ik}(\xi) = \frac{k!}{i!(k-i)} (\xi)^i (1-\xi)^{n-1} \quad . \tag{18.24b}$$

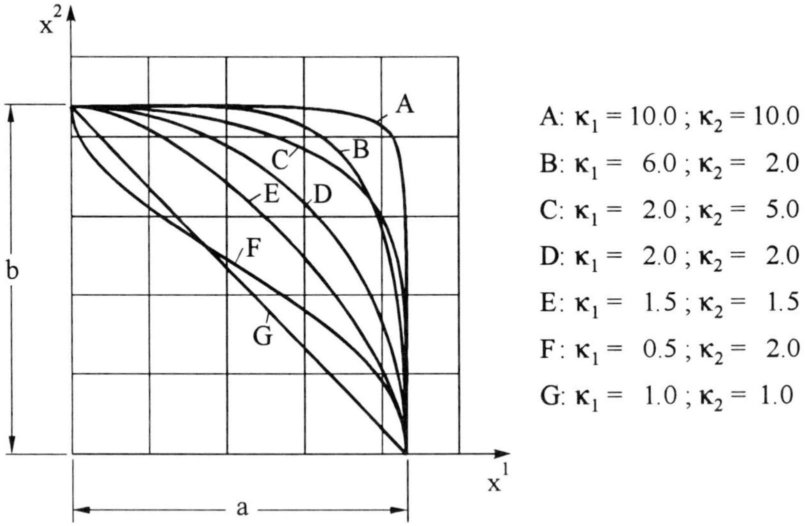

Fig. 18.5: Influence of the shape parameters on the modified ellipse [D.13]

Ellipse functions with variable exponents can successfully be employed in order to determine the shape of boilers or to find optimal notch configurations [D.13]. In mathematical terms these functions read:

$$\left(\frac{x^1}{a}\right)^{\kappa_1} + \left(\frac{x^2}{b}\right)^{\kappa_2} = 1 \qquad (18.25a)$$

with the shape parameters κ_1, κ_2 and the semi-axes a, b.

The parametrical representation of the ellipse equation reads:

$$x^1 = a(\sin\varphi)^{2/\kappa_1}, \quad x^2 = b(\cos\varphi)^{2/\kappa_2} \quad \text{with parameter } \varphi . \qquad (18.25b)$$

Fig. 18.5 illustrates the influence of the shape parameters κ_1, κ_2 on the shape.

18.3 Augmented optimization loop by additional strategies [D.3, D.12, D.40]

Fig. 15.4 presents the basic modules of an optimization model. The sensitivity analysis treated in Chapter 17 as well as the two optimization strategies multicriteria or and optimization (Chapter 18) contribute modules that are implemented into the optimization model, and they thus present important elements of an effective treatment of structural optimization problems. *Direct methods* are especially appropriate for solving multicriteria and shape optimization problems the basic procedure of which is shown in Fig. 18.6, while shape optimization problems can be processed by a mere augmentation of the design model, multicriteria optimization requires a special evaluation model. Fig. 18.7 illustrates how the optimization loop is augmented by these additional modules within the *Three-Columns-Concept* [D.12] discussed in Section 15.6.

In a similar manner, other modules can be implemented into the loop, e.g. for optimization with time-dependent parametric quantities, or for stochastic optimization problems. An important future demand on the optimization process will be the consideration of multidisciplinary aspects from the fields of fluid- or aerodynamics, thermodynamics, heat transfer, manufacturing, etc. In this context, the term *multidisciplinary optimization* has become general use [D.42].

Vector Optimization
"Min" $f(x)$, $x \in \mathbb{R}^n$
$h(x) = 0$
$g(x) \leq 0$
$x_l \leq x \leq x_u$

Shape Optimization
Min $F(R)$, $R \in \Gamma_2$
$H_i R = 0$, $i = 1,...,q$
$G_j R \leq 0$, $j = 1,...,p$
$R_l \leq R \leq R_u$

Direct Optimization Strategies (OS)

Preference Functions
$f(x) \rightarrow p[f(x)]$

Shape Approach Functions
$R \approx \widetilde{R}(\xi^\alpha, x)$

Parameter Optimization Problem

Min $p[f(x)]$
$h_i(x) = 0$, $i = 1,...,m$
$g_j(x) \leq 0$, $j = 1,...,n$
$x_l \leq x \leq x_u$

Min $F[\widetilde{R}(\xi^\alpha, x)]$
$h_i(x) = 0$, $i = 1,...,q$
$g_j(x) \leq 0$, $j = 1,...,p$
$x_l \leq x \leq x_u$

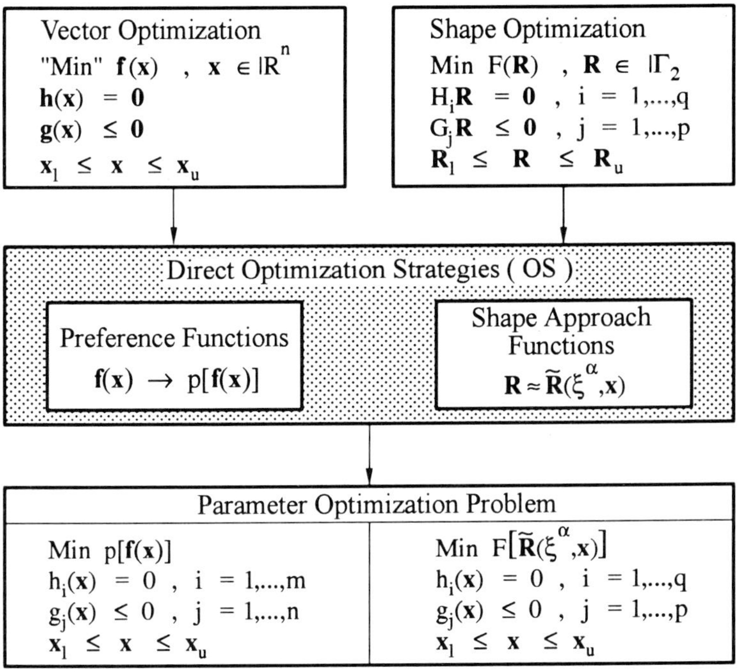

Fig. 18.6: Direct optimization strategies

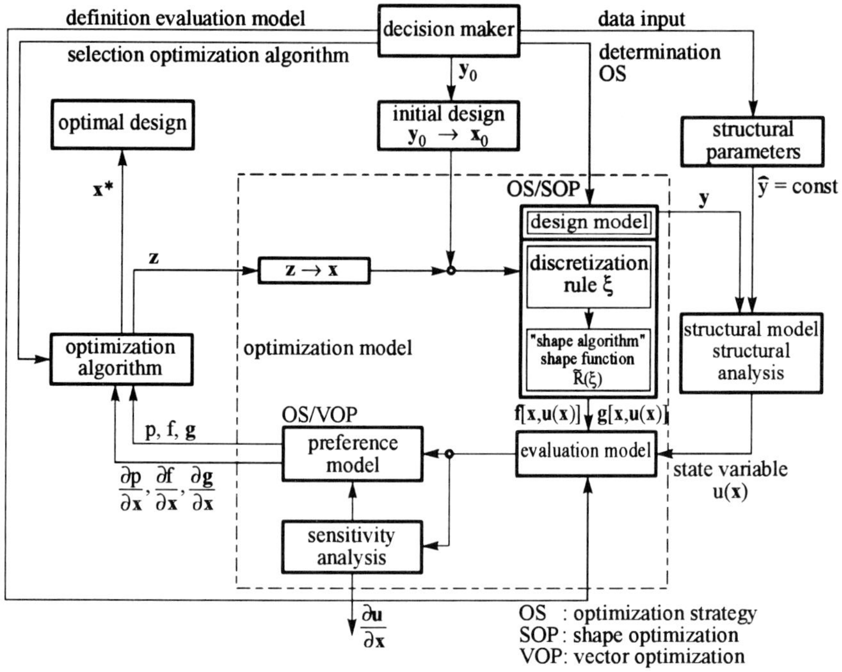

Fig. 18.7: Optimization loop augmented by multicriteria and shape optimization

D. 2 Exercises

Exercise D-15/16-1:

An unconstrained optimization problem is given by the objective function

$$f(x_1, x_2) = 12\,x_1^2 + 4\,x_2^2 - 12\,x_1 x_2 + 2\,x_1 \longrightarrow \text{Min} \quad , \quad x_1, x_2 \in \mathbf{R}^n.$$

a) Determine the minimum of this function using the necessary and the sufficient conditions.

b) Check the exact result by means of the POWELL-method, starting with $\mathbf{x}_0 = (-1, -2)^T$ as initial point.

c) Apply also the algorithm of conjugate gradients according to FLETCHER–REEVES to obtain the result. Let again $\mathbf{x}_0 = (-1, -2)^T$ be the starting point.

Solution:

a) We are confronted with an unconstrained optimization problem with a continuously differentiable objective function possessing an exact solution.

According to *(15.7)* the candidate minimum point is obtained from the necessary conditions

$$\frac{\partial f}{\partial x_1} = 24\,x_1 - 12\,x_2 + 2 \stackrel{!}{=} 0 \ ,$$

$$\frac{\partial f}{\partial x_2} = 8\,x_2 - 12\,x_1 \stackrel{!}{=} 0 \longrightarrow x_2 = \frac{3}{2}\,x_1 \ .$$

By substituting x_2 into the first equation one obtains:

$$24\,x_1 - 18\,x_1 + 2 = 0 \longrightarrow x_1^* = -\frac{1}{3} \ , \quad x_2^* = -\frac{1}{2} \ .$$

The corresponding function value becomes

$$f^* = -\frac{1}{3} \ .$$

The HESSIAN matrix is calculated from *(15.8)*

$$\mathbf{H}(\mathbf{x}) = \begin{bmatrix} 24 & -12 \\ -12 & 8 \end{bmatrix} = 48 \ .$$

This proves positive definiteness, i.e. a minimum solution has been found.

b) The starting vector for the POWELL-method is given as:

$$\mathbf{x}_0 = (-1, -2)^T \longrightarrow f_0 = 2 \ .$$

First cycle

For the first search direction we choose $\mathbf{s}_0 = (1,0)^T$. Thus, we obtain according to *(16.1a)*:

$$\mathbf{x}_1 = \begin{pmatrix} -1 \\ -2 \end{pmatrix} + \alpha \begin{pmatrix} 1 \\ 0 \end{pmatrix} = \begin{pmatrix} -1+\alpha \\ -2 \end{pmatrix} . \qquad (1)$$

By substituting (1) into the given function

$$f(\alpha) = 12(-1+\alpha)^2 + 4(-2)^2 - 12(-1+\alpha)(-2) + 2(-1+\alpha)$$

$$\frac{\partial f}{\partial \alpha} = 24(-1+\alpha) + 24 + 2 = 0 \quad \longrightarrow \quad \alpha = -\frac{1}{12} ,$$

which yields

$$\mathbf{x}_1 = \begin{pmatrix} -\frac{13}{12} \\ -2 \end{pmatrix} \quad \text{and} \quad f(\mathbf{x}_1) = 1.9167 .$$

As second search direction we choose $\mathbf{s}_1 = (0,1)^T$, which, in accordance with (1), leads to

$$\mathbf{x}_2 = \mathbf{x}_1 + \alpha\,\mathbf{s}_1 = \begin{pmatrix} -\frac{13}{12} \\ -2 \end{pmatrix} + \alpha \begin{pmatrix} 0 \\ 1 \end{pmatrix} = \begin{pmatrix} -\frac{13}{12} \\ -2+\alpha \end{pmatrix} . \qquad (2)$$

Substitution of (2) into the given function:

$$f(\alpha) = 12\left(-\frac{13}{12}\right)^2 + 4(-2+\alpha)^2 - 12\left(-\frac{13}{12}\right)(-2+\alpha) + 2\left(-\frac{13}{12}\right)$$

$$\frac{\partial f}{\partial \alpha} = 8(-2+\alpha) + 13 = 0 \quad \longrightarrow \quad \alpha = \frac{3}{8} .$$

One thus obtains

$$\mathbf{x}_2 = \begin{pmatrix} -\frac{13}{12} \\ -\frac{13}{8} \end{pmatrix} \quad \text{and} \quad f(\mathbf{x}_2) = 1.3542 .$$

An additional search direction is determined by means of *(16.1d)*

$$\mathbf{s}_2 = \mathbf{x}_2 - \mathbf{x}_0 = \begin{pmatrix} -\frac{13}{12} \\ -\frac{13}{8} \end{pmatrix} - \begin{pmatrix} -1 \\ -2 \end{pmatrix} = \begin{pmatrix} -\frac{1}{12} \\ \frac{3}{8} \end{pmatrix} . \qquad (3)$$

Then it follows

$$\mathbf{x}_3 = \begin{pmatrix} -1 \\ -2 \end{pmatrix} + \alpha \begin{pmatrix} -\frac{1}{12} \\ \frac{3}{8} \end{pmatrix} = \begin{pmatrix} -1-\frac{\alpha}{12} \\ -2+\frac{3}{8}\alpha \end{pmatrix} . \qquad (4)$$

Substitution into $f(x_1, x_2)$ and re-formulation yields

$$f(\alpha) = -\left(1 + \frac{\alpha}{12}\right)\left(14 - \frac{11}{2}\alpha\right) + 4\left(-2 + \frac{3}{8}\alpha\right)^2$$

$$\frac{\partial f}{\partial \alpha} = -\frac{1}{12}\left(14 - \frac{11}{2}\alpha\right) - \left(1 + \frac{\alpha}{12}\right)\left(-\frac{11}{2}\right) + 8\left(-2 + \frac{3}{8}\alpha\right)\frac{3}{8} \stackrel{!}{=} 0$$

$$\longrightarrow \quad \alpha = \frac{40}{49} \quad .$$

From (4) one determines

$$\mathbf{x}_3 = \begin{pmatrix} -\frac{157}{147} \\ -\frac{83}{49} \end{pmatrix} \quad \text{and} \quad f(\mathbf{x}_3) = 1.319728 \quad .$$

This conclude the first cycle.

Second cycle

The second cycle also proceeds from the search direction $\mathbf{s}_0 = (1, 0)^T$. We get

$$\mathbf{x}_4 = \mathbf{x}_3 + \alpha\,\mathbf{s}_0 = \begin{pmatrix} -\frac{157}{147} + \alpha \\ -\frac{83}{49} \end{pmatrix} , \qquad (5)$$

$$f(\alpha) = 12\left(-\frac{157}{147} + \alpha\right)^2 + 4\left(-\frac{83}{49}\right)^2 - 12\left(-\frac{157}{147} + \alpha\right)\left(-\frac{83}{49}\right) + $$
$$+ 2\left(-\frac{157}{147} + \alpha\right)$$

$$\frac{\partial f}{\partial \alpha} = 24\left(-\frac{157}{147} + \alpha\right) + 12 \cdot \frac{83}{49} + 2 \stackrel{!}{=} 0 \quad \longrightarrow \quad \alpha = 0.1377552$$

and finally from (5)

$$\mathbf{x}_4 = \begin{pmatrix} -0.9302720 \\ -1.6938775 \end{pmatrix} \quad \text{and} \quad f(\mathbf{x}_4) = 1.092008 \quad .$$

We then formulate

$$\mathbf{x}_5 = \mathbf{x}_4 + \alpha\,\mathbf{s}_2 = \begin{pmatrix} -0.9302720 \\ -1.6938775 \end{pmatrix} + \alpha \begin{pmatrix} -0.0833333 \\ 0.375 \end{pmatrix} , \qquad (6)$$

$$f(\alpha) = 12(-0.9302720 - 0.083333\,\alpha)^2 + 4(-1.6938775 + 0.375\,\alpha)^2 -$$
$$- 12(-0.9302720 - 0.083333\,\alpha)(-1.6938775 + 0.375\,\alpha) +$$
$$+ 2(-0.9302720 - 0.083333\,\alpha)$$

$$\frac{\partial f}{\partial \alpha} = 0 \quad \longrightarrow \quad \alpha = 0.438567 \quad .$$

From (6) follows

$$\mathbf{x}_5 = \begin{pmatrix} -0.9668191 \\ -1.5294147 \end{pmatrix} \qquad \text{and} \qquad f(\mathbf{x}_5) = 0.89566164 \quad .$$

The course of the optimization process clearly shows a very slow convergence towards the solution point. We therefore stop the treatment at this point and proceed to c).

c) Algorithm of conjugate gradients according to FLETCHER–REEVES

We again proceed from the starting point $\mathbf{x}_0 = (-1, -2)^T$. The starting direction is calculated from the gradient

$$\nabla f(\mathbf{x}_0) = \nabla f(\mathbf{x})\Big|_{\mathbf{x}_0} = \begin{pmatrix} 24\,x_1 - 12\,x_2 + 2 \\ 8\,x_2 - 12\,x_1 \end{pmatrix}\Bigg|_{\mathbf{x}_0} .$$

The steepest descent direction is

$$\mathbf{s}_0 = -\nabla f(\mathbf{x}_0) = \begin{pmatrix} -2 \\ 4 \end{pmatrix} .$$

Eq. *(16.6)* yields for the end of the first step

$$\mathbf{x}_1 = \mathbf{x}_0 + \alpha_0\,\mathbf{s}_0 = \begin{pmatrix} -1 \\ -2 \end{pmatrix} + \alpha_0 \begin{pmatrix} -2 \\ 4 \end{pmatrix} = \begin{pmatrix} -1 - 2\alpha_0 \\ -2 + 4\alpha_0 \end{pmatrix}$$

and thus

$$\begin{aligned} f(\alpha_0) = {} & 12(-1 - 2\alpha_0)^2 + 4(-2 + 4\alpha_0)^2 - \\ & - 12(-1 - 2\alpha_0)(-2 + 4\alpha_0) + 2(-1 - 2\alpha_0) \;, \end{aligned}$$

$$\frac{df}{d\alpha_0} = 0 \qquad \longrightarrow \qquad \alpha_0 = 0.048077 \quad .$$

The new point $\mathbf{x}_1$ reads

$$\mathbf{x}_1 = \begin{pmatrix} -1.0961 \\ -1.8077 \end{pmatrix} \qquad \text{and} \qquad \nabla f(\mathbf{x}_1) = \begin{pmatrix} -2.6140 \\ -1.3084 \end{pmatrix} .$$

The next search direction is calculated with *(16.7)*

$$\mathbf{s}_1 = -\nabla f(\mathbf{x}_1) + \frac{\left|\nabla f(\mathbf{x}_1)\right|^2}{\left|\nabla f(\mathbf{x}_0)\right|^2}\,\mathbf{s}_0 \;,$$

$$\mathbf{s}_1 = \begin{pmatrix} 2.6140 \\ 1.3084 \end{pmatrix} + \frac{(-2.6140)^2 + (-1.3084)^2}{(-2)^2 + 4^2} \begin{pmatrix} -2 \\ 4 \end{pmatrix} =$$

$$= \begin{pmatrix} 2.6140 \\ 1.3084 \end{pmatrix} + 0.4272 \begin{pmatrix} -2 \\ 4 \end{pmatrix} = \begin{pmatrix} 1.7596 \\ 3.0172 \end{pmatrix} .$$

The next point is determined from

$$\mathbf{x}_2 = \mathbf{x}_1 + \alpha_1 \mathbf{s}_1 = \begin{pmatrix} -1.0961 \\ -1.8077 \end{pmatrix} + \alpha_1 \begin{pmatrix} 1.7596 \\ 3.0172 \end{pmatrix}$$

and correspondingly $f(\alpha_1)$. The minimum condition

$$\frac{df}{d\alpha_1} = 0 \qquad \text{yields} \qquad \alpha_1 = 0.4334 \quad .$$

One obtains

$$\mathbf{x}_2 = \begin{pmatrix} -0.3334 \\ -0.5 \end{pmatrix} \qquad \text{and} \qquad \nabla f(\mathbf{x}_2) = \begin{pmatrix} 0 \\ 0 \end{pmatrix} \quad .$$

Thus, the condition

$$\mathbf{s}_0^T \mathbf{H} \mathbf{s}_1 = (-2 \ , \ 4) \begin{pmatrix} 24 & -12 \\ -12 & 8 \end{pmatrix} \begin{pmatrix} 1.7596 \\ 3.0172 \end{pmatrix} \approx 0$$

is fulfilled.

Fig. D-1 illustrates the single search steps for the FLETCHER–REEVES–method. It is obvious that this method converges much faster than the POWELL-method. By suitable modifications, however, convergence of the latter method can be improved.

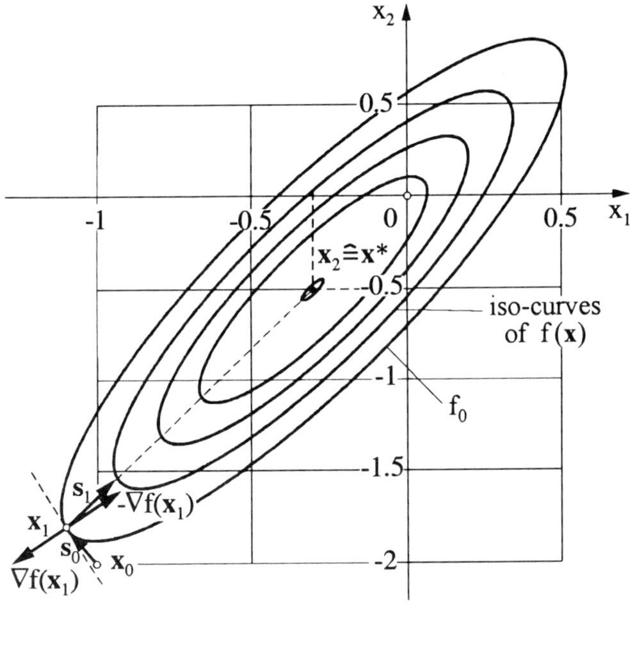

Fig. D-1: Search steps for the FLETCHER-REEVES-method

Exercise D-15/16-2:

The truss structure shown in Fig. D-2 consists of 13 steel bars with the cross-sectional areas A_i $(i = 1,\ldots,13)$ and 10 nodal points. A vertical force $F = 100\ kN$ acts at node 3.

Determine the cross-sectional areas in such a way that the weight of the structure is minimized. The stresses in the single bars must not exceed an admissible tensile stress of $\sigma_{t_{adm}} = +150\ MPa$, and an admissible compressive stress of $\sigma_{c_{adm}} = -100\ MPa$.

As further values are given:

$l = 1.0\ m$, $E = 2.1 \cdot 10^5\ MPa$.

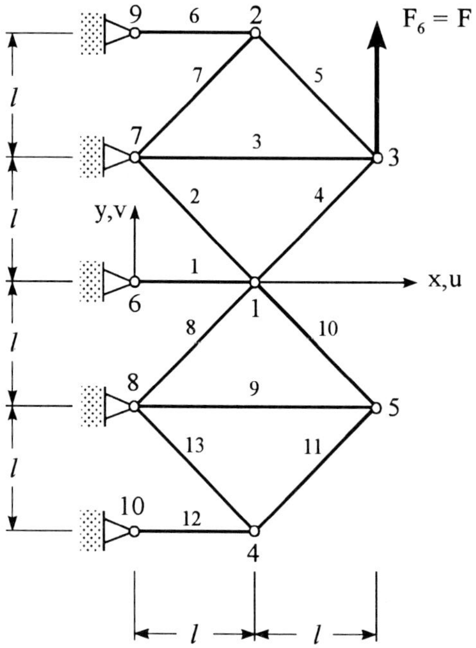

Fig. D-2: Plane truss structure

a) Formulate the structural model, and determine the solutions for displacements and stresses.

b) In order to formulate the optimization problem, define the objective function and the constraints when the cross-sectional areas are used as design variables $\mathbf{x} := \mathbf{A} = (A_i)^T$ $(i = 1,\ldots,13)$.

c) Determine the optimal solution of the constrained optimization problem by means of an external penalty function approach.

Solution:

a) The relation between the nodal forces and nodal displacements is established by means of the displacement method. This will be demonstrated for the forces F_3 and F_4 acting at node 2 and the corresponding displacements u_2 and v_2.

Equilibrium conditions:

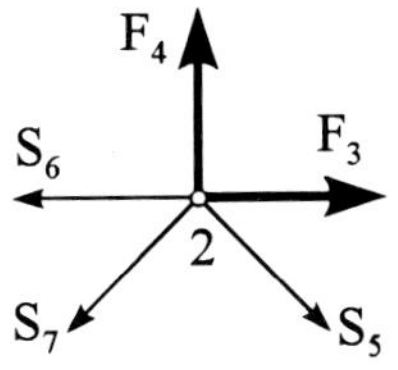

Equilibrium conditions give the external forces in terms of the bar forces at node 2:

$$\left.\begin{aligned} F_3 &= S_6 + \tfrac{1}{2}\sqrt{2}\,S_7 - \tfrac{1}{2}\sqrt{2}\,S_5 \ , \\ F_4 &= \tfrac{1}{2}\sqrt{2}\,S_7 + \tfrac{1}{2}\sqrt{2}\,S_5 \ . \end{aligned}\right\} \qquad (1)$$

Elasticity law:

$$\Delta l_6 = \frac{l}{EA_6}\,S_6 \ , \quad \Delta l_7 = \frac{\sqrt{2}\,l}{EA_7}\,S_7 \ , \quad \Delta l_5 = \frac{\sqrt{2}\,l}{EA_5}\,S_5 \ . \qquad (2)$$

Kinematics:

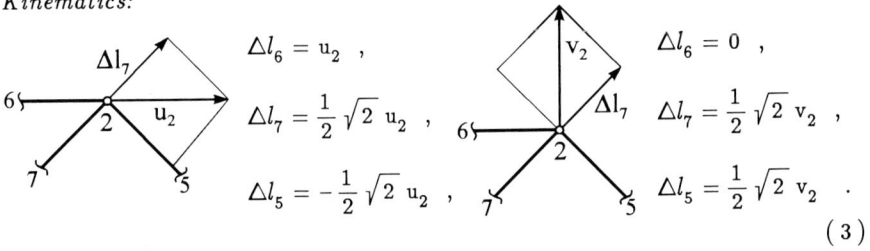

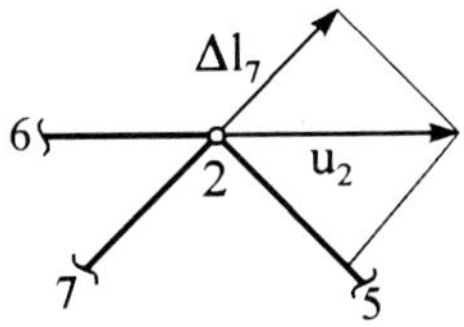

$$\Delta l_6 = u_2 \ , \quad \Delta l_7 = \frac{1}{2}\sqrt{2}\,u_2 \ , \quad \Delta l_5 = -\frac{1}{2}\sqrt{2}\,u_2 \ ,$$

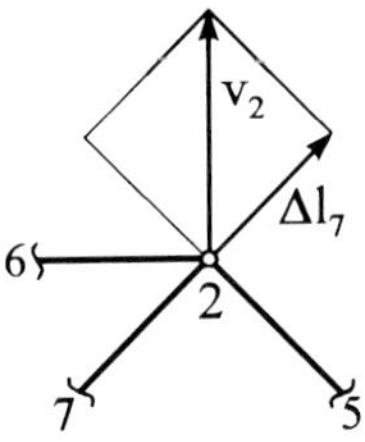

$$\Delta l_6 = 0 \ , \quad \Delta l_7 = \frac{1}{2}\sqrt{2}\,v_2 \ , \quad \Delta l_5 = \frac{1}{2}\sqrt{2}\,v_2 \ . \qquad (3)$$

Substitution of (2) and (3) into (1) yields

$$\begin{bmatrix} F_3 \\ F_4 \end{bmatrix} = \mathbf{K}^*_{22} \begin{bmatrix} u_2 \\ v_2 \end{bmatrix}$$

with the element stiffness matrix

$$\mathbf{K}^*_{22} = \frac{E}{2\sqrt{2}\,l}\begin{bmatrix} A_5 + 2\sqrt{2}\,A_6 + A_7 & -A_5 + A_7 \\ -A_5 + A_7 & A_5 + A_7 \end{bmatrix} = \frac{E}{2\sqrt{2}\,l}\,\mathbf{K}_{22} \ . \qquad (4)$$

Analogous relations can be derived for the other nodes.

The total stiffness matrix $\mathbf{K}$ consists of the single stiffness matrices of the bars; it relates the external forces to the displacements in the following linear equation:

$$\mathbf{f} = \mathbf{K}\,\mathbf{v} \qquad (5a)$$

with the displacement vector

$$\mathbf{v} = (u_1, v_1, u_2, v_2, u_3, v_3, u_4, v_4, u_5, v_5)^T \ ,$$

the force vector consisting of the 10 nodal forces

$$\mathbf{f} = (F_1, F_2, \ldots, F_{10})^T \quad ,$$

and the symmetric total stiffness matrix

$$\mathbf{K} = \frac{E}{2\sqrt{21}} \left[\begin{array}{c|c|c|c|c} \mathbf{K}_{11} & \mathbf{K}_{12} & \mathbf{K}_{13} & \mathbf{K}_{14} & \mathbf{K}_{15} \\ \hline \mathbf{K}_{21} & \mathbf{K}_{22} & \mathbf{K}_{23} & \mathbf{K}_{24} & \mathbf{K}_{25} \\ \hline \mathbf{K}_{31} & \mathbf{K}_{32} & \mathbf{K}_{33} & \mathbf{K}_{34} & \mathbf{K}_{35} \\ \hline \mathbf{K}_{41} & \mathbf{K}_{42} & \mathbf{K}_{43} & \mathbf{K}_{44} & \mathbf{K}_{45} \\ \hline \mathbf{K}_{51} & \mathbf{K}_{52} & \mathbf{K}_{53} & \mathbf{K}_{54} & \mathbf{K}_{55} \end{array} \right] \tag{5b}$$

with $$\mathbf{K}_{11} = \begin{bmatrix} 2\sqrt{2}\,A_1 + A_2 + A_4 + \\ + A_8 + A_{10} & -A_2 + A_4 + A_8 - A_{10} \\ -A_2 + A_4 + A_8 - A_{10} & A_2 + A_4 + A_8 + A_{10} \end{bmatrix} ,$$

$$\mathbf{K}_{22} = \text{see } (4) \ ,$$

$$\mathbf{K}_{33} = \begin{bmatrix} \sqrt{2}\,A_3 + A_4 + A_5 & A_4 - A_5 \\ A_4 - A_5 & A_4 + A_5 \end{bmatrix} ,$$

$$\mathbf{K}_{44} = \begin{bmatrix} A_{11} + 2\sqrt{2}\,A_{12} + A_{13} & A_{11} - A_{13} \\ A_{11} - A_{13} & A_{11} + A_{13} \end{bmatrix} ,$$

$$\mathbf{K}_{55} = \begin{bmatrix} \sqrt{2}\,A_9 + A_{10} + A_{11} & -A_{10} + A_{11} \\ -A_{10} + A_{11} & A_{10} + A_{11} \end{bmatrix} ,$$

$$\mathbf{K}_{13} = \mathbf{K}_{31} = \begin{bmatrix} -A_4 & -A_4 \\ -A_4 & -A_4 \end{bmatrix} ,$$

$$\mathbf{K}_{15} = \mathbf{K}_{51} = \begin{bmatrix} -A_{10} & A_{10} \\ A_{10} & -A_{10} \end{bmatrix} ,$$

$$\mathbf{K}_{23} = \mathbf{K}_{32} = \begin{bmatrix} -A_5 & A_5 \\ A_5 & -A_5 \end{bmatrix} ,$$

$$\mathbf{K}_{54} = \mathbf{K}_{45} = \begin{bmatrix} -A_{11} & -A_{11} \\ -A_{11} & -A_{11} \end{bmatrix} .$$

All remaining $\mathbf{K}_{ij}$ vanish.

Assuming non-singularity of the stiffness matrix, (5a) allows us to calculate the displacements of the nodal points u_i, v_i ($i = 1, \ldots, 5$):

$$\mathbf{v} = \mathbf{K}^{-1}\mathbf{f} \quad . \tag{6}$$

Thus, the displacements of the end-point of each single bar is established, and we can now, on the basis of the element stiffness matrices, determine the stresses within the bars by means of the matrix relation between stresses and displacements:

$$\boldsymbol{\sigma} = \mathbf{R}\,\mathbf{v} \quad . \tag{7}$$

Here, **R** is a (13×10)–matrix of the form

$$\mathbf{R} = \frac{E}{l}\begin{bmatrix} 1 & 0 & 0 & 0 & 0 & 0 & 0 & 0 & 0 & 0 \\ \frac{1}{2} & -\frac{1}{2} & 0 & 0 & 0 & 0 & 0 & 0 & 0 & 0 \\ 0 & 0 & 0 & 0 & \frac{1}{2} & 0 & 0 & 0 & 0 & 0 \\ -\frac{1}{2} & -\frac{1}{2} & 0 & 0 & \frac{1}{2} & \frac{1}{2} & 0 & 0 & 0 & 0 \\ 0 & 0 & -\frac{1}{2} & \frac{1}{2} & \frac{1}{2} & -\frac{1}{2} & 0 & 0 & 0 & 0 \\ 0 & 0 & 1 & 0 & 0 & 0 & 0 & 0 & 0 & 0 \\ 0 & 0 & \frac{1}{2} & \frac{1}{2} & 0 & 0 & 0 & 0 & 0 & 0 \\ \frac{1}{2} & \frac{1}{2} & 0 & 0 & 0 & 0 & 0 & 0 & 0 & 0 \\ 0 & 0 & 0 & 0 & 0 & 0 & 0 & 0 & \frac{1}{2} & 0 \\ -\frac{1}{2} & \frac{1}{2} & 0 & 0 & 0 & 0 & 0 & 0 & \frac{1}{2} & -\frac{1}{2} \\ 0 & 0 & 0 & 0 & 0 & 0 & -\frac{1}{2} & -\frac{1}{2} & \frac{1}{2} & \frac{1}{2} \\ 0 & 0 & 0 & 0 & 0 & 0 & 1 & 0 & 0 & 0 \\ 0 & 0 & 0 & 0 & 0 & 0 & \frac{1}{2} & -\frac{1}{2} & 0 & 0 \end{bmatrix}^{13 \times 10} \tag{8}$$

Substitution of (6) into (7) then yields the relation required for calculating the stresses:

$$\boldsymbol{\sigma} = \mathbf{R}\,\mathbf{K}^{-1}\mathbf{f} \quad . \tag{9}$$

The equations for the structural model that is required for the optimization have now beeen established.

b) In the following, the equations of the optimization problem shall be set up. In accordance with the problem formulation, the cross-sectional areas of the bars shall serve as design variables, i.e. we define

$$\mathbf{x} := \mathbf{A} \quad .$$

According to (5b), **K** depends on the design variables, i.e. $\mathbf{K} = \mathbf{K}(\mathbf{x})$. Given the same bar material, weight minimization is equal to volume minimization; the *objective function* of the structural volume is thus a linear function with respect to the design variables:

$$f(\mathbf{x}) := V(\mathbf{x}) = \boldsymbol{l}^T \cdot \mathbf{x} = \sum_{i=1}^{13} l_i x_i \tag{10}$$

with l_i denoting the bar lengths.

For the bar stresses we formulate the following constraints

$$g_1(\mathbf{x}) \quad \widehat{=}\, g_{ti}(\mathbf{x}) \;:=\; \sigma_i(\mathbf{x}) - \sigma_{t_{adm}} \;\leq 0 \qquad (i = 1,\ldots,13)\ , \qquad (11a)$$

$$g_2(\mathbf{x}) \quad \widehat{=}\, g_{cj}(\mathbf{x}) \;:=\; \sigma_{c_{adm}} - \sigma_j(\mathbf{x}) \leq 0 \qquad (j = 1,\ldots,13)\ . \qquad (11b)$$

Finally, we demand non-negativity for the cross-sectional areas of the bars:

$$x_i \geq 0 \qquad \text{for all} \quad i = 1,\ldots,13\ . \qquad (12)$$

c) The constrained optimization problem is solved using an external penalty function by means of which the task is transformed into an unconstrained problem. With *(16.11b)* we state

$$\Phi_i(\mathbf{x}, R_i) := V(\mathbf{x}) + R_i \sum_{j=1}^{26} \Big(\max[\,0, g_j(\mathbf{x})\,] \Big)^2 \qquad (i = 1,2,3\ldots)\ , \qquad (13)$$

where $$\max[\,0, g_j(\mathbf{x})\,] = \begin{cases} g_j^2(\mathbf{x}) & \text{in the infeasible domain}\ , \\ 0 & \text{in the feasible domain}\ . \end{cases}$$

Here, the choice of a suitable initial value for the penalty parameter R_i is crucial; for the present task we choose:

$$R_1 = 10^{-5}\ .$$

The unconstrained problem (13) can be solved by means of suitable algorithms; in the present case, the POWELL-method of conjugate gradients has been used, where a quadratic polynomium (LAGRANGE-interpolation) has proved sufficient for a one-dimensional minimization. In addition it could be shown that different initial designs ($A_i = 100,\ldots, 1000mm^2$) virtually lead to the same optimal solution.

The calculation, the scale of which requires the use of a computer, yields the result that the force F is mostly carried via bars 4 to 8 into the supports 7, 8, 9 (denoted by bold lines in Fig. D-3). Consequently, the remaining bars need very small cross-sectional areas only.

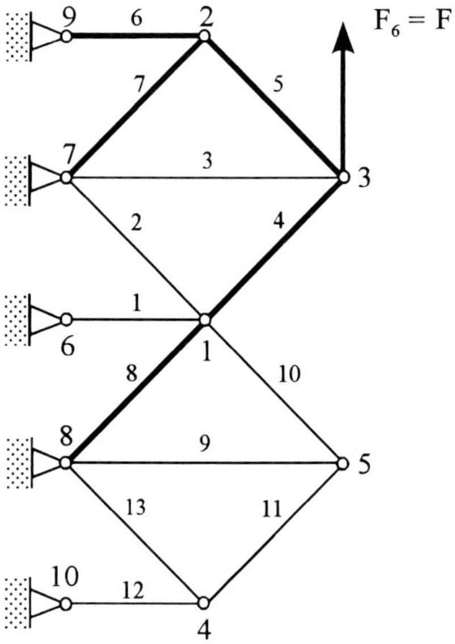

Fig. D-3: Optimized truss structure by changing the cross-sections of the bars

Exercise D-15/16-3:

Fig. D-4 shows a section of a circular cylindrical shell C with a ring stiffener S. The considered part of a boiler is subjected to a constant internal pressure p and has a constant inner temperature Θ_{iC}.

The temperature distribution within boiler and stiffener has been determined by measurements; for the cylindrical section C we assume a linear temperature distribution over the thickness with the gradient ${}^1\Theta_C$ = const in the longitudinal direction

$$\Theta_C(z) = {}^0\Theta_C + z\,{}^1\Theta_C$$

with
$$ {}^0\Theta_C = \frac{\Theta_{iC} + \Theta_{oC}}{2} \quad , \quad {}^1\Theta_C = \frac{\Theta_{oC} - \Theta_{iC}}{t} \quad .$$

The temperature distribution in the ring stiffener is assumed to be constant over the thickness, and is approximated in the mid-plane by a second-order polynomium in r with the following form:

$$ {}^0\Theta_S(r) = \Theta_{iC}\left[1 + \frac{(r-1)(r+1-2\omega)}{(1-\omega)^2}\left(1 - \frac{\Theta_{oS}}{\Theta_{iC}}\right)\right]$$

with $\quad r = \xi_s / b \quad , \quad \omega = a/b \quad .$

Choosing as two design variables the *half thickness h of the ring stiffener* and the *boiler thickness t*, the section is to be dimensioned with respect to minimum weight, subject to the condition that the maximum reference stresses in the ring and the stiffener must not exceed a prescribed value.

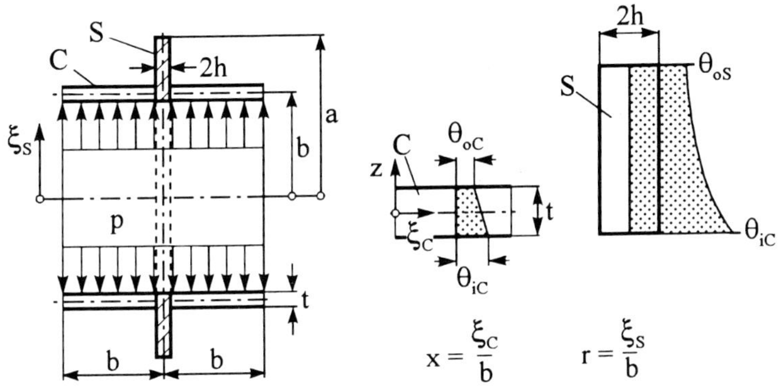

Fig. D-4: Section of a ring stiffened circular cylindrical boiler under pressure und thermal load

a) Determine the stress curves by means of the *Theory of Structures.*

b) Formulate the expressions required for the optimization (objective functions and constraints), and determine the design domain. The design variables $x_1 = h$ and $x_2 = t$ are restricted by upper bound values of 20 and 40 *mm*, respectively. State the wall-thicknesses of the optimal design.

Numerical values:

Geometry: $b = 0.5\,m$, $a = 0.65\,m$;

Loads:

$$\Theta_{iC} = 170\ ^oC\ ,\ \Theta_{oS} = 50\ ^oC\ ,\ {}^1\Theta_C = 1\ ^oC/mm\ ,\ p = 2\,MPa\ ;$$

Material:

$$\alpha_{TC} = \alpha_{TS} = \alpha_T = 1.11 \cdot 10^{-5}/^oC\ ,\ E_C = E_S = E = 2.1 \cdot 10^5\ MPa\ ,$$

$$\varrho_C = \varrho_S = \varrho = 0.785 \cdot 10^4\ kg/m^3\ ,\ \nu = 0.3\ ,\ \sigma_{C,S_{adm}} = 200\ MPa\ .$$

Solution:

a) – *Structural model and structural analysis*

The stress state of the given stiffened boiler structure can be most conveniently calculated by applying the compatibility between the single parts. Since the respective steps for establishing the structural equations have already been described in detail (see C.13.1/2), only the most important aspects will be treated here.

In a first step, we separate the two semi-infinite cylindrical shells from the ring stiffener. Owing to the different deformations of boiler and stiffener at the interface point, the required compatibility is induced by yet unknown boundary forces R and boundary moments M. Each of the substructures shows deformations caused by temperature and pressure loads (state " 0 "), and by the forces R acting at the boundaries (state " 1 "), and the moments M (state " 2 "). In the present case, the parts of the boiler can be idealized as circular cylindrical shells subject to axisymmetric loads (pressure, temperature, boundary force R, and boundary moment M), and the ring stiffener can be treated as a circular disk subject to internal pressure, temperature and the radial force R. The boundary moment M of the circular cylindrical shell does not effect the stiffener.

The deformations are calculated from the basic equations for the circular cylindrical vessel and for the circular disk (see **C-13-2**). After determination of the deformations at the points of the substructures, we formulate the compatibility conditions

$$w_C = w_C^{(0)} + w_C^{(1)} + w_C^{(2)} \stackrel{!}{=} u_S^{(0)} + u_S^{(1)} = u_S\ , \qquad (1a)$$

$$\chi_C = \chi_C^{(0)} + \chi_C^{(1)} + \chi_C^{(2)} \stackrel{!}{=} 0\ , \qquad (1b)$$

where w_C and u_S denote the expansions of the vessel and the radial displacements of the stiffener, respectively; χ_C are the corresponding angles of rotation. Conditions (1a,b) constitute a linear system of equations for determining the unknown boundary quantities R and M. After some calculation one obtains:

$$R = \frac{-\alpha_{TC}\,{}^0\Theta_C + 2\alpha_{TS}\,\Theta_{iC}\,\dfrac{\vartheta(\omega)-\vartheta(1)}{\omega^2-1} - \left[\dfrac{b}{E_C t} + \dfrac{1}{E_S}\left(\dfrac{1+\omega^2}{1-\omega^2} - \nu\right)\right]p}{\dfrac{b^2}{4K_C\,\kappa^3} - \dfrac{1}{E_S h}\left(\dfrac{1+\omega^2}{1-\omega^2} - \nu\right)}\ , \qquad (2a)$$

$$M = -\frac{1}{2}\frac{b}{\kappa}R - K_C(1+\nu)\alpha_{TC}\,{}^1\Theta_C \tag{2b}$$

with

$$\vartheta(r) = \frac{1}{\Theta_{iC}}\int r\,{}^0\Theta_S(r)\,dr = \frac{r^2}{12}\left[6 + \frac{3r^2 - 8\omega r + 12\omega - 6}{(1-\omega)^2}\left(1 - \frac{\Theta_{oS}}{\Theta_{iC}}\right)\right] ,$$

$$\vartheta(\omega) = \frac{\omega^2}{2} - \frac{\omega^2(5\omega^2 - 12\omega + 6)}{12(1-\omega)^2}\left(1 - \frac{\Theta_{oS}}{\Theta_{iC}}\right) ,$$

$$\vartheta(1) = \frac{1}{2} + \frac{4\omega - 3}{12(1-\omega)^2}\left(1 - \frac{\Theta_{oS}}{\Theta_{1C}}\right) ,$$

$$K_C = \frac{E_C t^3}{12(1-\nu^2)} , \qquad \kappa^4 = 3(1-\nu^2)\left(\frac{b}{t}\right)^2 .$$

Refer to **C-13-2** for further details of determining the curves of stress resultants and deformations.

- *Stresses within the parts of the boiler*

Cylindrical shell C

- Longitudinal stresses

$$\sigma_{xx}(x) = \pm\frac{6}{t^2}\left\{\left[\frac{b}{\kappa}R\sin\kappa x + \left(M + (1+\nu)\alpha_{TC}K_C\,{}^1\Theta_C\right)\cdot\right.\right.$$
$$\left.\left.\cdot(\sin\kappa x + \cos\kappa x)\right]e^{-\kappa x} - (1+\nu)\alpha_{TC}K_C\,{}^1\Theta_C\right\} . \tag{3a}$$

- Circumferential stresses

$$\sigma_{\varphi\varphi}(x) = \frac{bE_C}{2\kappa^2K_C}\left\{\frac{b}{\kappa}R\cos\kappa x + \left[M + (1+\nu)\alpha_{TC}K_C\,{}^1\Theta_C\right]\cdot\right.$$
$$\left.\cdot(\cos\kappa x - \sin\kappa x)\right\}e^{-\kappa x} + \frac{b}{t}p . \tag{3b}$$

- Reference stress according to VON MISES´ hypothesis

$$\sigma_{rC} = \sqrt{\sigma_{xx}^2 + \sigma_{\varphi\varphi}^2 - \sigma_{xx}\sigma_{\varphi\varphi}} . \tag{3c}$$

Ring stiffener S (disk)

- Radial stresses

$$\sigma_{rr}(r) = \frac{p - \frac{R}{h}}{\omega^2 - 1}\left(1 - \frac{\omega^2}{r^2}\right) +$$
$$+ E_S\alpha_{TS}\Theta_{iC}\left[\vartheta(1) + \frac{\vartheta(\omega) - \omega^2\vartheta(1)}{\omega^2 - 1}\left(1 - \frac{1}{r^2}\right) - \frac{1}{r^2}\vartheta(r)\right] . \tag{4a}$$

- Circumferential stresses

$$\sigma_{\varphi\varphi}(r) = \frac{p - \frac{R}{h}}{\omega^2 - 1}\left(1 + \frac{\omega^2}{r^2}\right) + E_S\,\alpha_{TS}\,\Theta_{iC}\Big[\vartheta(1) + \\ + \frac{\vartheta(\omega) - \omega^2\vartheta(1)}{\omega^2 - 1}\left(1 + \frac{1}{r^2}\right) + \frac{1}{r^2}\vartheta(r) - {}^0\Theta_S(r)\Big] \quad . \tag{4b}$$

- Reference stresses

$$\sigma_{rS} = \sqrt{\sigma_{rr}^2 + \sigma_{\varphi\varphi}^2 - \sigma_{rr}\,\sigma_{\varphi\varphi}} \quad . \tag{4c}$$

The reference stresses provide the basis for defining the constraints for the optimization.

b) Definition of the *Optimization model*

In order to illustrate the design domain, only two design variables are considered in the following: the half thickness of the stiffener ring $x_1 =: h$ and the shell thickness $x_2 =: t$, both of which are combined in the *design variable vector*:

$$\mathbf{x} = (x_1, x_2)^T \quad . \tag{5}$$

As stated in the problem formulation, a pure weight minimization problem is to be solved. We thus require the *objective function* to be the sum of the weights of the two parts of the boiler:

$$f(\mathbf{x}) \triangleq W(\mathbf{x}) = \rho g\big[V_C(\mathbf{x}) + V_S(\mathbf{x})\big] \tag{6}$$

with the volumes of cylinder and ring stiffener given by

$$V_C(\mathbf{x}) = 4\pi b(b - x_1)x_2 \quad ,$$

$$V_S(\mathbf{x}) \approx 2\pi x_1\big[a^2 - b^2\big] \quad \text{for } x_2 \ll b \quad .$$

We now consider the *constraints* that at each point x or r of the two boiler parts, the reference stresses σ_r have to be smaller than the maximum admissible stress values:

Cylinder C

$$\sigma_{rC_{max}} := \max_x \sigma_{rC}(x, \mathbf{x}) \le \sigma_{C\,adm}$$

$$\Longrightarrow \quad g_1(\mathbf{x}) = \frac{\max_x \sigma_{rC}(x, \mathbf{x})}{\sigma_{C\,adm}} - 1 \le 0 \quad , \tag{7a}$$

Stiffener S

$$\sigma_{rS_{max}} := \max_r \sigma_{rS}(r, \mathbf{x}) \le \sigma_{S\,adm}$$

$$\Longrightarrow \quad g_2(\mathbf{x}) = \frac{\max_r \sigma_{rS}(r, \mathbf{x})}{\sigma_{S\,adm}} - 1 \le 0 \quad . \tag{7b}$$

The two design variables are restricted to the intervals

$$0 < x_1 \leq 20 \; mm \quad , \quad 0 < x_2 \leq 40 \; mm \quad . \tag{8}$$

Now, the following structural optimization problem *(15.4)* with the scalar objective function (6) and the inequality constraints (7a,b) shall be solved:

$$\underset{\mathbf{x} \in \mathbf{R}^n}{\text{Min}} \left\{ f(\mathbf{x}) \,\middle|\, \mathbf{g}(\mathbf{x}) \leq \mathbf{0} \right\} \quad .$$

In order to solve this constrained problem, an algorithm can be chosen from MP-algorithms of zeroth, first, and second order. In the case of the actual non-convex problem (Fig. D-5), the algorithm should perform as simply and robust as possible; here, one of the penalty function methods (e.g. internal penalty function) or the COMPLEX algorithm by BOX are very suitable zero-order methods (see [D.24]).

Since only two design variables are considered, the determination of the optimal design of the current problem can be carried out analytically. As shown in Fig. D-5, the feasible domain X of the design domain is determined by the active constraints of the reference stresses in the vessel and the stiffener ring (7a,b), and by bounding the wall-thicknesses of vessel and stiffener (8). In addition, the iso-lines of the objective function f($\mathbf{x}$) ($\widehat{=}$ total weight W of the considered parts of the boiler) are depicted in the diagram.

Fig. D-5 displays the optimal values for the design variables as

$$\mathbf{x}_{opt} = (4.2 , 11.8)^T \quad ,$$

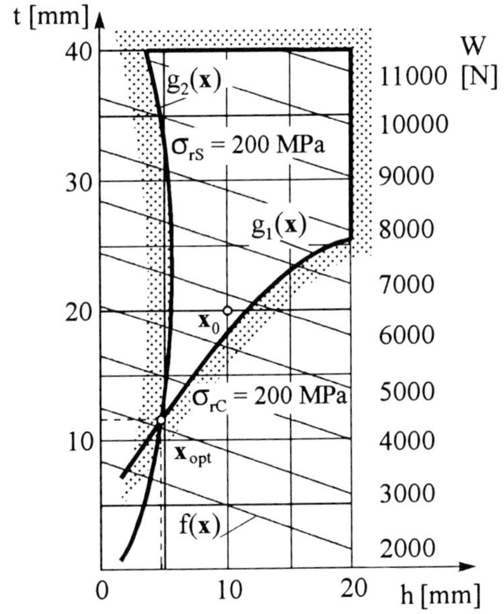

Fig. D-5: Design domain of the ring-stiffened boiler

which according to equation (6) allows us to determine the optimal weight as:

$$W_{opt} = 3185.5 \; N .$$

The inequality constraint functions $g_1(\mathbf{x})$ and $g_2(\mathbf{x})$, i.e the reference stresses σ_{rC} and σ_{rS}, respectively, become equal to the admissible value ($\sigma_{adm} = 200 \; MPa$) at the optimal point, and thus the material is utilized optimally. If we start the optimization calculation with an initial design

$$\mathbf{x}_0 = (h, t)^T = (10, 20)^T [mm] ,$$

we obtain a weight reduction of approximately 43% at the optimum point.

Exercise D-18-1:

Perform a mapping into the criteria space for a vector of the two objective functions (criteria)

$$\mathbf{f}(x) = \begin{bmatrix} f_1(x) \\ f_2(x) \end{bmatrix} = \begin{bmatrix} x^2 - 4x + 5 \\ \frac{1}{2} x^2 - 5x + \frac{29}{2} \end{bmatrix} .$$

a) Show the graphs of the individual objective functions in the design space, and determine the domain of the functional-efficient set of solutions in the design space.

b) Determine the curves of the functional-efficient solutions in the criteria space, using a constraint-oriented transformation (trade-off method).

Solution:

a) Presentation of the objective functions in the design space:

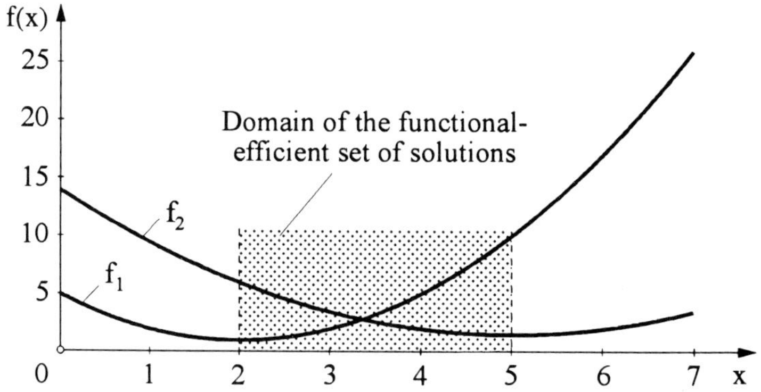

Fig. D-6: Objective functions and domain of the functional-efficient set of solutions

Fig. D-6 shows that the curves have slopes of opposite signs in the dotted area; according to Def. 1 in Section 18.1 there exist functional-efficient (or PARETO-optimal) solutions of the two functions.

b) Functional-efficient set of solutions in the criteria space

The Vector Optimization Problem *(18.1)* can be transformed into a scalar, constrained optimization problem by minimizing only one of the objective functions, for instance $f_1(x)$, and by imposing upper bounds on the remaining ones *(18.7)*, e.g.

$$f_1(x) \longrightarrow \text{Min} \qquad \forall \quad x \in X \,, \tag{1}$$

subject to

$$f_j(x) = \overline{y}_j \qquad \forall \quad j = 2, \ldots, m \,,$$

where f_1 is denoted the main objective, and $f_2, \ldots, f_m$ secondary objectives. The present task can be interpreted in such a way that, when minimizing f_1, all remaining components of the objective function are to attain prescribed values $\overline{y}_2, \ldots, \overline{y}_m$. These constraint levels illustrate the preference behaviour.

If one is to precisely achieve the constraint levels in (1), the given task corresponds to a minimization of the respective LAGRANGE–function *(15.9)*:

$$L(\mathbf{x}, \boldsymbol{\beta}) := f_1(\mathbf{x}) + \sum_{j=2}^{m} \beta_j [f_j(\mathbf{x}) - \overline{y}_j] \implies \text{Min} \tag{2}$$

with the necessary optimality conditions *(15.10)*

$$\frac{\partial L}{\partial x_i} = \frac{\partial f_1}{\partial x_i} + \sum_{j=2}^{m} \beta_j \frac{\partial f_j}{\partial x_i} \stackrel{!}{=} 0 \qquad (i = 1, \ldots, n) \,, \tag{3a}$$

$$\frac{\partial L}{\partial \beta_j} = f_j(\mathbf{x}) - \overline{y}_j \stackrel{!}{=} 0 \qquad (j = 2, \ldots, m) \,. \tag{3b}$$

The optimal values for $x_1, \ldots, x_n$ and the corresponding LAGRANGEAN multipliers $\beta_2, \ldots, \beta_m$ are then calculated from (3a,b).

For the present problem holds that

$$\mathbf{f}(x) = \begin{bmatrix} f_1(x) \\ f_2(x) \end{bmatrix} = \begin{bmatrix} x^2 - 4x + 5 \\ \frac{1}{2}x^2 - 5x + \frac{29}{2} \end{bmatrix} ,$$

and we thus choose according to (1)

$$f_1(\mathbf{x}) \longrightarrow \text{Min} \,,$$

subject to

$$f_2(\mathbf{x}) = \overline{y}_j \quad (j = 2, \ldots, 6) \,.$$

Using the LAGRANGE–function (2)

$$L(x, \beta) = f_1(x) + \beta_j [f_2(x) - \overline{y}_j] \longrightarrow \text{Min} \,,$$

by means of (3a,b), the optimal values are determined as

$$\frac{\partial L}{\partial x} = 2x - 4 + \beta_j (x - 5) = 0 \quad \longrightarrow \quad \beta^*_{j\,1,2} = \frac{2x^*_{j1,2} - 4}{5 - x^*_{j1,2}} , \tag{4}$$

$$\frac{\partial L}{\partial \beta_j} = \frac{1}{2} x^2 - 5x + \frac{29}{2} - \bar{y}_j = 0 \quad \longrightarrow \quad x^*_{j\,1,2} = 5 \pm \sqrt{2\bar{y}_j - 4} . \tag{5}$$

Finally, results are listed for different values of $\bar{y}_j$:

$$\bar{y}_2 = 8 \longrightarrow x^*_{2\,1,2} = 5 \pm 2\sqrt{3} , \quad \beta^*_{2\,1,2} = \begin{cases} -3.73 \\ -0.27 \end{cases} , \quad \mathbf{f}(x^*) = \begin{pmatrix} 42.78 / 1.22 \\ 8.0 \end{pmatrix} ,$$

$$\bar{y}_3 = 6.5 \longrightarrow x^*_{3\,1,2} = 5 \pm 3 , \quad \beta^*_{3\,1,2} = \begin{cases} -4.0 \\ 0 \end{cases} , \quad \mathbf{f}(x^*) = \begin{pmatrix} 37.0 / 1.0 \\ 6.5 \end{pmatrix} ,$$

$$\bar{y}_4 = 4.0 \longrightarrow x^*_{4\,1,2} = 5 \pm 2 , \quad \beta^*_{4\,1,2} = \begin{cases} -5.0 \\ 1.0 \end{cases} , \quad \mathbf{f}(x^*) = \begin{pmatrix} 26.0 / 2.0 \\ 4.0 \end{pmatrix} ,$$

$$\bar{y}_5 = 2{,}0 \longrightarrow x^*_5 = 5 , \quad \beta^*_5 = \infty , \quad \mathbf{f}(x^*) = \begin{pmatrix} 10 \\ 2 \end{pmatrix} ,$$

$$\bar{y}_6 = 1.0 \longrightarrow x^*_{6\,1,2} = 5 \pm i\sqrt{2} , \quad \text{no real solution} .$$

This proves that only the constraint level of $2 \le \bar{y}_j \le 6.5$ leads to unique functional-efficient solutions. Fig. D-7 presents the β^*_j-values belonging to the different constraint levels $\bar{y}_j$ in the criteria space. The efficient boundary ∂Y^* (solid line) of Y is valid for non-negative values of β^*_j.

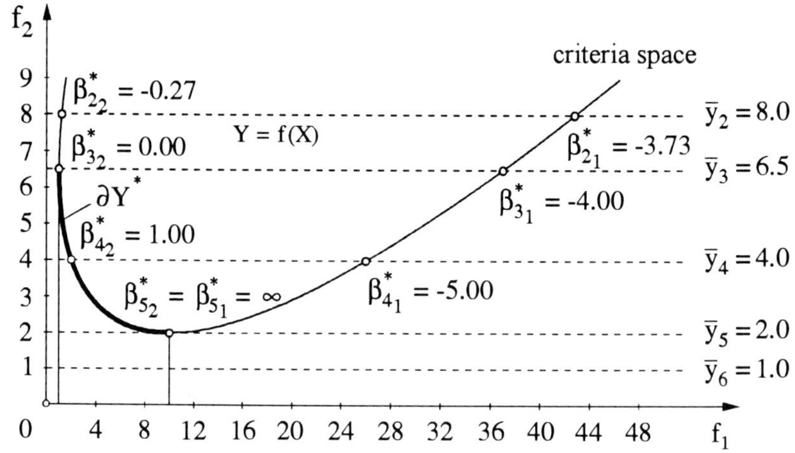

Fig. D-7: Functional-efficient boundary in the criteria space

The reader should check whether use of f_2 as the main objective and $f_1(x) = \bar{y}_j$ as a constraint leads to similar results.

Exercise D-18-2:

A simply supported column as shown in Fig. D-8 has variable, circular cross-sections (radius $r(x)$) and is subjected to buckling. The length l and the total volume V_0 are given.

a) Set up a functional which governs the problem of maximizing the buckling load F_{crit}.

b) Derive the optimality criterion for the problem.

c) Maximize the buckling load for the given volume V_0. Derive an equation for the optimal cross-section law $r = r(x)$.

d) Compare the optimal buckling load for a column with variable cross-section with the buckling load of a column with the same volume and constant cross-section.

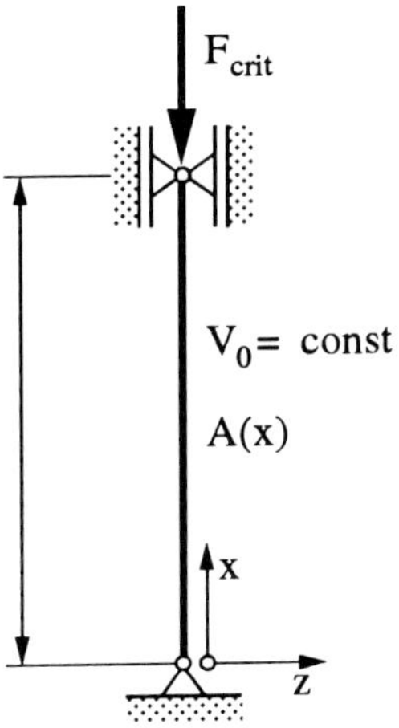

Fig. D-8: Simply supported column

Solution:

a) In order to establish a functional, we start with the following expressions describing the problem:

Volume $\longrightarrow$ $$V = \int_0^l \pi [r(x)]^2 \, dx = V_0 = \text{const} \quad . \tag{1}$$

Differential equation for column buckling $\longrightarrow$ $$w_{,xx} + \frac{F_c}{EI_y(x)} w = 0 \quad . \tag{2}$$

With $I_y(x) = \frac{\pi [r(x)]^4}{4}$ follows $$w_{,xx} + \frac{4 F_c}{\pi E} \frac{w}{r(x)^4} = 0 \quad . \tag{3}$$

Geometrical boundary conditions: $$w(0) = 0 \quad , \quad w(l) = 0 \quad . \tag{4}$$

The relations (1) to (3) are transformed into an integral expression of the form

$$I = \int F(x) \, dx \longrightarrow \text{Extremum} \tag{5}$$

with $F(x)$ as the basic function (see *(6.33)*).

Eq. (3) yields

$$w_{,xx} + \mu^2 \frac{w}{r(x)^4} = 0 \longrightarrow [r(x)]^4 = -\mu^2 \frac{w}{w_{,xx}} \tag{6}$$

with $$\mu^2 = \frac{4F_c}{\pi E} \quad .$$

Substituting (6) into (1) and considering that μ is independent of x, we can write

$$\frac{V_0}{\pi\mu} = \int_{x=0}^{l} \sqrt{-\frac{w}{w_{,xx}}}\, dx \quad . \tag{7}$$

Due to V_0 = const , minimization of the left-hand-side term yields the maximum value for the force F :

$$I = \frac{V_0}{\pi\mu} = \int_{x=0}^{l} \sqrt{-\frac{w}{w_{,xx}}}\, dx \longrightarrow \text{Min} \quad . \tag{8a}$$

With (8a) we have established an *unconstrained mathematical form* of our originally constrained optimization problem.

b) The basic function according to (8b) reads :

$$F(x, w, w_{,xx}) = \sqrt{-\frac{w}{w_{,xx}}} \quad . \tag{8b}$$

In accordance with the rules of the calculus of variation one obtains EULER´s differential equation as the necessary condition :

$$\frac{\partial F}{\partial w} + \left(\frac{\partial F}{\partial w_{,xx}}\right)_{,xx} = 0 \quad . \tag{9}$$

With $\quad \dfrac{\partial F}{\partial w} = \dfrac{1}{2}\left(-\dfrac{w}{w_{,xx}}\right)^{-1/2}\left(-\dfrac{1}{w_{,xx}}\right)$

and $\quad \dfrac{\partial F}{\partial w_{,xx}} = \dfrac{1}{2}\left(-\dfrac{w}{w_{,xx}}\right)^{-1/2}\left(\dfrac{w}{w_{,xx}^2}\right) = \dfrac{1}{2}\sqrt{-\dfrac{w}{w_{,xx}^3}}$

$$\Longrightarrow \quad \sqrt{-\frac{w_{,xx}}{w}}\left(-\frac{1}{w_{,xx}}\right) + \left(\sqrt{-\frac{w}{w_{,xx}^3}}\right)_{,xx} = 0 \quad .$$

Multiplication by w leads to

$$-\sqrt{-\frac{w}{w_{,xx}}} + \left(\sqrt{-\frac{w}{w_{,xx}^3}}\right)_{,xx} w = 0 \quad . \tag{10a}$$

Augmentation of the first term of (10a) by ($w_{,xx}$) yields

$$-w_{,xx}\sqrt{-\frac{w}{w_{,xx}^3}} + \left(\sqrt{-\frac{w}{w_{,xx}^3}}\right)_{,xx} w = 0 \quad . \tag{10b}$$

Eq. (10b) constitutes the optimality criterion for the present problem.

c) Based on (10b), the optimal cross-sectional radius function $r = r(x)$ shall now be determined.

Using the abbreviation $v = \sqrt{-\frac{w}{w_{,xx}^3}}$ we obtain from (10b):

$$-w_{,xx}\,v + v_{,xx}\,w = 0 \quad \longrightarrow \quad (v_{,x}\,w - v\,w_{,x})_{,x} = 0 \quad . \tag{11}$$

Taking into consideration that $w(0) = v(0) = 0$, (11) can be written

$$v_{,x}\,w - v\,w_{,x} = 0 \quad \longrightarrow \quad \left(\frac{v}{w}\right)_{,x} = 0 \quad . \tag{12}$$

After integration of (12) we have

$$\sqrt{-\frac{w}{w_{,xx}^3}} = c\,w \quad \longrightarrow \quad c^2\,w^2 = -\frac{w}{w_{,xx}^3} \quad . \tag{13a}$$

Since $w(x)$ can only be determined up to a multiplying factor ($w(x) \mathrel{\widehat{=}}$ eigenmode), one can choose $c = 1$, and thus

$$w = -\frac{1}{w_{,xx}^3} \quad . \tag{13b}$$

For the subsequent calculations (13b) is reformulated in the following way:

$$w_{,xx} = -w^{-1/3} \quad \longrightarrow \quad 2\,w_{,x}\,w_{,xx} = -2\,w_{,x}\,w^{-1/3}$$

or

$$(w_{,x}^2)_{,x} = -3\left(w^{2/3}\right)_{,x} \quad . \tag{14}$$

Integration of (14) with the integration constant a^2 yields:

$$w_{,x}^2 = 3(-w^{2/3} + a^2) \quad \longrightarrow \quad w_{,x} = \sqrt{3}\,\sqrt{a^2 - w^{2/3}} \quad . \tag{15}$$

After transformation of (15) we obtain:

$$\int dx = \int \frac{dw}{\sqrt{3}\,\sqrt{a^2 - w^{2/3}}} \quad . \tag{16a}$$

Now, introducing

$$w = u^3 \quad , \quad dw = 3\,u^2\,du$$

and integrating (16a), we get

$$x = \sqrt{3} \int \frac{u^2\,du}{\sqrt{a^2 - u^2}} + C \quad . \tag{16b}$$

Solution of the right-hand-side integral yields

$$x = \sqrt{3}\left[\frac{a^2}{2}\arcsin\frac{u}{a} - \frac{u}{2}\sqrt{a^2 - u^2}\right] + C \quad .$$

Re-substitution and factoring out leads to

$$x = \frac{\sqrt{3}}{2}\,a^2\left\{\arcsin\frac{\sqrt[3]{w}}{a} - \frac{\sqrt[3]{w}}{a}\sqrt{1-\left(\frac{\sqrt[3]{w}}{a}\right)^2}\,\right\} + C \quad . \tag{17}$$

The boundary conditions (4) provide the constants a^2, C:

$$x = 0: \quad w = 0 \longrightarrow \quad C = 0 \quad ,$$

$$x = l: \quad w = 0 \longrightarrow \quad \frac{\sqrt{3}}{2}\,a^2 = \frac{l}{\pi} \quad .$$

According to (6) we have $\quad r^4 = -\mu^2\,\dfrac{w}{w_{,xx}} \quad .$

Eq. (13b) leads to $\quad w = -\dfrac{1}{w_{,xx}^{\,3}} \quad$ or $\quad w_{,xx} = -w^{-1/3}$

$$\Longrightarrow \qquad r^4 = \mu^2\frac{w}{w^{-1/3}} = \mu^2 w^{4/3} \quad . \tag{18}$$

Decreasing the power of r in (18) to 3, one obtains

$$r^3 = \sqrt{\mu^3}\;w \quad . \tag{19}$$

Substitution of (19) into (17) then yields the implicit form of the equation for the optimal cross-sectional radius function

$$x = \frac{l}{\pi}\left\{\arcsin\frac{r}{r_0} - \frac{r}{r_0}\sqrt{1-\left(\frac{r}{r_0}\right)^2}\,\right\} \quad \text{with} \quad r_0^4 = \frac{16}{3\pi^3}\,\frac{l^2 F_{crit}}{E} \quad . \tag{20}$$

r_0 is the largest radius at $x = l/2$. If we solve the transcendental equation (20) with respect to r, we obtain

$$r = r_0 f(x) \quad . \tag{21}$$

For the given volume V_0, one obtains r_0 from (21) and (1)

$$r_0 = \sqrt{\frac{V_0}{\pi\int_0^l f^2(x)\,dx}} \quad . \tag{22}$$

Eq. (20) finally gives the buckling load

$$F_{crit} = \frac{3\pi^3 E r_0^4}{16\,l^2} \quad . \tag{23}$$

d) Comparison of the optimal buckling load according to (23) with the buckling load of a column with the same volume but uniform circular cross-section.

Proceeding from (22) we obtain

$$r_0^4 = V_0^2\left(\pi\int_0^l f^2(x)\,dx\right)^{-2} \quad . \tag{24}$$

With (24) we obtain from (23)

$$F_{crit} = \frac{3\pi^3 E}{16\, l^2} \frac{V_0^2}{\left(\pi \int_0^l f^2(x)\, dx\right)^2} \quad . \tag{25}$$

It is demanded that the volume V_0 be identical for both the column with constant and with variable cross-section. Thus, $V_0 = \pi r_k^2 l$ be valid for the column with constant cross-section.

The area moment of inertia for $r_k = \text{const} \longrightarrow I_y = \frac{\pi}{4} r_k^4$.

Thus, we can write

$$V_0^2 = 4\pi I_y l^2 \quad . \tag{26}$$

By substituting (26) into (25), we determine the critical load as follows

$$F_{crit} = \underbrace{\frac{3}{4} \frac{1}{\left(\frac{1}{l} \int_0^l f^2(x) dx\right)^2}}_{\varphi} \cdot \underbrace{\frac{E I_y \pi^2}{l^2}}_{F_{crit_{const}}} = \varphi \cdot F_{crit_{const}} \quad .$$

If the cross-sectional radius function is chosen according to (20) or (21), respectively, the buckling load increases by 36 % in comparison to the critical load with constant cross-section. Fig. D-9 shows the column designs.

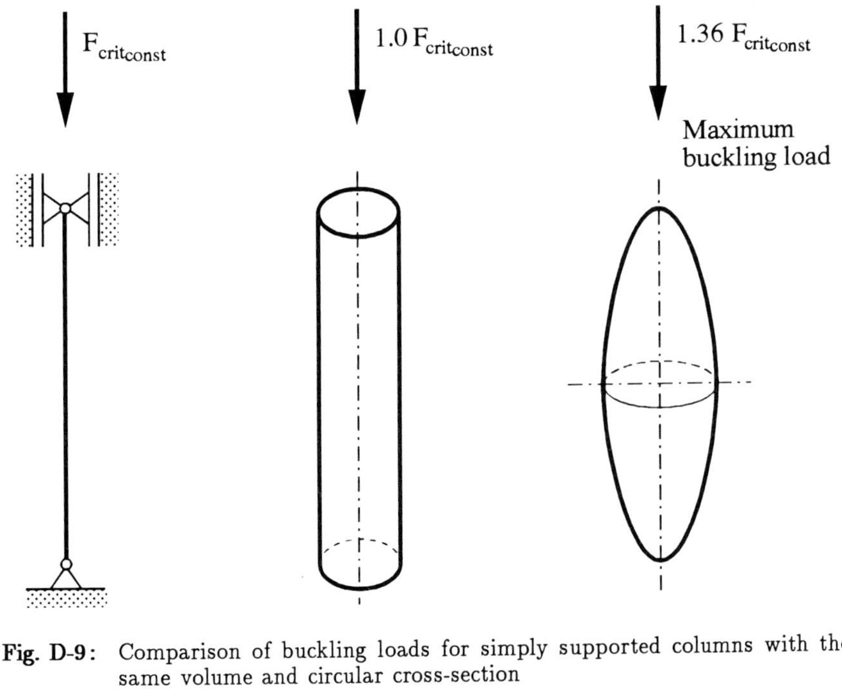

Fig. D-9: Comparison of buckling loads for simply supported columns with the same volume and circular cross-section

Exercise D-18-3:

The essential components of a conveyor belt drum are the belt, the supporting rollers, as well as the drive and guide drum (Fig. D-10a,b). The single drums consist of a drum casing (1) and a drum bottom (2). For the present type of construction, the bottom is connected with the shaft (4) via a tension pulley (3).

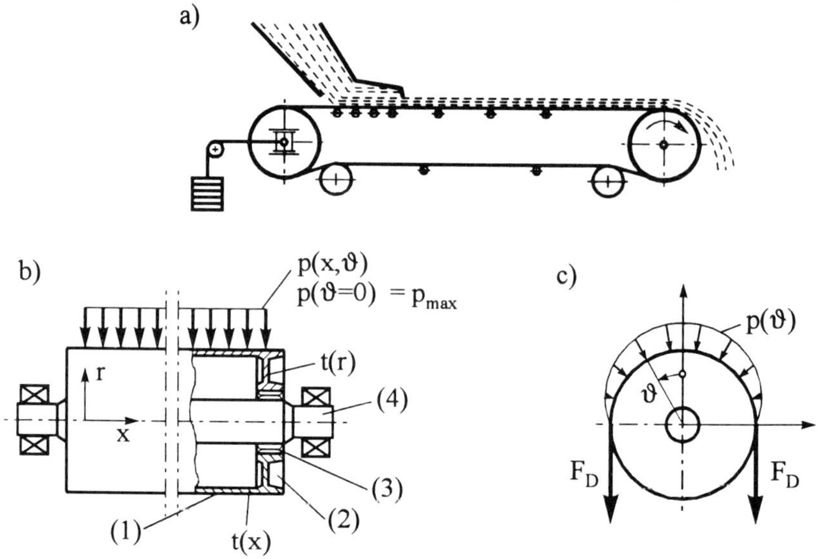

Fig. D-10: Belt conveyor a) integrate system
b) conveyor belt drum
c) surface load

The drum forces F_d of the conveyor belt induce a surface load $p(\vartheta)$ (see Fig. D-10c), where the pressure in the direction of the drum axis is assumed to be constant as a first approximation. The pressure distribution in the circumferential direction is defined as a load depending on the circumferential angle ϑ. The maximum pressure occurs at $\vartheta = 0$, and smaller values occur at the points $\vartheta = \pm \pi/2$.

The coefficients of the chosen pressure distribution

$$p(\vartheta) = p_0 + p_1 \cos\vartheta + p_2 \cos 2\vartheta$$

result from the conditions that the resulting pressure forces in the guide area correspond to the drum forces, i.e.,

$$F_d = r_a\, l \int_0^{\pi/2} p(\vartheta) \cos\vartheta \, d\vartheta \quad , \qquad (1a)$$

and from the condition that the load for the remaining area attains a minimum via a root mean square formulation.

Thus, the drum force $F_d = 650$ kN leads to the load:

$$p(\vartheta) = (0.2117 + 0.3326\cos\vartheta + 0.1411\cos 2\vartheta)\ [MPa] \quad . \tag{1b}$$

The shape of the mid-surface of the drum consists of portions of the drum bottom with constant thickness (idealized as a circular ring plate) and of the drum casing (circular cylindrical shell). The unknown wall thickness distribution $t(\varphi)$ at the transitions is defined by section-wise linear and constant approximation functions, using the *design variables* t_1, t_2, t_3, t_4 (see Fig. D-11):

$$\begin{aligned} t(\varphi, t_1, t_2) &= t_1 - (t_1 - t_2)\frac{\varphi - \varphi_1}{\varphi_2 - \varphi_1} & \varphi_1 \le \varphi \le \varphi_2 \;, \\ t(\varphi, t_3, t_4) &= t_3 - (t_3 - t_4)\frac{\varphi - \varphi_2}{\varphi_3 - \varphi_2} & \varphi_2 \le \varphi \le \varphi_3 \;, \\ t(\varphi, t_4) &= t_4 & \varphi_3 \le \varphi \le \frac{\pi}{2} \;. \end{aligned} \tag{2}$$

We also have to consider upper and lower bounds for the wall-thicknesses:

$$t_l(\varphi) \le t(\varphi) \le t_u(\varphi) \quad . \tag{3a}$$

$$\begin{aligned} 10\text{ mm} &\le t_1 \le 200\text{ mm} \;, & 10\text{ mm} &\le t_2 \le 150\text{ mm} \;, \\ 10\text{ mm} &\le t_3 \le 150\text{ mm} \;, & 10\text{ mm} &\le t_4 \le 100\text{ mm} \;. \end{aligned} \tag{3b}$$

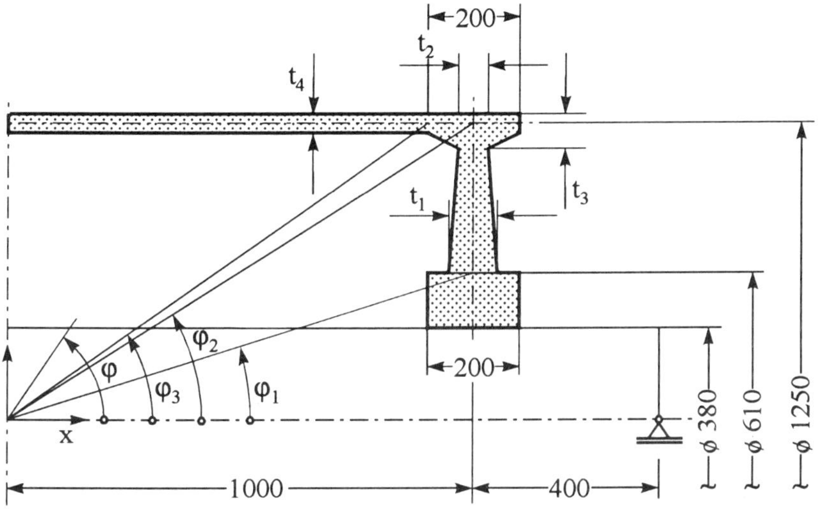

Fig. D-11: Shape function for a conveyor belt drum

a) Establish the optimization modeling relations for the task of designing a conveyor belt drum, when this task is treated as a multicriteria optimization problem with the objectives of minimizing the *weight* W and the *maximum reference stress* $\sigma_{r_{max}}$ ($\sigma_{r\,adm} = 30\ MPa$).

b) Determine the optimal wall-thickness distribution of the conveyor belt drum according to Fig. D-11.

Solution:

a) The objective functionals read as follows:

$$F_1(\varphi) = \int_V \rho g \, dV \mathrel{\widehat{=}} W \ , \tag{4a}$$

$$F_2(\varphi) = \max\left[\sigma_r(\varphi, \vartheta)\right] \ , \tag{4b}$$

where the reference stress is calculated by means of the VON MISES hypothesis:

$$\sigma_r = \sqrt{\sigma_{\varphi\varphi}^2 + \sigma_{\vartheta\vartheta}^2 - \sigma_{\varphi\varphi}\,\sigma_{\vartheta\vartheta} + \tau_{\varphi\vartheta}^2} \ . \tag{5}$$

The scalarized objectives *dead-weight* and *maximum reference* stress result from (4a,b) as vector functions of the variable shape parameters $\mathbf{x} = (t_1, t_2, t_3, t_4)$:

$$f_1(\mathbf{x}) = W(\mathbf{x}) = \rho g V(\mathbf{x}) \ , \tag{6a}$$

$$f_2(\mathbf{x}) = \max_{\substack{\varphi_1 \le \varphi \le \pi/2 \\ 0 \le \vartheta \le \pi}} \left[\sigma_r(\mathbf{x}, \varphi, \vartheta)\right] \ . \tag{6b}$$

The present multicriteria optimization problem is treated by means of the constraint-oriented transformation according to *(18.7)*. For this purpose, the secondary optimization objective (minimization of the maximum reference stresses) is substituted by the following constraints:

$$\left.\begin{aligned}
g_1(\mathbf{x}) &= \max_{\substack{\varphi_1 \le \varphi \le \varphi_2 \\ 0 \le \vartheta \le \pi}} \left[\sigma_r(\mathbf{x}), \varphi, \vartheta\right] - \sigma_{r\,adm} \le 0 \ , \\
g_2(\mathbf{x}) &= \max_{\substack{\varphi_2 \le \varphi \le \varphi_3 \\ 0 \le \vartheta \le \pi}} \left[\sigma_r(\mathbf{x}), \varphi, \vartheta\right] - \sigma_{r\,adm} \le 0 \ , \\
g_3(\mathbf{x}) &= \max_{0 \le \vartheta \le \pi} \left[\sigma_r(\mathbf{x}), \varphi = \varphi_3, \vartheta\right] - \sigma_{r\,adm} \le 0
\end{aligned}\right\} \tag{7}$$

with $\sigma_{r\,adm} = 30\ MPa$.

In the structural analysis, the drum bottom is treated as an uncoupled disk-plate problem, and the drum casing is considered as a circular cylindrical shell. For this purpose, a special transfer matrix procedure has been used according to Section 13.2. The results were additionally verified by control computations by means of an FE-software system [A.21].

b) The present shape optimization problem has been solved by means of the optimization algorithms SLP and LPNLP (see 16.2.2 a,b). Fig. D-12 a,b illustrates the efficiency of the above algorithms when using the constraint-oriented transformation as optimization strategy for the treatment of multicriteria optimization problems. Sequential linearization shows fast convergence; if active constraint limits are imposed, about six to ten linearization steps are necessary, where the gradient evaluations require the highest computational effort. Fig. D-12 b shows that the rate of convergence of the LPNLP-algorithm is lower than that of the SLP-algorithm.

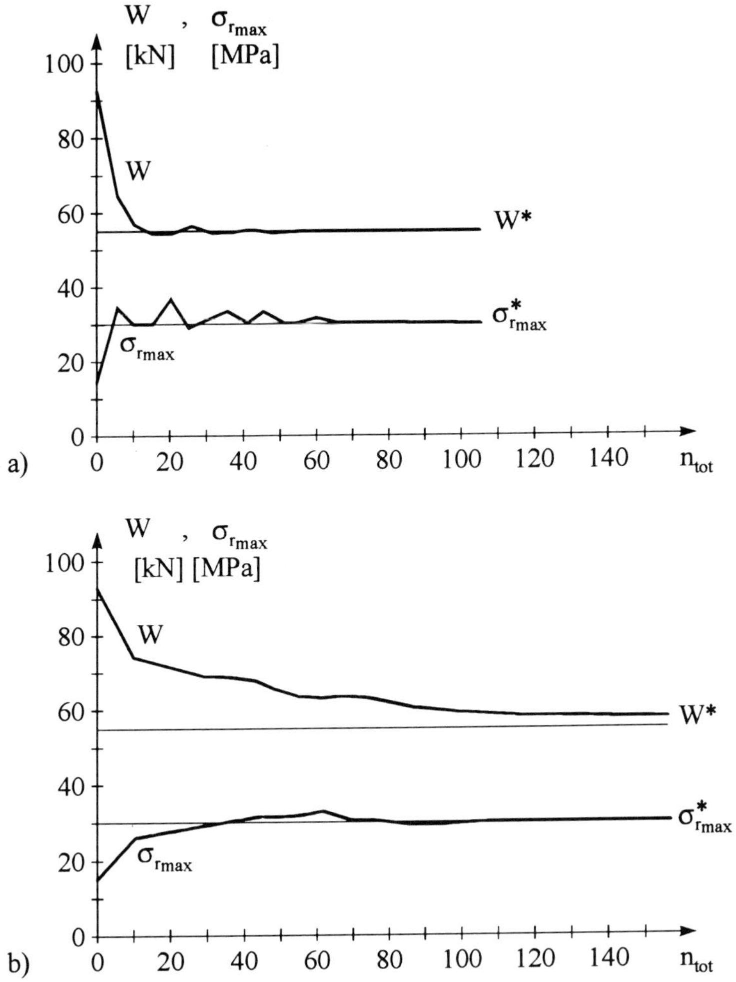

Fig. D-12: Optimization graphs in dependence on the number of function evaluations

a) Sequential Linearization SLP

b) LAGRANGE-multiplier-method LPNLP

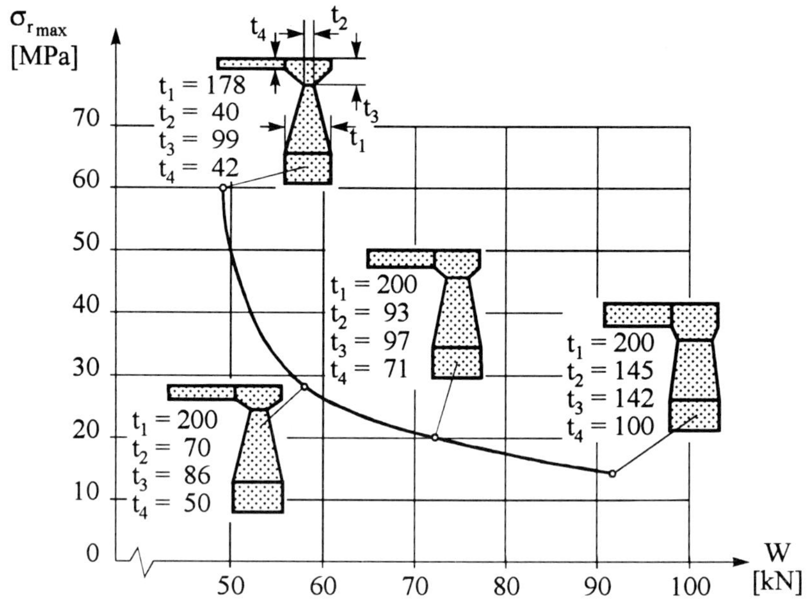

Fig. D-13: Functional-efficient solutions for a conveyor belt drum

Fig. D-13 illustrates the functional-efficient solutions which clearly show the influence of different admissible stresses on the shape of the optimized conveyor belt drums. Proceeding from the weight-optimal design characterized by active stress constraints, the increase of the variables t_1 and t_2 leads to a substantial decrease of the maximum reference stress with only a small increase of weight. Only in those cases where the range of t_1 has been used to its full potential, the remaining variables gain influence on further stress reductions. Variable t_4 in particular causes a strong increase of weight without reducing the stresses in a decisive manner.

Exercise D-18-4:

Component optimization plays an important role especially in space technology. As a typical example, a satellite that is to be brought into its orbital position should have an extreme lightweight design for saving transportation costs; even small weight savings for single components result in a substantial cost reduction. One of these components is the *fuel tank* of the satellite which stores the fuel for the position control rockets over the entire life-time of the satellite.

In the present example, the calculation and optimization of a thin-walled, satellite tank subjected to constant internal pressure shall be dealt with (a quarter section of the components can be considered for reasons of symmetry (Fig. D-14)).

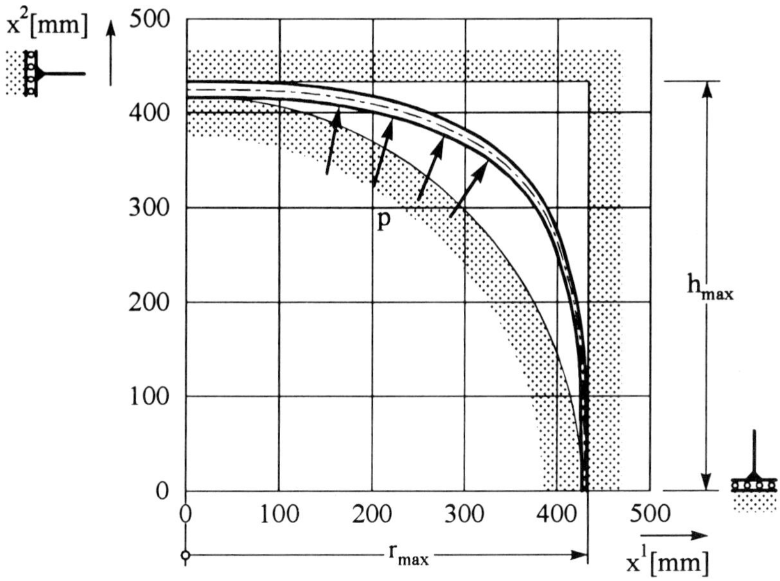

Fig. D-14: Principle sketch of the contour of a satellite tank with boundary conditions

The following design specifications and strength verifications are given:

- The construction space allows for a maximum outer radius of r_{max} = 436.9 mm. The tank height h_{max} must not exceed a value of 433 mm.
- The tank is subjected to an internal pressure p = 34.4 bar. The dead-weight is to be disregarded.
- The half-tank must be able to store a volume V which is larger than a minimum value $V_{min} = V_0 = 215.2$ liter. The following volume constraint is specified:

$$g_1 = 1 - \frac{V}{V_0} \leq 0 \quad .$$

- The tank is made of titanium alloy with specified material characteristics:

Density	$\varrho = 4.5 \cdot 10^3\ kg/m^3$,
YOUNG's modulus	$E = 1.1 \cdot 10^5\ MPa$,
POISSON's ration	$\nu = 0.3$,
Breaking strength	$\sigma_B = 1080\ MPa$.

- The strength verification is performed depending on the sign of the principal stresses in meridional and circumferential direction and in accordance with the following stress hypotheses of the state of plane stress:

  1) If the principal stresses σ_1, σ_2 have the same sign, the reference stress is calculated from the maximum stress:

  $$\mathrm{Max}(\sigma_1, \sigma_2) \leq \sigma_{\mathrm{r\,adm}} \quad .$$

  2) If the principal stresses have different signs, the VON MISES equal stress hypothesis is to be used:

  $$\sqrt{\sigma_1^2 + \sigma_2^2 - \sigma_1 \sigma_2} \leq \sigma_{\mathrm{r\,adm}} \quad .$$

The required thickness t_r can be calculated from the resultants and from the admissible reference stress $\sigma_{\mathrm{r\,adm}}$. For this purpose we define the following thickness constraint:

$$g_2 = 1 - \frac{t}{t_r} \leq 0 \quad .$$

The task is to minimize the weight of the satellite tank subject to the given constraints by simultaneously determining a suitable meridional contour and a thickness distribution of the tank.

Solution:

Structural Analysis

In the following, some general remarks shall be made concerning the *structural analysis*. The minimum volume of the tank already occupies more than 80% of the given construction space. This fact demands tank contours that smoothly follow their boundaries both at the poles and at the equator. At the equator, the shape of the tank approaches a cylindrical shell curved in the circumferential direction; at the pole, the radii of curvature increase to such a degree during optimization that a very shallow shell emerges. It appears that linear calculations produce large displacements in proximity of the pole, exceeding the wall-thickness by far. As the displacements do not occur constantly over the arc length, the radii of curvature of the deformed structure change substantially. According to the above theory, the pole area shows a decisively larger curvature in the state of deformation, which results in a violation of the conditions of equilibrium that were originally formulated for the undeformed element. Thus, we used an augmented approach for the structural analysis according to Section 13.2.

Shape Optimization

The following shape optimization requires a mathematical description of the tank shape as a function of free parameters by means of shape functions. The description should be characterized in such a way that a large number of admissible shapes can be achieved with a relatively small amount of parameters. The shape functions have to comply with the following requirements:

- The tank shape should not exceed the specified fitting space.
- The meridional contour of the curve must be determined in such a way that the given minimum volume is attained.
- The tangent at the pole must be perpendicular to the axis of rotational symmetry; the tangent at the equator must be parallel to the axis of rotational symmetry.
- A curvature undercut (change of sign) is not admissible.

Simple shape functions can be achieved by using a circle or an ellipse as meridional contour. These functions are, however, not suitable for the present problem because the required tank capacity cannot be fulfilled, and because the equator curve remains arched. A further disadvantage is the invariability of the curve shape. The same applies if a so-called CASSINI-curve is used since it does not possess any free parameters either. The shape is uniquely determined by the volume, and thus optimization calculations for finding a more suitable contour cannot be carried out. Further investigations were carried out using cubic spline-functions as shape functions. These third order polynomials define a continuous curve up to the second derivative, i.e. the derivative conditions at the pole and the equator are fulfilled. The splines, however, are disadvantageous in as far as changes of curvature can easily occur, and because the prescribed dimensions of the construction space cannot be complied with. In addition, they often cause problems in the structural analysis.

In order to avoid the above difficulties, we here choose a modified ellipse function according to *(18.22a)* as shape function (see Fig. 18.5). The use of a modified ellipse has the advantage that meridional shapes always exist for $\kappa_1 > 1$ and $\kappa_2 > 2$ which satisfy the demands made with respect to the curvature shape and to the tangent position. In the pole area, the ellipse function is replaced by a polynomium of fourth order.

Treatment as a Multicriteria Optimization Problem

In the following, some results shall be presented for a pure shape optimization and for a simultaneous shape and thickness optimization. When optimizing the tank, one has to address two conflicting objectives: The weight W of the tank shall be minimized, and the internal storage volume V shall be maximized. This multicriteria optimization problem, too, can efficiently be solved by using the constraint-oriented transformation according to *(18.7)*. For the present problem, the volume is introduced into the optimization model (g_1) as an additional constraint (secondary objective). For various desired volumes V_0, a scalar weight optimization is carried out, and thus the functional-efficient boundary is determined. The following design variables will be considered:

	$x_1 = \varkappa_1$	1st ellipse parameter ,
	$x_2 = \varkappa_2$	2nd ellipse parameter ,
	$x_3 = t$	thickness of the shell ,
or	$x_{3i} = t_i$	thickness of the i-th shell element .

The thicknesses t_i of the shell elements are used as additional design variables in the transfer matrix procedure. Since the computational effort increases substantially with the number of design variables (> 200), only the geometry variables are transferred to the optimization algorithm. The optimal thickness distribution is determined within each functional call, i.e. for each shape design, employing a *Fully-Stressed-Design (FSD)-strategy*, the use of which fulfills *a priori* the stress constraints and thus the thickness constraints. By this, the problem can be reduced to the following design variables and constraints:

$$x_1 = \varkappa_1 \qquad \text{1st ellipse parameter} \quad ,$$

$$x_2 = \varkappa_2 \qquad \text{2nd ellipse parameter} \quad ,$$

$$g_1 = 1 - \frac{V}{V_0} \leq 0 \qquad \text{volume constraint} \quad .$$

The procedure of the Generalized Reduced Gradients was used as optimization algorithm according to [D.1].

Fig. D-15 compares the functional-efficient boundaries of the simultaneous optimization to those of a pure shape optimization (without variation of the thickness). It can be shown that the integration of an FSD-strategy into the structural analysis leads to a substantial improvement of the designs. Three functional-efficient solutions (I: V_0 = 200 l; II: V_0 = 215.21; III: V_0 = 230 l) are depicted separately, and Fig. D-16 shows the corresponding designs of the shape and the cross-sections, of the radial displacements, of the membrane forces, and of the meridional bending moments. It is in the responsibility of the decision-maker to choose the most appropriate design.

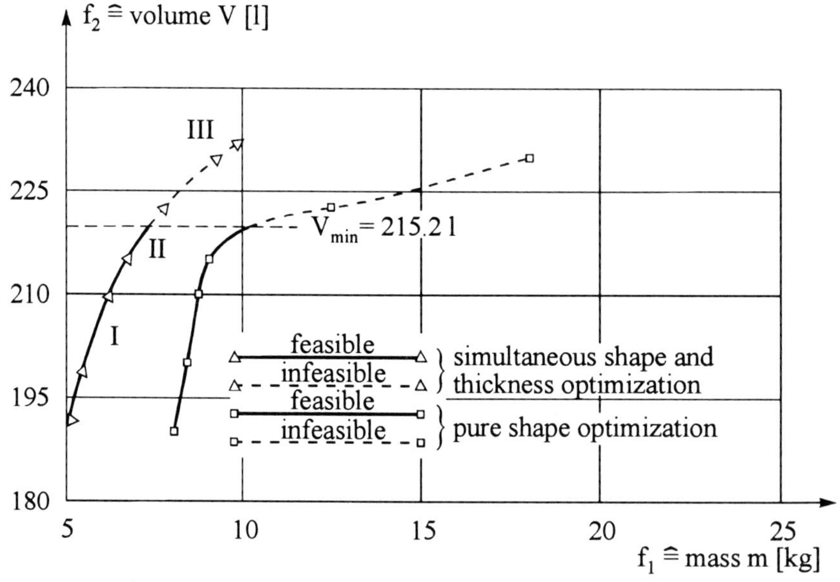

Fig. D-15: Functional-efficient boundaries for a pure shape optimization and for a simultaneous shape-thickness-optimization

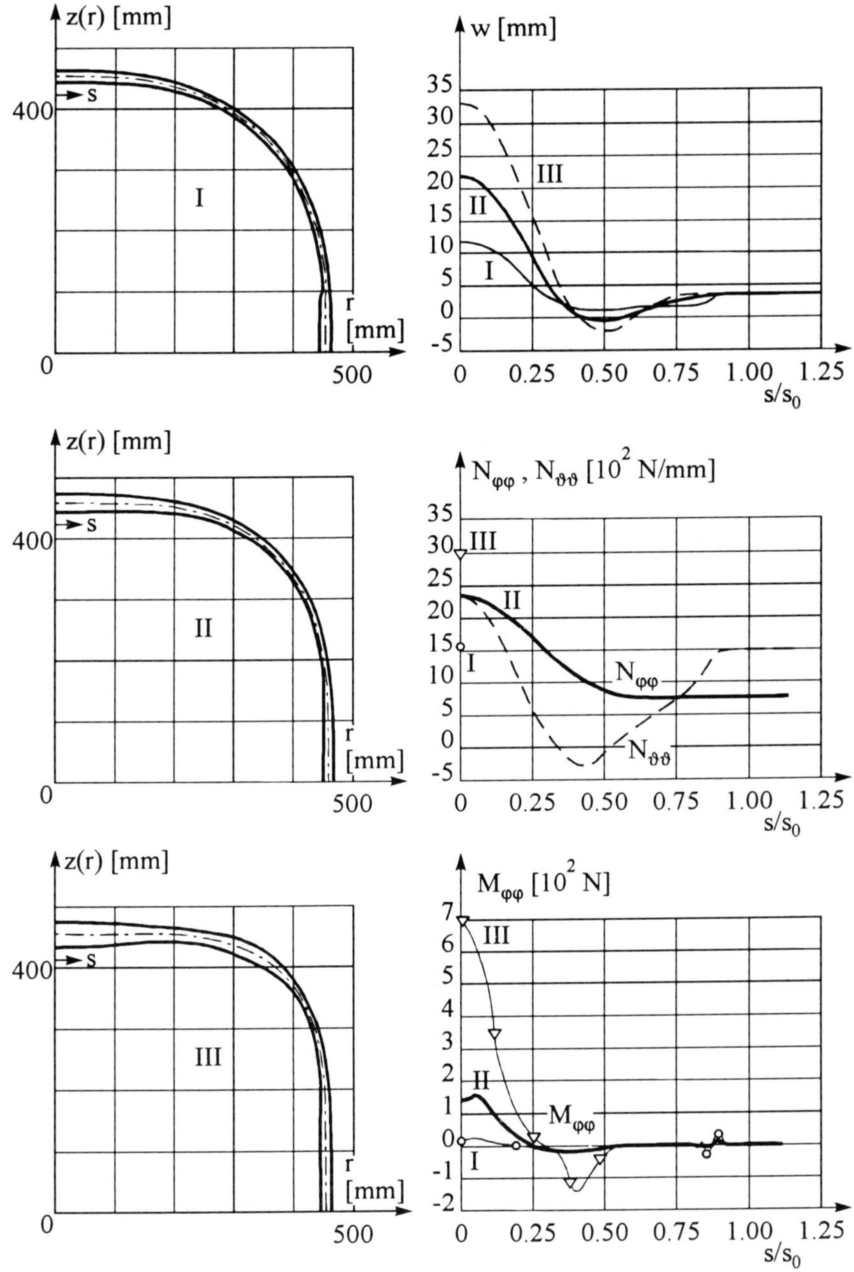

Fig. D-16 : Optimal meridional contours and their respective displacements, membrane forces, and meridional bending moments (s_0 = entire meridional arc length)
(For the shapes I and III only the maximum membrane forces are shown)

Exercise D-18-5:

Fig. D-17 shows a spatial sketch of a parabolic radiotelescope reflector with circular aperture. The reflector is assembled from single panels with sandwich structure, each of which consists of an aluminium honeycomb core and top layers made of Carbon Fibre Reinforced Plastics CFRP (see Fig. D-18).

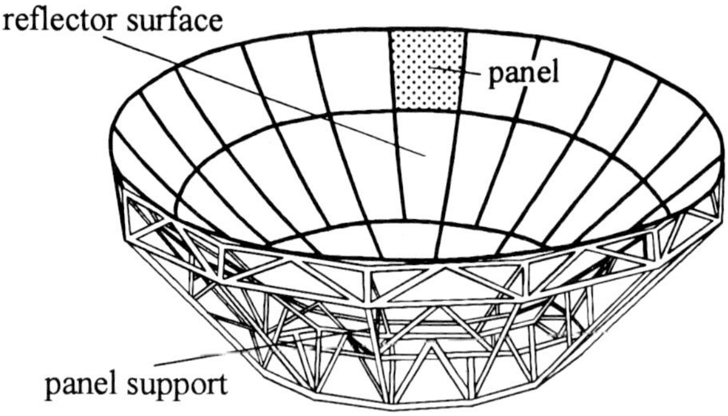

Fig. D-17: Sketch of a parabolic reflector with circular aperture and panel structure

The panel treated in the following is assumed to be plane and rectangular, and at several points it is supported at the rear truss structure by means of adjusting devices (see Fig. D-18). The number n of point-supports predominantly depends on the desired panel accuracy, i.e., on the maximum transverse displacement.

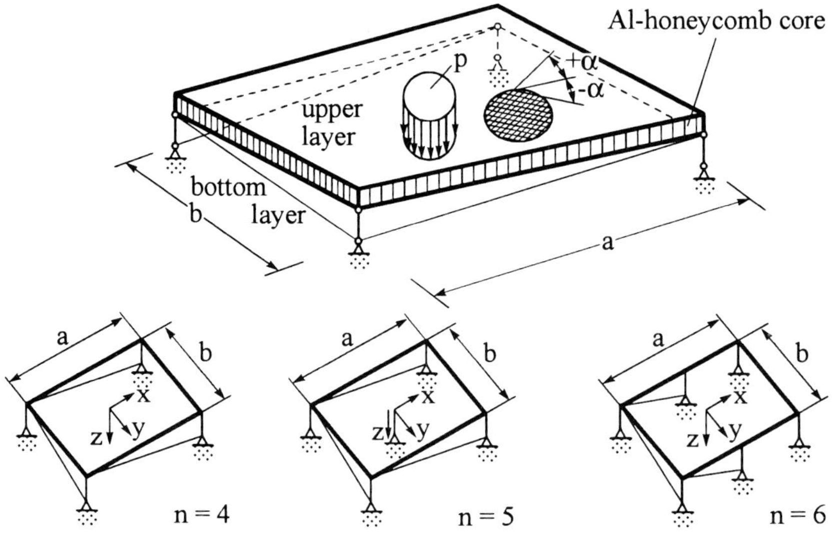

Fig. D-18: Point-supported, rectangular sandwich panel made of CFRP

In addition to the load cases deadweight, wind pressure, and concentrated forces, the layout must also take temperature effects into account. The optimization objctives consist in finding a construction which, for reasons of dynamics, is as light and stiff as possible, in order to increase the lowest eigenfrequency of transverse vibrations.

A special rectangular CFRP-sandwich panel design with four, five, and six point supports which is subjected to a constantly distributed wind load of $p = 1.384 \cdot 10^{-4}\,MPa$ according to Fig. D-18, shall be investigated in the form of a multicriteria optimization problem.

Solution:

Optimization modeling

In order to find *optimal* compromise solutions, we choose, according to *(18.1)*, two objective functions, namely weight $f_1(\mathbf{x}) := W$ and maximum displacement $f_2(\mathbf{x}) := w_{max}$. The present multicriteria optimization task will treated using preference functions like *(18.5)* to *(18.8)*. As a main strategy, however, a constraint-oriented transformation (Trade-off method) will be applied *(18.7)* that defines the weight as the primary objective, and the maximum displacement as the secondary objective, i.e. as constraint:

$$p[\mathbf{f}(\mathbf{x})] = f_1(\mathbf{x}) \mathrel{\widehat{=}} W \quad \text{with} \quad f_2(\mathbf{x}) \mathrel{\widehat{=}} g_1(\mathbf{x}) \leq \bar{y}_2 = w_{max} \quad . \tag{1}$$

The design variable vector $\mathbf{x}$ here defines the fibre angle α_i, the ply thicknesses t_i, the core height h_c, and the sides ratio b/a:

$$\mathbf{x}^T = (\alpha_1, \ldots, \alpha_n; t_1, \ldots, t_n; h_c; \frac{b}{a}) \quad . \tag{2}$$

The panel weight is the sum of the layers and of the core:

$$f_1(\mathbf{x}) := W = g\,a\,b\left\{h_C \rho_C + \sum_{i=1}^{n} t_i\left[\rho_{F_i} \varphi_{F_i} + \rho_{M_i}(1 - \varphi_{F_i})\right]\right\} \tag{3}$$

with $\rho_{F,M,C}$ defining the density of fibres, matrix, and core, respectively, and φ_{F_i} denoting the fibre volume fraction.

The optimization modeling is also augmented by a number of inequality constraints like a fibre breakage criterion, a fibre bonding criterion for the single layers as well as a shear failure criterion for the core [B.9, B.10]. In addition, the design variables are bounded by the following upper and lower constraint values:

$$\left.\begin{aligned} 0^\circ \leq \alpha_k \leq 90^\circ \quad &, \\ t_{min} \leq t_n \leq t_{max} \quad &, \\ h_{C_{min}} \leq h_C \leq h_{C_{max}} \quad &. \end{aligned}\right\} \tag{4}$$

Structural analysis

The maximum displacement w_{max} as the *secondary objective function* is here determined from the following system of equations, using the FE-program system ANSYS [A.21]:

$$\mathbf{K}(\mathbf{x})\mathbf{u} = \mathbf{r} \tag{5}$$

with

$\mathbf{K}(\mathbf{x})$	symmetrical total stiffness matrix as a function of the design variable vector $\mathbf{x}$,
$\mathbf{u}$	vector of displacements ,
$\mathbf{r}$	load vector .

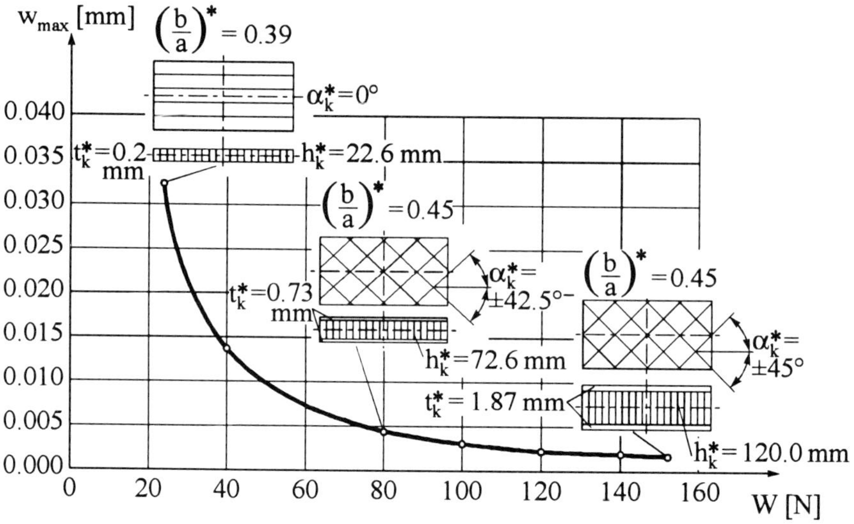

Fig. D-19: Functional-efficient solutions of a panel supported at six points

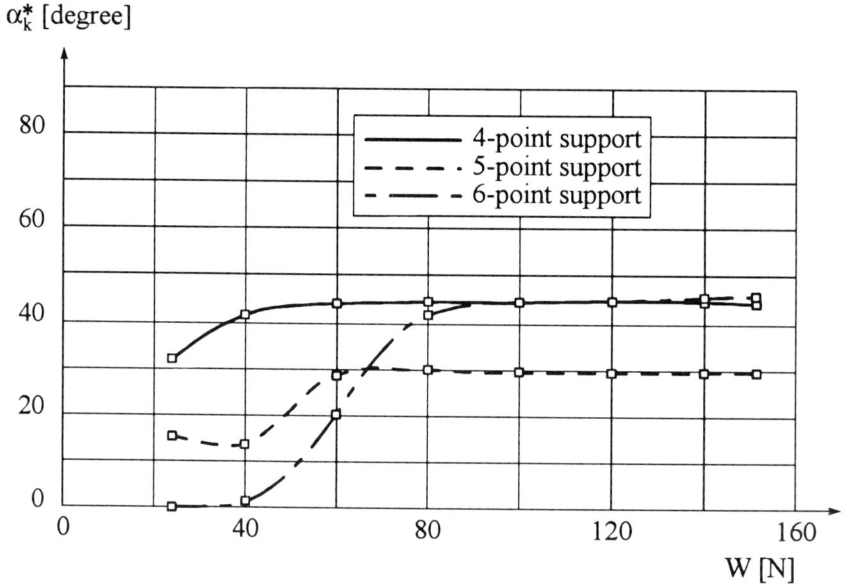

Fig. D-20: Optimal fibre angle α_k^* as a function of *weight* W

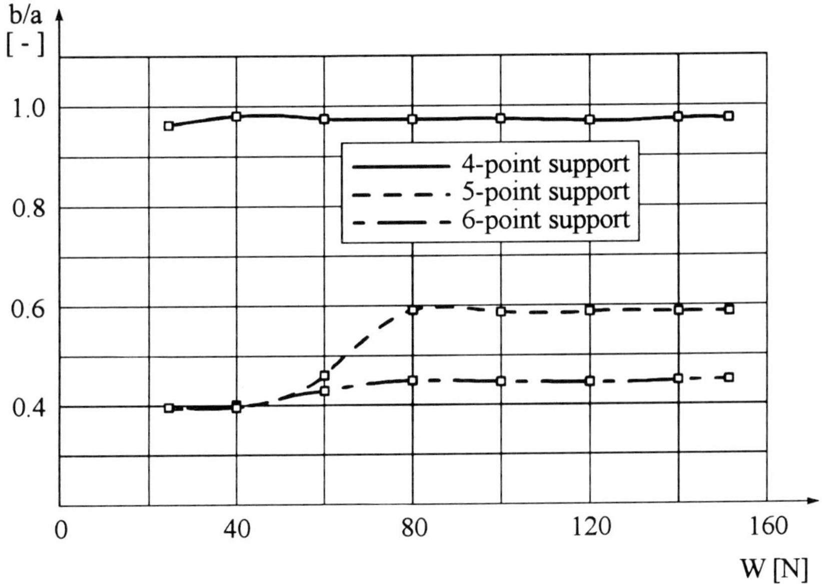

Fig. D-21: Optimal ratio b / a as a function of *weight* W

Results

Some optimization results shall be presented and interpreted in the following. Fig. D-19 shows the functional-efficient solutions of a panel supported at six points, including different fibre orientations as well as layer - and core thicknesses. Fig. D-20 illustrates a panel with one core thickness only, where the optimal fibre angles depend on the weight. It can be shown that, for a weight > 80 N, the fibre angle is nearly constant for all types of support. The fibre angle is equal to about 45° in the case of four or six supporting points, whereas it is 30° for five supports. According to Fig. D-20, a substantial change of the fibre angle occurs at the panels with five and six point-supports and with a weight less than 60 N and 80 N, respectively. The dependence of the optimal ratio b/a on the weight is shown in Fig. D-21. For each optimal weight, the panel with four supporting points displays an almost constant ratio of ≈ 1, while b/a ranges between 0.4 and 0.6 for the other cases of support of the panel. The above results illustrate the importance of the optimization investigations as they present an important decision tool to the engineer for choosing a *best possible* design.

References

The present work is based on the contents of the following two volumes (in German):

[ET1] ESCHENAUER, H.; SCHNELL, W.: Elastizitätstheorie I – Grundlagen, Scheiben und Platten. 2. Auflage, Mannheim, Wien, Zürich: BI–Wissenschaftsverlag 1986, 277 pages.

ESCHENAUER, H.; SCHNELL, W.: Elastizitätstheorie II – Schalen. Mannheim, Wien, Zürich: BI–Wissenschaftsverlag 1983, 269 pages.

[ET2] ESCHENAUER, H.; SCHNELL, W.: Elastizitätstheorie – Grundlagen, Flächentragwerke, Strukturoptimierung. 3. vollständig überarbeitete und erweiterte Auflage, Mannheim, Leipzig. Wien, Zürich: BI–Wissenschaftsverlag 1993, 491 pages.

ESCHENAUER, H.; SCHNELL, W.: Elastizitätstheorie – Formel– und Aufgabensammlung, Mannheim, Leipzig. Wien, Zürich: BI–Wissenschaftsverlag 1994, 279 pages.

A Fundamentals of elasticity
– Chapter 2 to 7 –

[A.1] BATHE, K.–J.: Finite Element Procedures in Engineering Analysis. Englewood Cliffs: Prentice Hall 1982

[A.2] COOK, R.D.; MALKUS, D.S.; PLESHA, M.E.: Concepts and Applications of Finite Element Analysis. 3rd ed. New York: Wiley & Sons 1989

[A.3] COURANT, R.; HILBERT, D.: Methoden der Mathematischen Physik (in German). Berlin, Heidelberg, New York: Springer 1968

[A.4] FLUEGGE, W.: Tensor Analysis and Continuum Mechanics. Berlin, Heidelberg, New York: Springer 1972

[A.5] FUNG, Y.C.: Foundations of Solid Mechanics. Englewood Cliffs: Prentice Hall 1965

[A.6] GALERKIN, B.G.: Reihenentwicklungen für einige Fälle des Gleichgewichts von Platten und Balken (in Russian). Petrograd: Wjestnik Ingenerow 1915

[A.7] GREEN, A.E.; ADKINS, J.E.: Large Elastic Deformations. Oxford: Clarendon Press 1970

[A.8] GREEN, A.E.; ZERNA, W.: Theoretical Elasticity. 2nd ed. Oxford: Clarendon Press 1975

[A.9] LANGHAAR, H.L.: Energy Methods in Applied Mechanics. New York, London: John Wiley 1962

[A.10] LOVE, A.E.H.: A Treatise on the Mathematical Theory of Elasticity. 4th ed. New York: Dover Publ. 1944

[A.11] LURIE, A.I.: Räumliche Probleme der Elastizitätstheorie. Berlin: Akademie-Verlag 1963

[A.12] MARGUERRE, K.: Ansätze zur Lösung der Grundgleichungen der Elastizitätstheorie (in German). ZAMM 35 (1955) 242-262

[A.13] NOWACKI, W.: Thermoelasticity. London: Pergamon Press 1986

[A.14] RITZ, W.: Über eine neue Methode zur Lösung gewisser Variationsprobleme der mathematischen Physik (in German). J. Reine Angewandte Mathematik 135 (1908) 1

[A.15] SOKOLNIKOFF, I.S.: Mathematical Theory of Elasticity. New York, Toronto, London: McGraw-Hill 1956

[A.16] TIMOSHENKO, S.; GOODIER, J.N.: Theory of Elasticity. 3. ed. New York, St. Louis, Toronto, London: McGraw-Hill 1970

[A.17] WANG, Ch.-T.: Applied Elasticity. New York, Toronto, London: McGraw-Hill 1953

[A.18] WASHIZU, K.: Variational Methods in Elasticity and Plasticity. 3rd ed. Oxford, New York, Toronto, Sydney, Braunschweig: Pergamon Press 1982

[A.19] WEMPNER, G.: Mechanics of Solids. New York, Toronto, London: McGraw-Hill 1973

[A.20] ZIEGLER, F.: Mechanics of Solids and Fluids. New York, Heidelberg, Berlin: Springer 1991

[A.21] N.N.: ANSYS Uscr's Manual for Revision 5.0. Swanson Analysis Systems, Inc. Johnson Road, Houston, TX / USA, ...

B Plane load-bearing structures

– Chapter 8 to 10 –

[B.1] ESCHENAUER, H.: Thermo-elastische Plattengleichungen; Beulen einer Kragplatte. TH Darmstadt: Dr.-Thesis 1968 (in German)

[B.2] GREEN, A.E.; ZERNA, W.: Theoretical Elasticity. 2nd ed. Oxford: Clarendon Press 1975

[B.3] JAHNKE, E.; EMDE, F.; LÖSCH, F.: Tafeln höherer Funktionen. 7. Aufl. Stuttgart: Teubner 1966

[B.4] MALVERN, L.E.: Introduction to the Mechanics of a Continuous Medium. Englewood Cliffs: Prentice Hall 1969

[B.5] MARGUERRE, K.; WOERNLE, H.-T.: Elastic Plates. Waltham, Toronto, London: Blaidsdell 1969

[B.6] MUSKHELISHVILI, N.I.: Einige Grundaufgaben zur mathematischen Elastizitätstheorie. München: Hase 1971

[B.7] PLANTEMA, F.J.: Sandwich Constructions. New York, London: Wiley & Sons 1966

[B.8] TIMOSHENKO, S.; WOINOWSKY-KRIEGER, S.: Theory of Plates and Shells. 2nd ed. New York, Toronto, London: McGraw-Hill 1959

[B.9] TROITSKY, M.S.: Stiffened Plates – Bending, Stability and Vibrations. Amsterdam, Oxford, New York: Elsevier Scientific 1976

[B.10] TSAI, S.W.; HAHN, H.T.: Introduction to Composite Materials. Westport, Conn.: Technomic Publishing 1980

[B.11] VINSON, J.R.; SIERAKOWSKI, R.L.: The Behavior of Structures Composed of Composite Materials. Dordrecht, Boston, Lancaster: Martinus Nijhoff Publ. 1986

C Curved load–bearing structures
– Chapter 11 to 14 –

[C.1] AXELRAD, E.L.; EMMERLING, F.A.: Flexible Shells. Heidelberg, New York, Tokio: Springer 1984

[C.2] DIKMEN, M.: Theorie of Thin Elastic Shells. Boston, London, Melbourne: Pitman 1982

[C.3] DONNELL, L.H.: Beams, Plates, and Shells. New York: McGraw–Hill 1976

[C.4] FLUEGGE, W.: Stresses in Shells. New York, Heidelberg, Berlin: Springer 1973

[C.5] FLUEGGE, W.: Tensor Analysis and Continuum Mechanics. Berlin, Heidelberg, New York: Springer 1972

[C.6] FUNG, Y.C.; SECHLER, E.E.: Thin–Shell Structures. New York: Prentice Hall 1974

[C.7] GECKELER, J.W.: Zur Theorie der Elastizität flacher rotationssymmetrischer Schalen (in German). Ing.–Arch. 1 (1930) 255 – 270

[C.8] GIBSON, J.E.: Thin Shells, Computing and Theory. Oxford, London, New York, Paris: Pergamon Press 1980

[C.9] GOLDENVEIZER, A.L.: Theory of Elastic Thin Shells. Oxford, London, New York, Paris: Pergamon Press 1961

[C.10] GOULD, P.L.: Analysis of Shells and Plates. Berlin, Heidelberg, New York: Springer 1988

[C.11] GREEN, E.; ZERNA, W.: Theoretical Elasticity. 2nd ed. Oxford: Clarendon Press 1975

[C.12] KOITER, W.; MIKHAILOV, K.G.: Theory of Shells. Amsterdam, New York, Oxford: Proc. IUTAM Symp., North–Holl. Publ. Comp. 1980

[C.13] KRÄTZIG, W.B.: Thermodynamics of Deformations and Shell Theory. Ruhr–Uni Bochum, Inst. f. Konstr. Ing. Bau (1971) Techn. Wiss. Mitt. 71 – 3

[C.14] KRÄTZIG, W.B.: Introduction to General Shell Theory. In: Thin Shell Theory – New Trends and Applications, ed. by W. OLSZAK. Wien, New York: Springer 1980, 3 – 61

[C.15] MARGUERRE, K.: Zur Theorie der gekrümmten Platte großer Formänderung (in German). Proc. 5th Int. Congr. Appl. Mech. (1939) 93 – 101

[C.16] MORLEY, L.S.: An Improvement on DONNELL's Approximation for Thin–Walled Circular Cylinders. Quart. Journ. Mech. and Appl. Math. XII (1959) 89 – 99

[C.17] NAGHDI, P.M.: Foundations of Elastic Shell Theory. In: Progress in Solid Mechanics, Vol. IV ed. by SNEDDON/HILL. Amsterdam: North-Holl. Publ. Comp. 1963

[C.18] NIORDSON, F.I.: Shell Theory. Amsterdam, New York, Oxford: North-Holland Series in Appl. Math. and Mech. 1985

[C.19] NOVOZHILOV, V.V.: The Theory of Thin Shells. Groningen: Noordhoff 1970

[C.20] REISSNER, E.: Stresses and Small Displacements of Shallow Sherical Shells. J. Math. Phys. 25 (1946) 80-85, 279-300; 27 (1948), 240; 38 (1959), 16-35

[C.21] SHIRAKAWA, K.; SCHNELL, W.: On Some Treatment of the Equations of Motion for Cylindrical Shells Based on Improved Theory. Ing.-Arch. 53 (1983) 275-63

[C.22] SINARAY, G.C.; BANERJEE, B.: Large Amplitude Free Vibrations of Shallow Spherical Shell and Cylindrical Shells. Int. J. Non-linear Mech. 20 (1985) 69-78

[C.23] SOEDEL, W.: Vibrations of Shells and Plates. New York, Basel: Marcel Dekker Inc. 1981

[C.24] TIMOSHENKO, S.; WOINOWSKY-KRIEGER, S.: Theory of Plates and Shells. 2nd ed. . New York, Toronto, London: McGraw-Hill 1959

[C.25] ZIENKIEWICZ, O.C.: The Finite Element Method in Engineering Science. New York, London, McGraw-Hill Vol. 1: Basic formulation and linear Problems 4th ed. 1988 Vol. 2: Solid and fluid mechanics, dynamics and nonlinearity 4th ed. 1991

D Structural optimization
– Chapter 15 to 18 –

[D.1] ABADIE, J.; CARPENTIER, J.: Generalization of the WOLFE Reduced Gradient Method to the Case of Nonlinear Constraints. In: FLETCHER, R. (Ed.): Optimization. New York: Academic Press 1969, 37-48

[D.2] ARORA, J.S.: Introduction to Optimum Design. New York, Toronto, London: McGraw-Hill 1989

[D.3] ATREK, E.; GALLAGHER, R.H.; RAGSDELL, K.M.; ZIENKIEWICZ, O.C. (Eds.): New Directions in Optimum Structural Design. New York, Toronto: John Wiley & Sons 1984

[D.4] BANICHUK, N.V.: Optimality Conditions and Analytical Methods of Shape Opimization. In: HAUG, E.J./CEA, J. (eds.): Optimization of Distributed Parameter Structures. Alphen aan den Rijn: Sijthoff und Noordhoff, 1987

[D.5] BANICHUK, N.V.: Introduction to Optimization of Structures. New York, Berlin, Heidelberg: Springer 1990

[D.6] BENDSØE, M.P.; OLHOFF, N.; TAYLOR, J.E.: A Variational Formulation for Multicriteria Structural Optimization. J. Struct. Mech. 11 (1983) 523-544

[D.7] BERNADON, M.; PALMA, F.J.; ROUSSELET B.: Shape Optimization of an Elastic Thin Shell under Various Criteria. J. Structural Optimization 3 (1991) 7 – 21

[D.8] BRAIBANT, V.; FLEURY, C.: Shape Optimal Design using B–splines. Comput. Methods Appl. Mech. Eng. 44 (1984) 247 – 267

[D.9] CARMICHAEL, D.G.: Structural Modelling and Optimization. New York, Toronto: John Wiley & Sons 1987

[D.10] CHENEY, E.W.; GOLDSTEIN, A.A.: NEWTON–Method and Convex Programming and CHEBYSCHEV Approximation. J. Num. Math. 1 (1959) 253 - 268

[D.11] DAVIDON, W.C.: Variable Metric Method for Minimization. A.E.C. Research and Development Report ANL–5990, 1959,

[D.12] ESCHENAUER, H.: The *Three Columns* for Treating Problems in Optimum Structural Design. In: BERGMANN, H.W. (ed.): Optimization: Methods and Applications, Possibilities and Limitations. Berlin, Heidelberg, New York: Springer 1989, 1 – 21

[D.13] ESCHENAUER, H.: Shape Optimization of Satellite Tanks for Minimum Weight and Maximum Storage Capacity. J. Structural Optimization 1 (1989) 171 – 180

[D.14] ESCHENAUER, H.A.; KOSKI, J., OSYCZKA, A.: Multicriteria Design Optimization. Berlin, Heidelberg, New York: Springer 1990

[D.15] ESCHENAUER, H.A.; SCHUHMACHER, G.; HARTZHEIM, W.: Multidisciplinary Design of Composite Aircraft Structures by LAGRANGE. In: Computers & Structures, 44, 4 (1992), 877 – 893

[D.16] ESCHENAUER, H.; WEINERT, M.: Structural Optimization Techniques as a Mathematical Tool For Finding Optimal Shapes of Complex Shell Structures. In: GIANESSI, F. (Ed.): Nonsmooth Optimization Methods and Applications. Switzerland, Australia, Belgium: Gordon and Breach 1992, 135 – 153

[D.17] FIACCO, A.V.; McCORMICK, G.P.: Computational Algorithm for the Sequential Unconstrained Minimization Technique for Nonlinear Programming. J. Management Sci. 10 (1964) 601 – 617

[D.18] FLETCHER, R.; POWELL, M.J.D.: A Rapidly Convergent Descent Method for Minimization. Computer J. 6 (1963) 163 – 168

[D.19] FLETCHER, R.; REEVES, C.M.: Function Minimization by Conjugate Gradients. Computer J. 7 (1964) 149 – 154

[D.20] GRIFFITH, R.E.; STEWART, R.A.: A Nonlinear Programing Technique for the Optimization of Continuous Processing Systems. J. Management Sci. 7 (1961) 379 – 392

[D.21] HAFTKA, R.T.; GÜRDAL, Z.; KAMAT, M.P.: Elements of Structural Optimization. Dordrecht, Boston, London: Kluwer Academic Publishers, 2nd ed. 1990

[D.22] HAUG, E.J.; ARORA, J.S.: Applied Optimal Design. New York, Toronto: John Wiley & Sons 1979

[D.23] HESTENES, M.R.; STIEFEL, E.: Methods of Conjugate Gradients for Solving Linear Systems. J. Res. Nat. Bur. Std. B49 (1952) 409

[D.24] HIMMELBLAU, D.M.: Applied Nonlinear Programming. New York: McGraw-Hill 1972

[D.25] KIEFER, J.: Sequential Minimax Search for a Maximum. Proc. Am. Math. Soc. 4 (1953) 105 - 108

[D.26] KUHN, H.W.; TUCKER, A.W.: Nonlinear Programming. In: NEYMAN, J. (Ed.): Proceedings of the 2nd Berkeley Symposium on Mathematical Statistics and Probability. University of California, Berkeley 1951, 481 - 492

[D.27] LUKASIEWICZ, S.: Local Loads in Plates and Shells. Alphen aan den Rijn: Sijthoff and Noordhoff 1979

[D.28] MROZ, Z.; DEMS, K.: On Optimal Shape Design of Elastic Structures. In: ESCHENAUER, H.; OLHOFF, N. (eds.): Optimization Methods in Structural Design. Mannheim, BI-Wissenschaftsverlag 1983, 224 - 232

[D.29] OLHOFF, N.; TAYLOR, J.E.: On Structural Optimization. J. Appl. Mechanics, 50 (1983) 1139 - 1151

[D.30] OLHOFF, N.; LUND, E.; RASMUSSEN, J.: Concurrent Engineering Design Optimization in a CAD Environment. In: HAUG, E.J. (ed.): Concurrent Engineering Tools and Technologies for Mechanical System Design. New York: Springer-Verlag 1993, 523 - 586

[D.31] OLHOFF, N.; LUND, E.: Finite Element Based Engineering Design Sensitivity Analysis and Optimization. In: HERSKOVITS, J. (ed.): Advanced in Structural Optimization. Dordrecht, The Netherlands: Kluwer Academic Publishers 1995, 1 - 45

[D.32] OSYCZKA, A.: Multicriterion Optimization in Engineering. New York, Toronto: John Wiley & Sons 1984

[D.33] PAPALAMBROS, P.Y.; WILDE, D.J.: Principles of Optimal Design. Cambridge, New York, Melbourne, Sydney: Cambridge University Press 1988

[D.34] PARETO, V.: Manual of Political Economy. Translation of the French edition by A.S. SCHWIER (1927). London-Basingslohe: The McMillan Press 1972

[D.35] PARKINSON, A.; WILSON, M.: Development of a Hybrid SQP-GRG Algorithm for Constrained Nonlinear Programming. Proc. ASME Design Engineering Technical Conferences, Columbus/Ohio 1986

[D.36] PIERRE, D.A.; LOWE, M.J.: Mathematical Programming via Augmented Lagrangians. London: Addison Wesley 1975

[D.37] POWELL, M.J.D.: An Efficient Method for Finding the Minimum of a Function of Several Variables without Calculating Derivations. Computer J. 7 (1964) 155 - 162

[D.38] POWELL, M.J.D.: VMCWD: A FORTRAN Subroutine for Constrained Optimization. University of Cambridge, Report DANTP 1982/NA4

[D.39] ROZVANY, G.I.N.: Structural Design via Optimality Criteria. Dordrecht, Boston, London: Kluwer Academic Publishers 1989

[D.40] SCHITTKOWSKI, K.L.; HÖRNLEIN, H.: Numerical Methods in FE-Based Structural Optimization Systems. ISNM-Series, Zürich: Birkhäuser 1993

[D.41] SCHMIT, L.A.; MALLET, R.H.: Structural Synthesis and Design Parameters. Hierarchy Journal of the Struct. Division, Proceedings of the ASCE, Vol. 89, No. 4, 1963, 269 - 299

[D.42] SOBIESZCZANSKI-SOBIESKI, J.: Multidisciplinary Optimization for Engineering Systems: Achievements and Potentials. In: BERGMANN, H.W. (Ed.): Optimization: Methods and Applications, Possibilities and Limitation. Berlin, Heidelberg, New York: Springer 1989, 42 – 62

[D.43] STADLER, W.: Preference Optimality and Application of PARETO-Optimality. In: MARZOLLO/LEITMANN (Eds.): Multicriterion Decision Making. CISM Courses and Lectures. New York: Springer 1975, 125 – 225

[D.44] STADLER, W.: Multicriteria Optimization in Engineering and in the Sciences. New York, London: Plenum Press 1988

[D.45] SVANBERG, K.: The Method of Moving Asymptotes – A New Method for Structural Optimization. Int. J. Num. Methods in Eng. 24 (1987) 359 – 373

[D.46] VANDERPLAATS, G.N.: Methods of Mathematical Optimization. In: BERGMANN, H.W. (Ed.): Optimization: Methods and Applications, Possibilities and Limitations. Berlin, Heidelberg, New York: Springer 1989, 22 – 41

[D.47] ZIENKIEWICZ, O.C.; CAMPBELL, J.S.: Shape Optimization and Sequential Linear Programming. In: GALLAGHER/ZIENKIEWICZ (Eds.): Optimum Structural Design. Chichester, New York, Brisbane, Toronto: John Wiley & Sons 1973, 109 – 126

[D.48] ZYCZKOWSKI, M.: Recent Advances in Optimal Structural Design of Shells. European J. of Mechanics, A/Solids 11, Special Issue (1992) 5 – 24

Subject index

D

E

F

T

U

V

W

Y

Printing: COLOR-DRUCK DORFI GmbH, Berlin
Binding: Buchbinderei Lüderitz & Bauer, Berlin